T0252787

Handbook of Meta-analysis in Ecology and Evolution

HANDBOOK OF META-ANALYSIS IN ECOLOGY AND EVOLUTION

Edited by
Julia Koricheva, Jessica Gurevitch,
and Kerrie Mengersen

PRINCETON UNIVERSITY PRESS

Princeton and Oxford

Copyright © 2013 by Princeton University Press
Published by Princeton University Press, 41 William Street,
Princeton, New Jersey 08540
In the United Kingdom: Princeton University Press, 6 Oxford Street,
Woodstock, Oxfordshire OX20 1TW

press.princeton.edu

All Rights Reserved

Library of Congress Cataloging-in-Publication Data

Handbook of meta-analysis in ecology and evolution / edited by Julia Koricheva,
Jessica Gurevitch, and Kerrie Mengersen.
pages cm
Summary: "Meta-analysis is a powerful statistical methodology for synthesizing research
evidence across independent studies. This is the first comprehensive handbook of meta-
analysis written specifically for ecologists and evolutionary biologists, and it provides
an invaluable introduction for beginners as well as an up-to-date guide for experienced
meta-analysts. The chapters, written by renowned experts, walk readers through every
step of meta-analysis, from problem formulation to the presentation of the results. The
handbook identifies both the advantages of using meta-analysis for research synthesis and
the potential pitfalls and limitations of meta-analysis (including when it should not be used).
Different approaches to carrying out a meta-analysis are described, and include moment
and least-square, maximum likelihood, and Bayesian approaches, all illustrated using
worked examples based on real biological datasets. This one-of-a-kind resource is uniquely
tailored to the biological sciences, and will provide an invaluable text for practitioners from
graduate students and senior scientists to policymakers in conservation and environmental
management. Walks you through every step of carrying out a meta-analysis in ecology and
evolutionary biology, from problem formulation to result presentation   Brings together
experts from a broad range of fields   Shows how to avoid, minimize, or resolve pitfalls such
as missing data, publication bias, varying data quality, nonindependence of observations, and
phylogenetic dependencies among species   Helps you choose the right software   Draws on
numerous examples based on real biological datasets "— Provided by publisher.
Includes bibliographical references and index.
ISBN 978-0-691-13728-5 (hardback) — ISBN 978-0-691-13729-2 (paperback)
1. Ecology—Statistical methods. 2. Evolution—Mathematical models. 3. Meta-analysis.
I. Koricheva, Julia, 1967– II. Gurevitch, Jessica. III. Mengersen, Kerrie, 1962–
QH541.15.S72H36 2013
576.8—dc23
2012041108

British Library Cataloging-in-Publication Data is available

This book has been composed in Times New Roman

Printed on acid-free paper. ∞

Printed in the United States of America

10 9 8 7 6 5 4 3 2 1

To Ransom Myers (1952–2007), whose enthusiasm for meta-analysis and what it could reveal in ecology, passion for the natural world, and demand for rigorous science and statistical methodology, pervade this book

CONTENTS

Contents

PREFACE

YOU ARE HOLDING IN your hands the first handbook of meta-analysis specifically written for ecologists and evolutionary biologists. While meta-analysis has a longer history in disciplines like the medical and social sciences, it was introduced to ecology and evolution only in the early 1990s (Järvinen 1991, Gurevitch et al. 1992) and is still coming of age in these fields. However, despite this relatively short history, several hundreds of meta-analyses in ecology and evolution have already been published, and Gurevitch et al. (2001) concluded that at the turn of the twenty-first century meta-analysis had begun to have a substantial impact on the way data are summarized in these disciplines. They suggested that incorporation of meta-analysis as a routine and familiar approach will fundamentally change the nature of ecology and evolutionary biology.

Several factors contributed to the growth of meta-analysis in ecology and evolution. First, accumulation of an enormous amount of published and unpublished research has made the traditional narrative approach to comprehensive research synthesis increasingly unfeasible. Second, the importance of research synthesis in ecology and evolution is now particularly high because of the growing pressure on researchers to provide accurate quantitative assessments, predictions, and practical solutions to pressing environmental issues (e.g., biodiversity losses, biotic responses to global climate change). Third, sophisticated methods for meta-analysis have been developing in medicine and the social sciences since the late 1970s, and thus ecologists and evolutionary biologists could adapt these methods without starting from the very beginning.

An important role in the promotion of the quantitative approach to research synthesis in ecology has been played by the National Center for Ecological Analysis and Synthesis (NCEAS), established in Santa Barbara, California in 1995; the center has been funded by the United States National Science Foundation (Grant #EF-0553768), the University of California, Santa Barbara, and the State of California. The mission of NCEAS is to advance the state of ecological knowledge through the search for general patterns and principles, and to organize and synthesize ecological information in a manner useful to researchers, resource managers, and policy makers addressing important environmental issues. In 1996, a working group was created at NCEAS to help evaluate and guide the application of meta-analysis to ecological questions. A series of articles resulting from this workshop was published in a special feature issue of *Ecology* in 1999 (Gurevitch and Hedges 1999, Hedges et al. 1999, Osenberg et al. 1999). In late 2004, the National Evolutionary Synthesis Center (NESCent) was established in Durham, North Carolina, with the aim of promoting synthetic research in evolutionary biology. NESCent works in a manner similar to NCEAS by providing funding for working groups, postdoctoral researchers, and sabbatical fellows. Both NCEAS and NESCent have supported the development of quantitative research synthesis methods in ecology and evolution, as well as numerous applications of these methods to specific problems.

The work on this book was conducted as a part of the Meta-analysis in Ecology: Lessons, Challenges and Future Working Group supported by NCEAS. The working group consisted of 14 scientists from the United States, Canada, Australia, and the United Kingdom with expertise in statistics and/or meta-analysis application in ecology, evolutionary biology, medicine, and the social sciences. Four group meetings were held in Santa Barbara during 2006–2008. The topics to be covered were agreed on in the first meeting, and the remaining meetings were used to develop the individual book sections and chapters. The resulting handbook differs from the usual edited volume because it was a genuinely collaborative venture; the number of participants was small, most contributed to more than one chapter, and the contributors worked closely together on the material. The working group format also resulted in all the contributors being more aware of the content of other chapters, which we hope has resulted in a more coherent treatment of the subject.

RATIONALE FOR THIS HANDBOOK

The decision to write a handbook on meta-analysis for ecologists and evolutionary biologists was prompted by several factors. First, meta-analytic techniques are either not covered or barely mentioned in the standard statistical books widely used by ecologists and evolutionary biologists (e.g., Sokal and Rohlf 1995, Quinn and Keough 2002) and are seldom included in the syllabi of standard statistical courses. Therefore, training in these techniques is not readily available to ecology and evolution students, and those ecologists and evolutionary biologists who have mastered meta-analysis are nearly all self-taught.

Second, although several reviews on meta-analysis in ecology and evolution have been published (Arnqvist and Wooster 1995a, Gurevitch et al. 2001, Hillebrand 2008, Stewart 2010, Harrison 2011), and software for ecological meta-analysis has been developed (Rosenberg et al. 2000), none of these publications describe in sufficient detail the entire procedure of conducting a meta-analysis from the problem formulation stage through data collection and analysis to the reporting of the results, nor do they address many of the issues and problems involved. On the other hand, the available handbooks on meta-analysis (Hedges and Olkin 1985, Lipsey and Wilson 2001, Borenstein et al. 2009, Cooper et al. 2009) are aimed at scientists from other disciplines (medicine and social sciences). As a result, they contain approaches, terminology, and examples that are unlikely to resonate with ecologists and evolutionary biologists.

Third, although the statistical issues and problems confronting meta-analysts in ecology and evolutionary biology share much in common with medicine and the social sciences, they also differ in important and substantive ways. For example, the dichotomous response variables that are common in medicine (e.g., dead/alive) are less frequent in ecology and evolutionary biology where continuous response variables prevail. Medicine and the social sciences seek to understand only a single species (*Homo sapiens*) whereas ecological and evolutionary meta-analyses usually strive to reveal patterns across many species. This introduces a number of issues, including the challenging problems of heterogeneity, and potential nonindependence among studies due to shared phylogeny. Most importantly, the types of questions asked and the data structure in meta-analyses in ecology and evolution differ from those in other fields. For example, meta-analyses in ecology and evolution typically are unconcerned with grand means, and are most concerned with the explanation of heterogeneity; meta-analyses in medicine tend to be much smaller and more narrowly focused, and the grand mean is the statistic of primary interest. For all these reasons, we felt that there was a compelling need for a handbook on meta-analysis aimed specifically at ecologists and evolutionary biologists.

In addition, the growing popularity of meta-analysis in the absence of proper training and guidelines results in highly variable standards for the application of meta-analysis to ecology and evolutionary biology; these standards fall short of those in other fields, such as medicine (Gates 2002, Roberts et al. 2006). Unfortunately, methodological shortcomings in some of the published ecological and evolutionary meta-analyses sometimes trigger criticism of the meta-analytic approach as a whole, and are used as an argument to resort to more primitive and statistically flawed methods of research synthesis, such as vote counting. There is clearly a considerable scope to improve the rigor of quantitative research synthesis in ecology and evolutionary biology, both by better educating ecologists and evolutionary biologists about available methodological tools and by taking advantage of recent developments in meta-analysis in other disciplines.

This handbook is aimed at addressing the above problems. It should facilitate and promote the thoughtful and critical use of meta-analysis for research synthesis in ecology and evolutionary biology, and increase the scope of its application in these disciplines. This handbook should generally improve the robustness of meta-analysis in ecology and evolutionary biology. This book will, we hope, put powerful tools and concepts for research synthesis directly into the hands of those who need them, and make them available in one place.

THE INTENDED AUDIENCE

This handbook should be useful to anyone who is involved in conducting or interpreting meta-analyses in ecology or evolution; for example, researchers (ranging from graduate students to senior researchers), statisticians, and decision makers. In addition to academic scientists in the fields of ecology and evolutionary biology, we believe it will be useful to people in more applied areas, such as researchers in forestry, conservation, and other nonacademic areas. Furthermore, this book will be helpful to managers and policy makers who are not themselves conducting meta-analyses, but need to be able to evaluate and use information derived from published meta-analyses in their areas of interest. In addition to teaching meta-analytical techniques to beginning researchers, this handbook will also provide an up-to-date guide for more experienced meta-analysts by expanding the array of available techniques and by tackling some of the current problems in meta-analysis application to ecological data. Finally, the various chapters discuss and provide examples of the main applications of meta-analysis in ecology. This book will thus be ideal for both those who are new to meta-analysis, as well as those familiar with some of the techniques but who are interested in learning more about the background, assumptions, and additional methods in order to advance their own research.

LAYOUT OF THIS BOOK

The first section of this handbook defines the place of meta-analysis among the other methods of research synthesis (Chapter 1) and summarizes the process and procedure of meta-analysis in a nutshell (Chapter 2).

The second section describes the first steps in planning and initiating a meta-analysis, such as defining the problem and developing the protocol (Chapter 3), collecting data (Chapter 4), extracting and critically appraising the data (Chapter 5), and deciding on the metrics of effect size (Chapters 6 and 7).

The third section describes the main types of statistical models and approaches to statistical inference in meta-analysis (Chapter 8), including moment- and least squares-based approaches (Chapter 9), maximum likelihood methods (Chapter 10), and Bayesian analysis (Chapter 11). It also provides an overview of statistical software available for meta-analysis (Chapter 12).

The fourth section deals with specific statistical issues and problems in meta-analysis, including missing data (Chapter 13), publication bias (Chapter 14), and temporal changes in effect sizes (Chapter 15). Nonindependence is discussed when it is due to multiple measures per study (Chapter 16) or shared phylogeny (Chapter 17). In addition, this section covers meta-analysis of primary data (Chapter 18) and results from multisite experiments (Chapter 19).

The fifth section covers issues related to the interpretation and presentation of the results of meta-analysis. These include quality standards for meta-analysis (Chapter 20), graphical presentation of the results (Chapter 21), power statistics for meta-analysis (Chapter 22), and the role of meta-analysis in interpreting scientific literature (Chapter 23) and in testing ecological and evolutionary theory (Chapter 24).

Finally, the last section of this book describes the contribution of meta-analysis in ecology and evolution by reviewing the history of the approach and comparing its progress in ecology and evolution with that in the medical and social sciences (Chapter 25). Its contribution is also described by discussing the application of meta-analysis to environmental management and conservation (Chapter 26) and by summarizing the current status of meta-analysis in ecology and evolution. Future challenges are also considered (Chapter 27). We have additionally included the section of frequently asked questions about meta-analysis.

BOOK DEDICATION AND ACKNOWLEDGMENTS

We dedicate this book to the memory of our friend and colleague Ransom Myers, who was an active and enthusiastic member of the NCEAS working group and played a key role in planning the statistical part of the book. Sadly, he did not live to see it completed. His untimely death has left an enormous void, and his wonderful personality and great passion for meta-analysis and conservation are sorely missed. This book is part of his legacy.

Working on this book as part of a highly interdisciplinary working group at NCEAS has been a wonderful, illuminating, and rewarding experience. We are very grateful to all the contributors for their cooperation and patience in putting up with the endless editorial comments, demands, and deadlines, and for their support and friendship. We also express our deepest thanks to the staff at NCEAS for their hospitality, technical assistance, and continuous supply of coffee and snacks.

Drafts of all the individual chapters were reviewed internally by NCEAS working group members as well as externally. The external reviewers, who greatly helped improve the chapters, were:

Dean Adams (Iowa State University, US)
Joseph Bailey (University of Tennessee, US)
Kasey Barton (University of Hawaii, US)
Daniel T. Blumstein (University of California, US)
Geoffrey D. Borman (University of Wisconsin, US)
Michael Brannick (University of South Florida, US)
James S. Clark (Duke University, US)
Michael Dietze (University of Illinois, US)
Aaron M. Ellison (Harvard University, US)
Helmut Hillebrand (University of Oldenburg, Germany)
Laura Hyatt (Rider University, US)
John P. A. Ioannidis (Stanford University, US)
Khalid Khan (Birmingham Women's Hospital, UK)
Roosa Leimu (University of Oxford, UK)
Mark W. Lipsey (Vanderbilt University, US)
Arne Mooers (Simon Fraser University, Canada)
Emily V. Moran (Duke University, US)
Christa P. H. Mulder (University of Alaska, US)
Stephan B. Munch (Stony Brook University, US)
Shinichi Nakagawa (University of Otago, New Zealand)
Kiona Ogle (University of Wyoming, US)

Therese D. Pigott (Loyola University Chicago, US)
Hugh Possingham (University of Queensland, Australia)
Jennifer Schweitzer (University of Tennessee, US)
Alexander J. Sutton (University of Leicester, UK)
Tom Tregenza (University of Exeter, UK)
Jeffrey C. Valentine (University of Louisville, US)
Jennifer Verdolin (Stony Brook University, US)
Jacob Weiner (University of Copenhagen, Denmark)

We are very grateful to Emily Rollinson for her help with the technical editing of this book and to Megan Higgie for redrawing the figures for Chapter 24. Last but not least we would like to thank the Princeton University Press team, particularly Sheila Dean, Nathan Carr, Quinn Fusting, Stefani Wexler, Robert Kirk, and Alison Kalett for their continuous support and advice through the process of preparing this handbook.

Julia Koricheva
Jessica Gurevitch
Kerrie Mengersen

SECTION I

Introduction and Planning

Place of Meta-analysis among Other Methods of Research Synthesis

Julia Koricheva and Jessica Gurevitch

IN THE MOST GENERAL terms, meta-analysis is one method of research synthesis. Research synthesis may be defined as a review of primary research on a given topic with the purpose of integrating the findings (e.g., for creating generalizations or resolving conflicts). Research synthesis is central to the scientific enterprise. Without it, the evidence for various alternative hypotheses cannot be properly evaluated and generalizations cannot be reached, thus the advance of the scientific field as well as any potential practical applications are inhibited. Research synthesis can be performed either qualitatively, in the form of a narrative review, or quantitatively, by employing various statistical methods for the integration of results from individual studies.

Research reviews in ecology and evolutionary biology have traditionally been carried out either in the form of narrative reviews, or by "vote counting," where the number of statistically significant results for and against a hypothesis are counted and weighed against each other (see below). In other fields, it has been widely recognized that neither of the above methods is adequate to address current needs for quantitative research synthesis, and we believe that this is also true in ecology, evolution, and conservation biology. Narrative reviewing offers expert interpretation and perspective, but it is inherently subjective and nonquantitative. Vote counting has very poor properties as a statistical procedure, as discussed below. Neither procedure is able to provide critically important information on the magnitude of the effects or the sources of variation in outcomes among studies.

The lack of training in formal, rigorous protocols for research synthesis in ecology and evolutionary biology contrasts markedly with the standard quantitative training that scientists in these fields receive for primary research, including training in experimental design and analysis. Fortunately, this asymmetry in the tools employed for primary versus synthetic investigations began to change dramatically in the past two decades as meta-analytic methods have started to be introduced and incorporated into standard practice. The summarization and integration of research results across studies is increasingly viewed as a scientific research process in its own right in ecology and evolutionary biology, as it has come to be recognized in other disciplines. Like primary research, research synthesis follows the scientific method and is held to scientific standards. Meta-analysis employs specialized techniques for data gathering and analysis developed specifically for the purposes of research synthesis. Below, we briefly review the variety of approaches to research synthesis and discuss their relative merits, advantages, limitations, and drawbacks.

"TEXTBOOK EXAMPLES"

The simplest method of presenting the results of research on a particular topic is the "textbook example." While not a formal method of research synthesis, it is one with which every biologist is familiar, because that is the way we learn and subsequently teach our fields of study. This ubiquitous method picks out particular case studies which seem to best illustrate the evidence for a particular phenomenon, and uses these examples to explain the phenomenon, to provide the evidence that it exists, and to summarize the findings on that phenomenon. The practice of using "textbook examples" or case studies as a summary of scientific findings in a research field is based on a widespread but erroneous belief that a single primary study, particularly a well-designed experiment, is able to provide the ultimate test for a hypothesis and hence resolve an issue. This is a mistaken idea. Research results are probabilistic and subject to artifacts such as sampling error and measurement error; thus, the findings of any single study may have occurred simply by chance and may be refuted by subsequent research (Taveggia 1974, Schmidt 1992, Ioannidis 2008). Therefore, in the majority of current methods for research synthesis, and in meta-analysis in particular, any individual primary study is considered to be one of a population of studies on a given question, rather than a single definitive and conclusive piece of evidence capable of fully resolving the question.

NARRATIVE REVIEWS

Traditionally, research synthesis in ecology and evolutionary biology has been carried out in the form of narrative reviews; numerous examples can be found in journals such as *Annual Review of Ecology, Evolution and Systematics*, *Trends in Ecology & Evolution*, *Biological Reviews*, or *The Quarterly Review of Biology*. These narrative reviews are generally invited contributions written by senior scientists who are recognized experts in their fields. Some of these reviews are narrower in their scope while others are more comprehensive, and their bibliographies may contain dozens to hundreds of primary studies. The structure of these reviews differs profoundly from that of primary research papers. There is usually no "material and methods" section where the author explains search methods or selection criteria for the inclusion of primary studies into the review. An analysis of 73 review articles on conservation and environmental management topics published during 2003–2005 found that 63% of the reviews were narrative. Of these, only 30% reported details of the sources used to acquire studies for review, 34% defined inclusion criteria for identification of relevant studies, and fewer than 20% reported the search terms used (Roberts et al. 2006). This lack of methodological and reporting rigor can increase the degree of subjectivity in which papers are chosen for review and how their results are interpreted. For example, two authors could review the literature on the same topic and, by including or excluding different subsets of studies, arrive at opposite conclusions. Narrative reviews are opaque in the sense that the methods used for searching the literature are seldom laid out explicitly, and this means that the review cannot be repeated or updated systematically as more information becomes available.

Another very practical limitation of narrative reviews is that they are inefficient in handling a large number of studies and ill equipped for dealing with variation in outcomes among studies. At best, the results of the studies may be presented in the form of large tables, which can be difficult to interpret. At worst, large proportions of the studies are ignored, while the results of a small number of "exemplars" are emphasized. This places narrative reviewing uncomfortably close to "textbook examples" or case studies as a method of summarizing research. If a large amount of research has been done on a particular topic, chance alone dictates that studies exist that report inconsistent and contradictory findings, particularly when sample sizes and the statistical power to detect significant outcomes differ greatly among studies. When the number of reviewed studies is relatively small, potential explanatory variables could be suggested by comparing various study attributes in a table. As the number of studies increases, however, the task of sifting through many potential explanatory variables becomes unmanageable. Consequently, the results of studies in narrative reviews may appear to be "inconsistent," "inconclusive," or "conflicting"; these types of results are uninformative for resolving intellectual or methodological conflicts. They are also insufficient to address needs for making practical decisions, which is becoming an increasingly important task in applied ecology and conservation biology.

Narrative reviews can serve important functions by presenting perspectives, historical development of ideas, and other conceptual contributions. But their inherently high subjectivity and low repeatability, and their poor suitability for dealing with variation in results among studies greatly limit their utility for research syntheses, including evaluation of whether hypotheses are supported by the existing data. This can lead to a paradoxical situation that occurs when many hypotheses exist to explain some phenomenon or process, and all of them have been repeatedly tested, but none has ever been rejected, and they all coexist by virtue of supporting evidence in one system or another (e.g., Berenbaum 1995).

Ecologists and evolutionary biologists have often been criticized for having long-term debates which seemingly persist for generations (e.g., on the role of competition in structuring communities, or on the relationship between the diversity and stability of communities). We argue that the reliance of ecologists and evolutionary biologists on "textbook examples" and narrative reviews as ways of informing conclusions and synthesizing the results of published literature is at least partly responsible for this pattern.

VOTE COUNTING

Vote counting is a quantitative method for research synthesis that has been discredited and largely abandoned in other scientific fields but is unfortunately still all too commonly employed in ecology and evolutionary biology. In its simplest form, the available studies are sorted into three categories: those that report significant results in the predicted direction, those that yield significant results in the opposite direction, and those that yield nonsignificant results. Sometimes the latter two categories are combined because both types of studies are considered as evidence not in support of the hypothesis (e.g., Watt 1994). The studies are counted up and the category into which most studies fall is declared the "winner," and is considered to provide the best evidence about the direction of the effect. The relative number of studies "voting" for or against the effect may be considered evidence for its magnitude, strength, and consistency. Alternatively, outcomes of studies could be classified into more than three categories. For example, in a review on the relationship between productivity and diversity (Mittelbach et al. 2001), the shape of the relationship in each of the studies being reviewed was classified into five categories: linear positive relationships, linear negative, humped (maximum diversity at intermediate productivity values), U-shaped, and no significant relationship. The percentages of studies displaying the above shapes were then compared. Authors sometimes use goodness of fit or other statistical tests to compare the expected and observed frequencies of positive and negative outcomes (Waring and Cobb 1992, Daehler 2003, Colautti et al. 2006), with the null hypothesis being that the number of positive and negative outcomes is equal.

The advantages of vote counting are simplicity and seemingly broad applicability; regardless of design and the statistical approach used in the primary study and the way the results are reported, the outcomes are likely to be broadly classifiable into one of the vote-counting categories. Although counting up significant results seems to be logically straightforward, it is unfortunately seriously flawed as a statistical technique for research synthesis. The results of a vote count are statistically biased and often misleading. More importantly, they do not provide the information most necessary and relevant in synthesizing the results of different studies. There are several reasons for this. First, the vote-counting procedure gives one vote to each study regardless of its sample size and the statistical precision with which the outcome was tested. Thus, a study testing an effect with three replicates, for example, receives the same weight as a study with 100 or 1000 replicates. A second limitation of vote counting as a method for quantitative research synthesis is that the number of studies reporting statistically significant versus nonsignificant outcomes tells us nothing about the magnitude of the effect of interest, which we would argue is more likely to be biologically meaningful. One could partially deal with the problem of accounting for differences among studies in the precision with which an effect is tested by comparing the number of positive and negative outcomes regardless of their statistical significance. However, this still does not tell us anything about the magnitude of the effect in question. Also, the direction of the outcome may not be reported for nonsignificant results.

Vote counting may also lead to a wrong conclusion about the overall outcome across studies. Because vote counting is based on the statistical significance of the research findings, it has low power for effects of relatively small magnitude (which might still be biologically or otherwise significant); this is due to the statistical power of small studies possibly being too low to detect an effect. Effect sizes and sample sizes in ecology and evolution are often rather small. For example, Møller and Jennions (2002) have shown that across 43 published meta-analyses on various ecological and evolutionary topics the average effect size reported for Pearson's correlation coefficient was between 0.180 and 0.193. The sample size required to detect such a small effect with a power of 80% is over 200 replicates, which is considerably larger that the sample size of most studies in ecology and evolutionary biology. Moreover, Hedges and Olkin (1980) demonstrated that the power of the vote-counting procedure decreases as the number of studies integrated increases—a strange and undesirable property for a statistical test.

Surprisingly, despite the serious statistical drawbacks outlined above, which have been detailed in the ecological literature for well over a decade, vote counting is still used, published, and accepted as a "valid" method of research synthesis in ecology and evolution. For example, in a recent paper in *Science* on the shape of the productivity/plant species richness relationship, Adler et al. (2011) counted the number of study outcomes in five categories (nonsignificant, positive linear, negative linear, concave up, or concave down) at local, regional, and global scales. The most common relationship was nonsignificant, and the authors concluded that there was no clear relationship between productivity and plant species richness. While their results may very well be correct, a vote count is not the best way to investigate this question, and the authors did not attempt to justify this approach. We return to the reasons for the persistence of vote counting in the ecological literature below (see "Choosing the Method for Research Synthesis").

Vote counting can be more pervasive and subtle than is sometimes realized. Narrative reviews may rely upon a kind of "virtual vote count" in that they report on findings based on the statistical significance of their outcomes, and may then base conclusions on the preponderance—that is, the number—of statistically significant findings. Informally, one's assessment of the literature read on a question may likewise be based on the number of studies that report statistically significant findings. Formal systematic reviews (below) and meta-analyses bring a more rigorous, scientific approach to understanding the available evidence.

COMBINING PROBABILITIES AND RELATED APPROACHES

Another approach to quantitative research synthesis commonly used in the social sciences, but quite rarely in ecology and evolution, is based on combining probability values (significance levels) from statistical tests in individual studies that are testing the same scientific hypothesis. Methods for combining probabilities across studies have a long history dating at least from procedures suggested by R. A. Fisher in his book *Statistical Methods for Research Workers* (1925). As many as 18 different methods of combining probabilities currently exist (Becker 1994) depending on the statistical distribution employed. The method most commonly applied in the social sciences is the sum of Z's method based on normal distribution (Becker 1994, Cooper 1998). Another common approach to combining probabilities is Fisher's sum of logs methods, as described by Sokal and Rohlf (1995), which uses the inverse of the chi-square distribution to determine significance.

Similar to vote counting, the advantage of the combining-probabilities method is its broad applicability (any kind of statistical test can be used). Combining probabilities is also less problematic than vote counting because it uses exact probabilities, hence there is no deep gap between studies reporting $P = 0.04$ and $P = 0.06$ (note that in vote counting the above two studies would have been placed in different categories, nonsignificant and significant). However, similar to vote counting, combining probabilities is not very informative. The null hypothesis for the test of combined significance is that the effects of interest are not present in any of the studies (Becker 1994). If the null hypothesis is rejected, all the synthesist can tell is that in at least one of the studies the effect is not zero. This approach also provides no information about the sign or magnitude of the effect. Small P values may arise because of large effects or because of large sample sizes. The alternative hypothesis can be made more specific; for example, one-tailed tests are often used to estimate the probability that effects of interest are positive, negative, or larger than a specified value in at least one study (Becker 1994).

A practical problem with the application of the method of combining probabilities is that exact probability values are often not reported in primary studies. Commonly, significance values of statistical tests are reported as $P > 0.05$, $P < 0.05$, $P < 0.01$, or $P < 0.001$. In addition, Cooper (1998) pointed out that the method of combining probabilities is too liberal; if many tests exist and at least one of them provides a very low P value, the null hypothesis is practically always rejected.

The combining-probabilities method is seldom used for research synthesis in ecology and evolution. However, Koricheva et al. (2004) used the sum of Z's method in a meta-analysis testing for the potential trade-offs between constitutive and induced plant defenses against herbivores. If induced response decreases with increasing constitutive defense, as predicted by the trade-off hypothesis, the slope of the regression line of induced defense on constitutive defense will be < 1. The authors combined the one-tailed probabilities from individual t-tests of slopes of the linear regression of induced defense on constitutive defense. The combined Z value for 14 tests yielded 3.9394, which corresponds to the overall $P < 0.00005$, and thus results in rejecting the null hypothesis that the slope of the linear regression was < 1 in none of the 14 tests. A weighted meta-analysis of the same studies using Pearson's correlation coefficient between induction ratios and constitutive defenses provided further support for the existence of the trade-offs between plant allocations to constitutive and induced defenses against herbivores (Koricheva et al. 2004). Both approaches were used because correlations between induction ratios and constitutive defenses may be spurious, as the ratio is correlated with its own denominator. The combination of two approaches thus reinforced the conclusions of the meta-analysis. Combining probabilities is often used in combination with weighted meta-analysis in the social sciences (Kraus 1995).

A related approach, unfamiliar in ecology, is the combination of *t*-statistics. This approach has the advantage of considering the magnitude of effects in the individual studies. It also takes the sign of the effect into account, unlike the combination of probabilities. However, it suffers from other statistical drawbacks (described by Becker and Wu 2007, among others), particularly in terms of the inferences that it is possible to make from the outcome.

META-ANALYSIS

The term "meta-analysis" was coined by Glass (1976) in reference to "the statistical analysis of a large collection of analysis results from individual studies for the purpose of integrating the findings." The above definition is broad and covers all the techniques used in quantitative research synthesis including vote counting and combining probabilities, as described earlier. In this book, we define meta-analysis more narrowly, as a set of statistical methods for combining the magnitudes of the outcomes (effect sizes) across different data sets addressing the same research question. The methods of meta-analysis were originally developed in medicine and various social sciences (Glass et al. 1981, Hedges and Olkin 1985); they were introduced in ecology and evolutionary biology in the early 1990s (Järvinen 1991, Gurevitch et al. 1992, Arnqvist and Wooster 1995a).

Meta-analysis provides a powerful, informative, and unbiased set of tools for summarizing the results of studies on the same topic. It offers a number of advantages over narrative review, vote counting, and combining probabilities (Table 1.1). Meta-analysis is based on expressing the outcome of each study on a common scale. This measure of outcome, an "effect size," includes information on the sign and magnitude of an effect of interest from each study. In many cases the variance of this effect size can also be calculated (see Chapters 6 and 7). These effect size measures can then be combined across studies to estimate the grand mean effect size and

TABLE 1.1. Comparison of methods of research synthesis.

Characteristics of the review type	Narrative review	Vote counting	Combining probabilities	Meta-analysis
Imposes restrictions on the type of studies that can be used in review	No	No	No	Yes
Interprets study outcome based on its statistical significance	Yes	Yes	Yes	No
Takes into account sample size and statistical power of the individual studies being combined	No	No	Yes	Yes
Assesses statistical significance of the mean (overall) effect (i.e., whether it is significantly different than zero)	No	No	Yes	Yes
Assesses the magnitude of the mean effect	No	No	No	Yes
Allows analysis of sources of variation among studies	No	No	No	Yes

its confidence interval, and to test whether this overall effect differs significantly from zero. In many cases in ecological meta-analysis, it is of interest to examine heterogeneity among outcomes and model the relative contribution of different factors to the magnitude of the effect sizes. Hence, unlike other methods of research synthesis described above, meta-analysis allows the research synthesist to estimate the magnitude and sign of the grand mean effect across studies, assess whether the confidence interval around the effect includes zero, and examine sources of variation in that effect among studies.

A key aspect of modern approaches to meta-analysis is accounting for unequal precision in the magnitude of the effect among studies by weighting each study's effect size by the inverse of its variance. Meta-analysis offers an improved control of type II error rates (Arnqvist and Wooster 1995a), because the low power of individual studies to detect an effect is "corrected" by the accumulation of evidence across many studies. This is particularly important in areas where failure to reject false null hypothesis may have large detrimental impacts, as in conservation biology (Fernandez-Duque and Vallegia 1994) and medicine. Meta-analysis can potentially allow the detection of an effect even in situations when none of the studies included in the analysis show statistically significant results because of low statistical power (Arnqvist and Wooster 1995a). An important contribution of meta-analyses can be to identify gaps in the literature where more research is needed, and also to identify areas where the answer is definitive and no new studies of the same type are necessary. When the data are already sufficient to resolve a question, it is a waste of time and money to keep accumulating more of the same kind of information; meta-analysis can be an invaluable tool in this regard, and has been used in this way in ecology and other fields.

The first meta-analysis in ecology was published in 1991 (Järvinen 1991) and the number of meta-analyses in ecology increased greatly over the next two decades, reaching 119 publications by 2000 (Gurevitch et al. 2001), and exceeding 500 publications per year by 2010 (see Fig. 25.1C in Chapter 25). The first general review of meta-analysis in ecology and evolution, published in 1995 (Arnqvist and Wooster 1995a), stated: "Meta-analysis is still rare in our domain, but the first applications show that it can successfully help address a variety of questions." The second review was published in 2001 (Gurevitch et al. 2001) and concluded that at the turn of the century, meta-analysis had started to have a substantial impact on the way data were summarized in ecology and evolutionary biology. The authors predicted that incorporation of meta-analysis as a routine and familiar approach in ecology and evolution will fundamentally change the nature of these scientific disciplines. It is too early yet to assess the accuracy of their prediction.

More widespread adoption of meta-analysis may have a number of subtle but profound effects on ecologists' perspective of research. Most importantly, individual studies contributing to meta-analysis are seen as members of a population of studies that all provide information on a given effect, rather than as isolated and presumably definitive examples. More controversially, meta-analysis can influence one's interpretation of results by reducing the emphasis on the statistical significance of results and moving the focus to the magnitude, direction, and variance in effects (Chapter 23). Variability of results among studies may become more a source of hypothesis generation regarding the nature of the sources of variation rather than being seen as a hindrance to understanding and interpreting a given phenomenon. While scientists have always emphasized biological importance over statistical significance, meta-analysis makes this view more compelling, and shifts the interpretation of biological importance from case studies and textbook examples to "the weight of evidence" across all of the literature on a particular question. Meta-analysis helps us to advance from seeking complete answers based on individual experiments, however creative and elegant, to combining evidence across existing studies, each of them perhaps imperfect in some way, but which together provide a wealth of information. Meta-analysis is thus much more than just a new method of research synthesis; it is a new approach to scientific research which requires major changes in our views of the role of individual studies and the acceptance of the cumulative nature of scientific knowledge (Schmidt 1992).

SYSTEMATIC REVIEW

An important method of research synthesis that has become ubiquitous in the medical literature (Khan et al. 2003), and is beginning to be used in conservation and environmental management (Davies et al. 2008, Newton et al. 2009, Stewart et al. 2009), is the "systematic review." Systematic review is research synthesis on a precisely defined topic using explicit methods to identify, select, critically appraise, and analyze relevant research. The crucial element of the systematic review that distinguishes it from an ordinary narrative review is an a priori protocol, which describes the methodology, including detailed search strategy and inclusion criteria. This review protocol makes the review process rigorous, transparent, and repeatable. Details and examples of review protocols are discussed in Chapter 3.

Systematic reviews may or may not include meta-analysis or other methods of quantitative research synthesis. Systematic reviews without meta-analyses are used to identify the current state of knowledge, including gaps when insufficient data exists to conduct meta-analysis. On the other hand, meta-analyses are also produced without fully defined systematic reviews, but this practice may lead to biased or erroneous results if systematic methods have not been used to obtain and synthesize the data. Throughout this book, we shall retain the formal distinction between systematic review as the systematic collation and analysis of data and meta-analysis as the statistical methods for combining effect sizes.

In medicine, systematic reviews in combination with meta-analyses provide critical information needed for evidence-based medicine (Khan et al. 2003; Chapter 25). Use of systematic reviews for supporting decision making in conservation and environmental management has been promoted by the Centre for Evidence-Based Conservation at Bangor University, UK (http://www.cebc.bangor.ac.uk/). Guidelines for systematic reviews in conservation and environmental management have been developed (Pullin and Stewart 2006) and completed systematic reviews on these topics can be found at http://www.environmentalevidence.org/Reviews .html. For published examples of systematic reviews in conservation and management see Davies et al. (2008), Newton et al. (2009), and Stewart et al. (2009).

CHOOSING THE METHOD OF RESEARCH SYNTHESIS: IS META-ANALYSIS ALWAYS THE BEST CHOICE?

Meta-analysis has many advantages over other methods of research synthesis (Box 1.1). The rest of this handbook focuses specifically on meta-analysis and aims to promote its correct and thoughtful use as a part of systematic reviews in ecological and evolutionary research. Does this mean that meta-analysis is always the preferred method of research synthesis, and that narrative reviews, combining probabilities, and vote-counting procedures have to be abandoned altogether? Are there situations when it is not advisable or feasible to conduct a meta-analysis? Below we evaluate these questions, examine some common objections to the use of meta-analysis in ecology and evolutionary biology, and suggest situations where the use of other methods of research synthesis in combination with meta-analysis can be advantageous.

BOX 1.1.
Why should ecologists and evolutionary
biologists learn about meta-analysis?

(1) Meta-analysis provides a more objective, informative and powerful means of summarizing the results from individual studies as compared to narrative/qualitative reviews and vote counting.

(2) Applications of meta-analysis in ecology are becoming increasingly more common, and thus even if you are not planning to conduct your own meta-analyses, you need to understand the method to follow and evaluate the literature in your field.

(3) Application of meta-analysis to applied fields (e.g., conservation and environmental management) can make results more valuable for policy makers.

(4) Learning the basics of meta-analysis can dramatically improve standards for data reporting in primary studies so that the results can be included in subsequent research synthesis on the topic.

(5) Conducting meta-analysis changes the way you read and evaluate primary studies; it makes you acutely aware that the statistical significance of the results depends on statistical power, and in general improves your abilities to critically evaluate evidence.

Meta-analysis is more demanding than other research synthesis methods in terms of the format of the data required for analysis (Table 1.1). In ecology and evolutionary biology this problem is further exacerbated by uneven and poor reporting standards. Many primary studies in ecology and evolution do not report essential data needed to calculate effect sizes, or needed for critical appraisal of the results (e.g., standard deviations or other measures of variation, and sample sizes). As a result, those studies have to be excluded from conventional weighted analyses, often resulting in a dramatic reduction in the number of available studies and a consequent loss of information. This is sometimes used as an argument to justify the use of vote counting instead of meta-analysis in ecology and evolution (Heck et al. 2003). However, the problem of poor reporting of primary data can sometimes be solved in more satisfactory ways than by resorting to the statistically flawed vote-counting procedure. Two approaches have been suggested by Gurevitch and Hedges (1999) for situations where lack of variance or sample size information prevents the use of weighted meta-analysis; these are randomization tests (Adams et al. 1997), or unweighted standard parametric statistical tests, such as ANOVA or least-squares regression. Examples of the use of these approaches in ecology are provided by Johnson and Curtis (2001) and Coleman et al. (2006). The advantage of randomization tests and unweighted parametric methods over vote counting is that they provide an estimate of the magnitude of the effect, which vote counting lacks. In addition, resampling methods are free from normality assumptions of the parametric statistical tests. Note, however, that both randomization tests and unweighted parametric methods assume homogeneity of variances; this assumption is likely to be seriously violated when combining studies, because different studies typically have vastly different sample sizes and vary greatly in precision (Gurevitch and Hedges 1999).

Another possible way to avoid the loss of information when only a subset of studies provide data needed for meta-analysis is to conduct a proper weighted meta-analysis on studies that provide enough information to calculate effect sizes, with a vote count on the remaining studies. This approach is common in ecology (e.g., Borowicz 2001, Huberty and Denno 2004, Liu and Stiling 2006, Attwood et al. 2008); it allows all studies to be used, and the results of the meta-analysis can be compared with those of vote counting. Good agreement between the methods is observed in some cases, thus reinforcing the conclusions of the research synthesis, but discrepancies are also common (Liu and Stiling 2006). Bushman and Wang (1996, 2009) have proposed a procedure that combines estimates based on effect size and vote-counting procedures to obtain an overall estimate of the population effect size. To our knowledge, this approach has not been used yet in ecological and evolutionary meta-analyses. In general, we concur with the recommendation by Bushman and Wang (2009) that vote-counting procedures should never be used as a substitute for effect size procedures and should never be used alone. Combining probabilities may also be used in combination with weighted meta-analysis (Kraus 1995). Additional quantitative and qualitative sources of information like expert opinions can be incorporated into research synthesis by using a Bayesian approach (Kuhnert et al. 2005, Newton et al. 2007, Choy et al. 2009).

Some ecologists argue that meta-analysis is ill advised in situations when the data are very heterogeneous (e.g., in terms of research methods, experimental design, response variable measured, and organisms studied) and suggest that vote counting offers a more cautious and conservative approach to research synthesis in these circumstances (Daehler 2003, Heck et al. 2003, Ives and Carpenter 2007, Tylianakis et al. 2008). This argument refers to the "apples and oranges dilemma" and essentially repeats the argument of Eysenck (1994) that "meta-analysis is only properly applicable if the data summarized are homogeneous—that is, treatment, patients, and end points must be similar or at least comparable." While it is true that sometimes collected studies are too heterogeneous to allow meaningful synthesis (Markow and Clarke 1997), using this argument to justify a vote count instead is problematic. Vote counts of a truly heterogeneous body of studies offer no advantages, and considerable disadvantages, over meta-analysis. If the studies are considered to be too heterogeneous to be sensibly combined by meta-analysis, why is it justifiable then to lump them together in a vote count, or even a narrative review? Meta-analysis at least provides a way to explicitly analyze the extent of variation in effect sizes among studies and to reveal the causes of this heterogeneity whereas both vote counting and narrative reviews are unsuited to doing so. The scope of the review has to be carefully considered before any research synthesis is undertaken, and generalizations have to be sought on a biologically meaningful level. If the original question is too broad and results in a set of primary studies that is too heterogeneous, the question has to be redefined more narrowly to allow meaningful synthesis, regardless of the synthesis method used. Sometimes, combining heterogeneous data can be valid, depending on the level of generalization one wishes to make. For example, meta-analysis has a very important role in generalizing across species, ecosystems, and other larger-scale entities, beyond the scope of individual studies (Chapter 23).

SUMMARY AND CONCLUSIONS

Meta-analysis alone or in combination with other methods of research synthesis should be used whenever the estimate of the magnitude of an effect and an understanding of sources of variation in that effect is of interest, and when at least some of the primary studies gathered provide sufficient data to carry out the analysis. However, if few studies on the question exist, and the aim is largely to make the reader aware of an emerging research field or a new direction in the established field (as in *Current Opinion in . . .* and *Trends in . . .* journal series), meta-analysis combined with a systematic review of the topic may be unnecessary and a short narrative review may suffice.

If ecologists and evolutionary biologists wish to bring scientific methodology to bear on using the "weight of evidence" to inform policy making in conservation and environmental management, moving toward more scientific methods of synthesizing available data will become increasingly necessary. This is already widely accepted practice in medicine and in social policy. As the need for ecologists to be heard by policy makers and the public in addressing critical issues, such as biodiversity loss and climate change, becomes ever more urgent, mastery and implementation of the "evidence-based" tools of systematic review and meta-analysis becomes all the more compelling.

The Procedure of Meta-analysis in a Nutshell

Isabelle M. Côté and Michael D. Jennions

IT IS SAID THAT a picture is worth a thousand words. Taking this at face value, we offer two figures to summarize the entire meta-analytic process. In Figure 2.1 we cover Part I, the initial stage of formulating a question and systematically searching the primary literature for suitable studies. In Figure 2.2 we cover Part II, the stage at which you extract data from publications, run statistical tests, and present and interpret your results. The two figures explicitly link each step to the relevant chapters (indicated by a circled number). These indicate where Chapters 3 to 27 fit into the process of meta-analysis. You will notice teardrop-shaped symbols next to each step. These are not intended to indicate the tears of frustration that will be shed as you proceed (although they might!). Instead, they are meant to reflect the beads of sweat, or hard work, that should be expected. Our reason for including them is to reassure the novice that the hardest part of a meta-analysis is not mastering complex statistics (statistical procedures for meta-analysis, as we will see, may range from very simple to advanced) or software, but the labor involved in gathering, evaluating, and assimilating papers. In particular, meta-analysis requires the additional work of extracting data, which is often frustrating and challenging, and is not typically part of an old-fashioned narrative review. The key information to extract is an estimate of the strength of a relationship in the form of an effect size. Effect sizes are a common currency into which the outcomes of all papers are translated so that they can be combined and compared (e.g., Hedges' d is the difference between a control and treatment group expressed as the number of standard deviations by which they differ, or Pearson's r is the correlation between two variables of interest). For now you can think of an effect size as a P-value corrected for sample size with the direction of the relationship also provided (for details, see Chapter 6). If two studies report the relationship between two variables as $P = 0.05$, $n = 20$, and $P = 0.05$, $n = 50$, the relationship must be stronger—hence Pearson's r larger—in the first study. Why? With less data the confidence interval on the estimate of r is wider, so the estimated mean value of r must be greater for the lower limit of the 95% confidence interval to be above the null value of 0 (recall that $P = 0.05$ when the 95% confidence interval for a test statistic just touches the null value).

Performing a meta-analysis is in fact conceptually no different from an empirical study in the sense that sometimes statistical problems bog you down. But researchers usually design a study with their own statistical abilities in mind, or they follow an established design that allows them to replicate a standard analytic approach. The difference between a good and a bad empirical study often boils down to whether an interesting question is being asked, and to the quality and quantity of the data collected using an unbiased sampling technique. The same is true of meta-analysis. Statistical elegance is wonderful, but it is subservient to the quality/quantity of data collected. (After all, you can always reanalyze data.) If you are a novice you will be reassured to know that anyone who uses a basic Windows-driven statistical package, such as Minitab, SPSS, or JMP, has the ability to do the statistics necessary to produce a publishable meta-analysis. Even tricky issues often have simple solutions if you are prepared to use less powerful statistical approaches (e.g., use the mean of a set of nonindependent measures rather than a repeated-measures model; see Chapter 16). Another alternative is to collaborate with someone with advanced statistical skills when you wish to delve into more complex models. If like many biologists, you consider yourself "mathematically challenged," do not be deterred; you likely already have most of the skills, training, and insights needed to conduct a top-notch meta-analysis.

SOME CAVEATS ABOUT THE ROAD MAP

The two figures offer a road map depicting the usual sequence of events that make up a meta-analysis. You should, however, be aware that the exact content of the activities in each box can vary considerably among meta-analyses.

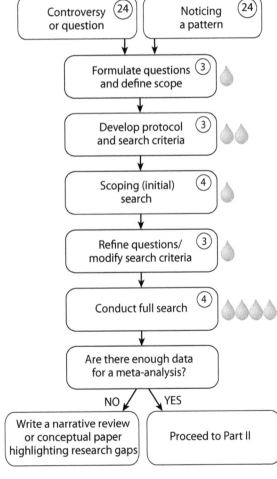

Figure 2.1. Part I of conducting a meta-analysis. The teardrop shapes represent (in our experience) the relative effort (drops of sweat) associated with each step of the process. Circled numbers refer to relevant chapters in this book.

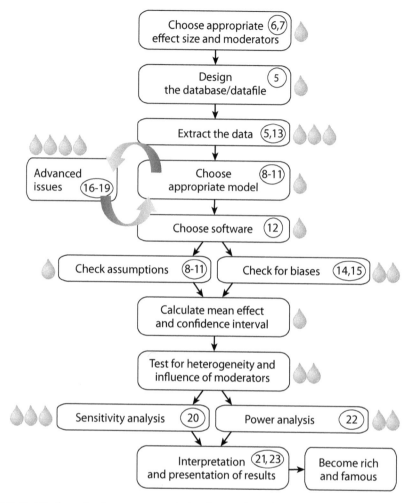

Figure 2.2. Part II of conducting a meta-analysis. The teardrop shapes represent (in our experience) the relative effort (drops of sweat) associated with each step of the process. Circled numbers refer to relevant chapters in this book.

First, the extent of the initial (scoping) search and the rigor of the final full search vary among published meta-analyses. The ideal scenario is to search systematically for all available data from all possible sources (Chapter 3), but very few ecologists and evolutionary biologists carry out this type of full-scale systematic review. In fact, current practice leans rather strongly toward trying to find all published (and some unpublished) papers on the question of interest, with less attention given to gray literature (government reports, conference proceedings, and so on). The extent to which the gray literature needs to be searched is likely to vary widely depending on the type of meta-analysis. It is expected to be important when tackling a question about conservation or environmental policy (e.g., the effect of traffic on frog populations), but unlikely to provide useful data when conducting a meta-analysis on a highly theoretical evolutionary topic (e.g., the effect of sperm competition on mutation rates).

Second, while our outline makes the process seem linear, there is often uncertainty that leads to some steps being repeated. In particular, in Part I you might have to repeat several cycles of scoping searches, adjusting study questions and modifying the protocol and search criteria until you are confident that a full search will generate the appropriate amount of data (i.e., enough to produce meaningful results, but not so much as to make it logistically impossible to extract data from all relevant sources before your retirement).

Third, you will usually find that once you start to extract data from papers, your database has to be redesigned. This is usually because you have overlooked a factor that is a potential source of variation in study outcomes (i.e., effect sizes). For example, perhaps only after reading 20 papers on photosynthesis do you realize that there could be important differences in the results because researchers use two different methods to estimate average daily light intensity. This means that you can end up reading some papers several times to fully encode them into your database, and then you will also have to update your protocol.

Fourth, although the best practice is to fully test all model assumptions, look for publication bias, perform a sensitivity analysis, and calculate the power of every test you conduct, few published ecology and evolution meta-analyses do this. Almost none, for example, report statistical power. In our view, such shortcomings do not invalidate a meta-analysis any more than, for example, failure to analyze subsets of the data invalidates a primary empirical study. In short, while you should strive to be as thorough as possible, do not be put off by fears that you will be tackling some elements of the process less thoroughly than others (due to lack of expertise and/or time). The bottom line is that if you have a clearly defined protocol you will produce a data set whose limitations can be interrogated further by others, or yourself, long after you have published your meta-analysis.

A FEW WORDS ON EACH STEP

As the two figures show, each step in the process is covered in detail in one or more relevant chapters. Here we simply provide a short description of what each step involves. We also offer tips based on our personal experiences. Someone once said, "advice is just a set of anecdotes about what worked for the giver," so you can take what we say at face value, or be more circumspect and see where our views fall relative to the expert advice offered by our colleagues in the rest of this book.

Formulate question and define scope

This is the fun bit. What do you want to know?

(1) Do you want to evaluate a theoretical hypothesis? If so, you will need studies that use experiments to test for causality. You should be careful not to mix up studies that report an observed relationship and relationships identified through experimental manipulation (i.e., you should encode the study type). (Note that in evolutionary biology, "natural experiments" and "phylogenetic correlations" are sometimes considered to provide insights on causality, especially with respect to identifying selective forces; see Shadish et al. 2002).

(2) Do you simply want to test whether there is a consistent relationship between two or more variables (i.e., a biological rule)? If so, observational and experimental data can probably be combined (although it is still advisable to encode study type and test for its influence).

(3) When formulating a question, think carefully about how widely you wish to generalize and what size data set you can manage. For example, it would be great to know the general relationship between genetic heterozygosity and fitness, but do you have the resources to examine all studies published on this subject? Or should you confine yourself to a specific aspect of fitness (e.g., mating success or longevity) and/or taxonomic group (e.g., only birds or all vertebrates)? The answer will partly depend on what your initial search of the literature reveals; if there are too many studies, you will have to be less ambitious in the scope of your questions (or get more funding for library assistance). Our anecdotal tip is that the closer the questions you ask are to your own area of expertise, the easier your life will be. When a research field is distantly related to your own, it becomes more difficult to read papers and extract data, and the likelihood grows that you will overlook studies (e.g., due to use of unsuitable search terms, and unfamiliarity with key narrative reviews and the identity of researchers) or miss some important potential moderators of effect sizes.

(4) In most cases, the key question in ecological and evolutionary meta-analyses (as in many other fields) can be distilled to the following: What is the mean effect? Does it differ from the null expectation? Can we explain variation (heterogeneity) in the outcome of different studies? The last question requires that you think about factors (moderators) that differ among studies that might affect the effect size estimate. There is much opportunity here for creative insights (see Griffin et al. 2005 for a great example from evolutionary biology). Our practical tip to get you started is to encode studies based on the *population* studied, with the most obvious groupings of taxonomy, geography, habitat or ecotype, functional type (e.g., predator/herbivore, tree/shrub), and organism size; the *methodology* used (e.g., lab or field study, duration of study); how the *outcome* is measured (e.g., sperm count/ejaculate volume, leaf area/leaf mass); and the population to which a treatment is *compared*, that is, the baseline level of the measured outcome (e.g., rate of disease in control plants). You should note that in some cases the comparison will simply be to a null value (e.g., $r = 0$). Remember that moderators of effect sizes can be discrete factors or continuous variables. A final tip is that there is no point having a very large number of moderators, or many levels of a discrete moderator. If the number of papers per group is small you will have very low statistical power. Be prepared to pool groups before conducting your final statistical analyses (e.g., you might code "bird," "mammal," "insect," and "spider," but end up using the codes "vertebrate" and "invertebrate" instead). You should be keenly aware of confounded moderators, and decide how you want to handle this. For example, if all studies on woody plants are longer than one year and all studies on herbaceous plants are shorter than four months, then one cannot test for study duration without also comparing functional groups and vice versa.

Develop protocol and search criteria

Once you have formulated your questions, you need to write up a protocol that will (1) formally specify the questions you intend to ask, (2) specify an objective search strategy, and (3) establish study inclusion criteria. Specifying the questions you intend to ask (1) will include being specific about potential sources of heterogenity in effect sizes.

Specifying an objective search strategy (2) entails not biasing data collection toward relevant papers that you are already familiar with; otherwise, this could affect the outcome as it is often easier to remember papers with significant results (Chapter 14). Primarily, this involves making a list of which electronic databases you will search and what search terms you will use. Chapter 4 includes excellent practical tips on how to search. Secondarily, an objective search strategy will involve a decision about how much effort to expend searching the "gray literature." For example, will you write to colleagues asking for unpublished data, and if so, who and why?

Once you have conducted a search and compiled a list of potential papers, you have to establish study inclusion criteria (3). These criteria are often fairly obvious, and include the following: Does the study fit the scope of your questions? (For example, is it on a marine ecosystem?) Does the methodology fit with how your question is defined? (For example, was there an experimental manipulation?) If so, was it of sufficient magnitude or duration? Does the study contain extractable data, that is, is there sufficient information to extract an effect size, its variance, and the sample size used? Your inclusion criteria will sometimes have to take into account study quality. This is far more difficult to assess than the criteria we have listed above, but it can be just as important (see Chapter 5).

To date, most published ecological and evolutionary meta-analyses have not presented a de-tailed protocol. This is not a good reason to do the same; quality standards for meta-analysis are rising in ecology and evolutionary biology. It is worth noting, however, that as in primary re-search, your protocol for searching literature and extracting effect sizes will almost certainly be modified as you proceed. The reality is therefore that, in many respects, your final protocol will end up describing what you did rather than what you ideally wanted to do. Our best advice is to keep your eye focused on two goals. First, you must tell the reader how you collected your data. So, just as in primary research, you provide the reader with enough information on data collec-tion and analysis to allow your review to be repeated and updated in the future. Second, you must have a protocol that forces you to evaluate continually whether your sampling is biased. A protocol increases the objectivity with which you compile data, but it should not blind you to the reality that the process of meta-analysis involves numerous subjective decisions; these are most apparent when trying to decipher the results of a given paper and deciding whether you can extract the necessary data for your synthesis. If more than one person is collecting the data, a well-described and tested protocol is very important to ensure uniformity in data extraction and coding decisions about moderators (Chapter 4).

Scoping (initial search), refining questions, modifying search criteria, and deciding on an effect size measure

Sometimes, if you are confident that most studies will be confined to a few key sources, you might only search a limited set of journals. This was how almost all research syntheses were done originally, before online databases were available (e.g., Gurevitch et al. 1992). No one uses this approach any more because of the explosion of data accessibility (at least for those with access to scientific journals). However, the question of whether to delve into unpublished or "gray" sources remains an important one. In ecology and evolution, using only easily ac-cessed published literature is arguably justifiable when dealing with a fundamental topic with no obvious applications. This work is almost always published in peer-reviewed journals. For example, studies on strategic investment in sperm by males in response to female mating status are unlikely to be featured in government reports and, due to the specificity of the topic, it is fairly easy for an experienced researcher to know which papers will be irrelevant based purely on the journal of publication. In other cases, however, such as questions about the loss of rain-forest or coral reefs, it is likely that suitable and sometimes crucial data can be extracted from many sources (e.g., journals, conservation reports, ecological impact assessments). If so, you need to give some consideration to how many effect sizes each source is likely to provide. This can be difficult to estimate in the absence of prior experience. It matters though, because these and other issues, such as whether non-English language literature is included, can have a real bearing on how unbiased your analysis will be (Chapter 14).

Remember that during a scoping search you are *not* trying to find every study or obtain a preliminary estimate of the mean effect. You main goals are to (a) estimate how much data there is so that you can decide whether to broaden or narrow your study questions; (b) work out what factors vary among studies that you might encode as potential moderators; (c) decide what criteria mark a study as obviously irrelevant (e.g., if your search identifies 2000 papers to read in full, you will definitely have to make some exclusion decisions based on the title, abstract, and/or place of publication); (d) work out what criteria each potentially relevant study must fulfill before you try to extract an effect size; (e) establish the format of your data extraction form/spreadsheet; and (f) decide upon the most suitable measures of outcome (effect sizes). This last decision will often depend on whether data are reported as a relationship between two continuous variables, in which case the effect size r is the most popular choice in ecology and evolution. Alternatively, the decision may involve a comparison between two groups, in which case there is a range of options depending on whether the response variable is discrete (e.g., alive/dead) or continuous (e.g., body size) (Chapters 6 and 7). Our quick tip is that it is sometimes easiest to conduct separate meta-analyses dividing studies on the basis of the most appropriate effect size. For example, in a review of the effect of a treatment on animal mortality, you might want to first do a meta-analysis using the log odds ratio for studies that measured whether animals were alive or dead at the end of the study, and then do a second meta-analysis using Hedges' d for those that measured lifespan.

Conduct a full search

Once you have established your protocol you will proceed to do a full search. This will generate a large number of studies, but many will be discarded as obviously irrelevant using criteria based on the study's title, abstract, or place of publication. The remaining "potentially relevant" studies will need to be read more closely and divided into relevant and irrelevant. This process can lead to a large reduction in the number of papers at each step (Fig. 4.2). Be prepared for a large number (often the majority) of studies that you initially identify as relevant turning out to be unsuitable for the meta-analysis. This reflects in part the low publication standards in ecology and evolutionary biology (e.g., variance estimates or sample sizes are often missing) and, in part, the diversity of approaches used by biologists to address specific questions. The final step is to try and extract the information you need (effect sizes and moderators) from relevant papers.

A practical tip we offer is that there is a trade-off between building up a pile of relevant papers and then returning to them to extract effect size once you have a finalized data spreadsheet versus extracting data from a paper as you read it. The advantage of the former is that you can be more confident that, having read all the relevant papers once, your spreadsheet contains all the information that you want to extract. The advantage of the latter is that you do not have to read a paper in depth twice. It can be surprisingly hard to establish exactly how a study was designed and which are the relevant data to extract an effect size. Unless you keep good notes (which we strongly recommend), it is often no easier on a second reading. If you are confident that you have a good understanding of the main features of the relevant studies, then you might consider designing a database and extracting data as soon as you classify a paper as relevant. The caveat, of course, is that you might still have to return to these papers if you later discover that you need to encode an additional moderator term or adjust your study inclusion criteria.

Extracting information on the initial reading is most feasible when dealing with studies that closely follow a specific and commonplace experimental design.

Designing the database/datafile and extracting data

Designing a database is an art, and Chapter 5 offers concise and practical advice. We recommend you read it. The basic rules are the same as those for an empirical study; you should ensure the datasheet contains all the information you need and is set out logically so that it is easy to complete and difficult to enter data in the wrong place. A trial run can help to tweak the format. Our practical tip is to use the longest, most complicated looking papers you have in order to trial test your datasheet. If it can handle them, you are off to a good start.

The main difference between an empirical study and a meta-analysis is that you need to record which subjects you did and did not collect data from. In short, keep a bibliographic library of studies and explain why some were excluded (e.g., irrelevant, missing key information needed to satisfy inclusion criteria, not possible to extract an effect size and/or variance estimate). Again, Chapter 5 has sound practical advice on organizing a bibliographic library.

With a good protocol it is fairly easy to encode information on study moderators. In contrast, extracting effect sizes is among the most challenging parts of meta-analysis. It can lead to self-doubt, especially during your first meta-analysis. To extract effect sizes you often have to make subjective decisions (albeit based on procedures outlined in your protocol) about such seemingly easy questions as the relevant sample size (often used to calculate the effect size variance). You will be shocked to discover how poorly results are often presented in the ecological and evolutionary literature (this is a favorite topic among meta-analysts meeting over beer!). In some cases you might simply give up on a paper and place it in the "not possible" category. Do not be put off by this "failure"; everyone in this field who conducts a meta-analysis has the same experience. Take it as a sign that your meta-analysis will be a valuable contribution. You are helping others identify papers with results that are too confusing to be appropriately cited as showing a given effect. These papers were presumably previously cited based on qualitative statements made in the abstract or discussion, which you now know to be difficult to substantiate based on the results presented.

Finally, have a protocol in place to deal with studies that report multiple effect sizes. Specifically, if treatment effects are measured repeatedly over time, you need criteria to determine which comparisons you will use. Likewise, if there is a single control and several treatments varying in magnitude, will you extract all treatment-control comparisons, or only the one between the control and the most extreme treatment? The statistical issues surrounding nonindependence of effect sizes are dealt with in Chapters 16 and 17. If you want to keep life simple, the easiest solution is to design a protocol that only extracts one effect size per study. Be aware, however, that you are sacrificing information and statistical power for ease of analysis. For an excellent basic introduction to complex data structures within a study and the similarities and differences among multiple independent subgroups, multiple outcomes (including time points), and multiple comparisons, we can highly recommend Borenstein et al. (2009, 215–45). They provide very simple, fully worked examples of how to calculate effect sizes and variance in each case. These should help you to decide on the best protocol for your meta-analysis.

One complicating issue in ecological and evolutionary meta-analyses is that there might be nonindependence between studies due to a shared phylogeny history. This can occur when studies of the same species or closely related species produce more similar results (effect sizes) than those from distantly related taxa. Statistical techniques to incorporate analysis of correlations among studies due to shared phylogeny are covered in Chapter 17. However, to be pragmatic we should note that most published meta-analyses do not fully control for this form of nonindependence. At best, they treat studies of the same species as the equivalent of multiple independent subgroups in a study, such that the final analysis is based on the mean effect size for each species. This may be a serious limitation for some meta-analyses (e.g., those in evolutionary morphology), and little or no problem for other meta-analyses (e.g., those investigating many ecological traits). Note, however, that this relaxed approach will undoubtedly change soon because phylogenies are becoming more readily available and a user-friendly software including phylogenetic data in meta-analyses has recently been developed (Lajeunesse 2011a). For now, however, we think that it is reasonable to publish meta-analyses that do not fully control for phylogeny (as long as this limitation is acknowledged), because when the data are presented in full they can always be reanalyzed in the future by other researchers.

Statistical models, software, calculating means, and looking for moderators

There is no easy way to summarize the many different statistical models for meta-analyses in a few sentences (for details, see Chapter 8). However, we can provide at least some reassurances for the novice meta-analyst. First, in the same way that you use statistics to test empirical data, even though their formal mathematical basis may be a distant memory from your undergraduate years, you can use meta-analysis software without having to recall every detail of the underlying mathematics. However, you do need to be aware of the underlying assumptions and understand at least the gist of the mathematics involved to ensure that you choose the right model for the task at hand. As with conventional statistics, interpreting your results correctly (which requires a good conceptual understanding of what has actually been tested) is vital. (Again, we can highly recommend Borenstein et al. (2009) who provide a very clear explanation of the main statistics used in meta-analysis. The authors take the reader through worked examples in sufficient detail so that each step is easy to follow. By "copying" their approach it is easy to ensure that you appropriately analyze your own data set and know which approaches are needed to answer standard questions and resolve generic problems common to many data sets.)

A second reassurance is that the main tests you will conduct have the same basic logic as those used in conventional statistical analyses. For example, in a primary empirical study you might want to know whether the difference between two treatments is greater than zero (the null value) (i.e., a t-test that is equivalent to an ANOVA with two levels), and whether there is a confounding variable that explains why the difference is bigger in some cases than others (i.e., a two-way ANOVA if the confounding variable is discrete, or an ANCOVA if it is continuous). Likewise, in a meta-analysis you usually want to calculate the mean effect size and see whether it differs from the null value. For example, for the effect size r, does the 95% confidence interval overlap zero? You also usually want to test for more variation among studies in effect sizes than expected by chance. In other words, you are testing for heterogeneity, so that you know whether there is a reason to look for confounding variables. You then test whether the effect size covaries with a continuous moderator variable, or whether the mean effect varies significantly among groups defined by a discrete moderator variable.

Third, as with analyses of empirical data, the real issue is whether your data on effect sizes conform to the assumptions of the statistical model you use, and whether you can interpret the output. For example, it matters that you recognize how a major outlier can change the P-value of a test, or whether nonsignificance is hard to interpret when sample sizes are small and statistical power is low.

Finally, as with conventional statistics, the models become more challenging when there are more predictor variables, when response variables are not statistically independent, and when there are nested, split, or crossed designs. Comparable solutions to those used with empirical data can be applied in a meta-analysis. They can range from the path of least resistance (e.g., use of the mean value when you have repeated measures from the same subject), to the more sophisticated (e.g., use of the most powerful possible statistical design and collaboration with a colleague who has greater statistical expertise).

Chapter 12 offers help in deciding which software to use. It is also useful to consult published meta-analyses by colleagues in your own or allied research areas to see what packages they have used. As a practical tip, we have mainly used MetaWin (Rosenberg et al. 2000). It has limitations (the most obvious being that you can only look at one moderator variable at a time), but the advantage for the novice is that it has a "point and click" format that is easy to use. You should be able to calculate mean effect sizes and test for differences among groups (or for an effect of a continuous moderator variable) within the first hour of using it.

Rigor: Check assumptions, publication bias, and power and sensitivity analysis

As with any scientific study it is always advisable to be self-critical. In the case of a meta-analysis, this can be distilled down to (a) checking that your data fit the assumptions of the statistical model you use, (b) testing for publication bias, (c) calculating the statistical power of key tests, and (d) testing how sensitive your results are to the inclusion/exclusion of data (e.g., whether the removal of effect sizes from unpublished studies or studies you classify as "lower-quality" changes your main conclusions). In general, it is possible to follow simple recipes to conduct these analyses (Chapters 14, 22, and 24), but the more you understand your data and the more you have thought about them, the higher the quality of your meta-analysis is likely to be. Also, as already noted, while it is best to be as thorough as possible, failure to, for example, report statistical power is no more likely to invalidate your meta-analysis than it would a primary study.

Presentation and interpretation of results

Once you have compiled your data set and obtained your results, you need to write up your paper and publish it. Chapter 20 contains a comprehensive checklist that you can work through to maximize the quality of your meta-analysis and publication. In our view, it is important to remember that, like a primary study, there is much benefit in presenting "raw data" showing the effect sizes associated with each study (note that this is now required by some funding sources), and to also offer a full list of studies that were included and excluded from your final data set. These can be lengthy documents, so they are often better presented as electronic appendixes than as tables in the main text. These raw data can considerably facilitate updates of a review as new studies become available and can provide an explanation for conflicting results of two reviews on the same topic. The other point we highlight is that it pays to provide a detailed account of your methods (i.e., your protocol). In some cases this could lead to difficulties with reviewers unfamiliar with systematic reviews because such descriptions may, in their eyes, take up too much space in the Methods section. Again, a simple solution is to be concise in the main text and use an electronic appendix to publish your protocol in its entirety. Finally, while it can be useful to consult published meta-analyses to gain a feel for how to present your results, you should remember that meta-analysis is still a relatively new technique in ecology and evolutionary biology. Many published meta-analyses are the first conducted by their authors, and the reviewers of the publication might have had no experience conducting meta-analyses and even limited experience reading them. In short, standards are certain to rise so you should not assume a meta-analysis is well presented simply because it is published. (That is, unless it was done by one of us, of course, but we will not say which one!)

CONCLUSIONS

So there you have it. Ultimately, like riding a bicycle, the only way to become comfortable with conducting a meta-analysis is to give it a try. We strongly encourage you to do so because there are many benefits (Chapter 1). You will find that thinking in terms of effect sizes makes you a better scientist who is more aware of the disparity between the outcome of individual studies and the sweeping generalizations that can ensue (Chapter 23), and of the perils of relying solely on P-values to classify studies (Chapters 1 and 23). You will also find that if you persevere and publish your meta-analysis it will be a far longer-lasting (and we hope more widely cited) publication than an equivalent narrative review. If nothing else, we can assure you that simply attempting to extract effect sizes from seemingly relevant papers will teach you some valuable lessons. You will discover that many researchers do an appalling job of presenting their statistical results. (Where are the degrees of freedom? What is the sample size? What is the direction of the difference between the two treatments? How did this paper ever get past peer reviewers and editors?) This will help you improve both the presentation of your own primary research and how you review manuscripts. You will realize that many studies do not show what they are cited as demonstrating. This can be a sobering experience. Meta-analysis forces you to examine the primary literature more critically. Finally, you will find that there are often far fewer papers with usable data than you originally assumed. This can be a useful insight allowing you to decide which primary studies are worthwhile and/or to convince a reviewer that, in fact, your own primary research is not testing a repeatedly confirmed hypothesis. So think of a question and start formulating your search protocol. Good luck!

Initiating a Meta-analysis

First Steps in Beginning a Meta-analysis

Gavin B. Stewart, Isabelle M. Côté,
Hannah R. Rothstein, and Peter S. Curtis

"Do nothing in haste; look well to each step; and from the beginning think what might be the end."

Edward Whymper (1986)

THIS CHAPTER IS CONCERNED with initiating the process of systematic research synthesis. Whymper's advice following the death of his companions on the first ascent of the Matterhorn is as relevant to those synthesizing mountains of data as to those climbing real peaks. Failure to carefully define the problem and methods can result in serious, sometimes fatal errors and terminal falls for meta-analyses. Without a systematic approach to defining, obtaining, and collating data, meta-analyses may yield precise but erroneous results, with different types of sampling error (biases) and excess subjectivity in choice of methods and definition of thresholds; these devalue the rigor of any statistical approaches employed. Formal a priori definition of the problem and methods using an explicit protocol has become *de rigueur* in medicine, psychology, and other evidence-based disciplines. The Cochrane (http://www.cochrane.org) and Campbell (http://www.campbellcollaboration.org) Collaborations provide centralized repositories of reviews and protocols in public health and social sciences, respectively (Chapter 25); an international registry of protocols for all health-related synthesis has been established (http://www.crd.york.ac.uk/PROSPERO) together with an agreed minimum data set for protocol registration (Booth et al. 2011). A repository of environmental management reviews and protocols is available at http://www.environmentalevidence.org, but many syntheses either do not have or do not report protocols, especially if they do not consider an intervention.

While the submission of review protocols may not always be required in ecology and evolutionary biology, researchers should embrace the philosophy of the approach and develop their own protocols to help them plan and report their meta-analyses. The increasing availability of online supplementary material provides the potential for all published syntheses to include a protocol and discuss deviations from planned analysis in full. Those interpreting the output of meta-analyses that are not accompanied by protocols should be very cautious to avoid a tumble down a precipice themselves!

This chapter considers exactly the same issues that face an ecologist designing a field experiment. What's the question? How can I define my sampling universe? How should I collect my data? What analyses should I undertake? How should I interpret my results robustly? These questions are considered in the context of research synthesis below.

FRAMING QUESTIONS FOR RESEARCH SYNTHESIS

The first step in planning and initiating a meta-analysis is to define the question and develop a protocol. The development and definition of postulates as hypotheses is widely recognized as central to scientific methodology for primary research (Ford 2000, Quinn and Keough 2002). However, ecologists often fail to use a formal, rigorous process when carrying out reviews (Gates 2002, Roberts et al. 2006). It is especially important to define questions prior to synthesis when reviewing existing data because this activity is retrospective by nature (Light and Pillemer 1984), and is therefore especially vulnerable to data mining. However, post hoc rationalization may be valid where relationships were not fully understood prior to the initial analyses (see below). Where questions are not explicitly defined a priori, research syntheses can become biased, as there will inevitably be disagreements regarding which studies are relevant, let alone how they are to be combined and weighted relative to one another. It therefore becomes difficult to distinguish between data mining and post hoc rationalization. Failure to fully define questions (and methods) can make interpretation of results very difficult, and severely constrains the utility of a research synthesis.

Reasons for conducting meta-analyses vary. For example, you might be interested in evaluating the direction and magnitude of a pooled effect, or you might want to explore sources of variation among studies, evaluate the evidence for various hypotheses, compare results from small studies with those of large studies, test hypotheses that cannot be or are difficult to test within primary studies (e.g., exploring large geographic trends). Or perhaps you want to identify gaps in research, test the effectiveness of various management strategies, or combine data from large databases. Depending on the specific aim of the meta-analysis, the scope will be different and this will in turn affect protocol.

There are four critical components to questions regarding effectiveness posed in a research synthesis: (1) the subject (population or subpopulation to be examined), (2) the intervention (treatment or other independent variable of interest), (3) the outcome(s) or responses of interest, and (4) where applicable, the comparison or control group. Some syntheses also specify the nature of study designs that will be acceptable, while others do not limit themselves in this way. Formulation of the question for a research synthesis often starts with identifying the above four components; this is known as the PICO approach (the acronym PICO stands for population, intervention, comparator, and outcome). However, the PICO components may not all be relevant in any particular situation. Different approaches have been taken to question setting in systematic reviews that either do not consider an intervention at all, or that consider very broad questions encompassing multiple interventions. These are active areas of methodological research in medicine and more widely, and merit further exploration in an ecological setting.

Medical prognostic studies may define a response variable as disease status and a prognostic factor (or more commonly multiple factors) as explanatory variables usually with defined cut points and populations (Centre for Reviews and Dissemination 2009). In these situations the treatment or intervention component is largely irrelevant, although it may receive some attention because the prognosis may only be relevant to a group of patients receiving a particular intervention. Similarly, ecologists and evolutionary biologists considering an impact may wish to adapt the PICO components, and define response and explanatory variables explicitly by considering definitions in terms of the limits of relevant variation. This may involve specifying the spatial or temporal domains of specific variables, or the manner in which they are measured or may relate to taxonomic definitions.

Reviews of public health questions also often adapt the PICO approach because they are usually broad and multifaceted, often seeking to address wide policy-based enquiries where a range of specific interventions exist. Broad questions can be split to provide a narrow focus where a PICO approach becomes feasible, but this may limit direct policy relevance. Alternatively, the context within which the intervention is delivered results in the inclusion of an additional C in the PICO acronym (Petticrew and Roberts 2006)—or more rarely, the use of a logic framework. The development and recording of a conceptual framework in the protocol involves mapping the hypothesized causal relationships between determinants (environmental, social, biological) and outcomes. The framework is then used to identify links between determinants, outcomes, possible interventions, and strategic points at which to intervene. Once mapped in this way, decisions can be made about which interventions to include (Centre for Reviews and Dissemination 2009). Such logic maps usually take the form of diagrams, but the use of Bayesian belief networks would provide a more formal framework both to compare initial stakeholder positions and as an outline for subsequent synthesis. However the question components are defined, the key point is that a systematic approach remains paramount, with a focus on critical thought, transparency, and explicitness about methods.

On occasion, meta-analyses conducted at approximately the same time by investigators with access to the same data, reach different or even contradictory conclusions (Rosenfeld and Post 1992, Williams et al. 1993, Kerlikowske et al. 1995, Smart et al. 1995). This is illustrated in ecology by the debate regarding the stress-gradient hypothesis where analysis of studies with different gradient lengths led to divergent results (Maestre et al. 2005, 2006; Lortie and Callaway 2006). These problems can be overcome, or at least made transparent, if questions are properly framed by defining the question elements and the methods used to address the question. For example, definition of the range of gradient lengths, or use of meta-regression to analyze the impact of gradient length in the initial analyses, could have explained the heterogeneity giving rise to the discrepancy in results.

Framing questions for research synthesis is a process that comprises many judgments and decisions. The definition of subject scope is particularly problematic, and meta-analysts are frequently accused of combining studies that do not deal with the same constructs and relationships (Lipsey and Wilson 2001). This can result in being criticized for attempting to compare "apples with oranges" (Wachter 1988). However, a combination of apples and oranges may be germane if the domain of interest is fruit (Rosenthal 1994, Rosenthal and DiMatteo 2001). This is illustrated in the first modern meta-analysis, which examined the effectiveness of psychotherapy (Smith and Glass 1977). This analysis was derided as "mega-silliness" (Eysenck 1978), because it combined findings from distinctly different therapies, including cognitive, behavioral, and psychodynamic. However, as Smith and Glass were interested in the overall effectiveness of all types of psychotherapies and in making comparisons between the different types, a broad sampling universe was essential. Proponents of meta-analysis therefore argue that heterogeneity of the measured effects should not be a deciding factor regarding the validity of a meta-analysis (Fernandez-Duque and Valeggia 1994), because within- and between-study variability, and the precision of effects, can all be informative even when the pooled mean is hard to interpret.

A focused reductionist question has the advantage of reduced heterogeneity, but this may be at the expense of widespread applicability, scope, and power, and is clearly inappropriate if a primary aim of the analysis is investigating reasons for variation. Broad questions and broadly defined syntheses may therefore often be appropriate in ecology, although researchers should beware of gathering and collating diverse data if they cannot ascertain how these data can be combined to answer a question.

Because deciding on scope remains subjective and strangely controversial, given the (over) reliance of ecologists on hypothesis testing (Anderson et al. 2000, Kirk 2007), meta-analysts should provide explicit definitions of the domain of interest (sampling universe). The utility of subsequent syntheses may be questioned, but as long as they are transparent and repeatable, readers can judge their value for themselves. Clear definition of question components provides this transparency and also ensures that the subsequent analyses are clearly focused and relevant (Lipsey and Wilson 2001, Khan et al. 2003, Pullin and Stewart 2006, Higgins and Green 2011).

General questions are typically broken down into specific hypotheses that are then structured using question components. For example, a general policy question regarding how we should conserve marine biodiversity could be expressed as specific hypotheses about the impacts or effectiveness of marine protected areas on specified biota, or the impacts of specific fishing practices. It is well-recognized that framing questions is a complex process; therefore most serious analysts devote substantial time and effort to question development before embarking on the search for data (Khan et al. 2003). The question is critical to the process because it defines the relevant domain for data collection (i.e., it generates the literature search terms for reviews) and determines relevance criteria. The relationship among the question components is complex (see below), and decisions made at this point can have a major influence on the conduct and outcome of the work.

It is also important to bear in mind the resources available for the synthesis. Clearly addressing all facets of a broad question will require more resources than addressing selected facets of a narrower question. Scoping searches will be required to aid decision making (Chapter 4). Scoping allows researchers to ascertain the quantity and nature of the information available, which may have a bearing on the way that the questions are framed. For example, where scoping suggests that little material is available, broadening the question may be appropriate. Alternatively, a narrowly focused question can be retained and a knowledge gap identified. The latter generally leads to the production of a systematic review without a meta-analysis in ecology, while the former may result in multiple analyses or a narrative review.

QUESTION COMPONENTS

Structured approaches to framing questions generally use four main components to define the hypotheses of interest, with additional criteria sometimes defined (Lipsey and Wilson 2001, Khan et al. 2003, Pullin and Stewart 2006, Higgins and Green 2011). The key components are the populations of interest, comparators, treatments, and responses related to the problem addressed by the synthesis. Additional definitions of study designs, subgroups, covariates, and magnitudes of differences are defined to greater or lesser degrees depending on available knowledge and the desired degree of rigour, but best practice requires clear definition of all of these factors (Booth et al. 2011). Each component is considered below.

Key components
Subject or population

The unit of study (e.g., ecosystem, habitat, species) is defined in terms of the subject(s) to which the treatment will be applied or the dependant variable. For example, systematic review of windfarm impact on birds (Stewart, Pullin, and Coles 2007) defined the subject explicitly as birds, with groups of particular concern being raptors (order Falconiformes, families Pandionidae, Accipitridae, Falconidae), breeding waders (uplands), nonbreeding swans, geese, coastal waders, sea ducks in general, and common scoters at sea. Search terms employed in the review to retrieve the data were based on multiple taxonomic organizational levels (see Chapter 4). The population was also defined geographically with a global focus. In this instance, the taxonomic subdivision of the population was based on an unstated combination of policy concern regarding these species and ecological functional grouping. The best practice is to explicitly define the rationale for choice of population(s); for example, see Cassey et al. (2005). Identifying potential reasons for variation between populations can also aid the interpretation of results, which may have important policy ramifications.

Treatment or intervention (often absent in observational studies and almost always absent in impact studies)

In an applied context the treatment or intervention is the proposed management regime, policy, or action. In an ecological or evolutionary context, this could be any independent variable of interest. Variation in the intensity, duration, and ecological context of the intervention may be associated with variable effects. Such factors should be considered and defined. For example, application of the herbicide asulam was defined as the intervention in a review of bracken control, with dose, season, and frequency of application defined as reasons for heterogeneity (Stewart, Pullin, and Tyler 2007).

Response or outcome

The response includes all the relevant objectives of the proposed intervention that can be reliably measured, or the relevant measures of the dependent variable. It is not uncommon to list multiple outcomes, but it should be explicitly acknowledged where it is known that they are not independent (Chapter 16). Some or indeed all relevant outcomes may not be measured by existing studies, and it is useful to identify such deficiencies in the evidence base. Surrogate or proxy outcomes are sometimes defined (often post hoc) in such circumstances. For example, a review of burning as a form of habitat management used species richness as a surrogate for habitat condition (Stewart et al. 2005) because high-quality habitat was assumed to have a high species richness. This masked the impact of increases in "undesirable" species, and did not allow consideration of other important habitat condition elements, such as structural heterogeneity; thus, cautious interpretation was required, decreasing the utility of the synthesis.

Additional question components
Comparator (often absent from nonexperimental studies)

Use of a comparator (a "control," baseline, or other means of comparison between groups of interest or points in time) may be a prerequisite for high-quality quantitative synthesis, just as controls are considered fundamental elements of many experimental studies (Ford 2000). The term "comparator" rather than control is used, because on occasion ecologists may be interested in how treatment one compares to treatment two, rather than to no treatment. Other elements of study design are also important (see below). The generation of effect sizes for continuous dependent outcome measures relies on the existence of comparators and associated variance measures, while binary outcomes also require comparators to generate risk ratios, odds ratios, or risk difference metrics. Lack of comparators therefore may preclude standard inferential meta-analyses (an exception is the use of correlations); however, meta-analysis of the outcome using other parameters as moderators (covariates) is possible (Chapters 8 to 10), and the generation of probability distributions allows Bayesian approaches to be employed (Chapter 11). Often, more than one comparator may be of interest. For example, Coltman and Slate (2003) examined correlations between phenotypic variation and two measures of genetic variation at microsatellite loci; these were multilocus heterozygosity (MLH) and mean d^2. Effect sizes reported using mean d^2 were smaller than those using MLH, although analyses of paired effect sizes that reported using both measures from the same data did not differ significantly. Direct and indirect comparisons often differ, and care must be taken to explain this variation where multiple comparators are used. Mixed treatment comparisons (or network meta-analysis) offer alternative ways of combining or exploring variation between direct and indirect comparisons (Salanti, Higgins, et al. 2008; Salanti, Kawoura, and Ioannidis 2008).

Study design

Study design (or data collection methods) must be accounted for in research syntheses, as the strength of the inferences made by a synthesis depends on the integrity of the studies that are combined (Higgins and Green 2011). When diverse methods are used in original studies it is wise, prior to pooling, to test for differences in outcome in relation to data collection method, either to justify pooling (Côté et al. 2005) or as sensitivity analyses (Stewart, Pullin, and Coles 2007; Stewart, Pullin, and Tyler 2007). Given that the power may either not be high enough to detect a difference, or so high that biologically meaningless differences are statistically significant, means and variances of subgroups should be displayed to allow readers to assess the implication of pooling decisions.

Bayesian techniques can be used to synthesize disparate types of evidence (Chapter 11) and to explore any discrepancies between expert opinion and evidence, where expert opinion has been elicited and contrasted with empirical data (Newton et al. 2007). However, large quantities of "low-grade" evidence do not equate to a high-grade synthesis. This poses considerable problems given that a high proportion of individual studies are erroneous (Hurlbert 2004, Ioannidis 2005c). Subjective decisions are required to determine if a quality threshold is appropriate, and which elements of study design should be considered in critical appraisal (Chapter 5).

Subgroups and/or moderators

Numerous moderators (covariates) may influence effect sizes; these may be related to any of the key question components. It may be useful to consider ecological variables (often related to the subject), methodological variables (related to the nature of the data), or treatment-related variables (e.g., the intensity or duration of the intervention). Where the data are discrete, subgroups should be defined.

Different subgroup classifications may be relevant and produce different results. For example, should taxa be combined by genera, families, phylogenetic relatedness, or functional group (Chapter 17)? Multiple subgroups may be defined where there is doubt, but multiple comparisons are associated with loss of power and the widening of confidence intervals (or credibility intervals) when interpreting the results. The number of moderators will also be limited by the number of studies. It is not advisable to test the same data set many times, each time with a different moderator (Chapter 20).

Large numbers of covariates also increase the complexity of data extraction, requiring the generation of complex rule sets to maintain independence and to address aggregation bias (also called "ecological bias"). This bias occurs where some subgroups or individuals within a group display very different responses as compared to the rest of the group, thus making the average response aggregated over the entire group misleading (Berlin et al. 2002). Minimizing the number of covariates and focusing on the most important ones are pragmatic ways of decreasing the risk of these problems. Separating components of the question is another approach, allowing the analysis to be split into a robust hypothetico-deductive analysis of a few variables, and an exploratory or descriptive multivariate analysis. Despite the problems of examining large numbers of covariates, there are advantages to extracting the maximum possible amount of information as studies are coded. This has the benefit of maximizing information retrieval and can save time if extra information is desired later because it avoids the need to read and extract data all over again.

Magnitude of differences

It is good practice to define statistical significance levels for pooled effects and covariates. For example, when multiple comparisons are undertaken, will Bonferroni adjustment be performed or will information criterion approaches be used to define the most parsimonious combination of covariates? It is also important to consider what responses are biologically significant. For example, is a 50% increase in abundance biologically meaningful in the context of the synthesis? Are absolute values required or do relative measures provide enough information? Obsession with statistical significance and P-values can lead ecologists and evolutionary biologists to overlook the magnitude of impacts (both in primary research and research synthesis). It is worth remembering that with sufficient statistical power, any two biological groups will be significantly different in a statistical sense, and a definition of the magnitude of difference for ecological significance is required. Stakeholders may disagree on magnitudes, which should be noted, as should variations in the effect sizes contained within frequently cited publications on the topic of interest. Sometimes, this may simply necessitate an a priori value judgement; for example, a 10% change in species richness is considered important (scientifically or from a policy perspective). It is often easier to define such magnitudes where research is asking basic science questions, as effects can be defined in terms of the existing scientific debates. Policy-orientated reviews may be severely hampered if insufficient efforts are made to define the operational variables of interest and how they relate to one another. Biodiversity in ecology and pain in medicine are two examples of potential outcome measures, where the policy implications of a unit change in effect are context dependent. It is important that policymakers understand the importance of this when commissioning reviews, and that scientists and statisticians are able to articulate policy implications in the appropriate range of meaningful contexts.

DEVELOPING A REVIEW PROTOCOL
What should a protocol include and why?

The definition of the question components provides the structure for the protocol, as well as for the subsequent systematic data synthesis. In addition, the protocol outlines each phase of the review process, and describes what the researchers will do at each step. The major issues addressed are how the search for relevant studies will be conducted; how the identified studies will be appraised and retained, or excluded from the review; what data will be extracted from the included studies; and what methodology will be used to summarize the evidence extracted from the included studies.

The main goal of the protocol is the production of an unbiased and repeatable synthesis that addresses a meaningful question. It does this by providing guidance to the researchers as they work their way through the decisions and judgments that will have to be made in the course of the review, consequently limiting the degree of subjectivity that is introduced. The protocol also creates transparency so that other researchers and interested parties can see what decisions were made at each choice point. This is necessary because meta-analysis alone does not reduce bias but only reduces imprecision (Sackett et al. 1997). It is perfectly possible to produce a highly precise but erroneous meta-analysis. The use of a review protocol to define the question and the manner in which it will be addressed provides a mechanism for reducing (but not eliminating) bias. It makes the process of research synthesis as rigorous, transparent, and well defined as possible (Light and Pillemer 1984).

Given that the question itself may be a focus of criticism, a clear a priori rationale becomes especially important. Beside a formal presentation of the question and its background, a review protocol sets out the strategy for obtaining data and defines relevance criteria for data inclusion or exclusion (Khan et al. 2003, Pullin and Stewart 2006). The subject, intervention, and outcome elements defined in the question-setting stage provide a priori inclusion criteria and demonstrate that data were not included or excluded for spurious reasons, for example, to increase statistical significance or to promote a particular outcome. Similarly, definition of subgroups, covariates, and the manner of their analysis guards against the possibility of generating results of dubious value by repeatedly analyzing the same data in different ways until statistically significant patterns emerge. Examples of formal protocols regarding the effectiveness of environmental interventions are available at http://www.environmentalevidence.org. PROSPERO (http://www.crd.york.ac.uk/PROSPERO) provides a range of medical examples.

The protocol also acts as a document that all stakeholders can agree upon (or at least comment on) in policy-related meta-analyses. Relevant decision makers, interested stakeholders, editorial referees, and the wider scientific community can provide feedback. This may prevent unanticipated problems arising during the review process and ensures that the final synthesis meets defined minimum standards. Differences of opinion may of course not always be possible to resolve, but it is useful if this uncertainty is acknowledged early in the process. For example, meta-analyses of the effectiveness of marine protected areas assume that site comparisons are not confounded, although this is controversial. The assumption has implications for the relevance of study design, as only BACI studies would be appropriate if confounding was proven.

Additional protocol considerations

The details of searching are dealt with in Chapter 4. Here, we briefly consider the resource implications of searches. If insufficient resources exist to retrieve a broad pool of diffuse literature, question modification may be a better approach than undertaking a review with a deficient search strategy. Scoping searches (Chapter 4) are useful tools to determine how much literature is available and how accessible it is, but ultimately the decisions about sensitivity/specificity, the necessity to retrieve large quantities of gray literature, and the need for inclusion of non-English material are subjective. Subject experts and authors are useful to ensure that key material has not been missed, but they have been shown to selectively cite material with positive results (Gotzsche 1987, Ravnskov 1992), and should therefore only be used to augment searches, not to replace them. The details of the search strategy should be documented in detail in the protocol. Once searching is complete, relevant articles must be identified without wasting resources examining irrelevant articles in detail (Chapter 4). To determine the level of confidence that may be placed in selected data sets, each should be critically appraised to determine the extent to which its research methodology is likely to prevent systematic errors or bias (Moher et al. 1995). The approach to critical appraisal should be specified a priori in the protocol. The manner in which data will be extracted and utilized also requires explanation in the protocol. Details regarding data extraction and critical appraisal are provided in Chapter 5.

Data synthesis includes both qualitative synthesis and quantitative analysis. Qualitative synthesis allows informal evaluation of the effect of the intervention and the manner in which it may be influenced by measured study characteristics and data quality. Formal quantitative analysis can be undertaken to generate overall point estimates of the effect size and, where appropriate data exist, to analyze reasons for heterogeneity in the effect of the intervention. These techniques are discussed in detail in Section III. With regard to protocol development, it is important to specify the methodological approach in as much detail as possible. Good scoping searches (Chapter 4) and careful assessment of retrieved data should enable researchers to determine if sufficient "high"-quality material exists for quantitative synthesis and the approximate nature of the data. It should then be possible to at least suggest which effect sizes are most appropriate (Chapters 6 and 7), how they will be combined (Section III), how heterogeneity will be investigated, and how results will be interpreted. Failure to specify methodology a priori leads to the potential for bias, because different meta-analytical techniques produce different results (Lajeunesse and Forbes 2003).

Interpretation of results and next steps

Although a priori definition of the review question and methods reduces the probability of bias, post hoc rationalization is not uncommon and is often sensible (Lipsey and Wilson 2001, Khan et al. 2003, Pullin and Stewart 2006, Higgins and Green 2011). Questions are developed without detailed knowledge of much relevant data, despite scoping searches and collaboration with subject experts. It may therefore become necessary to modify either the question or the approach in light of the accumulated research. Such modifications are germane provided they are supported by a sound scientific rationale, and are based on alternative ways of defining question elements that were not initially considered. This may entail revision of the protocol, undertaking further searches, revising inclusion criteria, or running additional analyses. Reviewers should be fully transparent about modifications, which should be listed under substantive amendment of the protocol. The final work should clearly distinguish a priori reasoning from post hoc rationalization, and unambiguously state the justification for changes. Inevitably, post hoc rationalizations will carry less weight in the overall interpretation of the results (Chapter 20). It is interesting to note that peer reviewers may ask for post hoc analyses (for example, water quality covariates to explore variation in the effectiveness of instream devices), particularly where conventional beliefs are challenged by a meta-analysis.

SUMMARY: KEY POINTS

- Research syntheses should be accompanied by protocols (published or internal) to reduce the probability of bias. These should provide definitions of the question and the manner in which it will be addressed.

- Characteristics of the subjects, differences in interventions (where relevant), variation in outcome, and study design all contribute to heterogeneity in the relationship between outcome and intervention. The impact of these factors and the methods of exploring them require careful consideration and should be specified.

- Reasons for including or excluding data should be explicit and unambiguous.

- The manner in which studies will be critically appraised, weighted, and combined should be specified in as much detail as possible.

- Alternative ways of defining question elements may become apparent during the work. It is reasonable to alter the original questions when this occurs, but the modifications should not be driven by the knowledge of results of studies or data sets.

- Meta-analyses must provide detailed definitions of their scope, how data were included, excluded, weighted, and combined; they must explicitly distinguish between a priori and post hoc reasoning if they are to be robust. Sensitivity analyses exploring the impact of variation in key definitions or sources of data are recommended.

- Systematic reviews that are guided by formal protocols and use meta-analysis to integrate the results of included studies represent best practice. They should be used to inform policy or develop evidence-informed practitioner guidance where applied ecology is concerned.

- The danger of producing highly precise but erroneous meta-analyses where systematic review methodology is not followed should not be underestimated.

ACKNOWLEDGEMENTS

We would like to thank a large number of people who have supported the development of an evidence-informed approach to ecology and have helped us to critically appraise our thoughts on the matter. Special thanks to Helen Bayliss, Kevin Charman, John Hopkins, Khalid Khan, Terry Rowell, Emma Stewart, Lesley Stewart, Bill Sutherland, Rod Taylor, Adrian Newton, Centre for Reviews and Dissemination colleagues, and our NCEAS working group editors and coauthors.

Gathering Data: Searching Literature and Selection Criteria

Isabelle M. Côté, Peter S. Curtis,
Hannah R. Rothstein, and Gavin B. Stewart

PEOPLE SOMETIMES THINK THAT doing a meta-analytic review will give them an easy or rapid publication. Think again! There is no such thing as a good, quick and dirty meta-analysis, particularly when carrying out a systematic review. Gathering data for such a project can be a long and tedious affair—the most arduous and time-consuming stages of the review begin once you have formulated your question clearly and developed your protocol (Chapter 3). These stages consist of identifying and retrieving relevant sources, evaluating the retrieved information against specified selection criteria, and extracting and appraising data from the studies that survive this screening.

There is an important relationship between how thorough and unbiased the search for relevant data is and the validity of the resulting meta-analysis. Many reviewers fail to uncover citations to documents relevant to their project because of inadequate search tools or strategies. In this chapter, we cover literature searching and information retrieval, as well as the application of study selection (inclusion) criteria. Those undertaking systematic reviews and meta-analyses may also want to look at *The Oxford Guide to Library Research* (Mann 2005), *Library Use: Handbook for Psychology* (Reed and Baxter 2003), or the *Cochrane Handbook for Systematic Reviews of Interventions* (Higgins and Green 2011) for more extensive information on search processes and use of research libraries. Oddly enough, graduate training in ecology, evolutionary biology, and the environmental sciences almost always omits formal training in this area, and assumes that these skills are somehow inborn or caught on the fly. While they can be learned independently, this is a haphazard way to develop these important skills.

INITIAL SCOPING

As noted in Chapter 3, it is increasingly advised that a scoping search should be performed prior to a full-scale systematic review, in order to determine the feasibility and value of conducting the review (Khan et al. 2003, Lipsey and Wilson 2001). A scoping search is essentially analogous to a pilot study. It allows you to determine the size and nature of the literature on the question of interest, as well as whether any review of the topic has already been carried out. It will also help identify the major research questions of interest in the field, and may thus help with refining the focus of the synthesis. A scoping search will provide important information about how much literature exists on a particular topic, its quality, the major terms or key words used, and which databases are most likely to yield relevant studies. It also provides an idea of how much time, effort, and money will be needed to conduct the full systematic review. At the end of a scoping search, you may decide that the question you had in mind has already been reviewed adequately enough or that not enough literature is available to justify a meta-analysis.

A scoping search can be carried out using one or a few large electronic databases that are known or believed to include studies on the topic of interest. These might include the Web of Science or Web of Knowledge, as well as one or more of a number of other large databases and repositories available in ecology, evolution, and environmental biology (Box 4.1). Different institutions and individuals have access to different resources and it is not necessary to search them all during the scoping phase, but a few key resources should be examined as a sampling exercise. You may also have to consider searching for material that has not necessarily been subject to peer review using the World Wide Web. Meta-search engines, such as www.dog pile.com, are useful in this context. Government and nongovernmental organizations, subject experts, and existing traditional narrative reviews of the question of interest, if they exist, may also provide useful input to a scoping search. Keep in mind that existing reviews may be inadequate for various reasons (e.g., those based on a biased sample of literature, improper methods of meta-analysis, etc.), so do not rely exclusively on them. Different combinations of keywords must be used to retrieve the information, and it is worth remembering that the search functionality of the electronic resources varies considerably (see Box 4.2 later in this chapter for database searching tips). It is as important at the scoping stage as it is during the full review that the procedures used to identify, retrieve, and evaluate potentially relevant research are objective, systematic, and transparent, and that the description of these procedures is recorded.

The results of a pilot search will frequently lead to modification of the research question or search terms (e.g., broader or narrower), changes in study selection criteria, or even in a decision that the time is not yet right to conduct a systematic review. If the scoping search reveals that a systematic review of the proposed question has already been carried out, consider whether the previous review could be updated, or if you should refine/refocus your question. For example, a review of the impacts of aquaculture on marine nutrient loadings may be extended to consider biological response metrics as well as chemical parameters. Extending existing analyses by adding more recent data, related taxa, or coverage of previously missing geographical areas may also be considered. Even if you reach the disappointing conclusion that there are not enough primary studies to conduct a systematic review, your identification of specific gaps in the literature can be used as a basis for suggesting where new primary research is most needed. In an applied context, the identification of recognized research gaps may be a very important outcome, particularly if a lack of knowledge in a particular area is not currently recognized by the policy or scientific communities. Conversely, if there are large numbers of hits on your search terms, then it may indicate that your question is ripe for a systematic quantitative summary, or perhaps that the question is too broad. However, the existence of lots of relevant references can be misleading. It is not uncommon to find that despite a large and relevant literature base, there is a paucity of robust data suitable for the generation and/or pooling of effect sizes (Pullin and Stewart 2006). The quality and nature of the retrieved studies therefore also require consideration during the scoping phase of review.

DATA SOURCES

It is perhaps self-evident that the widest possible range of sources should be consulted in the quest for data. This will reduce the possibility that the systematic review will contain a biased subset of research on the topic of interest. Generally, data for a systematic review will come

---

BOX 4.1.

Some of the electronic databases, repositories and search engines relevant to ecology, environmental biology, conservation, and evolution.

AGRICOLA: Database of agricultural and forestry literature, with some conservation ecology papers not easily retrieved elsewhere.

Biological Abstracts: Database of life science topics from botany to microbiology to pharmacology, with six million records extracted from five thousand journals, from 1969 to present.

BIOSIS Previews: Database of journal articles, meeting and conference reports, books, and patents relating to life sciences.

CAB Abstracts: Comprehensive database of journals, books, and conference proceedings in applied life sciences (agriculture, animal and veterinary sciences, environmental sciences, human health, food and nutrition, leisure and tourism, microbiology and parasitology, and plant sciences).

Conference Proceedings Citation Index: Bibliographic information and author abstracts from papers delivered at international scientific conferences, symposia, seminars, colloquia, workshops, and conventions.

CSA: Dozens of bibliographic and full-text databases and journals in natural sciences and technology. Includes, among others, Agricola, Aquatic Sciences and Fisheries Abstracts, Animal Behavior Abstracts, Ecology Abstracts, Environmental Impact Statements, and Zoological Record.

Current Contents: Bibliographic information from over 8000 scholarly journals and more than 2000 books.

Dissertation Abstracts Online: Subject, title, and author guide to virtually every American dissertation accepted at an accredited institution since 1861; includes selected Masters theses accepted since 1962.

Google Scholar: Search engine for access to peer-reviewed papers, theses, books, abstracts and articles, from academic publishers, professional societies, preprint repositories, universities, and other scholarly organizations.

Index to Theses: Comprehensive listing of theses with abstracts accepted for higher degrees by universities in Great Britain and Ireland since 1716.

JSTOR: Not a database but a useful archive of important scholarly journals spanning many scientific disciplines. Useful for accessing digitized older literature.

ScienceDirect: Scientific, medical, and technical information online, including 2000 peer-reviewed journals and hundreds of book series, handbooks, and reference works.

Scirus: Search engine for science-specific Web pages.

Scopus: Abstract and citation database of research literature and quality web sources.

Web of Science: The largest database, with more than 8700 journals covered, and one of the most widely used, with cited reference searching and a growing selection of pre-1950 literature. Be aware that there is little overlap between different electronic databases and that although Web of Science is the largest, it does not replace the smaller databases.

WorldCat: Network of library content and services.

from three main sources: published data, gray literature, and unpublished data. We discuss each of these in turn.

Published data

The primary method for data retrieval is a systematic literature search of published information. This is by far the easiest type of data to find. Searching for published information will initially be carried out using electronic databases. This is deceptively simple to do and the scoping search conducted at the beginning will have yielded many of the most recent reviews and primary papers. In addition to working back in time from the most recent papers, you can identify the seminal papers that have triggered the questions you are addressing, and then you can search electronic databases for papers citing these key references.

It might be tempting to use a single database, such as the widely available Web of Science. However, no database is complete, and the use of multiple databases will make the search more comprehensive (thereby reducing selection bias). The identification of additional, potentially useful databases can be made relatively easy by using sources such as the *Gale Directory of Databases*.

It is important to know something about the search capacity of the databases used and your librarian can be extremely useful here. Many databases do not have full Boolean search capacity and are sensitive to word order and number. The help pages of individual resources provide some guidance on search functionality, which may improve search efficiency. However, it is well worth speaking early on to the librarian at your institution to discuss this and more general issues of resource availability. For many topics, the involvement of a librarian or other information search specialist can greatly improve the quality of your search; librarians are generally more likely to be knowledgeable and up to date about sources of information and searching techniques than you are. Although librarians usually will be able to vet the sources of information for you, they actually *know* keywords (see "Search Strategy," below), and will not have to merely guess at them. They are often enthusiastic about taking part in systematic reviews and make great research allies.

The studies unearthed by electronic searches may sometimes be just the tip of the iceberg. Additional relevant studies can be found by consulting the cited literature sections of these papers. Some will have already been found by the initial electronic search, but many older ones may not. The amount of "prehistoric" (i.e., pre-1980) scientific literature available on the web is increasing quickly (e.g., in archival repositories such as JSTOR), but access to a good university library and to interlibrary loan is needed. If you do not have access to such resources, consider proposing a collaboration with someone who is likely to be interested in the topic and who does have this access. Be aware that the literature cited in papers may be a biased sample of what is available. These citations may be limited to those studies that support the findings, so the search should definitely not be limited to studies cited in reference lists (Gotzsche 1987, Leimu and Koricheva 2005b).

In addition to following up leads from published papers, it may be fruitful to conduct manual cover to cover searches ("hand searches") on issues of journals that are particularly relevant to the question. Publishers' web pages often provide electronic tables of contents for more recent material that can be searched if paper copies are unavailable. This may be necessary because not all studies are included in electronic databases, and even if they are, they may not have been indexed with keywords that match the ones selected in the search strategy. Hand searching is most useful for older (pre-1980) issues, which have relatively poor coverage in many ecological databases, but this often yields additional studies even for later years. For example, Turner et al. (2004) were looking for randomized experiments in education and found that electronic searches identified on average only one-third of the trials identified by hand searching. Findings in the health sciences also show that electronic searches retrieve fewer studies than manual searches of the same volumes (Higgins and Green 2011).

Gray literature

The most formal definition of gray literature is, "that which is produced on all levels of government, academics, business and industry in electronic and print formats not controlled by commercial publishers" (Auger 1998). For the purposes of a systematic review, however, the key feature of gray literature is that it is hard to retrieve. According to Weintraub (2000), "scientific grey literature comprises newsletters, reports, working papers, theses, government documents, bulletins, fact sheets, conference proceedings and other publications distributed free, available by subscription or for sale." Others include book chapters as "nearly gray" literature, since they are generally not indexed in databases. "Deep gray" or "black" literature may also exist in an applied context as uncollated monitoring reports or field notes. We also include here information available on the World Wide Web.

Gray literature comprises an increasing proportion of the information relevant to research synthesis, particularly in some study areas, such as applied ecology. Searching for data in the gray literature is, however, likely to be a time-consuming task, and the reporting standards in this literature are extremely variable. Despite the increased time and money that are likely to be needed to locate and retrieve gray literature, as the standards for publication of data syntheses in ecology continue to increase and approach those in medicine and other fields, you may find it hard to justify conducting a synthesis that excludes this evidence, particularly if there is subsequent evidence of publication bias (Chapter 14).

The optimal extent of a gray literature search, however, will depend on both the subject matter and the resources available for the search, and must be decided on a case by case basis. It is also true that different sources of gray literature are likely to produce different yields. The four sources that seem to consistently produce "buried treasure" are conference proceedings, book chapters, theses and dissertations, and the World Wide Web. Conference proceedings tend to be the most productive source for identifying studies that have not been published as full journal articles (often for reasons other than quality, according to research in both health care and social science; see Cooper et al. 1997 and Scherer et al. 2007). In addition, the abstracts from these studies are not generally included in electronic databases. Although a few conference proceedings are available electronically from professional society websites, others exist only as compact discs or on paper, and some conferences do not even publish proceedings. In the latter case, a search of the conference program and subsequent contact with the abstract authors may be needed to retrieve potentially relevant studies and/or information missing from the abstracts. Book chapters have also been shown to provide a steady yield of studies that have not been published elsewhere, and manual searching of edited books on the topic of interest should therefore be considered. Dissertations and theses are more likely than other types of gray literature to be indexed (e.g., Index to Theses and Dissertation Abstracts databases, Box 4.1). Once identified, the dissertations or theses of interest must be ordered through interlibrary loan, purchased, or viewed at the institution of origin (assuming they are not data protected).

The World Wide Web is an increasingly important source of information about ongoing and completed research, particularly that which has not been formally published. Sources of gray literature that are accessed primarily through the internet include the websites of universities, major libraries, government agencies, nongovernmental organizations, professional societies, corporations, and advocacy groups. The internet is often particularly useful for identifying documents produced by government authorities, because in many cases these are published only online. Searching the World Wide Web, however, can be daunting, as much of what it contains is unreachable or invisible using standard search engines.

Note the danger that any data found in gray literature may have been published later, perhaps in a different format that will make identification of duplication difficult. Duplication could seriously bias a meta-analysis, particularly when sample sizes are small. To avoid this problem, careful notes must be taken of the methods (especially dates and locations of work) reported in the gray literature document for comparison with the published work that is suspected to stem from the same study. In case of doubt, contact the authors.

Unpublished information

Personal contacts can sometimes form an important part of assembling a database for systematic reviews. Contact can be made with individual researchers to gain access to totally unpublished data sets. There are many reasons why these data sets might not be published. They may have been collected very recently and have yet to be analyzed or published (Chapter 14). Alternatively, you might think that the studies were perhaps badly designed and that they failed the peer-review process leading to publication. However, you should not assume that this is the case. One of us, for example, found that there was a large number of good, unpublished data on coral cover on Caribbean reefs; these were collected during the course of behavioral and ecological studies of reef-dwelling organisms. Because these data served as background to other studies, they were simply never published as a time series of habitat information. In social and biomedical sciences, studies reporting nonsignificant results or small effect sizes have been shown to be less likely to be published than large significant results, even though they may be of equivalent quality; this is a case of "publication bias" (e.g., Dickersin 2005; Chapter 14), although evidence of this in ecolocgial studies, as well as in other fields, is mixed (Koricheva 2003, Møller et al. 2005, Stewart, Pullin, and Coles 2007).

You might also want to contact individuals to obtain unpublished information that would supplement data from published papers. There will often be important data missing from published sources (e.g., estimates of error or sample sizes, or key details of experimental design; see Chapter 13) or data will not be presented in a useful way for inclusion in the review, particularly if a meta-analysis is carried out. For example, means that are aggregated over several habitats or several species may be presented, but the research question specifically calls for disaggregated data.

It is important to manage personal contacts well. It can be useful in some cases to elucidate the relationships between individual researchers to ascertain who owns which data, and thereby prevent data duplication. Remember that individual researchers have many competing demands on their time and may be reluctant to share data or even acknowledge your request, so when asking for information, keep it short, polite, and to the point. Make a real effort to anticipate what your data requirements will be (to avoid making repeated requests of an individual) and to state your question and your data needs very clearly. Our own experience of contacting individuals to ask for methodological clarification or missing data is that the success rate hovers between 50% to 75% if you know the researcher personally, but is much lower if you do not and if the data sought are more than three to five years old. It may also be harder to retrieve raw data from authors if you do not have expertise in their specific field or are at an early stage of your scientific career. It may be useful to maintain an audit trail when contacting researchers for data, so that those who respond can be acknowledged. Evidence of failed attempts to retrieve data may also be useful if, for example, peer reviewers challenge the absence of certain data in an analysis. Checking for the existence of online supplementary material is also worthwhile because this presents more detail of methods and results than is reported in articles. Unfortunately, there are generally limited repositories of raw data in ecology, evolution, and the environmental sciences; however, this is changing rapidly, at least for published papers (e.g., http://knb.ecoinformatics.org/index.jsp; http://data.esa.org/esa/style/skins/esa/index.jsp; and http://datadryad.org).

Retrieving unpublished data presents a special challenge for several reasons. First, it can require a tremendous amount of effort from the person who collected the original data. Unpublished data have sometimes not even been entered in computerized spreadsheets and may lie in a rather disorganized fashion in a filing cabinet or at the bottom of a drawer. Success at obtaining these data is very likely to be contingent on your physical presence at the source. Second, you will need permission to use these data because they are not publicly available. Finally, it is difficult to ascertain the likelihood of bias in samples of unpublished data.

A thorough search of gray literature and unpublished information can seriously lengthen the amount of time needed to collate data for a meta-analysis. For example, a recent effort at collating all available information on densities of Caribbean coral reef fishes took one researcher, who was fully dedicated to the task, nearly two years! However, exhaustive searches of non-conventional sources can yield substantial material. For example, more than one-third (38%) of data included in a meta-analysis of rates of change in coral cover on Caribbean reefs originated from gray literature and unpublished data provided by personal contacts (Gardner et al. 2003). These additional data allowed the authors to extend both the spatial and temporal scales of their analyses considerably.

SEARCH STRATEGY

Your search strategy spells out exactly how you will go about searching for relevant information. The strategy is rooted in the clear and precise definition of the question of interest that you have delineated in your protocol (Chapter 3). It is at that early stage that you should have made key decisions about the independent and response variables, potential moderators, the subject population, and acceptable research designs. These become the criteria for inclusion or exclusion of the potentially relevant studies that you will identify as you search the data sources described above. Developing a search strategy begins with assessing the sources available to you and selecting those which are likely to offer the broadest and most unbiased survey possible (Box 4.2).

The search strategy will list all the search terms you anticipate using when searching electronic databases. The choice of keywords is critical. There are a number of tricks that can be used, such as the use of * or ?, to capture alternative spellings or multiple derivations of words of interest. The use of these "wild card" symbols can have a huge effect on the number of returns (Fig. 4.1). We give a few tips for selecting search words in Box 4.3.

A good approach to developing an electronic search strategy is to list multiple terms that describe the subject and the treatment of interest. For example, in a systematic review of the relationship between ectoparasite load and body size in fish, the subjects of interest could be described as [fish*] or [fish?] (to include fish and fishes) or [teleost*] (or [teleost?]—for simplicity, we continue this example using only *—while the treatments of interest could include [parasite*] or [ectoparasite*] and body size, body length, total length, standard length, or length.

If you are uncertain about the search capacity of the database you are using, the safest strategy is to search separately for every combination of subject and treatments, for example, [fish*

BOX 4.2.
Selection of information sources.

- Scope of search: Which fields should be searched? Do you need to go beyond Ecology/ Evolution/Environmental Science?

- Availability of databases: Which databases do you have access to at your institution?

- Format of databases: What format are they in? Online, print, web based?

- Date: How far back does each database go?

- Language: What is the language of the database? How can non-English material be located?

- Unpublished work: How can you access dissertations, reports, and other gray literature, or raw data?

AND parasite* AND body size], [fish* AND ectoparasite* AND body size], and so forth. In this case, there would be 2 × 2 × 5 possible keyword combinations, which would be time consuming to search and would be likely to yield study sets that overlap to a large extent. (Note that duplicates can be eliminated easily by using referencing software such as Endnote, which can receive direct input from electronic databases.) However, if the database has a full Boolean search capacity (which can be ascertained by consulting a librarian, for example), then a single search could combine all possible terms: [fish* OR teleost* AND parasite* OR ectoparasite* AND body size OR body length OR total length OR standard length OR length].

In ecology, resource-intensive searches of high sensitivity (i.e., capturing everything that might be relevant, including much irrelevant material) are particularly important, even though this is at the expense of specificity (i.e., capturing only what is relevant); this is because ecology lacks the sophisticated integrated databases of medicine and public health, which use controlled vocabulary such as Medical Subject Headings (MeSH). A high-sensitivity and low-specificity approach is necessary to reduce bias and increase repeatability. Typically, large numbers of retrieved references end up being rejected (see "Criteria for Inclusion of Studies" section, below).

You should be aware of the biases that might be introduced by the keywords used in your searches. Many prevalent types of biases are covered in Chapter 14, so we will not discuss them in depth here. However, the search could be inadvertently biased by not including synonyms, alternative spellings, and non-English language terms as keywords. It has been found, for example, that medical trials carried out in Germany that showed a significant effect were more likely to be published in English than in German (Egger, Zellweger-Zähner, et al. 1997). Similar biases may occur in ecological studies.

Devising a search strategy for unpublished data and gray literature is more difficult because there is a paucity of empirical evidence to guide these decisions in ecology. Unpublished and non-English language material has been incorporated in applied ecology meta-analyses, and these analyses would have yielded different results had this literature not been included (Côté et al. 2005; Stewart et al. 2005; Stewart, Pullin, and Coles 2007). However, ascertaining what proportion of the available resources is devoted to searching gray literature relies on value judgment and is likely to be topic specific. Funnel plot asymmetry (Chapter 14) may cast considerable doubt on the reliability of the pooled effect size if the gray literature search was cursory or nonexistent.

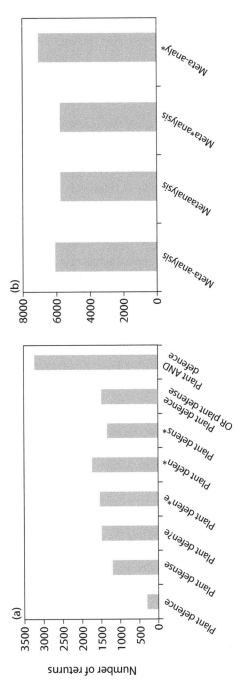

Figure 4.1. Examples of the effect of keyword choice on number of returns from an online scientific database.

BOX 4.3.
Database searching tips.

- Use "wildcard" symbols to capture alternative spellings or multiple derivations of words of interest.

  For example, in some databases searching [meta*anal*] or [meta?anal?] should capture meta-analysis, meta-analyses, meta-analysed, meta-analyzed, and their non-hyphenated alternatives.

- Do not truncate words excessively to avoid low specificity.

  For example, [rat*] will retrieve rat, ratio, rations, rate, rationale, etc. If interested in rats (the mammals), search using [rat] or [rats] or [rattus]. Alternatively, use NOT operator to exclude some areas (e.g., [phenol* NOT phenology] would yield literature on phenolic compounds, such as phenols and phenolics, but not phenology).

- Avoid or be wary of ambiguous search terms.

  For example, if searching with "fox" or "flowers," limit the search fields to subjects and titles to avoid retrieving articles by Dr. Fox or Dr. Flowers.

- Beware of common names which are applied across taxa.

  For example, parrots can be birds or fish, cucumbers can be vegetables or marine invertebrates, and a "pomme de terre" is not an apple.

- Beware of differences in American and British spelling (behavior vs. behaviour, defence vs. defense) (Fig. 4.1A).

- Beware of alternative nomenclature.

  For example, the name of the mangrove killifish *Rivulus marmoratus* was changed to *Kryptolebias marmoratus* in 2004. Searching both genus names would be necessary to capture all literature.

- Beware of terms with dual meanings.

  For example, translocation in genetics refers to part or all of a chromosome becoming attached to another chromosome. In conservation biology, it means moving plants or animals from one location to another. Deal with this using the NOT operator to exclude the pools of undesired literature.

As emphasized in Chapter 3, it is essential that the details of the search strategy be reported in as much detail as possible. You should keep track of the combinations of terms used in electronic searches, the range of years included in both electronic and hand searches, and the effort put into tracking down unpublished and gray literature. Information retrieval activities should be described in sufficient detail in a review so that the process is transparent and can be replicated. Box 4.4 includes examples of good and poor reporting of search strategies. It may be necessary to include the detailed information on searching methodology in an appendix (printed or online) if page limitations prevent including it in the body of the paper.

CRITERIA FOR INCLUSION OF STUDIES/DATA

Criteria for study inclusion are critical because they will influence the outcome of the systematic review. Study selection criteria are intimately linked to the research question. In the

BOX 4.4.
Good and poor reporting of search strategy.

GOOD REPORTING EXAMPLE: EFFECT OF WIND FARMS
ON BIRDS (STEWART, PULLIN, AND COLES 2007)

Multiple electronic databases and the internet were searched using a range of Boolean search terms in 11 languages. The databases searched on the internet were: Dogpile meta-search engine, SCIRUS, COPAC, and Web of Knowledge. Searches were performed in March 2002 on the complete range of references available at that time. Additional searches were performed on JSTOR, Index to Theses Online (1970 to present), and English Nature's "Wildlink" database.

Search terms were as follows: bird* AND wind turbine*, bird* AND windfarm*, bird* AND wind park*, bird* AND wind AND turbine*, bird* AND wind AND farm*, bird* AND wind AND park*, bird* AND wind AND installation*, raptor* AND wind*, wader* AND wind*, duck* AND wind*, swan* AND wind*, geese AND wind* and goose AND wind*. Although the search term "wind*" encompasses the terms "wind turbine*," "windfarm*," and "wind park*," initial trials proved that the number of hits became unmanageable unless the specificity of the terms was increased.

The Dogpile meta-search engine was used with the advanced search facility, and the terms "bird AND wind AND turbine." It was also searched using the following languages and terms: German "Vögel AND Windturbinen," French "oiseaux AND turbines AND éoliennes," Spanish "pájaros AND turbinas AND viento," Dutch "vogels AND windturbines," Norwegian "fugle AND vindkraft," Danish "fugle AND vindkraft," Finnish "lintu AND tuulivoimala," Swedish "fåglar AND vindkraft," Italian "uccelli AND vento AND turbina," and Portuguese "pássaros AND vento AND turbina."

These languages cover the following countries with wind energy developments: Australia, Austria, Belgium, Canada, Denmark, Finland, France, Germany, Ireland, Italy, Morocco, Netherlands, Norway, Portugal, Spain, Sweden, Switzerland, UK, US, and others with one of these languages in official use. Internet searches are unavailable in languages of other significant wind power nations including China, Greece, India, Japan, and Ukraine, although existing English language translations from these countries were accessible. For internet searches of relevant sites, the authors undertook "hand" searches (following links) or, where available, electronic site searches of the first 100 "hits" for each search engine within the meta-search. Articles identified by this process were assessed in the same manner as other articles.

The library of the Royal Society for the Protection of Birds (RSPB) was hand searched. In addition, bibliographies of articles accepted for full text viewing and those in otherwise relevant secondary articles were searched. The authors also contacted recognized experts and current practitioners in the fields of applied avian ecology and renewable energy technology to identify possible sources of data (including primary data) and to verify the thoroughness of our literature coverage.

POOR REPORTING

Imagine now how difficult it would be to replicate the search performed by Stewart, Pullin, and Coles (2007) if they had summarized their search strategy like this instead:

(Box 4.4. continued)

"[We] searched commonly accessible electronic databases, the World Wide Web, and bibliographies of relevant articles using various combinations of relevant search terms."

This brief description makes it impossible for readers to judge the depth and scope of the search. Effectively, the sampling universe is undefined, and inferences based on the sample must therefore be considered cautiously. If the search is comprehensive, sufficient detail must be included for readers to judge its validity. If the search is not comprehensive, then the potential for bias should be explicitly acknowledged or more preferably addressed by modifying the search.

---

wind farms study (Box 4.5; Stewart et al. 2005; Stewart, Pullin, and Coles 2007), 2845 bibliographical references were retrieved as a result of the search strategy outlined in Box 4.4. Initial assessment of study relevance was made on the basis of the title and abstract of each paper, which reduced the list to 124 papers (including papers with insufficient information in the title/abstract to reject). After examining the full text of these papers, only 20 were deemed to fulfill the inclusion criteria and provide all information needed. Once the independence of the data sets presented in each paper was ascertained, the papers were then subjected to data extraction and careful quality appraisal (Chapter 5). A generic study selection process, with rough estimates of number of references obtained at each step, is shown in Figure 4.2.

BOX 4.5.
Study inclusion criteria used in systematic review
of effect of wind farms on birds (see Box 4.4).

Studies were included if they fulfilled the following relevance criteria:

- *Subjects(s)* studied: Any bird species.

- *Treatment* used: Commercial wind harvesting installations in any country; wind farms and turbines.

- *Response(s)*: Population size or distribution, breeding success, population mortality rate, recruitment rate, turnover rate, immigration rate, emigration rate, demography, dispersal behavior, collision mortality, displacement disturbance, movement impeded, and habitat loss or damage.

- *Comparator*: Appropriate controls (e.g., reference areas) or predevelopment comparators.

- *Type of study:* Any primary studies with appropriate comparators and variance measures. Data from BACI designs was extracted preferentially, but site comparisons and before and after impact studies were also considered relevant. Stewart, Pullin, and Coles (2007) used a standardized mean difference effect size metric and therefore sought continuous data with means, sample sizes (pseudoreplicates), and variance measures.

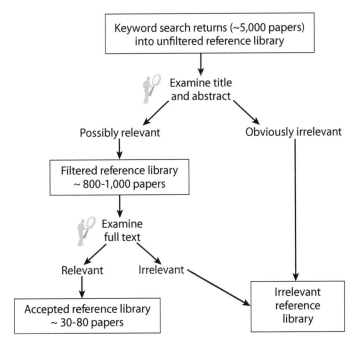

Figure 4.2. Schematic representation of the literature searching and study selection processes in a systematic review. The numbers of papers shown at each step are meant to reflect the ratio of attrition through the study selection process rather than to provide guidelines on the number of papers necessarily retrieved during a systematic search.

Study selection must be both unbiased and repeatable. Blinding researchers to the identity of authors of primary studies has been shown to have an impact on inclusion of medical studies (Berlin et al. 1997), but this method has not been empirically investigated in ecology. It is good practice to ask a second reviewer to go through the relevance assessment process on a random subsample of articles from the original list. The subsample should include at least 25% of the list, up to a predefined resource-limited maximum number of references (CEBC 2010). The two outputs can be compared using a kappa analysis (Box 4.6); this adjusts the proportion of records for which there was agreement, by the amount of agreement expected by chance alone (Cohen 1960, Edwards et al. 2002). Kappa can range from zero to one. If the results were generated entirely by chance, with neither reviewer showing judgment, the value would be zero. If the reviewers were in perfect agreement, the number of agreements would equal the number of trials and kappa would be one. A kappa rating of "substantial" (i.e., 0.6 or higher) is recommended to pass the assessment (Pullin and Stewart 2006, CEBC 2010). If the assessment fails, then the selection criteria should be refined and the process repeated.

Good record keeping is essential to provide evidence of the thoroughness of the search and of the lack of bias in the study selection. You should keep a record of all papers/data sets examined at every step of the search and selection processes, and when papers are rejected, make note of the reason for rejection (i.e., which selection criterion was not fulfilled). Note also any study not examined because you could not obtain a full version. These records will help to make these stages of the systematic review process completely transparent.

BOX 4.6.
Kappa assessment: A worked example.

The table below shows the number of references accepted and rejected by two reviewers in relation to one another.

		Reviewer 2		
		Reject	Accept	Total
Reviewer 1	Reject	20	19	39
	Accept	1	110	111
	Total	21	120	150

Agreement expected by chance is calculated as: (row total × column total)/overall total, providing a second matrix, shown below.

		Reviewer 2		
		Reject by chance	Accept by chance	Total
Reviewer 1	Reject by chance	5.46	33.54	39
	Accept by chance	15.54	95.46	111
	Total	21	129	150

Actual agreement is calculated as the number of times reviewers agreed (i.e., both reject or both accept, from first matrix): 110 + 20 = 130.
Similarly, expected agreement by chance is (from second matrix): 5.46 + 95.46 = 100.92.
Kappa is used to calculate the agreement above and beyond the agreement expected by chance.

K = (actual agreement − expected agreement)/(total number of trials − expected agreement)
  = (130 - 100.92) / (150 - 100.92) = 29.08 / 49.08 = 0.593

BEST PRACTICES

- Carry out an initial scoping study to assess how much literature is available and whether a systematic review and meta-analysis are possible.

- Develop an explicit search protocol which details exactly how you will go about searching the literature.

- Do not rely on a single electronic database—it is very unlikely to yield all articles of interest.

- Outline clear study selection criteria so that the reasons for inclusion or exclusion of studies are transparent.

- Keep good records at all steps of the literature search and study selection processes. This will ensure that your review is as thorough, representative, and replicable as possible.

ACKNOWLEDGEMENTS

We thank Helen Bayliss for setting us straight on many library issues, Michael Jennions and Laura Hyatt for provocative comments on the manuscript, and Julia Koricheva for providing both the inspiration and part of the data in Figure 4.1.

Extraction and Critical Appraisal of Data

Peter S. Curtis, Kerrie Mengersen, Marc J. Lajeunesse,
Hannah R. Rothstein, and Gavin B. Stewart

DATA EXTRACTION

THE EFFICIENT AND ACCURATE extraction of data from primary studies is an important component of successful research reviews. It is one of the most time-consuming parts of a research review and should be approached with the goal of repeatability and transparency of results. Careful definition of the research question (Chapter 3) and identification of the effect size metric(s) to be used (Chapters 6 and 7) are prerequisite to efficient data extraction. The components of the data extraction process may be simpler or more complex, depending on the scope of the meta-analysis (e.g., how many studies are to be included, how complex the analysis is likely to be) and the number of people involved in the process (i.e., individual investigator vs. large team effort). For more complex and large group efforts, separate data extraction and database spreadsheets may be used to initially code and then store data from individual sources. For a small, one-person analysis, a simple spreadsheet plus a bibliographic file may be sufficient. Eventually the data will be taken from these files to be analyzed using statistical software.

The data extraction spreadsheet

Although the diversity of ecology and evolutionary biology research resists any general description of the kinds of data that will require extraction, several general principles should be followed in any well-designed data extraction process. First is the development of an appropriate data extraction spreadsheet (Fig. 5.1). This constitutes the link between each primary study and the ultimate research review and as such, informs the meta-analysis model with data for effect size estimation and associated information; it also provides the basis for quality assessment and quality control (QA/QC) and hence the integrity of the review outcomes. Historically, data extraction was carried out exclusively using paper forms, which retain some advantages over the electronic spreadsheets that are now more widely used. For example, when assessments of the data extraction process are conducted by independent observers, it may be easier to compare some data entry fields on paper than on electronic forms. However, most data extraction is now done electronically. Some clear advantages of electronic extraction spreadsheets include the availability of flexible software for interfacing data entry with data management programs, the possibility for some data entry fields to be completed automatically, the reduction in coding errors when transcribing data from paper to computer, and the possibility of incorporating error checking software into the extraction spreadsheet. Furthermore, with the expansion of

CO_2 Meta-Analysis Project	PAGE	OF	ACCESSION NUMBER
FIRST AUTHOR		YEAR	

1. ACC#	2. PAGE#	3. PARAM	4. GENUS	5. SPECIES
6. PL FUNC GRP	7. TIME	8. POT	9. METHOD	10. X_TRT
11. LEVEL	12. SOURCE	13. X_AMB	14. SD_AMB	15. N_AMB
16. X_ELEV	17. SD_ELEV	18. N_ELEV	19. QUALITY	20. COMMENTS

Figure 5.1. Simplified example of a data extraction spreadsheet used in the meta-analysis of elevated CO_2 effects (Curtis and Wang 1998). Each paper is assigned an accession number (ACC#) and multiple entries from the same paper are given separate page numbers (PAGE#). Cells 3–20 contain details of the ecological parameter (outcome variable) being measured (PARAM); the taxa under study (GENUS, SPECIES, PL FUNC GRP); CO_2 exposure time (TIME); growth conditions and other experimental factors (POT, METHOD, X_TRT, LEVEL); the table or figure from which the data were extracted (SOURCE); the means (X_AMB, X_ELEV), standard deviations (SD_AMB, SD_ELEV), and sample sizes (N_AMB, N_ELEV) of the parameter in the ambient and elevated CO_2 treatments; the quality assessment score (QUALITY); and miscellaneous comments (COMMENTS).

electronically available journal holdings it is now possible to largely eliminate paper from the review process.

The specific elements of a data extraction spreadsheet are dictated by the research review questions and anticipated analyses as laid out in the review protocol (see Chapter 3). A common feature of any extraction spreadsheet is a study identifier linking that entry to a specific piece of primary research (e.g., the accession number in Fig. 5.1). It may be desirable to include other study identifiers in the extraction spreadsheet such as lead author or publication date, but typically separate full bibliographic databases also will be maintained (see "The Meta-analysis Database," below). In determining additional elements of an extraction spreadsheet, a general rule is to include those study characteristics that could affect the study results and help assess their applicability to the research review. Typically, study characteristics form the spreadsheet columns, while individual studies are represented in rows. While the studies being considered for data extraction will all have been previously examined for inclusion criteria, input of data into the spreadsheet provides a final check on suitability as well as allowing evaluation of study quality.

There are a number of other data entry elements that are common across extraction spreadsheets. For experimental studies, these include details of the research methods or experimental conditions that might affect the results (laboratory vs. field-based treatments, size or type of growth containers, etc.), the taxa or system being considered and any pertinent attributes (age, sex, etc.), and details of the experimental treatment (length, intensity, additional factors). For observational studies these might include location, site or population characteristics, and relevant conditions under which observations were made. For both types, an open-ended "Comments" field is useful. If a data entry element is not relevant for a particular study, that cell can simply be left empty or assigned a missing value code. For example, in the elevated CO_2 database, studies with no additional treatment crossed with the CO_2 treatment (X_trt = NONE; see Fig 5.2) would have a missing value designation in the cell coding the crossed treatment level (Level = ".").

The study results themselves, such as sample size, mean, and standard deviation, often represent a comparatively small number of entries in the extraction spreadsheet. To facilitate error checking and other forms of QA/QC it is useful to annotate, either electronically (e.g., by using "Highlight Text Tool" in PDF files) or on hard copy, each manuscript from which data were extracted. The annotated version should specify exactly where information was obtained, such as the specific numbers within a table, points on a graph, or data from the body of the paper. Highlighting descriptive text in the methods, and the extracted data from the reported results, will save considerable time later if questions arise regarding the accuracy of results obtained from a particular study. It is also useful to have a column in the data extraction spreadsheet which specifies the source of the data—for example a particular figure or table from the primary study (Figs. 5.1 and 5.2).

It is important to properly document not only the literature used in the meta-analysis, but all of the literature retrieved in the search process, recording those papers which were examined but not included in the analysis, and organizing the copies of the papers from which data were extracted. Each step in this process is important in establishing the breadth and freedom from bias of a literature search. Prior to the widespread availability of journal articles in electronic format, this part of a large meta-analysis could easily fill numerous filing cabinets. Fortunately for the meta-analysis enterprise, electronic article retrieval coupled with bibliographic software has vastly streamlined this part of the process. Figure 5.3 illustrates a flow chart and decision tree that begins with the filtered reference list (papers judged acceptable based on examination of title and abstract) and results in two bibliographic files (e.g., EndNote libraries). One library file contains citation information and electronic copies of papers (or permanent links to papers) found to not possess the necessary inclusion criteria. The other library file contains citation data and annotated copies of all papers included in the meta-analysis. It is important to maintain both libraries since it may be important to demonstrate why a particular study was not included in the analysis. For papers not included in the meta-analysis, a record should be kept of the reason for rejecting them. One of the most common reasons for rejecting studies in ecological and evolutionary meta-analyses is that essential data are missing (e.g., sample sizes and any measures of variance).

Data to be extracted may reside in any part of a paper, but are most commonly found in the "Materials and methods" section describing details of the experimental conditions, treatments, number of replicates, etc., and in the "Results" section describing the data needed for calculation of treatment effect size. Numerical data cited in the text or contained in tables are easily transferred to the data extraction spreadsheet. Data contained in figures can present a variety of extraction challenges depending on the manner in which they are plotted (e.g., scatterplots, histograms, trend lines, use of multiple or three-dimensional axes, etc.), the size of symbols or lines relative to the axis scale, and the size and printed quality of the figure itself. However, with a good-quality digital image of a data figure, it is relatively simple using graphical data-extraction software (e.g., freeware DataThief, Graphclick, or ImageJ) to obtain accurate numerical information. When symbol or line thickness is large relative to the axis scale it is usual to consider their midpoint as the correct numerical data point to record.

ACC	PG#	Param	Genus	Species	Pl Func Grp	Time	Pot	Method	X trt	Level	Source	X A	SD A	N A	X E	SD E	N E	Quality
44	1	PN	ALNUS	RUBRA	N2FIX	46	0.5	GC	FERT	HI	T3	11.8	1.43	5	23.2	10.31	5	3
44	2	PN	ALNUS	RUBRA	N2FIX	46	0.5	GC	FERT	CTRL	T3	11.7	2.59	5	25.9	3.31	5	3
44	3	TOTWT	ALNUS	RUBRA	N2FIX	47	0.5	GC	FERT	HI	T1	3945.0	1115.80	5	6816.9	1769.98	3	3
44	4	TOTWT	ALNUS	RUBRA	N2FIX	47	0.5	GC	FERT	CTRL	T1	2251.2	327.58	5	2596.1	667.47	3	3
121	1	PN	QUERCUS	PRINUS	ANGIO	76	2.6	GH	NONE	.	T2	2.6	1.05	6	5.4	1.42	6	3
121	2	TOTWT	QUERCUS	PRINUS	ANGIO	70	2.6	GH	NONE	.	T4	6.6	1.63	5	5.9	1.74	5	3
121	3	PN	MALUS	DOMESTICA	ANGIO	77	2.6	GH	NONE	.	T3a	8.2	1.13	6	12.1	1.59	6	3
121	4	TOTWT	MALUS	DOMESTICA	ANGIO	64	2.6	GH	NONE	.	T4	4.1	1.26	4	4.6	1.41	4	3
121	5	PN	ACER	SACCHARINUM	ANGIO	61	2.6	GH	NONE	.	T2	9.2	1.51	7	15.8	1.85	7	3
121	6	TOTWT	ACER	SACCHARINUM	ANGIO	50	2.6	GH	NONE	.	F2	6.4	2.03	3	10.8	1.16	5	3
159	1	TOTWT	CASTANEA	SATIVA	ANGIO	730	GRND	GC	NONE	.	T1	127.3	47.46	3	153.5	27.19	3	3
209	1	TOTWT	CASTANEA	SATIVA	ANGIO	365	24	GH	FERT	HI	F1	144.1	25.70	20	183.6	39.06	20	3
209	2	TOTWT	CASTANEA	SATIVA	ANGIO	365	24	GH	FERT	CTRL	F1	59.9	14.28	16	71.7	14.30	16	3
468	1	PN	CASTANEA	SATIVA	ANGIO	1095	24	OTC	NONE	.	T3	8.3	1.30	4	9.5	1.50	4	3
468	2	TOTWT	CASTANEA	SATIVA	ANGIO	1095	24	OTC	NONE	.	T2	136.0	35.00	5	146.0	10.00	5	3
502	1	PN	LIRIODENDRON	TULIPIFERA	ANGIO	840	GRND	OTC	NONE	.	T1	7.4	1.34	5	12.3	1.79	5	3
505	1	PN	QUERCUS	ALBA	ANGIO	168	2.6	GH	FERT	CTRL	F4a	6.8	2.75	5	11.2	3.60	5	3
505	2	TOTWT	QUERCUS	ALBA	ANGIO	168	2.6	GH	FERT	HI	T2	14.6	2.29	5	17.9	10.19	5	3
505	3	TOTWT	QUERCUS	ALBA	ANGIO	168	2.6	GH	FERT	CTRL	T2	11.2	5.75	5	14.7	5.01	5	3

ACC#	Authors & year	Title/Journal/Pages
44	Arnone, J.A., III, & J.C. Gordon. 1990.	Effect of Nodulation, Nitrogen Fixation and CO2 Enrichment on the Physiology, Growth and Dry Mass Allocation of Seedlings of Alnus rubra Bong. New Phytologist 116:55–66.
121	Bunce, J.A. 1992.	Stomatal Conductance, Photosynthesis and Respiration of Temperate Deciduous Tree Seedlings Grown Outdoors at an Elevated Concentration of Carbon Dioxide. Plant, Cell & Environment 15:541–549.
159	Couteaux, M.M., P. Bottner, H. Rouhier, & G. Billes. 1992.	Atmospheric CO2 Increase and Plant Material Quality: Production, Nitrogen Allocation and Litter Decomposition of Sweet Chestnut. In Responses of Forest Ecosystems to Environmental Changes (A. Teller, P. Mathy, and J.N.R. Jeffers, eds.), Elsevier, London.
209	El Kohen, A., H. Rouhier, & M. Mousseau. 1992.	Changes in Dry Weight and Nitrogen Partitioning Induced by Elevated CO2 Depends on Soil Nutrient Availability in Sweet Chestnut (Castanea sativa Mill.). Annales des Sciences Forestieres 49:83–90.
468	Mousseau, M. 1993.	Effects of Elevated CO2 on Growth, Photosynthesis and Respiration of Sweet Chestnut (Castanea sativa Mill.). Vegetatio 104/105:413–419.
502	Norby, R.J., C.A. Gunderson, S.D. Wullschleger, E.G. O'Neill, & M.K. McCracken. 1992.	Productivity and Compensatory Responses of Yellow-poplar Trees in Elevated CO2. Nature 357:322–324.
505	Norby, R.J., & E.G. O'Neill. 1989.	Growth Dynamics and Water Use of Seedlings of Quercus alba L. in CO2-enriched Atmospheres. New Phytologist 111:491–500.

Figure 5.2. Two linked database spreadsheets from the meta-analysis of elevated CO_2 effects (Curtis and Wang 1998). *Top panel:* main database that receives information from the extraction spreadsheet (Fig. 5.1). Each row corresponds to a unique set of response statistics, with the seven papers listed contributing from one to six independent sets of data. *Bottom panel:* reference spreadsheet containing citation information for each paper. Column headings are as described in Figure 5.1. In this example, the parameters, or outcome variables, in column 3 are photosynthetic rate (PN) and total plant weight (TOTWT).

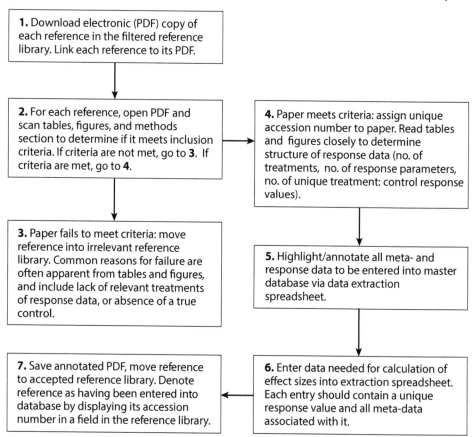

Figure 5.3. Flow chart and decision tree involved with managing the literature database in a meta-analysis. The filtered reference library in step 1 has passed an initial examination for relevance based on viewing the title and abstract, and hence each reference merits examination of its full text. This important initial process is detailed in Chapter 4. Stored electronic copies of the papers (the PDF files) are linked to their citation information within the bibliographic software. Extracted data are linked to the citation information and the PDF file through a unique accession number (step 4).

The data extraction protocol

Complementing the data spreadsheet itself is the set of instructions, or extraction protocol, which guides the coding decisions that will need to be made by anyone entering data. While this is especially critical for groups working together, it is also important for individuals working alone to increase accuracy and reduce potential bias. For example, we found that it was important to create uniform standards for the determination of rooting volume from reported pot dimensions to correctly code "pot size" on the CO_2 effects extraction spreadsheet (Fig. 5.1), or for handling missing data (Chapter 13). It may even be necessary to establish a protocol for identifying control and treatment groups. For example, Gurevitch et al. (1992) were interested in studies in which the densities of organisms were manipulated to determine responses to competition. They defined the "controls" in the meta-analysis to be the groups whose densities were close to natural densities in the field, and the "experimental" groups were those in which densities were higher or lower than natural densities. These definitions were not always congruent with those of the studies' authors. If quality assessments are made (see "Critical Appraisal of the Data," below), it is very important that clear guidelines for interpreting possible sources of bias within studies and assigning quality scores are provided. Accurate coding is critical and the extraction protocol should be concise enough to be practical yet provide sufficient detail to prevent erroneous decisions. The authors of the data extraction spreadsheet and protocol should practice using them and be involved in training any other people who will use them.

Establishing precisely which data points within a study are the appropriate ones to extract requires careful thought and explicit direction in the extraction protocol. This necessity for careful thought and direction is often encountered in studies reporting time series or other types of data exhibiting nonindependence. Working with nonindependent data is covered in detail in Chapter 16 and if a meta-regression model is appropriate, careful coding of measurement time, treatment duration, and so forth will be important. A common strategy in ecological and evolutionary meta-analyses has been to avoid nonindependence by extracting only a single data point per measurement time series; examples are at the end of the treatment or at peak standing biomass. There are two problems with this approach: studies with shorter durations may not be appropriate to combine with those having longer durations (results may be very different after two months and five years), and also there is a substantial loss of data. For example, trends over time may be the most interesting and valuable part of the results, and these will be lost if only a single point is extracted. Clear guidelines for selecting data from a time series would then be a necessary part of the extraction protocol. Similar considerations and the need for guidelines arise with single studies reporting response data from multiple treatments, species, locations, outcomes, and so forth.

Another common source of ambiguity in data extraction is determining the correct sample size associated with a reported statistic. Hence, rules for converting degrees of freedom to sample size need to be clear, as does the basis for determining sample size from a complex or ambiguously reported experimental design. The meta-analyst may be required to calculate necessary values, such as determining standard deviation from standard error, sample size from degrees of freedom, or conversion of test statistics to r (see Chapter 6). It is critically important for sample size, means and standard deviation to agree—that is, they must reflect the same group of sampling units. This must be specified and done in a way that is uniform across studies, and often may be performed within a separate database spreadsheet. One reason this may be difficult to do repeatedly in ecological and evolutionary studies is that data reporting in ecological and evolutionary journals often is haphazard, and because a wide range of very different and often complex experimental designs are used. Where subjective decisions are required in completing the extraction form, or if extracted data are transformed or otherwise altered from those that were reported, these should be noted in the "Comments" input cell.

The last stage in preparing for data extraction is pilot testing and modification of the extraction spreadsheet and protocol. Invariably, users will identify input that either is not needed or falls into a field that is missing from the spreadsheet. A small but representative sample of primary studies should be selected for this exercise with several data extractors taking part. In this way, both the completeness of the spreadsheet and the reliability of data extraction can be assessed. If several people will be inputting data it is quite important to evaluate the degree to which their assessments of the same study differ (e.g., with a kappa analysis; see Chapter 4) and to develop a plan for comparing data and resolving disagreements. Where the number of studies is manageable, it is possible to use two coders and double code every study, resolving any differences by discussion. This builds intercoder agreement directly into the data extraction procedure. Finally, changes to the spreadsheet and/or protocol may be needed to improve data collection reliability.

THE META-ANALYSIS DATABASE

The extraction spreadsheet may simply be appended to a growing database stored in a single spreadsheet (also known as "flat file database") (e.g., Microsoft Excel, Lotus, Quattro Pro), but it may be advantageous to develop relational databases (e.g., by using Microsoft Access, Paradox or dBase software), particularly for large or complex data. It should be kept in mind, however, that most meta-analytic software requires a "flat" data matrix of rows and columns, and thus if a relational database is used, a single spreadsheet will have to be created and used with the statistical software for the actual meta-analysis. One database model is illustrated in Figure 5.2 in which extracted data are stored in one spreadsheet, and citation information in another. A paper's accession number links the two databases. In Microsoft Access, this would be accomplished by assigning the accession number to the primary key field linking the separate databases. Specialized bibliographic databases such as EndNote or Mendeley can easily substitute for the spreadsheet-based reference database shown here. Where data syntheses involve examining many different subgroups or multiple effect sizes for different variables or times of measurement, it may be advantageous to organize the data in separate group-level or effect size-level files. Working within a relational database would facilitate the organization of complex analyses involving many subgroups. It may be desired to make the extracted database available online; however, doing so emphasizes the need for clear documentation of all input fields and points to the general advantage of simple data matrices over more complex designs and those requiring specialized software to read.

CRITICAL APPRAISAL OF THE DATA

During the process of data extraction the investigator has an opportunity for critical appraisal of data quality. Methods for assessing data quality have received considerable attention by meta-analysts in medicine and the social sciences, but how or even whether to incorporate quality measures in the meta-analysis is controversial (Herbison et al. 2006). We recognize that scientific studies vary in quality and therefore it seems reasonable to assume that combining poor-quality studies with those of higher quality will weaken the inferential capacity of any subsequent analysis. Put another way, a poorly designed or executed study is considered more likely to be influenced by systematic errors (low accuracy) although the random sampling error for such a study might be quite low (i.e., it can have high precision). In a research review, poor-quality studies can add to the unexplained variance among studies; if they are of sufficient weight or number, these studies can materially affect mean effect size estimates (e.g., EPA 1992).

One approach to quantitative assessment of study quality has been the use of numerical scales in which points are assigned to specific elements of the study and summed to produce an overall quality score (e.g., Jadad et al. 1996, Levine 2001). Although many such quality assessment tools have been developed, a variety of serious problems with their application have been identified, including the comingling of questions of methodological quality with those of reporting quality, reliance on "accepted criteria" from textbooks rather than empirical evidence of bias, and poor reporting of inter-rater reliability or concordance (Moher et al. 1995, Jüni et al. 1999, Sanderson et al. 2007). Herbison et al. (2006), in a study applying 45 different quality scales to 65 separate meta-analyses, concluded that none of the scales measured quality in a valid manner and that while study quality is important, summed scores as a method for assessing this should be abandoned. We concur with this assessment.

An alternative to summative scores is a more qualitative evaluation of the degree to which a given study is free from known sources of experimental bias (Pullin and Stewart 2006, Higgins and Green 2011). Some design features common to ecology and evolutionary biology may be readily identifiable as calling into question a study's quality, such as pseudoreplication, inappropriate controls, or highly variable sample size across treatments. Many others, however, will be specific to the research area or study, but nonetheless identifiable as potential sources of bias by knowledgeable practitioners. A high-quality study is defined as being free from these sources of bias whereas a lower-quality study might suffer from one or more potential sources of bias. This approach yields a relatively small number of categorical groups from which subgroup and sensitivity analyses can be conducted to test whether between-group heterogeneity exists among quality levels, and whether inclusion of studies with a high likelihood of biases (i.e., lower-quality studies) significantly affects the overall outcome of the review.

Although formal tests of the utility of quality groupings in ecology and evolutionary biology meta-analyses have not yet been undertaken, quality scales have been used in conjunction with sensitivity analyses within individual meta-analyses (e.g., Tyler et al. 2006; Stewart, Pullin, and Coles 2007). If these are used, it is critical that clear guidelines for evaluating sources of bias are articulated in the extraction protocol and that inter-rater reliability (degree of agreement among raters) is established if more than one person is coding for study quality. The latter can be achieved by a kappa analysis (see Chapter 4). It should be emphasized that poor reporting quality should not be confused with poor study quality although separating the two in practice can be challenging.

It also is possible to avoid any explicit quality ranking in favor of clearly defined study inclusion criteria combined with careful coding of methodological variation among included studies. Inclusion criteria initially can be set to exclude studies that lack specified design features, such as proper replication or appropriate controls, making transparent any a priori "quality" standard applied to a given review. Remaining methodological variation then is coded within the data extraction protocol and examined statistically using subgroup or sensitivity analyses. If specific methods are associated with different outcomes those studies can be reported separately. For example, in research on plant responses to elevated CO_2, a small pot size (i.e., small rooting volume) for experimental plants was considered by some (though not all) investigators to be a serious design flaw leading to biased results. Curtis and Wang (1998) coded all studies in their meta-analysis for pot size (Fig. 5.2) and found that plants grown in pots less than 0.5 L acclimated photosynthetically to elevated CO_2 much more than plants grown in larger pots. They used that result as a basis to exclude particular studies from further consideration. The goal of this approach, then, is to focus on testable relationships between specific methodological features and study outcomes, and away from attempts to assign quality measures to individual studies. This is our recommended strategy for handling perceived differences in study quality within the meta-analysis database.

CONCLUSIONS

It is not uncommon for those without experience building a meta-analysis database to suppose that this part of the review process will be relatively straightforward and proceed rapidly. Surely, extracting data from published sources must be faster and easier than collecting and analyzing it yourself! Unfortunately, this often is not the case. However, initial time spent developing clear data extraction protocols and testing data extraction spreadsheets will both speed the extraction process and reduce errors in the final database. While the thorny issue of distinguishing among high- and low-quality studies remains with us, use of transparent inclusion criteria and coding of specific method features allows for subsequent statistical tests of the impact of variable study quality on the outcome of the meta-analysis. Finally, electronic journal archives linked to bibliographic software have greatly simplified the process of documenting the data gathering process and retrieving papers for analysis, a clear boon for the meta-analyst. The end result should be a more transparent, repeatable, and error-free database, albeit in many cases its formation still comprises a large part of the time committed to the overall project.

Effect Sizes: Conventional Choices and Calculations

Michael S. Rosenberg, Hannah R. Rothstein, and Jessica Gurevitch

ONE OF THE FUNDAMENTAL concepts in meta-analysis is that of the effect size. It is this concept that allowed the development of modern meta-analysis. An effect size is a statistical parameter that can be used to compare, on the same scale, the results of different studies in which a common effect of interest has been measured. There is no universal effect size measure in meta-analysis; it depends on a number of considerations, including the nature of dependent variables (e.g., binary or continuous) and whether we are comparing two groups or looking at one group (e.g., over time). In experimental studies, the effect size is a measurement of the response of the subjects to an experimental treatment relative to a control group or another treatment. In observational studies, the effect size may be the relationship between two variables of interest.

The choice of effect size depends, in part, on the type of data available from primary studies and on the common ways of expressing outcomes in a given field. Different methods of estimating effect size have been developed for different questions and types of data; for some data sets multiple effect sizes may be possible, for others there is likely to be only one choice. For example, one may be interested in the effect of increased atmospheric CO_2 on plant growth (measured as total biomass). One might use the ratio of biomass production in ambient and elevated CO_2 conditions as a measure of effect; alternatively, this effect could be measured as the difference in mean biomass production in ambient and elevated CO_2 treatments, or as the slope of a regression of biomass production against a series of different CO_2 concentrations. All of these can be valid measures of the effect. In cases such as this, there are multiple ways an effect might be measured; in other cases only one measure may be appropriate. Where there are multiple possible measures, it is up to the investigators to determine which measure is most biologically relevant to the question and hypotheses they wish to test. The theoretical context of the questions being explored is the most critical component of determining the appropriate means of measuring and interpreting the effect size (Houle et al. 2011).

All effect size measures are a means of representing the results of primary research in a common way so that the results from individual studies can be compared and evaluated (Cooper 1998). In addition to the magnitude of the effect itself, it is important to have an index of precision (e.g., variance, standard error, or confidence interval). In individual studies, the precision serves to provide a context for the interpretation of the effect size. A standardized mean difference of 0.50 with a 95% confidence interval of 0.10 to 0.90 presents a very different picture than a mean difference of 0.50 with a confidence interval of 0.35 to 0.65. In the former case, the real effect could fall over a wide range of values, while in the latter case we have a much more precise estimate of the effect. In a meta-analysis the estimate of precision from each study is used to assign a weight to that study, so that studies in which the effect sizes can be estimated with greater precision are given higher weight than those in which the effect size is estimated with lower precision (Chapters 8 and 9).

This chapter describes the conventional effect sizes most commonly encountered in ecology and evolutionary biology, and the types of data associated with them. While choice of a specific measure of effect size may influence the interpretation of results, it does not influence the actual inference methods of meta-analysis (Chapters 8 to 11). Throughout this work, our general symbol for the estimate for the effect size is θ; while each specific metric of estimating effect size may have its own symbol, we are always trying to estimate θ. One critical point to remember, however, is that one cannot combine different measures of effect size in a single meta-analysis: once you have chosen how you are going to estimate θ, you need to use it for all of the studies to be analyzed. In general, different types of effect sizes cannot be directly compared to one another. A log response ratio and a standardized mean difference may be based on the same sort of data, but one is a ratio of means and one is a difference of means and these are not directly comparable. It is possible, however, to convert some metrics of effect size into others; this is described in detail in Chapter 13.

COMMON MEASURES OF EFFECT SIZE

This section describes standard measures of effect size used in the majority of meta-analyses and whose sampling properties are well understood. They are sorted by the type of data used to estimate the effect.

Pairs of Means

One common type of data in ecology results from the studies which compare two estimated means, $\overline{Y_1}$ and $\overline{Y_2}$, associated with sample sizes n_1 and n_2 and standard deviations s_1 and s_2. The two groups will often (but do not have to) represent treatment and control conditions. Two alternative measures of effect size are commonly used for comparing pairs of means: a standardized difference or a ratio.

The standardized difference in means has been one of the most common measures of effect in ecological meta-analysis (e.g., Järvinen 1991, Gurevitch et al. 1992, VanderWerf 1992, Poulin 1995, Koricheva et al. 1998). In principle, one could estimate the effect as the simple difference in the means. This is only tenable in the unrealistic situation where all studies have been measured in the identical scale and the variance among all studies and groups is assumed to be equal. Instead, one usually uses a standardized difference among the means to adjust for the differences in scale among studies. There are many ways of estimating the standardized difference among means, including Glass's (1976) original estimate Δ, a modification of this known as Hedges' g (Hedges 1981), and a similar alternate measure known as Cohen's d (Cohen 1969). These metrics (described in detail in Rosenthal 1994 and Rosenthal et al. 2000) differ in their assumptions about the homoscedasticity (equality) of the sampling variances in the contrasted groups and on how these are used to estimate the pooled variance for standardizing the effect difference. The most common (and preferred) metric in use today is known as Hedges' d (Hedges and Olkin 1985):

$$d = \frac{\overline{Y_1} - \overline{Y_2}}{\sqrt{\dfrac{(n_1 - 1)s_1^2 + (n_2 - 1)s_2^2}{n_1 + n_2 - 2}}} J,$$

$$\text{where } J = 1 - \frac{3}{4(n_1 + n_2 - 2) - 1}$$

is a correction for small sample size. (Note that there is some variation in terminology in different disciplines, and this term is sometimes referred to as Hedges' g.) The variance for Hedges' d is found as

$$v_d = \frac{n_1 + n_2}{n_1 n_2} + \frac{d^2}{2(n_1 + n_2)}.$$

Unlike some of the earlier standardized difference metrics, Hedges' d is not affected by unequal sampling variances in the paired groups, and includes a correction factor for small sample sizes; it thus works well when there are as few as five to ten studies. (Note: see Chapter 16 for a discussion of issues involving paired observations, e.g., premeasurements and postmeasurements of the same individuals.)

Interpretation of the magnitude of the standardized mean difference can be difficult in some cases. Cohen (1988) suggested rules of thumb that stated that values of 0.2 were small effects that would not be visible to the naked eye, 0.5 were medium effects apparent on casual observation, and 0.8 or more were large effects that are immediately obvious. Another way to interpret d is to convert it into the probability of the standard normal variate (Z-score), which indicates the percentage of scores in one group that are expected (given normally distributed data) to be exceeded by the average score in the other group. For example, Tonhasca and Byrne (1994) estimated the difference in insect densities between crop monocultures and polycultures as $d = 0.5$. By examining the probability associated with d converted to a Z-score, the result indicates that the average insect density in polycultures was lower than in 69% of the monocultures. Beyond interpreting the magnitude of the effect size in a meta-analysis, one also needs to consider its statistical significance. This raises the general issue of the difference between statistical significance and biological importance, which in meta-analysis, as in primary analyses, are not the same thing.

One must note that large standardized differences can result from either a large difference in the means (the numerator) or from a small estimate of the pooled variance (the denominator). Osenberg et al. (1997) and others have criticized the use of the standardized mean difference because it may confound the differences between treatment means with differences in variance among studies. For example, in the meta-analysis of predator impacts on prey density in streams, Wooster (1994) found that invertebrate predators have a larger impact on prey (as measured by Hedges' d) as compared to vertebrate predators. Osenberg et al. argued that it is possible that vertebrates reduce prey densities to the same degree as invertebrates, but that studies with vertebrates were characterized by a larger variance. This could arise for several reasons, such as greater individual variation in feeding behavior or the use of larger cages to enclose vertebrate predators, which might result in the cages being dispersed over a larger and more heterogeneous landscape. In these scenarios, the use of d would potentially confound the effect of the treatments with aspects of the sampling design and the degree of spatial heterogeneity. Finally, the sampling variances for the means are among the data most likely to be missing from published studies, occasionally making d more difficult to estimate than other effect size measures (see Chapter 13 for a discussion of this problem). It can also be argued, on the other hand, that although the individual deviations from the mean might be larger for vertebrates versus invertebrates, field experiments versus greenhouse experiments, and so forth, it is also very possible that experimenters adjust for this by using larger sample sizes in such studies to increase statistical power. In that case, there would be no systematic bias in the magnitude of the variances between such groups.

An alternate effect size to the standardized mean difference is the ratio of the means. The response ratio R (Hedges et al. 1999) is the ratio of the means of the two groups (e.g., experimental and control); it is useful when one wishes to compare the magnitudes of two means with the same sign. However, ratios generally have poor statistical properties so one generally transforms the ratio to a metric with more desirable properties using the natural log. Thus the natural log of the response ratio is

$$\ln R = \ln\left(\frac{\overline{Y_1}}{\overline{Y_2}}\right) = \ln \overline{Y_1} - \ln \overline{Y_2},$$

with variance

$$v_{\ln R} = \frac{s_1^2}{n_1 \overline{Y_1}^2} + \frac{s_2^2}{n_2 \overline{Y_2}^2}.$$

Both Hedges' d and the natural log of the response ratio range from $-\infty$ to $+\infty$, where 0 signifies no effect (no difference between the means). One usually reports the response ratio as the back-transformed values of $\ln R$ after analyses are completed since the ratio is easier to interpret than the log-transformed value (Box 6.1). The response ratio has been used in a number of ecological meta-analyses (e.g., Curtis and Wang 1998, Schmitz et al. 2000, Norby et al. 2001, Searles et al. 2001, Wan et al. 2001). Some advantages of this metric are that the effect size (although not its variance) can be estimated without knowledge of the sample sizes

---

BOX 6.1.
Ratios as effect sizes and the natural logarithm.

Several metrics of effect size are based on ratios, including the response ratio, the rate ratio, and the odds ratio. Ratios can be very useful measures but they generally have undesirable statistical properties, including highly asymmetric distributions. For example, ratios generally range from zero to infinity with the null value equal to one. While a ratio of 1/2 and a ratio of 2 reflect the same logical magnitude of change from the null value, to construct a direct average of these two values results in 1.25 (because of the skew, values above one are weighted more heavily than values below one in arithmetic calculations of the average).

A simple solution is to transform ratios using a natural (or other) logarithm. The logarithm of a ratio has many desirable statistical properties, including symmetric distributions with a null value of zero, with a range from $-\infty$ to $+\infty$. It is also often much easier to estimate the distributional properties of log-transformed ratios. These metrics can later be back-transformed for ease in interpretation. To back-transform a logarithm, one raises the base of the logarithm to the log-transformed value, thus $X = e^{\ln X}$. In the example above, if we average $\ln 1/2$ and $\ln 2$, the back-transformed mean is 1.

A consequence of using log-transformed ratios, however, is that the ratio must be positive (one cannot take a logarithm of a negative value or zero). For effect sizes derived from contingency matrices (the odds ratio and rate ratio, see below) this is not much of a problem since these ratios cannot logically be negative, but for contrasts of paired means it is quite possible to have a pair of means where one happens to be positive and the other negative. In this case, the response ratio is an inappropriate measure and Hedges' d is instead recommended. One should also not use a ratio as an effect size measure when either the numerator or denominator would be equal to zero; the transform is undefined and trying to adjust the values by adding a tiny fraction to the numerator and denominator usually results in abnormally large estimates of effect size.

or variances (allowing at the very least, unweighted meta-analysis); it may be easier to interpret than a standardized mean difference; in some fields of ecology results are commonly expressed as ratios between two groups. As discussed in Box 6.1, ratios are only meaningful when they are positive (i.e., the numerator and denominator have the same sign), thus the response ratio cannot be used if the mean of one group is positive and the other negative; in this case a standardized mean difference is appropriate.

Two × Two Contingency Data

An alternate form of data in ecology is based on a contingency table (Table 6.1). This compares two groups (again, often a treatment and a control) and the observed counts for two possible outcomes (e.g., alive or dead). This type of data is common in medical research and less common in ecology; up to this point, only a few ecological meta-analyses have made use of these types of data (e.g., Hyatt et al. 2003, Maestre et al. 2005). There are no standard effect size measures for tables larger than 2×2 (for a discussion of why this is, see Box 6.6).

Before effect sizes can be calculated for this sort of data, one must determine the rate of response for each group. The rate ranges from zero to one, and can be interpreted as the probability of a member of that group showing the response. The rate P is simply the number that shows the response divided by the total, or

$$P_1 = \frac{A}{n_1} \text{ and } P_2 = \frac{B}{n_2}.$$

Using these rates, there are three common effect sizes that can be calculated: the rate difference (RD), the rate ratio (RR), and the odds ratio (OR). Each of these metrics has certain advantages and disadvantages. Although the rate difference (DerSimonian and Laird 1986, L'Abbé et al. 1987, Berlin et al. 1989, Normand 1999) is relatively straightforward to calculate (Box 6.2) and interpret, a shortcoming of this effect size is that the range of variation of RD is greatly

TABLE 6.1. Hypothetical 2×2 contingency table representing the data from a primary study. In this table, A, B, C, and D are the number of observations in each cell.

	Group 1	Group 2	Total
Response	A	B	$A + B$
No Response	C	D	$C + D$
Total	$n_1 = A + C$	$n_2 = B + D$	$A + B + C + D$

BOX 6.2.
Calculation of the rate difference.

The rate difference is calculated as

$$RD = P_1 - P_2,$$

with variance

$$v_{RD} = \frac{P_1(1-P_1)}{n_1} + \frac{P_2(1-P_2)}{n_2}.$$

BOX 6.3.
Calculation of the rate ratio.

The logarithm of the rate ratio is calculated as

$$\ln RR = \ln\left(\frac{P_1}{P_2}\right) = \ln P_1 - \ln P_2,$$

with variance

$$v_{\ln RR} = \frac{(1 - P_1)}{n_1 P_1} + \frac{(1 - P_2)}{n_2 P_2}.$$

---

limited by the magnitudes of P_1 and P_2. That is, the possible values of RD are much greater when P_1 and P_2 are close to 0 and 1, and are constrained when P_1 and P_2 are both close to 0.5; this mathematical constraint may cause apparent heterogeneity among studies that is not present in the sample (Fleiss 1994).

The rate ratio (also known as the relative rate) (Greenland 1987, L'Abbé et al. 1987, Normand 1999; Box 6.3) and the odds ratio (also known as the relative odds, Box 6.4) are more commonly used measures. As with other ratios, it is preferable to analyze the natural logarithm of these metrics (Box 6.1).

The odds ratio is the most widely used effect size estimated from contingency tables. The odds of an event are the probability of the event occurring divided by the probability the event does not occur, relative to one. The odds ratio (or relative odds) estimates the odds of an event happening in one group relative to the odds of the same event happening in the other group (L'Abbé et al. 1987, Berlin et al. 1989, Sokal and Rohlf 1995, Normand 1999). Despite the common use of the odds ratio in meta-analyses, it should be noted that the rate ratio is more intuitive and tends to approximate the odds ratio as long as the event rate is not very low or very high.

Correlation Data

When association between two continuous variables is of interest, Pearson's correlation coefficient r is commonly used as an effect-size measure in ecological and evolutionary meta-analyses (e.g., Britten 1996, Bender et al. 1998, Møller and Thornhill 1998, Reed and Frankham 2001, Koricheva 2002). When the magnitude of Pearson's correlation coefficient approaches ± 1, its distribution becomes skewed. Therefore, rather than use the correlation directly, we transform it into a metric with desirable statistical properties. This is done using Fisher's z-transformation,

$$z = \frac{1}{2}\ln\left(\frac{1+r}{1-r}\right),$$

whose asymptotic variance estimate (Sokal and Rohlf 1995) is simply

$$v_z = \frac{1}{n-3}.$$

BOX 6.4.
Calculation of the odds ratio.

The odds of the response in group 1 (Table 6.1) is:

$$odds_1 = \frac{P_1}{1 - P_1} : 1.$$

The odds ratio is thus

$$OR = \frac{\left(\frac{P_1}{1 - P_1}\right)}{\left(\frac{P_2}{1 - P_2}\right)} = \frac{P_1(1 - P_2)}{P_2(1 - P_1)} = \frac{AD}{BC},$$

although we again prefer to work with the log odds ratio or $\ln OR$. Many formulas exist to estimate the variance of $\ln OR$ (Hauck 1979, Robins et al. 1986); the simplest is

$$v_{\ln OR} = \frac{1}{A} + \frac{1}{B} + \frac{1}{C} + \frac{1}{D}.$$

However, there is a better (although seemingly convoluted) way to estimate the $\ln$ OR and its variance for meta-analysis (Yusuf et al. 1985, Berlin et al. 1989, Haddock et al. 1998), known as the Mantel-Haenszel (1959) procedure. For Table 6.1, the observed response, O, from group 1 is simply A. The expected response $\hat{O}$ assuming no difference between groups 1 and 2, would be

$$\hat{O} = \frac{(A + B)(A + C)}{A + B + C + D}$$

The variance of the difference between the observed and expected values $(O - \hat{O})$ is

$$V = \hat{O}\left(\frac{A + C}{A + B + C + D}\right)\left(\frac{C + D}{A + B + C + D - 1}\right).$$

The $\ln OR$ can then be estimated as

$$\ln OR = \frac{O - \hat{O}}{V}$$

with variance

$$v_{\ln OR} = \frac{1}{V}.$$

Unlike the original formulation of the $\ln OR$ described above, this strange method of estimation turns out to be better for meta-analysis, particularly when samples sizes are small (Hauck 1989).

Note: The Mantel-Haenszel procedure is usually not presented in quite this way, but is mathematically equivalent to what we describe here. We use this presentation because it meshes more clearly with the remainder of our discussion of meta-analysis, effect sizes, and inference throughout the text.

BOX 6.5.
Back conversion of Fisher's z-transform.

For ease of interpretation, Fisher's z-transform can be back converted into a correlation coefficient using the hyperbolic tangent function, or

$$r = \tanh(Z_r) = \frac{e^{2Z_r} - 1}{e^{2Z_r} + 1}.$$

The variance of a correlation coefficient (not shown) is dependent on both sample size and the magnitude of the correlation itself; the variance of the transform is dependent only on sample size. The formula for back conversion of Fisher's z to r is provided in Box 6.5.

The reason this is such a popular effect measure is not simply because correlation coefficients are common in ecology and evolutionary biology (which they are), but because they are easily interpretable and because it is possible to transform a wide array of other statistics into correlation coefficients for use in a meta-analysis. A common problem in carrying out meta-analyses in these fields is that many primary research studies do not report sufficient information for estimating other effect sizes (e.g., Hedges' d requires means, sample sizes, and variances, Chapter 13). They may be more likely, however, to report the results of basic statistical tests such as Z-scores, Student's t, F-statistics, or χ^2. Significance values and test statistics from these basic metrics can be transformed into correlation coefficients using a variety of formulas (Chapter 13). Moreover, other commonly used correlation coefficients such as Spearman's and Kendall's rank correlations can also be converted to Pearson's correlation coefficient (Chapter 13). It is similarly possible to transform some of the other effect size measures into correlation coefficients and vice versa (see Rosenthal 1994 for details). See Chapter 13 for more details on how, why, and when one should transform among measures of effect sizes.

Mantel Correlations

The Mantel correlation (Mantel 1967) is another common metric in ecology and evolution, and is calculated as the correlation of corresponding elements in a pair of square matrices (excluding the diagonals). These $n \times n$ matrices contain pairwise contrasts representing distance (or similarity) between n objects, such as genetic distances or geographic distances (excluding the diagonal, there are $n \times (n-1) / 2$ contrasts in a symmetric distance/similarity matrix). Although it is calculated as a standard product-moment correlation, the Mantel correlation has a very different sampling distribution than a typical correlation coefficient because the data that go into the Mantel correlation lack independence due to pairwise contrasts within each matrix. While there is an asymptotic estimate of the variance of the Mantel correlation, significance is typically determined through a permutation test (Mantel 1967, Mantel and Valand 1970).

Mantel correlations should not be combined with typical correlation coefficients in a meta-analysis: they have very different meanings and sampling distributions. A set of Mantel correlations could be combined in a meta-analysis, however. One would still want to use Fisher's z-transform to improve the distributional properties of the Mantel correlation. For the estimate of precision, there are a few options. In the unlikely circumstance that the asymptotic variances are reported for all of the Mantel correlations to be combined (they rarely are since significance is generally estimated using permutation tests), these could be used directly. An alternative

---

BOX 6.6.

Conversion of F-statistics and χ^2 with more than 1 degree of freedom.

Meta-analysts often encounter F-statistics and χ^2 with more than 1 degree of freedom in the numerator (or a contingency table larger than 2×2) and want to convert these into correlation coefficients for use in a meta-analysis. One must consider the logic behind these analyses to understand why there is no simple formula for this conversion. The degrees of freedom in these tests are one less than the number of groups being compared. Thus a standard contrast between an experiment and a control group (for example) would yield 1 degree of freedom. A test with 2 degrees of freedom would have three groups, perhaps a control and two different experimental conditions. To convert the latter test directly into a correlation is difficult because the F-statistic (or χ^2) is evaluating not just the difference between the control and the experimental treatments, but also between the different experimental treatments. Converting a contrast of this sort into a correlation coefficient requires a very different logic than when one is contrasting a simple pair of groups. Tests with 1 degree of freedom are focused contrasts, while those with more than 1 degree of freedom are unfocused omnibus studies. It is still possible to convert the statistic into a correlation for use as an effect size, but one must first partition the test statistics into contrast and noncontrast subsets. For example, if one partitions an F-statistic into a contrast F with 1 degree of freedom (in the numerator) and a noncontrast F with $k - 2$ degrees of freedom (where k is the number of conditions in the original F, such that the original F has $k - 1$ degrees of freedom in the numerator), the effect size correlation coefficient can be determined as

$$r = \sqrt{\frac{F_{contrast}}{F_{contrast} + (k-2)F_{noncontrast} + df_{within}}}.$$

Specific procedures for partitioning test statistics are dependent on the exact experimental conditions and can be found in Rosenthal and Rosnow (1985, 1991) and Rosenthal et al. (2000).

---

would be to use the size of the matrix (n) as an estimate of precision. We suggest using n rather than n^2 or $n \times (n-1) / 2$, because the distances within the matrices lack independence and n is a better estimate of the independent size of the matrix. Again, we want to emphasize that one should not combine Mantel correlations and standard correlations into a single meta-analysis.

An expansion of the standard Mantel test is the partial Mantel test (Smouse et al. 1986, Legendre 2000), which examines the correlation of two matrices while holding a third (or more) constant. With respect to meta-analysis, these can be treated like standard Mantel correlations, although one would only want to combine partial correlations calculated under logically identical circumstances (e.g., the correlation of genetic distances and linguistic distances while holding geography constant).

Slopes from Simple Linear Regressions

In many cases one might be interested in using the slope from a simple linear regression ($y = a + bx$) in a meta-analysis. Slopes can be used directly as effect sizes, although the following three caveats and conditions must be kept in mind. First, the slope in every study must be measured in the same units. For example, if the slope in one study represents biomass per square meter and the slope in another study represents biomass per square kilometer, one or the other must be scaled so that the units match. Slopes measured in disparate units are not comparable. Second, one needs a measure of the variance of the estimate of the slope (s_b^2) for each study, which is rarely reported in published papers in ecology (this is different than the unexplained error variance of the regression, $s_{y \cdot x}^2$, which is more likely to be reported). If the variance of the slope is not directly provided, it may be possible to estimate it from other statistical values that are more likely to be reported (Box 6.7). Third, if some of the slopes need to be rescaled (as above), one must also remember to rescale the corresponding variances.

BOX 6.7.
Estimating the variance of a regression slope.

If the variance of a slope (s_b^2) is not directly provided in a paper, it may be possible to estimate it from other values which are more likely to be reported. The simplest way is to recognize that the significance of a slope (b) can be tested using Student's t-distribution as the ratio of the slope to its standard error

$$t_s = \frac{b}{s_b}$$

with $n - 2$ degrees of freedom (n is the sample size). Thus, one can work backwards from the significance of the slope by determining the value of t_s which would lead to the reported significance and solving the above equation for s_b. Alternatively, given the sample size, the slope, the variance of the independent variable (s_x^2), the unexplained variance ($s_{y \cdot x}^2$), and the R^2 of the regression,

$$s_b^2 = \frac{R s_{y \cdot x}^2}{(n-1) b (s_x^2)^2}.$$

Another alternative requires the sample size, the variance of both the dependent and independent variables and the variance explained by the regression ($s_{\hat{y}}^2$). With these one can find

$$s_b^2 = \frac{(n-1) s_y^2 - s_{\hat{y}}^2}{(n-1)(n-2) s_x^2}.$$

It may help to recall that the variance of the dependent variable (s_y^2) is the sum of the variance explained by the regression ($s_{\hat{y}}^2$) and the unexplained variance ($s_{y \cdot x}^2$). Also, if an F-statistic is used to test the significance of the regression, it is calculated as

$$F = \frac{s_{\hat{y}}^2}{s_{y \cdot x}^2}.$$

Thus, one can calculate any two of these values (F, s_y^2, $s_{\hat{y}}^2$, and $s_{y \cdot x}^2$) given the other two, which may aid in estimating s_b^2.

Alternative Measures of Effect Size

In principle, many other metrics could be used as measures of effect size (Chapter 7). The primary requirements are that, for each study, one can estimate an index which measures the magnitude of the effect and a variance which measures the precision of the index. In many cases this is much easier said than done, which is one reason why the standard metrics described above are most commonly used in meta-analyses in ecology and evolutionary biology, as well as in other disciplines.

Osenberg and colleagues (Osenberg et al. 1997, Osenberg and St. Mary 1998) have criticized the use of standard effect size metrics in ecology and evolutionary biology. Instead, they suggest that each time a meta-analysis is conducted the authors should model the biological process being studied as well as the spatial/temporal scales of the experiments; then based upon the model used, they should construct appropriate effect size metrics. To illustrate how the choice of effect size metric can potentially have a substantial impact on the results of a meta-analysis, Osenberg and St. Mary (1998) compared a meta-analysis of the same data set using three different metrics of effect size and found substantial and meaningful discrepancies among the results.

Gurevitch et al. (2001) raised two major objections to creating new metrics of effect size for each meta-analysis: first, lack of standardized measures makes evaluating the results of a meta-analysis very difficult (if a set of meta-analyses use different, nonstandardized measures, it is very difficult to compare their results). Second, for most ecologists and evolutionary biologists, determining the statistical properties of novel statistics is essentially out of reach. One can run into serious problems using novel metrics with unknown sampling distributions and unknown properties. In addition, for most meta-analyses in ecology and evolutionary biology there will not be a single way to model all studies one is interested in combining, and it is unlikely that they will all be on the same spatial or temporal scale. Gurevitch et al. (2001) suggest that a compromise between creating new metrics for each meta-analysis on the one hand and having a single metric for all meta-analyses on the other hand is to have a number of well-understood metrics (such as those described throughout this chapter) to choose from; these would represent a range of biologically meaningful types of comparisons between groups. New metrics can be introduced if their statistical properties are well understood and there is a compelling reason to use them.

SUMMARY

In conclusion, there are many different ways of estimating effect size. The six most common effect size metrics in meta-analysis vary both with respect to the type of data to which they can be applied and in their interpretation; thus, specific choice of a metric will depend not only on the nature of the data, but also on the specific questions being asked. Once a metric is chosen, it must be applied to all included studies; one cannot mix and match different measures of effect in a single meta-analysis.

Using Other Metrics of Effect Size
in Meta-analysis

Kerrie Mengersen and Jessica Gurevitch

META-ANALYSIS IN ECOLOGY AND evolutionary biology has generally been used to synthesize the results of independent experiments in order to assess overall results across studies, and to examine the causes of heterogeneity in those results due to modifying characteristics of the studies. The ability to do this is based on using standardized effect sizes that express the effect of interest on the same scale or in comparable terms across studies. In combining them, the effect sizes are weighted by their inverse sampling variances to account for differences in the precision of their estimates of the effect of interest, as well as for other desirable statistical properties. These effect sizes often involve a comparison of two groups, typically an experimental and control group. As discussed in Chapter 6, the use of such effect size metrics to compare two groups is not limited to experimental data; in many cases, analogous comparisons from nonexperimental data can also be used to construct such effect size measures. For example, particularly in evolutionary biology, it has been common to use Fisher's transform of Pearson's correlation coefficients (Chapter 6) as effect size measures in meta-analysis where the correlations between two parameters in observational data are combined across studies. Similarly, standardized mean differences may be used to construct effect sizes when one wishes to compare the relative performance of two groups in a series of observational studies (e.g., relative time invested in parental care for male and female birds).

However, in many other cases it is of interest to combine various types of noncomparative metrics (i.e., simple measurements of responses) of effect size across studies. The need to combine such ecological or evolutionary responses or parameters is characteristic of observational data, but may also be useful for experimental data. Such effects include parameters like heritability, diversity indices, rate of population increase, time to an event such as metamorphosis or death, age of first reproduction, respiration rate, and so on. In these cases, the response itself is combined across studies, rather than combining the comparison of the experimental and control groups' responses (as in Hedges' *d* or the log response ratio) across studies.

Information about such responses, or effects, is often available in a wider class of studies than just traditional experiments, and observational studies may provide a rich source of data. Perhaps even more than ecological experiments, these studies typically exhibit differences in design, conduct, analytic method, reporting, and control of confounders and biases. The analyst must therefore take great care to ensure that the effect estimates are standardized and truly comparable across studies, that an appropriate estimate of sampling variance is used for weighting, and that account is taken appropriately for differences in study characteristics. Similarly, it is important to understand the interpretation and inferential capability of the results:

What do they mean? How much do we trust them? How general are they? These issues will continue to become much more important in the future, particularly as it becomes increasingly necessary to broadly synthesize results from many different studies, whether observational or experimental, on responses of organisms and ecological systems to anthropogenic changes, from habitat fragmentation to changing climate.

As with familiar metrics such as standardized mean differences or odds ratios, meta-analysis of such parameters requires extraction of both a standardized effect estimate from each study, and a corresponding weight for this estimate, which is typically the inverse of its sampling variance. Thus there are two critical requirements. The first is making sure that the measure is really comparable across studies. The second is obtaining an appropriate sampling variance, or the necessary information from each study in order to calculate it.

There are many potential effects that might be of interest to a meta-analyst. For example, meta-analysis in ecology has been used to combine presence-absence matrices (Gotelli and McCabe 2002), spatial variation in recruitment (Hughes et al. 2002), similarity between communities (Soininen et al. 2007), fecundity and survival (Boyce et al. 2005), and field metabolic rate (Riek 2008). Importantly, although they illustrate the range of possibilities for meta-analysis, they also highlight some of the practical difficulties of properly synthesizing studies, as discussed below.

EFFECT ESTIMATES WITH KNOWN DISTRIBUTIONS AND/OR VARIANCES

Many effects of interest to a meta-analyst have known distributions. For example, the distribution of correlation coefficients is described in Chapter 6. For other effects, an appropriate distribution may be assumed based on the type of data and its characteristics. This is discussed and illustrated below for a range of data types. Some examples of effects of potential interest are also given in Table 7.1, with the corresponding effect estimate, the approximate distribution and variance of the effect estimate, and the possible transformation to a normal distribution.

Two points apply to all of the following. First, it is assumed that the underlying observations (i.e., study outcomes) are independent and identically distributed; that is, they all have the same underlying distribution and parameter values. Second, in most cases the suggested distributions are *approximate*. Thus, the analyst should carefully check the assumptions underlying a distribution to ensure that is indeed appropriate for the data. Unless the assumptions are met exactly, analysts should treat inferences made on the basis of a meta-analysis with caution. A sensitivity analysis can be conducted to assess the robustness of inferences to various assumptions. This is discussed in more detail below. As with all good statistical practice, any assumptions and data manipulations should be explicitly declared in the write-up of the meta-analysis.

Normally distributed data

Many effect estimates can be assumed to be approximately normally distributed. These include:

- sample means based on large sample sizes (i.e., large numbers of observations per study), based on observations from any distribution—by the central limit theorem
- regression parameter estimates (e.g., intercept, slope/trend) based on a regression with normally distributed residuals
- estimates of rates (e.g., rate of growth, rate of events) that are large in magnitude
- proportions that are not too close to zero or one, and are based on large sample sizes

TABLE 7.1. Examples of known distributions and possible effect estimates, assuming iid (independent, identically distributed) data.

Effect of interest	Effect estimate	Approximate distribution of effect estimate	Variance of the effect estimate	Transform to normal distribution
Population mean	sample mean $\overline{X}$ based on a sample size n	$\overline{X} \sim N(\mu, \sigma^2/n)$ if $X \sim N(\mu, \sigma^2)$ or if n is large (>30)	If $\overline{X}$ is normal, $\mathrm{Var}(\overline{X})$ is σ^2/n; estimated by s^2/n	not needed if X is normal or n is large
Rate of occurrence	count Y per unit (time period, spatial area, etc.)	$Y \sim \mathrm{Poisson}(\lambda)$; mean $\mu = 1/\lambda$	$\mathrm{Var}(Y)$ is equal to Y	$\log(Y)$
Probability of a "success," p	proportion $\hat{p}$ of successes X in n observations	X has Binomial distribution	$\mathrm{Var}(X)$ is $np(1-p)$; $\mathrm{Var}(\hat{p})$ is $p(1-p)/n$	$\log(\hat{p})$ or $\mathrm{logit}(\hat{p})$ or $\sin^{-1}\sqrt{\hat{p}}$
Logit of a probability (p), based on sample of size n	$Y = \ln(p/(1-p))$	Y has a Normal distribution	$\sqrt{\{(1/np) + 1/(n(1-p))\}}$	not needed
Effect of a covariate or modifier in a regression	Estimated regression coefficient, b	b has a Normal distribution if the regression residuals are Normal	obtained from regression analysis; must be reported	not needed
Growth rate, survival rate, etc.	estimated from sample	Gamma, Weibull, etc.	must be reported	log
Coefficient of variation (CV)	sample SD/ mean $S/\overline{X}$	if X is Normal, the inverse CV $\overline{X}/S$ is $N(0,1)$	variance of inverse CV is 1	not needed

For a meta-analysis, these effect size estimates need an associated variance of the estimate. These are sometimes reported— for example, as standard error of the mean (SEM) for a sample mean, or as standard error for a regression parameter estimate (however, see the discussion on variance of regression parameters in Chapter 9). Suggestions for what to do when they are not reported are given later in this chapter. Many measures of ecological interest fall into this category of effect estimates that can be assumed to be approximately normally distributed. For example, it may be useful to combine estimates of mean population sizes per unit area (where the means are calculated for each study from some number of plots or samples) across different studies or over time (Chapter 15). Mean sizes or amounts (e.g., mass, length) based on n samples per study are another common measure of this type.

It is essential for the researcher to ascertain that the measures in the different studies are on the same spatial/temporal scale. One important case is that of mean species richness (number of species) per sample (e.g., site, plot). While species richness appears to be a simple metric that is directly comparable across studies, there are serious limitations to combining this measure across studies directly because species-area curves are nonlinear and differ among sites and ecological communities (Scheiner et al. 2000). Failure to recognize this problem may lead to inaccurate and potentially misleading results.

Count data

It is not uncommon for ecological or evolutionary studies to report counts as an effect of interest. As examples, one might count numbers of organisms (such as animals sighted in a region), numbers of events (such as outbreaks of the marine cyanobacteria *Lyngbya* at a site in a year), and so on (e.g., Hamilton et al. 2007). Let T be the count reported in a study. If the observations are independent and have the same underlying probability of occurring—for example, if *Lyngbya* outbreaks occur independently with equal probability each time—then T could be assumed to have a Poisson distribution. The Poisson distribution depends on one parameter, the rate λ.

Sometimes the distribution is written in terms of the expected count (the mean) μ, where $\lambda = 1/\mu$. This can be useful for meta-analysis. For example, if a study reports the estimated rate instead of the count itself, such as the rate of koala sightings instead of the number of koalas sighted, we can convert the rate λ_i to an expected number μ_i, and combine these parameters in a meta-analysis, since this might be more biologically interpretable.

Another useful feature of the Poisson distribution is that the mean is equal to the variance. Thus the count gives us estimates of both the mean and variance. This can be advantageous in a meta-analysis; we can use a generalized linear model (GLM) in which the observed count is defined as Poisson with a study-specific parameter (e.g., μI) and these parameters are then combined. Thus, it is not necessary to provide a variance for the effect estimate, since the count defines both the mean and the variance, and the GLM formulation incorporates this variance into the analysis. This is described further in Chapters 8 and 9.

If the analyst does not wish to use a GLM, but wishes to use a model with normally distributed parameter estimates, then there are a number of options. First, a Poisson distribution tends to a normal distribution for large sample sizes, so if the counts are large (greater than 30, as a rule of thumb), then they can be assumed to be approximately normally distributed. Since the mean of the Poisson distribution is equal to the variance, then the variance associated with T can be estimated as T itself. Note that under this distribution, as the count increases, the variance also increases. This is contrary to the assumption of independence between the mean and variance for a normal distribution, so this approach is only approximate.

Second, the logarithms of the counts ($\log(T)$) are often assumed to be normally distributed. Note, however, that it is not straightforward to calculate the variance of these transformed values, since the mean-variance equality is no longer valid for the log-transformed values; we cannot simply take the log of the counts as an estimate of the variance of $\log(T)$. (That is, $\text{Var}(\log(T)) \neq \log(\text{Var}(T))$.) See the discussion below on this issue.

Third, the effect estimates (count data) from each study can be modeled as Poisson, and incorporated in a study-specific correlation or regression analysis; the correlation or regression parameters can then be assumed to have normal distributions and combined across studies. As an illustration, Cassey et al. (2005) used meta-analysis to quantify the establishment of species outside their natural geographical ranges, as an indicator of changes in global biodiversity. The 23 studies reported regional establishment success, but the definition of success varied across studies. In order to obtain a common metric, the authors converted the various reported effect estimates to a Pearson's correlation coefficient (see, e.g., Rosenthal 1994 for formulas for transformations).

In another study, Ren et al. (2007) investigated the effect of ozone on the association between temperature and human mortality across a wide array of studies. Counts of daily deaths from cardiovascular disease were obtained for 95 large United States communities during the summer months of 1987–2000. These were assumed to have a Poisson distribution with a different rate parameter for each community. A Poisson regression model was constructed for each study (i.e., community), with the daily deaths as the response, and temperature, ozone, and their interaction as explanatory variables. The regression coefficient estimates of the interaction between temperature and ozone, and their associated variance, for each community were assumed to be normally distributed, and were then combined (using a Bayesian meta-analysis model, Chapter 11) to estimate the overall ozone effect across the communities.

Note that in order to combine counts, they must be on the same scale. For example, the number of koalas sighted must be over the same area and the same time period. Estimates reported on different scales need to be adjusted to a common scale. This is by and large relatively easy to do. However, note that if the scale changes, then the variance also changes. For example, if one study reports 10 koala sightings over 6 months and another study reports 25 sightings over a year, and if a year is determined to be the common scale, then the count for the first study must be doubled to 20 koalas, and the estimated variance of the Poisson distribution will be 20 instead of 10.

Proportions (binary data)

Another common situation occurs when studies report proportions. Examples include proportion of seeds germinating out of a total number of seeds planted, proportion of females in a population, proportion of sites surveyed with koalas present, and so on. The underlying data are binary (germinated/not, female/male, presence/absence). Alternatively, for the same studies the number of "successes" (germination, female, koalas present) out of a total number of trials is reported. If each of the trials is independent with the same probability of a success, then the number of successes can be assumed to have a binomial distribution. This distribution depends on the probability of success, p, and the number of trials, n.

As is the case with count data, proportions can be analyzed directly using a GLM, as discussed in Chapters 8 and 9. Again, a variance term does not need to be specified since it will be incorporated as part of the analysis.

If the analyst prefers to fit a meta-analysis based on a normal distribution, then the following options are available.

If the sample size n is large and the estimated proportion $\hat{p}$ is not too close to 0 or 1, then $\hat{p}$ can be assumed to be normally distributed. For a binomial distribution, the variance of p is given by $p(1-p)/n$ which can be estimated by $\hat{p}(1-\hat{p})/n$, where $\hat{p}$ is the observed proportion; this can be used in the calculation of the weights required for the meta-analysis.

More commonly, the $\hat{p}$ values will have to be transformed to approximate a normal distribution. Usual transformations are $\log(\hat{p})$ or, preferably, $\sin^{-1}\sqrt{\hat{p}}$ or $\text{logit}(\hat{p}) = \log(\hat{p}/(1-\hat{p}))$. Again, if the values are transformed, then we lose the mean-variance relationship of the binomial distribution and the associated variances have to be calculated separately; see the discussion below.

Other common effects

Some common ecological metrics such as heritability, diversity, and population growth have corresponding variances that can be used as weights in a meta-analysis. Examples of these are discussed below:

- Heritability is defined as the ratio of genetic and phenotypic variation, where genetic variation can be broad sense (all genetic variation) or narrow sense (only additive genetic variation). The latter is commonly estimated via a regression of the offspring's phenotype on the average of the parents' phenotypes, and an associated variance can be computed depending on the assumptions made about the underlying population structure and distribution (Falconer and Mackay 1996).

- Diversity is estimated by metrics such as Shannon's and Simpson's indices. For example, the Shannon diversity index is defined as $H = -\Sigma p_i \ln(p_i)$, where p_i is the proportion of the ith species in the population, with corresponding large-sample and small-sample parametric variance equations (see, for example, Brower et al. 1998); alternatively, bootstrap estimates of the variance for either index are given by $\Sigma(g_i - g^*)/(k-1)$, where g_i is the diversity estimate from the ith bootstrap sample, g^* is the mean of these estimates, and k is the number of bootstrap samples (Dixon 1993, Efron and Tibshirani 1997). Note that diversity measures depend on the area sampled and other factors, and the results of different studies may or may not be appropriate to combine (e.g., Scheiner et al. 2000).

- Population growth measures include the rate of population increase and instantaneous rate of increase where the growth model and hence associated effect estimates and variances depend on the assumed population processes (birth and death rates, age distributions, patterns of dispersion, breeding periods, carrying capacity, etc.).

In these cases, given an estimate and its associated variance, a meta-analysis can proceed in the usual way.

An example of a meta-analysis of heritability of longevity in captive (zoo) populations of mammals and birds is reported by Ricklefs and Cadena (2008). The authors describe a number of ways of obtaining different estimates of heritability, including the use of ANOVA variance components and regression analysis of offspring age of death on the midparent age at death. Corresponding standard errors were obtained as part of the analyses and were used as weights in a standard meta-analysis model (Chapter 8).

Ricklefs and Cadena (2008) make an important point that inferences in a meta-analysis should acknowledge the quality of the inputs and the variability of the outputs: "Using meta-analysis to obtain a single overall estimate of additive genetic variation across species assumes that heritabilities are uniform across populations, which is probably not valid . . . we use this statistical technique only as a tool to address the question of whether, despite large SE values and low statistical power owing to small sample sizes, life span is influenced by genetic factors. Thus, rather than focusing on the cross-species point estimate of heritability, we focus on the confidence intervals (CI) around it, specifically on whether they include zero" (Ricklefs and Cadena 2008, 438). Thus, when interpreting the results, the authors acknowledge limitations in the data and take them into account.

It is important to recognize that even in these relatively straightforward cases, the comparability of the estimates obtained from different studies must be carefully considered. For example, there are a number of possible estimators of heritability, each with a corresponding formulation of the variance. If heritability is obtained from a regression equation, with an associated variance, then a normal distribution might be assumed or a bootstrap estimate of the density might be obtained. Similarly, diversity measures such as Shannon's and Simpson's indices are well known, but different diversity measures are not equivalent; indeed, there are variations even within an index. If disparate effect estimators are combined, the scientist must address the added question of the interpretation and applicability of estimates that arise from the meta-analysis.

PER CAPITA AND WHOLE-POPULATION PARAMETERS

Many of the parameters that are used in ecological meta-analyses can be categorized as those relating to individuals in the population ("per-capita" parameters, such as the number of births per female or the probability of a female giving birth). Others can be categorized as those relating to the whole population ("whole-population" parameters, such as the total number of births in the population or the change in population size per year).

We use the ideas detailed above to discuss how to deal with these two types of parameters. We first note that they have different statistical characteristics. This has an impact on the parameter estimates and, importantly, on the variances associated with these estimates. We then consider what this means in a meta-analysis context.

Per capita parameters

These are parameters (effects) that don't change as population size (N) changes. (We mean this in a simple, mathematical sense; biologically, of course, per capita ecological parameters can be density dependent, but we are not addressing that here.)

Consider the following examples:

(a) The number of outcomes per individual in the population.

Example: The number of births per female in the population.

This parameter can be between 0 and (technically) infinity. Under certain assumptions (e.g., a random sample), a statistic (effect estimate) that may be used to estimate this parameter is the number of outcomes per individual in the sample. The sample estimate could possibly be modeled with a Poisson distribution (depending on assumptions of independence, actual biologically meaningful maximum, etc.).

(b) The probability of an outcome in the population.

Example: The probability of a female giving birth in a population.

This parameter can be between 0 and 1. Under certain assumptions (e.g., constant birth conditions), this may be equivalent to the proportion of females who have given birth in the population. A statistic that may be used to estimate this parameter is the proportion of outcomes in the sample. The sample estimate for this parameter could possibly be modeled with a binomial distribution (depending on binomial distributional assumptions).

Note that in the case of example (b), the probability and the proportion are the same in the whole population. In a sample, we calculate the proportion and use this to estimate the population proportion— that is, the population probability.

Whole-population parameters

The per capita parameters described in (a) and (b) above can be multiplied by the population size N, to give the number of outcomes in the population. For example:

(c) The total number of births in the population (i.e., number of births per female times the number of females in the population).

This value can be between 0 and either infinity, for (a), or N, for (b). The sample estimate for this parameter could possibly be modeled with a Poisson distribution.

(d) The change in population size per year, or change in population size per year expressed as some multiple of the original population.

The bounds of this value depend on the way that the parameter is defined. For example, a change in population size per year could technically be between -max(N) and +infinity (or some large upper bound). The sample estimate for this parameter could possibly be modeled with a normal distribution.

Calculation of variances

For all of the above cases, the population parameter is typically estimated by a statistic (an effect estimate) based on a sample of size n ($n < N$). The variance of the effect estimates will depend on the sampling distribution (e.g., Poisson, binomial, normal), the sample size n, and for cases (c) and (d), the population size N.

Thus, if you are not given the variance, but you can categorize the parameter as illustrated above, then you may be able to estimate the variance as we have discussed earlier in this chapter. For example, for a Poisson and binomial distribution the mean and variance are equal; for a normal distribution, *if it can be assumed that the sampling variances are equal,* the variance of the effect size just depends on n.

This has the following implications:

- We could consider log and logit transformations for Poisson and binomial distributions, as we have already done in the chapter.

- If the parameter estimate is multiplied by the population size N, then the variance of the resultant estimate necessarily depends on N. Thus, it is not always reasonable to just take the population size N as a full proxy for the variance since it also depends on the distribution of the estimate; that is, Var(estimate $\times N$) = Var(estimate) $\times N^2$. Note, however, that since the N^2 often dominates Var(estimate), then using the population size as a proxy may in some cases be reasonable.

- The sample size can be used as a proxy for the variance in a normal distribution under the assumption that the sampling variances are equal, as we have already seen in this chapter. However, we again caution that in ecology (and most other areas), this assumption is almost never true.

EFFECT SIZE ESTIMATES WITH NONSTANDARD OR UNKNOWN DISTRIBUTIONS

In some cases, primary studies might report, or the meta-analyst might derive, "new" parameters of interest. For some such parameters, it may be possible to identify parallels between the parameter of interest and an effect with known properties. In any event, if the distribution of the corresponding effect estimates can be derived and if a "weight" that reflects the variance of the estimate can be obtained, then the meta-analysis methods described in the following chapters can be applied. Of course, the principles and statistical assumptions underlying the claimed distribution should also be followed.

If no such distribution is available, a different type of analysis might be considered, depending on the overall aim. For example, if the aim is to evaluate the effect of specified covariates or factors on a response, it might be reasonable to consider deriving a more standard effect size metric from each study, such as regression coefficients from a multiple regression. After deriving the study-specific effect estimates (such as regression coefficients), these can then be combined in a meta-analysis. In this case, regardless of the distribution of the effect estimates themselves, the derived regression coefficients (which provide measures of the effect of the covariate or factor) and associated hypothesis tests are valid if the residuals are normally distributed. Although there is certain robustness in linear regression to violations of the assumption of independent and identically normally distributed residuals, this assumption should be carefully checked using standard regression diagnostics. This is particularly important if the original effect estimates are not standard.

If none of these options is available to the meta-analyst, it is strongly recommended that a formal meta-analysis of the type described in this book NOT be conducted. It may (or may not) be possible to use an alternative (and statistically valid) method of synthesizing the information. Possible alternative analyses include general complex systems models, such as Bayesian networks (e.g., Hamilton et al. 2007) as well as discipline-specific models. Examples of the latter include genetic consensus maps (Rong et al. 2007), and genetic diversity models for subdivided models that also include explicit accommodation of within- and between-population variation (Toro and Caballero 2005). Alternatively, one might carry out a formal systematic review (Chapter 1) without quantitative meta-analysis. However, counting up and comparing the number of statistically significant outcomes—vote counting—is not a valid alternative strategy.

Further examples of other metrics of effect size with nonstandard distributions

Many other metrics of effect size have been used in meta-analysis in ecology and evolutionary biology. For example, Gotelli and McCabe (2002) carried out a meta-analysis of 96 published presence-absence matrices for the evaluation of models of species co-occurrence. Each study-specific matrix represented presence or absence of particular species (rows) within sites (columns), from which a number of indices of community structure were calculated. A standardized effect size was computed by comparing the observed index with the analogous index obtained from randomizations of the original matrices, with a corresponding variance also obtained from the randomizations. However, Gotelli and McCabe (2002) provide no explicit discussion of how these effect sizes were combined in the meta-analysis, or whether or how the study estimates were weighted. This makes it difficult to evaluate or repeat their methods.

Soininen et al. (2007) conducted a meta-analysis on the rate of decrease in community similarity with distance. The authors used two parameters: the Sorensen index of similarity between communities at a fixed distance of 1 km (the "initial similarity"), and the distance at which the community similarity was half of the initial similarity (the "halving distance"). These indices were derived from regressions of distance on similarity. Corresponding variances of these effect estimates would have been available from the regression analyses that could have been used to weight the effect estimates (Chapters 8 and 9), but these were not reported or used in the meta-analysis. The objective of the research synthesis was to hypothesize relationships between these ecological parameters and various ecological characteristics of the communities, including trophic level of the organism and the latitude of the communities.

There are also examples of the creation of metrics based on other metrics. Boyce et al. (2005) describe a series of five meta-analyses, reported over the period 1992 to 2004, of spotted owl demography; the authors focused on the problem of estimating vital rates of fecundity and survival. Vital rates were then combined into a demographic projection matrix for which the dominant eigenvalue was estimated to reflect the asymptotic trajectory of population size. The eigenvalue was then used as the effect estimate to be combined across studies.

Note that we have largely avoided discussing the method or model used by the respective authors to combine the effect estimates across the studies. Appropriate approaches are described in Chapter 8. We reiterate, however, that it is essential to be able to derive a variance (or equivalent) for the metric obtained in each study, and to use these to weight the effect sizes in the meta-analysis. Unweighted analyses produce biased estimates of overall effects. Moreover, it is essential for these metrics to be well described so that other readers can evaluate the resultant inferences and replicate the approach. If the approach cannot be replicated, the results cannot be updated with data from future studies. Moreover, the comparability of new measures with more established measures should be discussed from both statistical and ecological perspectives; then others can critically evaluate the conclusions of the meta-analysis in light of other published results. A lack of adequate description may not always be the "fault" of the authors; editorial policies may limit the description of methodology. With the advent of online appendices, such limitations should be possible to overcome.

ADDITIONAL ISSUES ARISING WITH EFFECT SIZES WITH NONSTANDARD AND UNKNOWN DISTRIBUTIONS

Transformations and other manipulation of effect estimates

A meta-analysis requires consistent, comparable effect estimates that, when combined, will provide overall estimates that are interpretable and meaningful. In many cases, it is necessary to transform or otherwise manipulate some of the effect estimates reported in the primary studies in order to make them comparable with effects from other studies.

It is often reasonable to assume normal distributions for the effect estimates and hence for the study-specific and overall parameters. Other statistics may be transformed to approximate normality under some assumptions (e.g., sufficiently large sample size and responses not too close to boundaries, such as zero). For example, a log transformation of a Poisson distribution or an arcsine or logit transformation of a binomial distribution will approximate a normal distribution.

Keep in mind, however, that transformation can alter the meaning and interpretation of an effect; for example, log transformation changes an additive effect into a multiplicative effect (Houle et al. 2011). Another challenge relates to calculating the weights for transformed effects (see below).

In many situations, however, such transformations may not be suitable. These situations occur when (1) the assumption of approximate normality are not reasonable, (2) the analyst prefers to model the data directly rather than through a transformation, (3) there is no transformation of the estimate that will make it "more" normal, and (4) the information required to make a normal transformation is not reported in the primary studies. In these situations, it is often better to avoid poor transformations or untenable assumptions, and combine the study estimates using the more natural nonnormal distributions directly. Rather than using simple meta-analysis models to analyze the data, however, use of a generalized linear model (GLM) provides a general, very flexible framework for carrying out a meta-analysis with nonnormal effect sizes. This model can be analyzed using maximum likelihood or Bayesian methods; see Chapters 10 and 11 for further discussion. This can be extended to a mixed model via a generalized linear mixed model (GLMM). Bolker et al. (2009) describe the interpretation and application of these methods in the context of ecology and evolution.

A related issue encountered in many meta-analyses is that the estimates from the individual studies are on different scales or use different measures (e.g., standardized mean difference, odds ratio, regression coefficient). In this case, the estimates must be transformed to a common measure if they are to be combined at all (Chapter 6). Examples of such transformations are given in Chapter 13 of this book as well as in Chapters 5 and 7 of Card (2012), Chapters 3 and 6 of Lipsey and Wilson (2001), and Chapters 7 and 34 of Borenstein et al. (2009).

Variance of effect estimates where distributions are unknown

As we have discussed, the meta-analyst needs to extract from each study an effect estimate and a corresponding weight for this estimate. This weight is typically based on the variance of the estimate. Knowing (or assuming) the distribution of the effect of interest is therefore important because it can tell us what information we need from each study in order to correctly weight the effect sizes in the meta-analysis. For example, if the data are binomial, then what we need to obtain from each study is the number of observations, samples, or trials (n), and either the number of "successes" (X, say) or the estimated proportion of successes ($\hat{p} = X)/n$). Similarly, if the data are normally distributed, then we need a sample mean ($\overline{X}$) and the variance of this estimate for each study. (Note that if the sample variance, S^2 of the observations and the sample size, n, are available, then the estimated variance of the sample mean is S^2/n).

If there is a known relationship between the mean and variance for the distribution in question, then it might be sufficient to extract just the effect estimate and the sample size. For example, for a binomial distribution the proportion p has a corresponding variance $p(1 − p)/n$, so if this information is extracted from a study, the analyst can construct the required weight for the effect estimate ($\hat{p}$). Similarly, for a Poisson distribution, the mean is equal to the variance. Hence, the analyst needs only to extract the relevant count (or alternatively, the estimated rate and the spatial or temporal scale over which the counts were taken) in order to have both the effect estimate (the count) and the associated variance (the count again). Then in the meta-analysis, the estimates from each study can be transformed to a common spatial or temporal scale, and the weight used for the Poisson effect estimate will be the inverse of the variance.

This trick is not applicable to the normal distribution, since the mean and variance are independent. Thus, knowing the mean does not provide information about the variance for this distribution. However, in the special case in which the true variance of the observations in each study can reasonably be assumed to be equal (and it is only within-study sampling properties that contribute to differences in observed variances), the inverse sample size might be justified as a weight in a meta-analysis of these effect estimates. This is because the variance of the sample mean is equal to s^2/n; so, if s^2 is equal across all studies, only n need be taken into consideration in the weights. The same reasoning can be extended to other metrics. For example, if two groups are being compared and the effect estimate is the difference between the group means (d), then if the variance in the two groups is equal, the corresponding variance of the effect estimate is $(n_1 + n_2)/(n_1 \times n_2)$; this last is the sum of the sample sizes in the two groups divided by their product (and hence the required weight is the inverse of this value).

Note that if the assumption of equal variances across studies is not true, basing the weights on the sample sizes alone can lead to biased overall effect estimates. The extent of the bias will depend on the magnitude of the inequality of the variances. Unfortunately, variances are almost never equal across studies, and so the outcome of the meta-analysis may be both biased (to an unknown extent), and potentially seriously misleading if sample sizes alone are used as weights. If it is possible to obtain or calculate the variances for many but not all of the studies, it may be reasonable to estimate the missing values using various approaches to imputation (e.g., Chapter 13).

It is important to be aware that if an estimate is transformed, the variance of the transformed estimate is not equal to the variance of the original estimate. If the transformation is simple, then the transformed variance can be computed from the original variance. For example, if a is a constant, then $\mathrm{Var}(aT) = a^2\mathrm{Var}(T)$. If a 95% CI for an estimate T is (a,b), say, then the 95% CI for $\log(T)$ is $(\log(a), \log(b))$. For computing variances with more complex transformations, the meta-analyst should consult a statistician or access the specialist literature. As noted above, some examples are also provided in the meta-analysis texts of Card (2012), Lipsey and Wilson (2001), Borenstein et al. (2009), and Sutton et al. (2000).

Meta-analysis of Corrected Effect Estimates

The need for comparable, unbiased, and standardized study-specific effect estimates in a meta-analysis has been highlighted above. In experimental studies, comparability is sometimes possible to achieve by "matching" for suspected influential covariates or factors, although it is in practice unlikely that different ecological experiments will use the same experimental factors. In nonexperimental studies the covariates or factors are almost never the same across studies. Thus, nonexperimental studies might report estimates that have been "corrected" or "adjusted" for any such covariates, or "corrected" effect estimates in which the covariates have been accounted for in the study-specific analyses, often through a form of regression. As an example, reported study estimates of population growth might correct for one, both, or neither of the two covariates, body weight and population size.

Although accounting for confounding in this manner is appealing, the choice of confounding factors, if any, and the statistical method by which adjustment is undertaken can vary considerably between studies. For the purposes of meta-analysis, these adjusted estimates then may not be directly comparable. For example, an estimate of extinction risk adjusted for species body length may be interpreted differently than the same estimate adjusted for species body length, geographic location, and competing species. Similarly, the impact of grazing on the presence/absence of a bird species may be estimated by a regression coefficient (Kuhnert et al. 2005), but this coefficient depends on the other factors in the (logistic) regression model; hence, even if two studies use the same form of regression analysis but include different factors in the model, the estimates of grazing impact may not be comparable across studies. If the raw data are not available, it may be impossible to combine parameters calculated and published in this manner.

The meta-analyst is thus faced with a dilemma. On the one hand, adjusted effect estimates are potentially more accurate study-specific measures of the ecological parameter of interest; on the other hand, they may not be directly comparable across studies.

We can describe this problem in a general context. Assume that there are two studies, A and B. Study A may have used the regression model $Y_j = \alpha + \beta_1 X1_j + \beta_2 X2_j + \beta_3 X3_j$ to obtain an adjusted effect of interest, where Y_j is the outcome for the jth unit in the study. Alternatively, study B may not have measured $X3$, or found it to be nonsignificant in the regression model, so the data are modeled using $Y_j = \alpha + \beta_1 X1_j + \beta_2 X2_j$, excluding covariate $X3$. If we assume that all else is equal between the two studies (even though this is unlikely, especially with observational studies), the study-specific effect estimates Y may not be comparable for synthesis in a meta-analysis. Similarly, the estimated relationships between the outcome and the covariates (e.g., β_1) will not be comparable because these values will depend on the other covariates in the model.

Greenland (2005) suggested a resolution to this problem by proposing a method of externally adjusting studies for certain covariates based on estimates from other studies in the meta-analysis. Suppose that we are interested in the effect of a particular factor (e.g., grazing) on a response (e.g., presence of bird species), and consider a study that has used logistic regression to estimate this effect. Let b_u be the estimate of the effect when a potentially important confounder is ignored in the analysis, and b_a be the effect estimate when this confounder is included. Then $U = \exp(b_u - b_a)$ is the "value" of adjusting for this confounder in the study. Assuming that the effect of adjusting for the confounder is the same in all studies in the meta-analysis, this value could be applied to another study that has not adjusted for the confounder, since $b_a = b_u - \ln(U)$. Greenland applies the standard error of b_u to an imputed b_a and in this manner, obtains a set of estimates (with standard errors) adjusted for the same confounders, which can be used as comparable, standardized effect estimates in a meta-analysis.

However, this general approach may be problematic because the number of covariates considered in adjustment analyses across all studies in a meta-analysis may exceed the number of studies in the meta-analysis. An alternative partial adjustment model may be constructed that includes a subset of covariates sufficient to account for most of the potential confounding in the study. The model will then adjust study estimates for these key covariates as well as a new covariate that encompasses the other covariates for which an individual study was adjusted.

Adjusting for a consistent subset of covariates, as described above, is not the same as obtaining unconfounded estimates. There may be potentially important covariates that are still unknown or unmeasured; these could influence the outcome. An associated important issue is that not all studies report adjusted estimates, and those that do may have selectively decided which estimates to report. While this decision is often objective, there is a concern that the estimates may have been selectively reported for some reason that could have an impact on the meta-analysis. Selective reporting within studies is a form of publication bias (Chapter 14; Hutton and Williamson 2000; Hahn et al. 2002; Chan, Hróbjartsson, et al. 2004; Chan, Krleza-Jeric, et al. 2004). For an example of within-study publication bias in evolutionary biology, see Cassey et al. (2005). Peters and Mengersen (2008) found evidence of selective reporting of adjusted estimates in medical meta-analyses. Based on a reanalysis of ten published meta-analyses and reading of over 100 primary studies, the authors cautioned that selective reporting of adjusted estimates may lead to a bias in some meta-analyses when adjusted study estimates are not reported because univariate analyses indicated a nonsignificant effect. Their recommendation is that both adjusted and unadjusted study estimates should be extracted for a meta-analysis and if the adjusted estimates cannot be obtained, the reasons for this should be investigated; sensitivity analyses could then be used to assess the impact on the meta-analysis.

Sensitivity analyses

As discussed above, the analyses suggested in this chapter are based on a number of assumptions about the statistical distribution of the effect estimates and the underlying data. These assumptions should be clearly stated in a meta-analysis report and, where possible, the robustness of the inferences (estimates, statements about statistical significance, etc.) should be assessed via a sensitivity analysis.

A sensitivity analysis can take a number of forms. One approach is to examine the results of multiple meta-analyses. Sometimes there is a discrete set of options to be considered in a sensitivity analysis. These include the following: two different types of distributions might be possible for describing the data (e.g., Poisson or normal); effect estimates can be adjusted or unadjusted; the variances associated with effect estimates could be as reported or alternatively inflated by some margin to allow for possible error of estimation; or the sample sizes could be used as a proxy for estimated variances. (This last option is based on the assumption that the variance of the observations in each study is the same and only the sample sizes differ, as discussed above.) If several options exist, a series of meta-analyses can be run using those different options, and the impact on the final inferences can be evaluated. This process can reveal, for example, how much the overall estimates change under the different options, and whether a change to the estimates alters any decisions that are based on the analyses.

CONCLUSIONS

The interests of ecologists and evolutionary biologists are very diverse, as are the organisms and systems that they study. Given the complexity of the world around them and of the problems that they are tackling, it is no surprise that ecologists and evolutionary biologists sometimes need to combine effects that do not conveniently fall under the range of common measures obtained from standard experimental studies or that are commonly employed in ecological and evolutionary meta-analyses (e.g., Osenberg et al. 1997; Chapter 6). This chapter has focused on some of the other effect size metrics that might be of interest to a meta-analyst in these fields. Some of these are quite standard in the particular discipline in which they arise, but may be nonstandard to others—for example, heritability is a very well established measure in evolutionary biology but may be unfamiliar (or nonstandard) to scientists outside this field. The aim of the chapter has thus been to discuss a range of approaches to combining these different measures of effects, in the hope that this will broaden the options available for meta-analysis and motivate prospective analysts to search their own literature for relevant ideas and approaches.

As stressed in this chapter, these measures of effects must be combined knowledgeably and wisely. The analyst must ensure that

- they understand the statistical distribution of the effect estimates that they wish to combine, and if not, they account for this limitation using one of the approaches described above;
- the assumptions underlying the distribution and the meta-analysis model are justified and supportable;
- the effect estimates are comparable and consistent across studies;
- the relevant estimates required for a meta-analysis are available from the individual studies; and
- the resultant estimates and inferences are interpretable scientifically as well as statistically.

Finally, these issues must be placed in the context of the wider range of issues that have to be considered in a meta-analysis, as discussed in other chapters in this handbook.

The use of meta-analysis to synthesize experimental and other data that compare two groups (such as an experimental and control group) has been an important and productive area of research in ecology and evolutionary biology. The availability of better statistical tools to synthesize effect sizes that summarize other sorts of biological parameters will offer opportunities to explore a broad range of new questions and a rich array of ecological and evolutionary data. These opportunities can only become more compelling in the future in both fundamental and applied research. The challenges facing scientists addressing these questions will be unevenness and limitations in the quality of the data available, awareness of the underlying statistical issues, and rigor in applying them so that results will be robust and credible.

Essential Analytic Models and Methods

Statistical Models and Approaches to Inference

Kerrie Mengersen, Christopher H. Schmid,
Michael D. Jennions, and Jessica Gurevitch

WE COME NOW TO the statistical aspects of meta-analysis— namely,

(1) the statistical model that describes how the study-specific estimates of interest will be combined;

(2) the key statistical approaches for meta-analysis; and

(3) the corresponding estimates, inferences, and decisions that arise from a meta-analysis.

The technical details of the statistical approaches for analysis and inference appear in later chapters. Here, the focus is on providing an introduction and overview of these three components. First, we describe common statistical models used in ecological meta-analyses and the relationships between these models, showing how they are all variations of the same general structure. From this perspective, more advanced models follow easily. We then discuss three main approaches to analysis and inference, again with the aim of providing a general understanding of these methods. Finally, we briefly consider a number of statistical considerations which arise in meta-analysis.

Note that in this chapter, we focus on estimation of univariate, independent outcomes. The analysis of dependent outcomes due, for example, to multiple estimates within publications, research teams, locations, species, and so forth, and the analysis of multiple outcomes of interest, are discussed in Chapter 16.

In order to illustrate the concepts described in this chapter, we consider the Lepidoptera mating example described in Appendix 8.1. This is a meta-analysis of 25 studies of the association between male mating history and female fecundity in Lepidoptera (Torres-Vila and Jennions 2005). The aim of the meta-analysis is to combine individual study estimates of this association (Hedges' *d*) in order to obtain an estimate of the overall standardized mean difference in reproductive output for females mated to virgin versus experienced males. The studies also differ with respect to suborder, family and % polyandry (percentage of females that mated more than once).

THE META-ANALYSIS MODEL
Model considerations

The most fundamental statistical component of a meta-analysis is the *model*. This reflects the aim of the meta-analysis, defines the way in which the study estimates will be combined, describes important sources of variation, and determines the type of overall estimates and inferences that will arise from the meta-analysis.

The meta-analysis model is thus influenced by many factors. We describe five of the key issues influencing the meta-analysis model in this section: (1) the aim of the meta-analysis, (2) the study-specific information available from the primary studies, (3) the treatment of between-study variation, (4) the statistical distribution of the estimates to be combined, and (5) the choice between a frequentist or Bayesian approach.

The aim of the meta-analysis is central to the decision about model structure. Three common aims in ecology and evolutionary biology are as follows: estimation of the magnitude of an overall effect that characterizes the population of studies being analyzed, and the corresponding uncertainty or variance associated with this estimation; assessment and description of the heterogeneity between the study-specific estimates of the effect of interest; and identification and characterization of factors that have an impact on the between-study heterogeneity and the overall effect estimate.

The overall effect could be a measure of the absolute size of an ecological quantity of interest; or a comparative measure, such as the impact of a treatment, defined as a standardized mean difference or relative risk; or the association between two factors, such as a correlation; and so on (see Chapters 6 and 7 for further details). Estimation of this overall effect can be followed by an assessment of whether it is significantly different from zero or some other value of ecological interest. This assessment is typically made on the basis of a hypothesis test or confidence interval, as described in Chapter 9.

The study-specific information will both inform and be dictated by the meta-analysis model. It is important to propose models that are appropriate for the type of information that is available from the primary studies. Similarly, it is important to extract relevant information from the studies in order to fit the meta-analysis model. Even for the most basic meta-analysis model, this will comprise

- the statistic that estimates the effect of interest for each study, such as a sample mean, standardized difference, or sample correlation; and

- a corresponding estimate of the variability of this statistic for each study (e.g., the standard error of the mean).

Other information will be required for more complex models, such as sample sizes, correlations, and factors or covariates that may explain heterogeneity among the studies and consequent differences between the study-specific effect estimates. Unfortunately, not all ecological and evolutionary studies report all of the statistics needed for a meta-analysis. In this case, it may be possible to obtain information to calculate them (such as deriving the standard error of the mean from the standard deviation and sample size, or vice versa) or to fit them as missing values. This is discussed briefly below and in more detail in Chapter 13.

Unlike primary data analysis of ecological experiments, in which heterogeneity of estimates from replicates is due to the sampling variance alone, meta-analysis typically involves consideration of a second source of heterogeneity. This is between-study variation. It is rare that individual studies are sufficiently similar and can be considered as simply repeat samples of "one big study." For example, they may have been conducted under different experimental or sampling designs or protocols; they may differ with respect to important covariates or modifying factors; or they may have accounted for potential biases or confounders in different ways.

These sources of variation determine whether the meta-analysis model is constructed as a *fixed-effects* or *random-effects* model, as described in detail below. If the aim of the meta-analysis is to identify and characterize the factors that have an impact on the magnitude (and/or variation) of the effect of interest, these factors must be explicitly included in the meta-analysis model. This is typically implemented through a *meta-regression* model (explained below).

The meta-analyst will also have to specify the statistical distributions to be used in the model. For example, standardized differences might be assumed to have a normal or *t*-distribution; counts might have a Poisson distribution; proportions might have a binomial distribution; or the residuals of a regression might have a normal distribution (Chapter 7). As in any good statistical analysis, the importance of these decisions on the results of the meta-analysis can be tested through a sensitivity analysis.

The choice between a frequentist or Bayesian approach to meta-analysis depends on a number of considerations, including philosophical inclinations, model complexity, availability of prior information, and type of inferences required from the meta-analysis. In a frequentist (sometimes called classical) setting, the analyst makes inferences about model parameters based on the *likelihood* of the data given these parameters. The "frequentist" label arises because the analyst considers the outcome of hypothetical repeated sampling from the unknown population. In a Bayesian setting, the analyst makes inferences based directly on the (posterior) distributions of the parameters themselves, given the data that have been observed. The posterior distribution combines both the likelihood (which provides information about the data) and priors (which provide any other information about the parameters). The priors can be "informative" in that they are based on a review of the relevant literature, previous phases of the same study, or expert opinion; or they can be "uninformative," and then inferences are based primarily on the observed data. The parameters of prior distributions can also have priors, leading to a natural way of describing complex models involving many parameters, hierarchies, sources of error or uncertainty in parameter estimates, nonnormal distributions, and so on. Moreover, because the Bayesian approach focuses on the distribution of the parameters themselves, many more inferences in addition to estimation of means and variances can be made. Examples of inferences that are difficult in a frequentist framework but easier in a Bayesian framework include ranking or ordering parameters from a set of studies or populations, and estimating the probability that a parameter is greater than some ecological or biological threshold.

While the prior in a Bayesian analysis can be "uninformative," thus placing all of the weight in the meta-analysis on the information in the studies themselves, it is rarely the case that a meta-analysis (or indeed a primary study) is conducted in ignorance; in ecology, we have some idea about the anticipated magnitude of an effect or at least the bounds on its size. We may also have some information about the uncertainty of estimates, expected sources of heterogeneity between studies, and study characteristics. This information may be drawn from previous analyses as described above, other published information, expert information, and so on, and can be formally included in the meta-analysis via priors in the Bayesian framework.

The elicitation of expert information and the formulation of this information as statistical priors is of current research interest in the Bayesian community in general (O'Hagan et al. 2006), and in ecology in particular. See, for example, Kuhnert et al. (2005), Martin et al. (2005), O'Leary et al. (2009), James et al. (2010), and Low Choy et al. (2010). While these papers do not address meta-analysis in particular, the use of prior information described by the authors can be easily applied to the meta-analysis context.

Further details about Bayesian modeling are given in Chapter 11. Ellison (2004) provides a clear exposition of Bayesian inference in ecology and makes philosophical and methodological comparisons with frequentist approaches. He describes the increasing use of Bayesian methods since the 1970s, with particular reference to ecological applications in the decade of 1993 to 2003. The issue of model choice in both frameworks is also discussed and illustrated in this paper.

Other issues will influence the meta-analysis model. For example, it is much more straight-forward to combine well planned studies of the same design, but it requires more care to com-bine survey, opportunistic, and experimental data; or to combine study estimates that are based on subsets of the primary studies, or trend estimates based on repeated measures studies.

We consider now how the five key issues described above relate to the Lepidoptera mating meta-analysis, described in the introduction to this chapter. The issues are as follows:

(1) The aims of the meta-analysis are to estimate the overall standardized mean difference in reproductive output for females mated to virgin versus experienced males, and to identify factors that impact on the overall effect.

(2) The study-specific information available for this analysis is the statistic that estimates the effect of interest; in this case, the statistic is Hedges' d, and its corresponding variance.

(3) Between-study variation is expected in this example due to study-specific differences in suborder, family, and % polyandry. Thus, a meta-analysis model will include both within-study variation, expressed through the variances of the Hedges' d statistics, and between-study variation, expressed through the explicit addition of covariates (suborder, family, % polyandry) in the model and/or an additional variance term. Thus the statistical model will need to allow for both within- and between-study variation. The model could also be expanded to include the specific covariates of suborder, family and % polyandry.

(4) The statistical distribution of the Hedges' d estimate is assumed to be normal. In the meta-regression, the residuals are assumed to be normally distributed.

(5) Given the lack of prior information in this data set and the fairly simple meta-analysis model that can be adopted, the choice between a frequentist or Bayesian approach is largely philosophical. The Lepidoptera example is analyzed in both frameworks in Chap-ters 9, 10, and 11.

Choosing a model

Suppose that we have estimates of an effect from k studies and that the aim of the analysis is to estimate an overall effect across all of the studies. We assume that each study-specific effect es-timate has an associated measure of uncertainty that reflects the sampling variability within the study. Thus in the Lepidoptera mating example described above and in Appendix 8.1, $k = 25$, the effect of interest is Hedges' d, and the variance of Hedges' d is the corresponding measure of uncertainty.

Meta-analysis models can be subject to numerous categorizations and nomenclature. Here we present a categorization based on the aim of the meta-analysis and the assumptions about the fixed or random nature of the effects.

Three main aims of meta-analysis were defined earlier: estimation of an overall effect, as-sessment and description of the heterogeneity between studies, and investigation of factors or covariates that may explain this heterogeneity and the effect. Simple estimation of an overall effect can be considered in both fixed- or random-effects framework. All of the models as-sume that there is within-study (sampling) variation. Models that attempt to explain heteroge-neity between studies include mixed effects, hierarchical, multifactorial, and meta-regression models.

A summary of the key characteristics of these models is given in Figure 8.1; note that these models apply to both frequentist and Bayesian settings. A Bayesian approach augments the models with priors (Chapter 11).

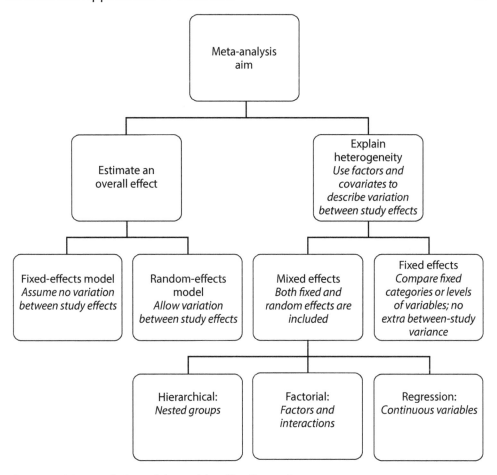

Figure 8.1. Meta-analysis models considered in Chapter 8.

Fixed-effects model

Suppose that we are only concerned about the particular studies in our meta-analysis and we wish to make comparisons among them, or we intend to generalize the results only to studies that have identical characteristics. Thus we consider the study effects as *fixed*, and we want to include sampling variability *within* each study but not across studies. In a simple *fixed-effects model*, we assume that there is negligible variation between studies, so that all the studies have the same true effect size and the estimates from the different studies vary only because of random sampling variation in the outcome. In this case the fixed-effects model is a natural contraction of the more general random-effects model, described below. Note that a fixed-effects model can be considered as the extreme of a random-effects model, in which the between-study variance is either zero (i.e., the studies have the same underlying parameters) or infinity (i.e., the study parameters are not measuring a common effect of any kind). In the latter case, a meta-analysis is neither mathematically nor biologically reasonable since the studies are not combinable, so we do not pursue this interpretation further in this chapter.

The simple fixed-effects model corresponds to an ANOVA-type situation in which the only variation in the overall population effect (where the "population" is the collection of effects from the different studies) is due to variability within studies, and there is no real between-study variation. However, unlike an ANOVA, a meta-analysis model allows for more flexibility in model specification and statistical inference, as described below and in the following chapters.

In the Lepidoptera mating example, a fixed-effects meta-analysis model would be appropriate if the meta-analyst believes or assumes that the 25 study-specific Hedges' d estimates (i.e., the population of effect sizes) differ only by sampling variation, with no additional variation between studies.

Random-effects model

In ecological and evolutionary meta-analyses, although we may acknowledge that the study estimates are measuring the same overall effect, we also typically want to account for additional variation in study-specific effects that are due to different experimental conditions, locations, and so on. In this case, we do not want to explicitly characterize or model these differences, but we do want to account for this additional random variation in the effects among studies in our model. One option is to allow for different study-specific effect sizes, and assume that the difference in the true effect sizes for the different studies is due to random variation around an overall mean effect, where the mean effect characterizes the population of studies. Thus, the model should encompass both the variance of effects *between* studies (due to random differences in their true effect sizes) as well as the variance of estimates *within* studies (due only to sampling variance or sampling error). This is typically called a *random-effects* model.

The random-effects model is a natural extension of the fixed-effects (and ANOVA) model, whereby we now allow two components of variance, those within and those between studies. Thus, in the Lepidoptera mating example, a random-effects meta-analysis model would include a between-study variance component to describe heterogeneity between the 25 studies. This is in addition to the within-study (sampling) variation of the Hedges' d estimates from each of the 25 studies. The ability to distinguish within- and between-study variation in meta-analysis models is one of key advantages of these models over ANOVA for combining the results of different studies.

In a random-effects model, we might be interested in "borrowing strength" from the combined set of studies to provide better estimates of the true effects of each of the studies, as well as estimating the overall mean effect. The conclusions of such a random-effects model are typically generalizable to a larger, unknown group of similar studies. Interest might also center on other features of the model, such as the absolute and relative sizes of the within- and between-study components of variance. In other words, is most of the variation among studies due to real differences in their outcomes, or to sampling error?

In practice, we may not know whether a fixed- or random-effects model is most appropriate. The random-effects model is in general conceptually applicable to most realistic meta-analysis contexts, apart from very carefully designed and similar experiments. Random-effects models are therefore the conceptual basis for more complex models in most ecological meta-analyses. However, this model does require estimation of an additional between-study variance parameter; poor estimation of this parameter due to small (within-study) sample sizes or a small number of studies might result in equally poor estimation of the other model parameters, and possibly in misleading inferences. Therefore, a general recommendation is to fit a random-effects model and assess the size and uncertainty of the estimate of between-study variation; if this is sufficiently small or too uncertain, a fixed-effects model might be adopted instead as a practical matter.

More complex models: Accounting for heterogeneity

We now go one step further in building the meta-analysis model; we acknowledge that there is variation among studies, but instead of simply including a general variance term to explain this heterogeneity, as in the random-effects model, we instead wish to identify and characterize the source of the variation. Thus there is a set of study characteristics that we wish to include in the meta-analysis model. For example, the habitat or species in a study might be important factors that affect differences in effect sizes among studies. In this case we could extend the simple random- and fixed-effects models above to a categorical or continuous model in which true differences between studies attributable to specific factors (such as habitat) can be taken into account. These models can include fixed, random, or both fixed and random effects (mixed effects). They are simply natural extensions of the primary fixed- and random-effects models, with the mean terms expanded to include factors, groups, or covariates. Obviously this categorization embraces a very wide range of meta-analysis models. We now consider some of the more common variants of these.

A *regression* meta-analysis model, or *meta-regression*, is a general term for a meta-analysis that includes factors (discrete or categorical variables) and/or covariates (continuous variables) to describe variation among study effects. This term embraces all of the models described below. (Sometimes *regression* is used to describe a model with only continuous covariates. In this chapter, regression is used to describe a model with any combination of continuous, ordinal, or categorical covariates.)

A *mixed model* is a form of meta-regression that encompasses both fixed and random effects, with fixed differences among categories of studies, and random variation among studies within categories (plus sampling error within studies, which is present in all models). For example, in the Lepidoptera mating example, a mixed model might include suborder as a fixed effect, with the aim of estimating a shared effect for studies within each suborder, as well as an overall mean effect across all studies. Variation among primary studies within suborders is treated as random effects, but the differences among suborders are considered to be fixed effects. Similarly, the model might include % polyandry as a continuous covariate. Although the "mixed model" terminology is common, this model is properly considered to be a form of random-effects meta-regression.

A *hierarchical* model is a form of regression meta-analysis (meta-regression) with nested subgroups, so that studies are combined into subgroups that may themselves be combined into larger groups, and so on. An example from the Lepidoptera mating model might be one in which studies are combined into subgroups according to species, and species are combined into groups according to suborder, and these subgroups are then combined into a single overall group.

A *factorial* model is a form of regression meta-analysis with (the same) categorical factors crossed within each study. For example, Gurevitch et al. (2000) combined studies in a factorial meta-analysis in which both competition and predation were manipulated within each study, while the overall effects of competition across studies, predation across studies, and the overall interaction between competition and predation were of interest.

STATISTICAL REPRESENTATION OF A META-ANALYSIS MODEL

In this section we consider a simple representation of each of the types of models described above, and then show how they might be extended. We write the models in two ways: the first as a linear model, and the second as a distributional model. The linear model is commonly seen in modeling and meta-analysis papers. However, the distributional framework is more easily extended to non-normal distributions and more elaborate meta-analysis models as described later in this chapter. The distributional framework also facilitates the use of different methods of estimation, as described later.

Suppose that each study provides an estimate of the effect of interest T_i, and that this estimate has a within-study sampling variance of σ_i^2, which is estimated by S_i^2. Thus, the T_i are the effect estimates from each study, and the meta-analysis model describes the combination of these estimates across studies. It is emphasized that σ_i^2 and S_i^2 represent the variance of the effect size estimate T_i; this is usually different from the variance of the data in the primary study. For example, if T_i is the estimated sample mean, then S_i^2 is the estimated standard error of this mean (not the sample variance of the data). (Note that the reader may have seen some of this terminology used to represent statistical parameters within primary studies, but we are using the terms to mean something different here.)

We denote the true effect from the ith study by θ_i; this is the unknown value that the statistic T_i is estimating. We then denote the overall effect by μ; this is the unknown value that the parameter θ_i is estimating. For exposition, suppose that T_i has a normal distribution. We discuss this assumption later in the chapter.

In the Lepidoptera mating example, T_i represents the Hedges' d estimate and S_i^2 is the corresponding variance of this estimate for the ith study, $i = 1, \ldots, 25$. Then θ_i represents the true standardized difference for the ith study and μ represents the overall mean standardized difference across all studies. In this case, the assumption of a normal distribution for T_i is reasonable, since this is the typical assumption for Hedges' d statistics.

Fixed-effects model

As discussed above, we might be intrinsically interested in the study effects themselves, rather than thinking of them as samples from an overall population. Since we are conducting a meta-analysis and wish to combine these studies, we must believe that they have the same true mean, which equals the overall mean. Hence $\theta_i = \mu$, $i = 1, \ldots, k$. Alternatively, we can think of the true study effects as distributed around an overall effect, but there is negligible between-study variation in these effects.

Recall that we are assuming normal distributions for concreteness and simplicity. Thus we want a model that says, "each study's estimate, T_i, is distributed normally with a true study-specific mean θ_i and variance σ_i^2, but the θ_i are all equal to an overall mean μ."

Therefore, in the Lepidoptera mating example, we assume that the Hedges' d statistic from each study is an estimate of the true standardized difference θ_i for that study, and that this statistic has a normal distribution with sampling variation σ_i^2 (estimated by S_i^2, the variance of the Hedges' d statistic). Moreover, under the fixed-effects model we assume that the true study-specific effects are all equal (to the overall mean standardized difference).

This fixed-effects model can be written as follows:

$$T_i = \theta_i + e_i \; ; e_i \sim N(0, \sigma_i^2)$$

$$\theta_i = \mu$$

or equivalently

$$T_i \sim N(\theta_i, \sigma_i^2)$$

$$\theta_i = \mu.$$

Here, e_i is the residual, the amount by which the estimate T_i differs from the true study value θ_i, induced by sampling variation within the ith study.

Random-effects model

A random-effects model accounts for both random sampling variation in estimating the true effect for each study and variation in the true effect among the studies. In the simplest case, the T_i vary around θ_i and these θ_i vary around the overall effect μ.

In this case, we want a model that says, "each study's effect size estimate, T_i, is distributed normally with a true mean θ_i and variance σ_i^2, and these study-specific means θ_i are normally distributed around an overall mean for all of the studies μ and a between-study variance τ^2."

Thus in the Lepidoptera mating example, we make the same assumptions about the distribution of the study-specific Hedges' d statistics, but under a random-effects model we allow the true standardized differences θ_i to vary from study to study. These θ_i have a normal distribution with mean μ (the overall true mean standardized difference) and variance τ^2.

This random-effects model can be written as follows:

$$T_i = \theta_i + e_i; e_i \sim N(0, \sigma_i^2)$$

$$\theta_i = \mu + \varepsilon_i; \varepsilon_i \sim N(0, \tau^2),$$

or equivalently

$$T_i \sim N(\theta_i, \sigma_i^2)$$

$$\theta_i \sim N(\mu, \tau^2).$$

Here, e_i is defined as above and ε_i is similarly defined as the amount by which the true study values θ_i differ from the overall value μ, induced by variation between studies.

Notice that if there is no between-study variation, $\tau^2 = 0$, so $\theta_i = \mu$ and the random-effects model reduces to the fixed-effects model.

Meta-regression model (regression meta-analysis)

A meta-regression allows explicit inclusion of important study characteristics in the model to explain the variability between studies and thereby reduce the between-study variance. We consider here examples of two simple meta-regression models.

First, suppose that there are p characteristics $X_1, X_2, \ldots, X_p$ available for each of the studies. For example, in the Lepidoptera mating example, it might be important to include suborder (X_1) and % polyandry (X_2). Hence, $p = 2$ and for the ith study, we would have additional data, a suborder value X_{1i} and a % polyandry value X_{2i}. The meta-regression model is written as follows:

$$T_i = \theta_i + e_i ; e_i \sim N(0, \sigma_i^2);$$

$$\theta_i = \beta_0 + \beta_1 X_{1i} + \beta_2 X_{2i} + \ldots + \beta_p X_{pi} + \varepsilon_i, \varepsilon_i \sim N(0, \tau^2),$$

or equivalently:

$$T_i \sim N(\theta_i, \sigma_i^2)$$

$$\theta_i \sim N(\mu_i, \tau^2)$$

$$\mu_i = \beta_0 + \beta_1 X_{1i} + \beta_2 X_{2i} + \ldots + \beta_p X_{pi}.$$

There are many variations of this model in terms of model construction and terminology; some examples follow.

- If the X's are continuous (for example, just % polyandry in the Lepidoptera example), the model is just like a linear *regression* meta-analysis with continuous terms.

- If the X's are factors that are manipulated or observed within each study, and we wish to compare effects across studies, the model reduces to a *factorial model*. This is akin to an ANOVA model with more factors leading to more sources of variation.

- If both random and fixed effects are included in the regression model, it may be referred to as a *mixed-effects model*. This is illustrated in the Lepidoptera mating example with Suborder as the fixed effect and random variation among studies within suborders included in the meta-analysis model.

- If subgroups exist so that study estimates are combined within subgroups which are then combined to obtain overall estimates, the model can be extended to a *hierarchical model*.

As a second example of a meta-regression, instead of fitting a regression model *across* studies as above, sometimes the regression model can be fitted *within* each study. A simple example is fitting a growth model to data within each study, with the aim of combining the study-specific parameters to obtain an overall estimate of growth. Of course, as described in the previous chapter, if the studies report the growth estimates (and appropriate variance), then these can be combined directly using a simple fixed- or random-effects model without recourse to regression. However, sometimes these estimates are not provided or are not reported in a sufficiently consistent manner to allow such a simple model.

Pursuing the growth example (but realizing that the same ideas apply if the data are not dependent on time), assume that the data now comprise a set of estimates T_{ij}, representing the average length of fish in study i, $i = 1, \ldots, I$, at time j, $j = 1, \ldots, J$. Assume for simplicity that a linear growth model is fitted within each study (see Chapter 16 for a nonlinear extension). Since the regression parameter (intercept and slope) estimates are correlated, they should be combined using a multivariate distribution. The model described above, a second example of a meta-regression model, is thus extended as follows:

$$T_{ij} = \theta_{ij} + e_{ij}; e_{ij} \sim N(0, \sigma_{ij}^2);$$
$$\theta_{ij} = \beta_{0i} + \beta_{1i}X_{1ij} + \varepsilon_{ij}, \varepsilon_{ij} \sim N(0, \tau_i^2)$$
$$\beta_i \sim MVN(\mathbf{B}, \Sigma).$$

Here, MVN represents a multivariate normal distribution, $\beta_i = (\beta_{0i}, \beta_{1i})$ represents the intercept and slope (growth) parameters for the ith study, $\mathbf{B} = (B_0, B_1)$ represents the overall intercept and slope values, and Σ is a 2×2 matrix denoting the variance-covariance matrix for the study-specific parameters. (Therefore, the diagonal terms indicate the variance of the overall intercept and slope parameters, and the off-diagonal term is the estimated covariance between these parameters.)

Note that the above model assumes that the study-specific T_{ij} estimates are summary statistics with an associated variance σ_{ij}^2, for which estimates (S_{ij}^2, say) are available. If instead, the T_{ij} are individual estimates (e.g., the length of a single fish), there is no variance associated with this value and the above model reduces to the following, as a third example of a meta-regression model:

$$T_{ij} = \beta_{0i} + \beta_{1i}X_{1ij} + \varepsilon_{ij}, \varepsilon_{ij} \sim N(0, \tau_i^2)$$
$$\beta_i \sim MVN(\mathbf{B}, \Sigma).$$

All of the above models are akin to standard regression and analysis of covariance (ANCOVA), or ANOVA with continuous and discrete sources of variation, as well as within- and between-study variation. As discussed above, however, the meta-analysis is typically based on summary data and the focus is on explaining the heterogeneity between studies and obtaining overall combined estimates.

In order to illustrate the generality of the modeling framework that has been constructed in this section, we shall now describe the mixed-effects and hierarchical models in more detail. The other forms of meta-regression models will then be revisited in later chapters.

Mixed-effects model

A mixed-effects model allows for both fixed and random effects. Suppose that the aim of the meta-analysis is to compare two groups. Studies are considered to be random within groups but the groups are considered to be fixed. In this case we want a model that says, "each study's effect size estimate, T_i, is distributed normally with a true mean θ_i and variance σ_i^2, and these study-specific means θ_i are distributed normally around group mean μ_1 or μ_2."

For example, in the Lepidoptera mating example, the aim might be to compare standardized differences in reproductive output for the two different suborders. Thus, the model allows for study-specific Hedges' d estimates to be combined within suborders, with the true standardized differences for each study assumed to be normally distributed around suborder means $(\mu_{Heterocera}, \mu_{Rhopalacera})$.

This mixed-effects model can be written as follows (with some abuse of notation):

$$T_i = \theta_i + e_i; e_i \sim N(0, \sigma_i^2)$$

$$\theta_i = \mu_j + \varepsilon_i; \varepsilon_i \sim N(0, \tau_j^2), j = 1, 2$$

or equivalently

$$T_i \sim N(\theta_i, \sigma_i^2)$$

$$\theta_i \sim N(\mu_j, \tau^2), j = 1, 2.$$

This is a form of random-effects meta-regression which can be expressed as a *hierarchical model*.

Hierarchical model

As discussed earlier, sometimes the regression model can be described by adding hierarchies to the random-effects model. Suppose that there are subgroups of studies that may be more homogeneous or for which we require group-specific effect estimates. Suppose that the study estimates T_i are normally distributed around study-specific true values θ_i, and that these are grouped into J subgroups with means and variances (ψ_j, ω_j^2), $j = 1, \ldots, J$. Furthermore, the subgroup-specific effects ψ_j are distributed around an overall effect μ. In this case, an additional hierarchy can be added to the random-effects model as follows:

$$T_i = \theta_i + e_{Ti}; e_{Ti} \sim N(0, \sigma_i^2)$$

$$\theta_i = \psi_j + e_{\theta i}; e_{\theta i} \sim N(0, \omega_j^2)$$

$$\psi_j = \mu + e_{\psi i}; e_{\psi i} \sim N(0, \tau_j^2),$$

or equivalently

$$T_i \sim N(\theta_i, \sigma_{Ti}^2)$$

$$\theta_i \sim N(\psi_j, \omega_j^2)$$

$$\psi_j \sim N(\mu, \tau^2).$$

This model says, "each study's estimate, T_i, is distributed normally with a true mean θ_i; these θ_i are normally distributed around subgroup means, ψ_j, and these subgroup means are normally distributed around an overall mean μ."

For example, in the Lepidoptera mating example, the study-specific Hedges' d estimates (T_i) are assumed to have a normal distribution (due to within-study sampling variance) with a study-specific true effect (θ_i), which are themselves normally distributed around suborder effects (ψ_j); these last are themselves considered to be a random sample of a larger set of suborder effects with overall mean μ.

This model can be extended to include more hierarchies. Note that these hierarchies could also represent geographic areas, specific experimental conditions, and so forth. As with the random-effects model, if $\tau^2 = 0$, this reduces to a hierarchical fixed-effects model.

APPROACHES TO INFERENCE IN META-ANALYSIS

So far, we have discussed the type of models that are typically considered in meta-analysis. The next step involves the decision about how to make inferences based on these models, that is, the method of analysis. *Statistical inference* means deriving estimates, testing hypotheses, and making decisions based on a model.

In this section we provide an overview of three main approaches: moment and least squares estimation, maximum likelihood estimation, and Bayesian estimation. This is intentionally a general and selective overview; statistical details of the approaches, including formulas, are provided in Chapters 9 to 11, and many more papers than we cite here provide good explanations of the models or give excellent examples of these approaches in practice. The reader is strongly encouraged to complement this reading by undertaking a literature search in their own field of interest for frequently cited and high-quality papers on these topics.

Moment and least squares inference

The vast majority of meta-analyses that have been conducted in ecology and related disciplines to date have been based on the method of moments (MM) and least-squares (LS) inference methods; they estimate global means and look at differences among subsets of studies broken into distinct groups. When multiple ways of explaining variation were testable, most meta-analyses in ecology have tended to analyze each explanation independently and without immediate regard to interaction. For example, in a meta-analysis of the Janzen-Connell hypothesis of distance-dependent seedling survival, Hyatt et al. (2003) independently examined at least seven factors which might have explained variation among the studies: (1) studies that explicitly looked at distance versus studies that used different habitat types, (2) tropical versus temperate species, (3) studies on seeds versus studies on seedlings, (4) trees versus shrubs vs. herbs, (5) the distance from the parent, (6) the density of seeds/seedlings in the depots, and (7) the duration of the studies. The first four of these were categorical contrasts, while the latter three were analyzed as linear regressions.

Although much less common, the least-squares inference approach has been used for more advanced modeling as well. For example, Gurevitch et al. (2000) examined the effects of both predators and competitors on growth/body size using Hedges' d in a 2×2 factorial design. In a more complex study, Verdú and Traveset (2004) used the general linear modeling approach (Chapter 9) to perform a meta-ANCOVA. Their effect size was the log odds ratio of the probability that a seed would germinate after passing through the gut of a frugivore (the control was germination rate for seeds which did not pass through a frugivore); explanatory variables included the size of the seeds (as a continuous variable), and whether the frugivores were mammals or birds (a categorical variable).

Both the MM and LS methods are appropriate if the aim of the analysis is to estimate moments (e.g., expected value and variances, regression coefficients), if these moments can indeed be estimated, and if the model is "not too complex"; see below for further discussion. In these cases, this approach offers analytical solutions (Chapter 9) and fast computation of estimates of interest.

In MM estimation, we match the theoretical moments with the corresponding estimates based on the data. This is described in detail in Chapter 9. For exposition here, consider the basic meta-analysis model for estimating a mean effect μ under normal distribution assumptions. The corresponding fixed-effects model is:

$$T_i \sim N(\theta_i, \sigma_i^2); \theta_i = \mu$$

for $i = 1, \ldots, k$, where k is the number of studies or data sets to be combined. The MM estimator of μ is a weighted average of the k study-specific means, θ_i, that is, $\Sigma w_i T_i / \Sigma w_i$, $i = 1, \ldots,$ k. Here, the weights, w_i are equal to the inverse of the variances σ_i^2; that is, $w_i = 1/\sigma_i^2$. Hence, study estimates with larger variance are downweighted and more precise study estimates (i.e., with larger values of w_i) are given more weight in the overall mean estimate. Remember that the variance σ_i^2 here relates to the variance of T_i (i.e., the study-specific estimates of effect, not the original observations from the individual studies). This is typically a function of the observed variance and the sample size in the primary study, so that as the sample size increases, the variance of the estimate decreases. Larger studies will thus typically have smaller values of σ_i^2 in the above equation, and contribute more weight to the overall estimate. This is conceptually reasonable.

The corresponding random-effects model is

$$T_i = \theta_i + e_i; e_i \sim N(0, \sigma_i^2)$$
$$\theta_i = \mu + \varepsilon_i; \varepsilon_i \sim N(0, \tau^2)$$

for $i = 1, \ldots, k$. The MM estimator of μ is again a weighted average of the k study-specific means, θ_i, that is, $\Sigma w_i T_i / \Sigma w_i$, $i = 1, \ldots, k$, where now $w_i = 1/(\sigma_i^2 + \tau^2)$. See Chapter 9 for further details. As discussed earlier in this chapter, the inclusion of the between-study variance τ results in more equal weighting of the studies, and if τ is zero (indicating no between-study variance), or relatively much larger than the study-specific variances (indicating large heterogeneity between studies), this model reduces to a fixed-effects analysis.

In practice, the variance σ_i^2 is almost always unknown and is estimated by a sample variance (S_i^2). In a MM analysis, this estimate is usually "plugged into" the above MM estimates and treated as known, without acknowledgement of the sampling variation of S_i^2. This is often argued to be reasonable since the variation contributed by this estimate is most often relatively small. Maximum likelihood (ML) and Bayesian approaches can appropriately accommodate this additional uncertainty.

In a Bayesian analysis, this is achieved by placing a prior distribution on σ_i^2; see Chapter 11 for details. For example, if T is normally distributed, then a possible prior is $\sigma_i^2 \sim v_i S_i^2 / \chi^2$ where v_i is the degrees of freedom related to S_i^2 (typically the sample size n_i minus 1), and χ^2 is a chi-squared distribution with v_i degrees of freedom. In a ML analysis, an iterative solution is used; see Chapter 10 for details.

As with MM, the popularity of LS is that it is a simple approach both conceptually and computationally when it is used in the appropriate models. Moreover, it is familiar to many practitioners because it mimics the very common least squares approach to regression that can be used for analysis of primary study data.

The MM and LS approaches to estimation are more problematic for increasingly complicated situations, and indeed may not be possible. For example, if the model has a hierarchical structure with more than two levels, likelihood-based or Bayesian approaches might be more appropriate. The formulas for the MM and LS approaches are given in Chapter 9. These approaches are common in most meta-analysis software packages or modules.

Maximum likelihood inference

As discussed above, moment-based estimation and least squares may not be possible or appropriate for more complex models. This may occur, for example, if data are nonnormal, models are mixed or nested, or more comprehensive inferences are required. A powerful alternative approach is the maximum likelihood (ML) inference. Instead of estimating parameters by matching moments or minimizing the squared residuals, with maximum likelihood we choose those parameters that maximize the probability of the data.

For example, in the simple meta-analysis situation described above, the aim would be to find the value of μ that maximizes the "likelihood" of obtaining the set of T_i's compiled from the studies. (For mathematical reasons it is often the *log-likelihood* that is maximized, since this is computationally easier and gives the same parameter estimates.) Since the values T_i are assumed to be normally distributed in the models described earlier, the maximum likelihood estimate (MLE) is the same as the moment-based estimate (and, indeed, equivalent to the posterior estimate obtained under Bayesian inference). This equivalence is not generally the case with more complex models or nonnormal distributions.

One of the great advantages of the ML (and the Bayesian) approach to inference is that we can simultaneously estimate all of the parameters in the model. Compared to the moment-based and least squares approaches, this gives a much clearer "picture" of the estimates and their uncertainty.

Variants of ML estimation have been developed for various situations. For example, as with the method of moments approach, if the true study-specific variances σ_i^2 are unknown, the meta-analyst can use the estimates S_i^2 and assume that these are the true values, but the MLE may be slightly biased. In this case, a restricted maximum likelihood (REML) can be used. This approach is also used to estimate variance components in a regression model, assuming that the residuals are normally distributed. The approach is modified for Bernoulli (binary) response variables, using, for example, marginal or penalized quasi likelihood (MQL and PQL) (see Chapter 10).

Bayesian inference

Recall that both moment-based and maximum likelihood approaches make inferences on the basis of the *likelihood*, that is, the probability of the data given the model parameters P (data | parameters). (Note that the "|" notation indicates "given" or "conditional on"; see Chapter 11 for details). In contrast, as discussed above, a Bayesian approach to meta-analysis makes inferences on the *posterior* distribution of the model parameters themselves, given the data P (parameters | data). This posterior distribution incorporates not only information about the data (through the likelihood), but also information about the parameters (through *prior* distributions).

Let all the unknown variables in the model be Θ; note that this can model parameters as well as other unknowns, such as missing data. For a simple random-effects model, $\Theta = (\theta_i, \sigma_i^2, \mu,$

$\tau, \ldots$). Let the observed data be denoted by T. This typically includes the study-specific effect estimates $(T_1, \ldots, T_k)$ and associated variances. Then under Bayes' rule, the posterior distribution for Θ, given the data T, is given by

$$P(\Theta \mid T) = P(T \mid \Theta)\, P(\Theta)/P(T)$$

where $P(T \mid \Theta)$ is the likelihood, $P(\Theta)$ is the set of priors on the components of Θ, and $P(T)$ is the probability of the data over all possible values of Θ.

Priors can be informative (representing information from literature reviews, previous experiments, expert opinion, etc.) or uninformative (so that inferences are based primarily on the observed data). Although priors can take any form, common distributions are often used, such as normal distributions for study-specific and overall means, and uniform distributions for within- and between-study standard deviations.

A wide range of inferences can be made based directly on the posterior distributions of the model parameters. For example, in addition to the usual estimation of the overall mean μ, we can use the posterior distribution of μ to estimate the uncertainty of this mean value, directly estimate a 95% interval for μ (by "chopping off" the 2.5% tails of the distributions), or calculate very easily the probability that μ is greater than some ecologically important threshold value. Other comparisons are equally straightforward to calculate and interpret, such as the probability that the study-specific mean from one study is larger than another (i.e., $P(\theta_1 > \theta_2)$). In an ecological context, this might translate to comparative statements such as, "the probability that plant species in unburned areas have longer average survival than species in burned areas." As another example, if the posterior distribution of a parameter is highly skewed, no single measure (e.g., a mean) can satisfactorily act as a summary, but multiple measures can be estimated if we have access to the distribution itself. It is also straightforward to predict an outcome of interest in a new study (e.g., a new site or population group) by simulating from the posterior distributions. Again, the Bayesian approach gives full probabilistic information about these forecasts (Lunn et al. 2000). Further examples of this are given in Chapter 11.

Sometimes estimates are substituted for parameters in the meta-analysis models in order to simplify the computation or reduce the model complexity. Common examples of this are the plug-in use of sample variances S_i^2 as "known" values of σ_i^2, and the plug-in use of an estimate of between-study variance as the "true" value of τ. The result is an *empirical Bayes* model. The advantage of this approach is that models may be able to be fit using standard statistical software such as SAS. However, since the uncertainty of the "plug-in" estimates is ignored in these models, the resultant posterior estimates may be biased and the precision of these estimates might be over optimistic. The extent of the bias and imprecision depends on the relative size of this uncertainty; this is typically not large for σ_i^2 but can be influential for τ.

However, these approximations are not necessary because full Bayesian models (with all of the parameters correctly specified as unknown) can be estimated using numerical methods such as the Markov chain Monte Carlo (MCMC). Under this approach, $P(\Theta \mid y)$ is "decomposed" into a series of distributions for each parameter (or set of parameters), conditional on the other parameters; for the random-effects model described above, these include

$$P(\theta_i \mid y, \mu, \tau^2, \sigma_i^2, \ldots),\ P(\mu \mid y, \theta_i, \tau^2, \sigma_i^2, \ldots),\ P(\tau^2 \mid y, \theta_i, \mu, \sigma_i^2, \ldots),\ P(\sigma_i^2 \mid y, \theta_i, \mu, \tau^2, \ldots),\ \text{etc.}$$

The MCMC algorithm then iteratively simulates from these distributions, resulting in a large number of values of each parameter of interest; these can then be used to construct the posterior distributions, find estimates, and so on. This approach is described further in Chapter 11. Some software packages are dedicated to fitting Bayesian models via MCMC estimation (e.g., Win-BUGS) and an increasing number of packages offer it as an option (MLwiN, Stata, and SAS).

Given the philosophical, modeling, and computational advantages of Bayesian modeling, there has been increased interest in and adoption of this approach for ecological meta-analysis (recognizing, of course, that the acceptance of Bayesian approaches is not universal). An example is given by Ren et al. (2007), who adopted a hierarchical Bayesian model to estimate the effect of ozone on the association between temperature and mortality. The effect estimates (regression coefficients and corresponding variances) were obtained for each of 95 United States communities. These estimates were then divided into seven regions and four levels of ozone. It was assumed that the regression estimates were normally distributed within region r and each ozone level l, with an overall effect that was itself normally distributed. Finally, these 28 estimates of overall effect were combined to provide a national estimate of the effect. Other examples are given in Chapter 11.

STATISTICAL ISSUES IN META-ANALYSIS

As we have seen in the previous sections, the construction of the meta-analysis model involves a number of statistical decisions and assumptions. In this section, we discuss three such issues that commonly arise in meta-analyses: appropriately weighting study estimates, testing for between-study heterogeneity, and including other information in the meta-analysis. Additional issues, including the choice of effect size metric, missing values, and the distributional assumptions for study estimates and model parameters, are discussed in Chapters 6 and 7.

Weighting studies

Once we have collected effect sizes for all of the studies and developed a model (or set of models) which tries to explain heterogeneity within and between studies, we then have to decide how to go about inferring basic properties, such as the mean overall effect. A naïve approach would be to calculate the simple arithmetic mean of the reported effect size estimates (adding them up and dividing by the number of such estimates). This effectively gives equal weight to all of the values. The problem with this approach is that in most cases, the studies are *not* contributing equal amounts of information; most obviously, the effect size estimates differ with respect to sample size and variance (also represented as precision, which is the inverse of the variance).

The solution of weighting is one of the fundamental aspects of meta-analysis. The meta-analysis model determines how study estimates will be weighted in the overall analysis. In general, it allows for effect estimates to be weighted by the variance terms included in the model. Thus, in a fixed-effects model, the study-specific estimates will be weighted by the precision (i.e., the inverse of the sampling variances) of the estimates, whereas in a random effects-model, this weighting will include both within- and between-study variance terms. As a consequence, more precise estimates will be more heavily weighted in analyses; a relatively large between-study heterogeneity will lead to greater uncertainty in estimated overall effects, and so on. Because the weights in a random-effects models include a common between-study variance term, they are more equal, so smaller studies are not as extremely down weighted as in a fixed-effects model. It is clear that if the between-study variance is very large relative to the within-study variance, the studies are effectively unweighted. This underlines the importance of taking care in choosing an appropriate meta-analysis model.

If the variance estimates are not available for each of the study-specific effect size estimates, alternative weighting approaches might be adopted based on the sample size, for example. The weights might also incorporate other measures such as quality scores (Chapter 7). It is recommended that one should consider an unweighted meta-analysis only when weights cannot be determined with any confidence or when there is extreme heterogeneity.

Note that maximum likelihood and Bayesian methods of estimation apply implicit weights, depending on the model. See Chapters 10 and 11 for details.

Testing for heterogeneity

For some, the choice of the meta-analysis model itself is a matter of inference rather than a decision based on the characteristics of the data or the problem at hand. Thus, one might undertake a "test of homogeneity" as part of the analysis. This tests the hypothesis that the studies arise from the same population by comparing the variance between study estimates with the variability of the estimates themselves (i.e., a comparison of "within" and "between" study variance). If the test is rejected, a random-effects model may be more appropriate, and the significant between-study variation is then included in the analysis. An inability to reject the test is interpreted as supporting a fixed-effects model whereby no between-study variation is allowed in the model. Common tests are based on the Q and I^2 statistics (Chapter 9) for moment-based approaches, likelihood ratio statistics (Chapter 10) for likelihood approaches, and posterior probabilities (Chapter 11) for Bayesian approaches. Homogeneity tests are usually not undertaken, and are not meaningful, in cases where a random-effects model has been used for conceptual reasons and/or because the meta-analyst recognizes in advance that there is substantial between-study variation. In these cases, it does not make sense to test for between-study variation, because we are already assuming it exists and has scientific meaning.

Although there are arguments in favor of choosing a fixed- or random-effects model on the basis of homogeneity tests, there are also strong arguments against this practice. In effect, the test is creating a discontinuity in the parameter that represents the between-study variation; a small change in the *estimate* of this parameter can make it "jump" from a positive value to zero. Moreover, the test has low power (Jackson 2006); that is, it is difficult to reject the null hypothesis of no heterogeneity even when there is indeed between-study variation, leading to the erroneous adoption of fixed-effects instead of random-effects models. Finally, the tests ignore the fact that I^2 (and Q) are estimated with error. (See Chapter 9 for confidence intervals for I^2.) For example, if $I^2 = 0.49$, those using a cutoff of 0.5 would call this a lack of heterogeneity, yet the upper bound of the confidence interval might include 1.0.

In practice, this test can lead to quite strange and potentially erroneous estimates. These types of estimates can be seen in the context of a cumulative meta-analysis (Chapter 15). Suppose that at a particular stage of the analysis the test of homogeneity supports a random-effects analysis. Hence, additional between-study variation is included in the model and reflected in widened confidence intervals on the parameters of interest. Suppose now that a new study is added to the analysis, and that this study increases the homogeneity of the study effect estimates so that the hypothesis of homogeneity is not rejected. Now the between-study variation is completely omitted from the model and the confidence intervals will be considerably smaller. Repeating this process, the confidence intervals and, indeed, the estimates may not provide a consistently smooth "story" of progressive learning about the parameter of interest.

Ideally the choice between random- and fixed-effects models should be made *before* undertaking any analysis. For example, if the aim of the meta-analysis is to compare and make inferences about the specific studies under consideration, without generalization, then a fixed-effects analysis might be appropriate. On the other hand, if the studies are considered samples or representatives of a larger group of studies (sites, populations, etc.), so that the results are more generalizable, then a random-effects model is automatically appropriate. Similarly, if the aim is to explain heterogeneity through categorical variables (forming subgroups of studies for which estimates and comparisons are interesting) then one might consider a fixed- or mixed–effects regression model. Finally, if the aim is to ascertain whether the studies under consideration could be considered to have the same population parameters, then a test of homogeneity might be appropriate (notwithstanding its faults as described above). However, if there is no good a priori reason for choosing one or the other model, then in almost all cases that we can envisage in ecology and evolutionary biology, a random-effects model should be adopted and the test of homogeneity should be abandoned as being meaningless.

In practice, the choice of model might also be determined as part of the process of analysis. For example, in the event that a fixed-effects model is fitted and there is a large between-studies variation component, the model could be changed by introducing a between-study variance term (leading to a random-effects model) or including moderators. Alternatively, this might induce a change in the interpretation of the results.

Including other information in a meta-analysis

We conclude our introduction to some of the statistical considerations in a meta-analysis by addressing the common desire to include in the analysis other information related to the outcome of interest. An immediately identified special case of this is a cumulative meta-analysis, which entails the integration of a set of current studies with the results of a previous meta-analysis of the same outcome (Chapter 15). This continuous updating of opinion about the effect or relationship of focus in the meta-analysis can be undertaken by a full reanalysis of past and present studies, or it can be achieved by formulating the outcome of the previous meta-analysis as a "prior," to be included with the new information (the "likelihood") in a Bayesian analysis (Chapter 11).

CONCLUSIONS

This chapter has focused on describing standard meta-analysis approaches and corresponding statistical models. The aim has been to present these as a coherent methodology, rather than a "grab-bag" of disjointed methods. This is important for the practicing meta-analyst. In particular, it provides a framework for a wide variety of methods, as well as for dealing with issues that are frequently encountered in practice, such as combining different types of data, coping with problems of bias and quality, allowing for missing data, and so forth.

APPENDIX 8.1: LEPIDOPTERA MATING EXAMPLE

"Do male virgins make better partners?" This question was addressed by Torres-Vila and Jennions (2005), who undertook a meta-analysis of the association between male mating history and female reproductive output in Lepidoptera. Female reproductive output was generally measured as lifetime fecundity (total number of eggs laid) or fertility (number of viable eggs or progeny produced).

The data set below (Table A8.1) is extracted from Torres-Vila and Jennions (2005). The studies are defined in terms of species, suborder (moths, Heterocera versus butterflies, and Rhopalocera), and family. The effect of interest, expressed as Hedges' d, reflects the standardized mean difference in reproductive output for females mated to virgin versus experienced males. Polyandry, which is defined as the percentage of females that were reported to have mated more than once, is included as a continuous covariate.

TABLE A8.1. Effect of male mating history on female reproductive output in Lepidoptera (data from Torres-Vila and Jennions 2005).

Species	Suborder	Family	%Polyandry	Hedges' d	Var (d)
Acrolepia assectella	Heterocera	Yponomeutidae	38	0.5385	0.0441
Agrotis segetum	Heterocera	Noctuidae	27.5	0.3762	0.0213
Busseola fusca	Heterocera	Noctuidae		0.4688	0.0819
Colias eurytheme	Rhopalocera	Pieridae	47	1.0125	0.2051
Chilo partellus	Heterocera	Pyralidae	5.5	−0.0411	0.0522
Choristoneura fumiferana	Heterocera	Tortricidae	27.5	0.1371	0.0479
Choristoneura rosaceana	Heterocera	Tortricidae	45	1.0281	0.0372
Diatraea considerate	Heterocera	Pyralidae	7.5	0.4577	0.1325
Epiphyas postvittana	Heterocera	Tortricidae	12.25	−0.1176	0.0895
Eurema hecabe	Rhopalocera	Pieridae		0.1105	0.2003
Helicoverpa armigera	Heterocera	Noctuidae	68	0.3043	0.1283
Jalmenus evagoras	Rhopalocera	Lycaenidae	0	0.3656	0.0714
Lobesia botrana	Heterocera	Tortricidae	22.5	0.0655	0.0463
Ostrinia nubilalis	Heterocera	Pyralidae	27.5	0.3585	0.0171
Papilio glaucus	Rhopalocera	Papilionidae	65	0.2317	0.0435
Papilio machaon	Rhopalocera	Papilionidae		0.2506	0.1551
Pectinophora gossypiella	Heterocera	Gelechiidae	81	0.809	0.0237
Pieris napi	Rhopalocera	Pieridae	70	1.169	0.255
Plodia interpunctella	Heterocera	Pyralidae	37.5	0.114	0.0458
Pseudaletia unipuncta	Heterocera	Noctuidae	55	−0.1524	0.0482
Sitotroga cerealella	Heterocera	Gelechiidae	68.5	0.5072	0.0217
Spodoptera littoralis	Heterocera	Noctuidae	61.5	0.1	0.1073
Thymelicus lineola	Rhopalocera	Hesperiidae	22	0.107	0.0781
Trichoplusia ni	Heterocera	Noctuidae	81	0.2631	0.075
Zeiraphera canadensis	Heterocera	Tortricidae	8	0.0155	0.0667

Moment and Least-Squares Based Approaches to Meta-analytic Inference

Michael S. Rosenberg

CHAPTER 8 INTRODUCED VARIANCE and structural models and various statistical inference approaches used in meta-analysis. This chapter describes the basic details behind the moment and least-squares approach to meta-analysis. This approach represents "classic" meta-analysis; it is the one most frequently found in meta-analytic introductions and used in ecological meta-analyses to date. As discussed in Chapter 8, this approach to meta-analytic inference has the advantage of using fairly simple formulas (for basic structural models) that can be easily calculated, and it is clearly and directly comparable to common statistical concepts, such as weighted means and sums of squares. The disadvantages of this approach is that it is less amenable to more complex modeling, particularly when considering features such as interaction effects; it also has certain limitations that in some cases reduce the applicability of random-effects models.

Throughout this discussion, we assume that one has already calculated effect sizes and their variance estimates (Chapters 6 and 7) for each of the studies that will be included in the analysis. As previously emphasized, all of the following approaches can be used with any measure of effect. For simplicity, we will use $\hat{\theta}_k$ to represent the effect measure of the kth study, with estimated study variance v_k. One of the fundamental aspects of meta-analysis is weighting (Chapter 8). Studies in the analysis are given different weights depending on the estimated reliability of the individual study. In general, this is based on the inverse of the sampling variance. Studies whose effect sizes are estimated with small variance are given larger weights, while studies whose effect sizes are estimated with large variance are given smaller weights. The only computational difference in statistical inference when using a fixed- or random-effects model is in the calculation of the weights. Recall from Chapter 8 that in fixed-effects meta-analysis, the general model can be written as $\hat{\theta}_k = \theta_k + e_k$, where $e_k \sim \mathrm{N}(0, v_k)$ and $\theta_k = \mu$. For this fixed-effects model, the weight for a study is simply the inverse of the study's variance, v_k (Table 9.1). The random-effects model is identical to the fixed-effects model, except that $\theta_k = \mu + \varepsilon_k$, where $\varepsilon_k \sim \mathrm{N}(0, \sigma^2)$. For this random-effects model, the weight for a study is the inverse of the sum of the study variance and the estimate of this additional random-effects variance, $\hat{\sigma}^2$.

As already discussed, v_k is estimated for each individual study as part of the estimate of effect size (Chapter 6); formulas for the moment/least-squares estimate of $\hat{\sigma}^2$ are specific to the structural model being used to explain heterogeneity and are described below. Other than this estimation of the study weight, fixed- and random-effects models use identical formulas for all moment estimators.

Throughout this chapter we will use a data set comparing female fecundity in Lepidoptera when mated with virgin or experienced males (described in detail in Appendix 8.1). The

TABLE 9.1. Moment estimators of weights in fixed- and random-effects model meta-analyses.

	Fixed-Effects Model	Random-Effects Model
Study Weight	$w_k = \dfrac{1}{v_k}$	$w_k = \dfrac{1}{v_k + \hat{\sigma}^2}$

standardized mean difference in fecundity (Hedges' d, see Chapter 6) among females mated with virgin or experienced males was used as the effect size for 25 species of moths and butterflies.

ESTIMATING THE GRAND MEAN ACROSS ALL STUDIES

The estimate of the mean effect size across all studies is the weighted mean of the individual study effect estimates,

$$\hat{\mu} = \frac{\displaystyle\sum_{k=1}^{K} w_k \hat{\theta}_k}{\displaystyle\sum_{k=1}^{K} w_k}$$

with variance

$$s_{\hat{\mu}}^2 = \frac{1}{\displaystyle\sum_{k=1}^{K} w_k}$$

The mean across all studies can be interpreted by comparison with a hypothetical null value, μ_0. The simplest way to examine this is through the use of a basic t-test, such that

$$t_s = \frac{\hat{\mu} - \mu_0}{s_{\hat{\mu}}},$$

whose probability can be determined from Student's t-distribution with $K - 1$ degrees of freedom (where K is the total number of studies). Typically μ_0 equals zero, and the test simplifies to

$$t_s = \frac{\hat{\mu}}{s_{\hat{\mu}}}.$$

Alternatively, one can construct confidence intervals around $\hat{\mu}$ with

$$CI = \hat{\mu} \pm s_{\hat{\mu}} t_{\frac{\alpha}{2}[K-1]},$$

where α is the desired significance level used to determine the confidence interval limits. The mean effect size is considered significantly different from μ_0 if its confidence interval does not include μ_0. If there are concerns that assumptions of the t-test are not met (e.g., the effect sizes may not be normally distributed), one can estimate the confidence interval for the mean using a bootstrap procedure that is discussed later in this chapter.

For the Lepidoptera data, the estimate of the average effect across all 25 studies was $\hat{\mu} = 0.3588$, with variance $s_{\hat{\mu}}^2 = 0.00203$. We can construct a 95% confidence interval around the average, which ranges from 0.2659 to 0.4517. Because this interval does not overlap with

---

BOX 9.1.
Student's *t*-distribution vs. the Normal distribution

Traditionally, most meta-analyses have tended to use the normal distribution for significance tests and the estimation of confidence intervals. Because the number of studies (the sample size of the overall test) in a meta-analysis is often small and the true variance of the test is unknown, we recommend use of the Student's distribution instead of the normal. The Student's distribution is more conservative, leading to broader confidence intervals when the number of studies is small. With a large number of studies, the difference between Student's and the normal distributions is minimal.

---

the expected null value of zero, this indicates that the effect size is significantly different than zero; that is, on average, across all studies, fecundity was significantly greater for females mated with virgin males than with experienced males. Alternatively, we could perform the *t*-test as described above and find that $t = 7.97$ with 24 degrees of freedom, corresponding to $P = 3.34 \times 10^{-8}$. We could also use a bootstrap procedure to estimate the confidence intervals, as discussed later in the chapter.

These calculations use a fixed-effects model where we assume that all variation among the studies is due to sampling error. That is, if the exact same experiment was carried out in the exact same way many times, the results would differ due only to chance; the variation is the sampling error in the fixed-effects model. If we assume there is additional variation due to genuine differences among the outcomes of the studies (e.g., because they are on different species), we might use a random-effects model instead. In this case, there are two sources of variation among studies: real differences in their outcomes (i.e., their effect sizes) and sampling error. Under the random-effects model, we estimate the pooled variance—that is, the variance in the true effect among studies—as 0.0486 (Box 9.2). Using this value to estimate new weights, and repeating

---

BOX 9.2.
Random-effects variance estimate when only estimating the mean.

The estimate of the additional variance component for the random-effects model when only estimating the overall mean is:

$$\hat{\sigma}^2 = \frac{Q_{E_{Fixed}} - (n-1)}{\sum\limits_{k-1}^{K} \hat{w}_k - \dfrac{\sum\limits_{k=1}^{K} \hat{w}_k^2}{\sum\limits_{k=1}^{K} \hat{w}_k}},$$

where n is the number of studies, $Q_{E_{Fixed}}$ is the error heterogeneity estimate from the fixed-effects model (for this model, $Q_E = Q_T$), and w_k is the weight of the individual studies under the fixed-effects model. All other calculations of effect, variance, and Q_T are identical as in the fixed-effects model, except for the use of the random-effects weights as described in Table 9.1. By accounting for excess heterogeneity using $\hat{\sigma}^2$, Q_T cannot be significant under a random-effects model.

the analyses with the new values, we find a new average effect size of $\hat{\mu} = 0.3308$, with a 95% confidence interval ranging from 0.1929 to 0.4686. The estimate of the mean is different than for the fixed-effects model because the means are weighted means, and the weights—based on the variances—are different in the two models. For this data set, the mean effect under the random-effects model is not all that different from that found in the fixed-effects model, but (as is typical and expected), the confidence interval around the estimate is wider.

HETEROGENEITY STATISTICS
Total heterogeneity

Heterogeneity in meta-analysis can be estimated as the weighted sums of squares, known as Q statistics (Hedges and Olkin 1985). The total heterogeneity of the study, Q_{Total} or Q_T, can be estimated as

$$Q_T = \sum_{k=1}^{K} w_k (\hat{\theta}_k - \hat{\mu})^2.$$

The null hypothesis is that all of the studies are a homogeneous sample from a population of studies with a true effect of μ. A test of homogeneity can be performed by comparing Q_T to a χ^2 distribution with $K - 1$ degrees of freedom. A significant value indicates that the estimated effect sizes are more heterogeneous than would be expected by chance. This calculation is only meaningful for a fixed-effects meta-analysis, where one assumes that the population is represented by a single fixed mean. In a random-effects meta-analysis, there is assumed to be additional variation in the population, accounted for by the additional variable $\hat{\sigma}^2$ (see Chapter 8). It is impossible for Q_T to be significant for a simple random-effects model because all of the additional variation is accounted for directly by the model.

For the Lepidoptera example, Q_T under the fixed-effects model is 46.55; when compared to a χ^2 distribution with 24 degrees of freedom, $P = 0.00380$. This indicates that the studies are more heterogeneous than would be expected by chance. Under the random-effects model, Q_T is not very meaningful by itself since the purpose of the addition of the random-effects variance is to account for all heterogeneity among studies. As expected, the Q_T value of 23.01 is not significant ($P = 0.519$) under this model.

Q statistics are not directly comparable among meta-analyses and their power is highly dependent on the number of studies. An alternative measure of heterogeneity in meta-analysis is

---

BOX 9.3.
Alternative formula for total heterogeneity.

The equation for Q_T is often written in a different form that lends itself to more efficient computation (particularly by hand), but that is less obviously related to the sums-of-squares nature of the Q-statistics.

$$Q_T = \sum_{k=1}^{K} w_k \hat{\theta}_k^2 - \frac{\left(\sum_{k=1}^{K} w_k \hat{\theta}_k \right)^2}{\sum_{k=1}^{K} w_k}$$

known as I^2 (Higgins and Thompson 2002). This metric quantifies the heterogeneity by comparing Q_T to its expected value under the assumption of homogeneity.

$$I^2 = \text{Max}\left[100 \times \frac{Q_T - (K-1)}{Q_T}, 0\right],$$

where the maximum is used to assure that the value cannot be less than zero (this occurs if Q_T is smaller than its degrees of freedom). This index can be interpreted as the percentage of total heterogeneity that can be attributed to between-study variance (Huedo-Medina et al. 2006). While Q_T can detect whether significant heterogeneity exists in a sample, I^2 can be used to quantify the extent of that variation and allows a direct comparison of the heterogeneity found in different meta-analyses. (Q_T's from different studies cannot be compared in any meaningful way because the magnitude of Q_T is dependent on the degrees of freedom.) These metrics can be overinterpreted; both have low power in detecting true heterogeneity when the number of studies is small and both may detect "significant" heterogeneity with a large number of studies, even when the true variability is quite negligible (Huedo-Medina et al. 2006).

For the Lepidoptera data, the fixed-effects meta-analysis yields an $I^2 = 48.44\%$, indicating that almost half of the observed variance can be attributed to differences among the studies.

BOX 9.4.
Confidence intervals for I^2.

The confidence interval for I^2 can be determined through a related metric known as H^2 (Higgins and Thompson 2002, Huedo-Medina et al. 2006). H^2 is the ratio of Q_T to its degrees of freedom or

$$H^2 = \frac{Q_T}{K-1}.$$

The standard error of the natural logarithm of H^2 can be determined by

$$S_{\ln H^2} = \frac{1}{2} \frac{\ln(Q_T) - \ln(K-1)}{\sqrt{2Q_T} - \sqrt{2K-3}} \qquad \text{if } Q > K,$$

or

$$S_{\ln H^2} = \sqrt{\frac{1}{2(K-2)}\left(1 - \frac{1}{3(K-2)^2}\right)} \qquad \text{if } Q \leq K,$$

allowing one to construct the confidence interval of $\ln H^2$ as $\text{CI} = \ln H^2 \pm s_{\ln H^2} z_{\alpha/2}$, where z indicates the critical value of the normal distribution for the desired α value. The upper and lower limits of the confidence interval of H^2 are simply the exponential of the corresponding limits of $\ln H^2$, $\text{CI}_{H^2} = e^{\text{CI}_{\ln H^2}}$. Because H^2 and I^2 are very similar metrics whose actual relationship is

$$I^2 = 100 \times \frac{H^2 - 1}{H^2},$$

the confidence interval for I^2 can be easily determined by applying this formula to the limits of the confidence interval for H^2.

Explained and unexplained heterogeneity

As discussed in Chapter 8, one may hypothesize that there is variation among studies beyond that due to sampling error or individual study characteristics, and then choose to model the effects of additional variables in explaining this variation. For meta-analysis with explanatory models (see below), the total heterogeneity, Q_{Total} (or Q_T), can be partitioned into heterogeneity from the structural model Q_{Model} (or Q_M) and unexplained (or sampling) heterogeneity Q_{Error} (or Q_E). Thus $Q_T = Q_M + Q_E$. In the simple model where all you are doing is estimating the heterogeneity among a single group of studies as described above, there is no additional explanatory model; in that case, $Q_M = 0$ and $Q_E = Q_T$.

The model heterogeneity, Q_M, describes the amount of heterogeneity which can be explained by the model structure. If the structural model describes a large portion of the total heterogeneity, Q_M will be significant. A significant Q_M does not indicate that a structural model is correct, merely that some of the variance can be explained by those factors. An alternate model might explain more or less of the variance. Q_M should always be evaluated in conjunction with Q_E.

The error heterogeneity, Q_E, describes the amount of heterogeneity which is left unexplained after the model is taken into account. A significant value of Q_E indicates that there is still significant heterogeneity unaccounted for by the model, thus a significant Q_E may indicate an inadequate model, even if Q_M is significant. While the significance of these statistics is generally tested by comparison to the χ^2 distribution, Q_M can also be tested using a randomization test described later in this chapter.

The ratio of Q_M to Q_T is the proportion of observed variance explained by the model; the ratio of Q_E to Q_T is the proportion of observed variance unexplained by the model. The ratio of Q_M to Q_T is functionally equivalent to the R^2 value from a typical linear regression (which is itself just a ratio of unweighted sums of squares).

SINGLE FACTOR EXPLANATORY MODELS

Thus far we have discussed models where all studies conceptually belong to a single group. Heterogeneity among studies may, however, be explainable by biologically important differences among groups. In fact, with meta-analysis in ecology, our primary interest is often in testing for hypothesized causes of differences in effect sizes. We may believe that there are differences among predefined groups of studies (e.g., studies on carnivores vs. herbivores) in the magnitude of the response to an experimental treatment, or due to a continuous variable (e.g., water depth). The following section describes the moment approach to meta-analytic inference for simple structural models of this sort, which attempt to explain variation using only a single factor. This factor can be either categorical (and thus logically analogous to a one-way ANOVA) or continuous (and thus logically analogous to linear regression).

Simple categorical ("ANOVA-like") structure

The simplest model which can be used to explain heterogeneity of effect sizes is one where the individual effects are split into two or more categories (or groups), much in the same manner that one attempts to explain variance using a single factorial ANOVA. The overall mean effect and total heterogeneity are estimated as above. The mean of the mth group is simply the weighted mean of the studies in that group, therefore

$$\hat{\mu}_m = \frac{\sum\limits_{k=1}^{K_m} w_{mk}\hat{\theta}_{mk}}{\sum\limits_{k=1}^{K_m} w_{mk}},$$

where K_m is the number of studies in the mth group and X_{mk} refers to the kth study in the mth group. The variance of the average effect of each group can be calculated as

$$s_{\hat{\mu}_m}^2 = \frac{1}{\sum\limits_{k=1}^{K_m} w_{mk}}.$$

Confidence intervals and differences from a null effect can be determined for each group as described above for the overall mean, while the heterogeneity within each group can be determined as

$$Q_{E_m} = \sum_{k=1}^{K_m} w_{mk}(\hat{\theta}_{mk} - \hat{\mu}_{mk})^2.$$

Once again, this value is expected to be χ^2 distributed with $K_m - 1$ degrees of freedom. At this stage, we have simply treated each group as if it were its own single population and calculated the mean, variance, heterogeneity, and confidence intervals as described earlier in this chapter.

To make comparisons among the groups, one takes an approach similar to that found in ANOVA. The model heterogeneity, Q_M (see Box 9.5) can be calculated as

$$Q_M = \sum_{m=1}^{M} W_m(\hat{\mu}_m - \hat{\mu})^2,$$

where M is the number of groups, and W_m is the sum of the weights in group m, or $\sum_{k=1}^{K_m} w_{mk}$. The error heterogeneity, Q_E, can be calculated as

$$Q_E = \sum_{m=1}^{M} Q_{E_m} = \sum_{m=1}^{M} \sum_{k=1}^{K_m} w_{mk}(\hat{\theta}_{mk} - \hat{\mu}_m)^2.$$

Note that these are simple weighted sums of square differences between the means at one level and the next (hierarchical) level; that is, Q_E is between individual studies and their group mean, Q_M, is between group means and the global mean. For a nested structure with additional levels in the hierarchy (see below), the Q for each intermediate level is determined in the same way.

BOX 9.5.
Alternative terms for partitioning heterogeneity.

Because of its similarity to ANOVA, Q_M and Q_E have often been referred to by different terms in the ecological (and other meta-analytic) literature. Specifically, Q_M is often referred to as $Q_{Between}$ or Q_B, while Q_E is often referred to as Q_{Within} or Q_W. While "Between" (as well as the rarely used "Among") and "Within" have specific meaning in an ANOVA-like structure, they are not terms of general use when other explanatory models (such as regression) are used, and we prefer to use the more generic "Model" and "Error" labels when partitioning the heterogeneity into its explanatory and unexplained components.

TABLE 9.2. Partitioning of heterogeneity for a simple categorical explanatory model. The elements in each column of the bottom row ("Total") are the sum of the elements in the two preceding rows.

Source of Heterogeneity			df
Model	Q_M	$\sum_{m=1}^{M} W_m(\hat{\mu}_m - \hat{\mu})^2$	$M - 1$
Error	Q_E	$\sum_{m=1}^{M} \sum_{k=1}^{K_m} w_{mk}(\hat{\theta}_{mk} - \hat{\mu}_m)^2$	$n - M$
Total	Q_T	$\sum_{k=1}^{n} w_k(\hat{\theta}_k - \hat{\mu})^2$	$n - 1$

As with total heterogeneity, Q_M and Q_E should both be χ^2-distributed with $M - 1$ and $n - M$ degrees of freedom, respectively (n is the total number of studies across all groups). The relationships of these values can most easily be seen by setting them into an ANOVA-like table.

The significance of Q_M can be tested not only by comparison to the χ^2 distribution, but it can also be examined using a randomization procedure, as described later in this chapter.

Returning to the Lepidoptera example, we might believe the reason for the high degree of heterogeneity among studies is that the experiments were conducted on organisms that are taxonomically quite distinct. We might wish to split the studies into two groups based on suborder; these are the studies of Heterocera (moths, 18 studies) and studies of Rhopalocera (butterflies and skippers, 7 studies). Using a fixed-effects model, the mean effect size for moths was 0.363 (95% confidence interval from 0.260 to 0.465), while for butterflies it was 0.336 (95% confidence interval from 0.043 to 0.628). There was significant heterogeneity among studies on moths, $Q_E = 40.3206$ ($P = 0.00117$), but not among those on butterflies, $Q_E = 6.1872$ ($P = 0.40255$). Keep in mind that the lack of significance in butterflies may be due to small sample size and low power, rather than a true absence of heterogeneity. We can calculate I^2 for the individual groups as well; for moths $I^2 = 57.84\%$, while for butterflies $I^2 = 3.03\%$.

Now, let us examine the overall fit of our model to the data. Recall that our model consists of two parts: a single factorial division of the data into moths or butterflies, and a fixed-effects model to describe the variation (or lack) within each group (Table 9.3A).

Our model failed to explain a significant amount of variation in the data ($Q_M = 0.04$, $P = 0.834$), and significant variation is still unexplained ($Q_E = 46.51$, $P = 0.003$). This should not be surprising; the mean effect sizes for butterflies and moths were very similar (0.363 and 0.336) with highly overlapping confidence intervals.

What if we change our model to allow for random-effects variation within each group (the so-called mixed model)? In this case we are fixing the group differences (moths vs. butterflies) but are not supposing that the effect within each group is perfectly fixed. Our estimation of the random-effects variance (Box 9.6) is 0.0526. (This is different from the estimate for the earlier random-effects model, which did not include our factorial design because estimation of the pooled random-effects variance is dependent on the entire model.) As expected, the estimates for the means of butterflies and moths have changed, although not substantially (results not shown). Table 9.3B shows the new model summary. Assuming random effects within groups, our model still does not explain a significant amount of heterogeneity. There do not appear to be differences between the taxa.

One thing to note, however, is that the magnitude of the total heterogeneity, Q_T, is different than what we found above when we did not categorize the studies into suborders. The total

TABLE 9.3. Summary of results and variation explained through a wide variety of models for the Lepidoptera data. Each model is explained more fully in the text. There is some variation in the number of studies across models because data was not always available for all species when additional factors were considered.

Source of Heterogeneity			df	$P(\chi^2)$
A. Fixed-effects categorical model: moths vs. butterflies. P (Rand) = 0.9047				
Model	Q_M	0.04	1	0.83400
Error	Q_E	46.51	23	0.00258
Total	Q_T	46.55	24	0.00380
B. Random-effects categorical model: moths vs. butterflies. P (Rand) = 0.7892				
Model	Q_M	0.0833	1	0.77292
Error	Q_E	22.0897	23	0.51484
Total	Q_T	22.1730	24	0.56894
C. Fixed-effects categorical model: families.				
Model	Q_M	12.87	5	0.02463
Error	Q_E	32.14	16	0.00960
Total	Q_T	45.01	21	0.00173
D. Fixed-effects nested model: families and suborders.				
Suborders	$Q_{M\text{-}Ord}$	0.07	1	0.79828
Families (in Suborders)	$Q_{M\text{-}Fam}$	12.80	4	0.01230
Error	Q_E	32.14	16	0.00960
Total	Q_T	45.01	21	0.00173
E. Fixed-effects continuous analysis. Regression explains 21% of the variation.				
Model	Q_M	9.4576	1	0.00210
Error	Q_E	36.5623	20	0.01320
Total	Q_T	46.0199	21	0.00127
F. Random-effects continuous analysis. Regression explains 17% of the variation.				
Model	Q_M	4.1667	1	0.04123
Error	Q_E	20.4476	20	0.43026
Total	Q_T	24.6143	21	0.26425
G. Fixed-effects GLM with continuous and categorical components.				
Model	Q_M	9.4807	2	0.00874
Error	Q_E	36.5392	19	0.00906
Total	Q_T	46.0199	21	0.00127
H. Random-effects GLM with continuous and categorical components.				
Model	Q_M	4.3074	2	0.1161
Error	Q_E	19.2236	19	0.4426
Total	Q_T	23.5310	21	0.3163

BOX 9.6.
Random-effects variance estimate when using a single categorical explanatory variable (the so-called mixed model).

The estimate of the additional variance component for the random-effects model when using an ANOVA-like single categorical explanatory model is

$$\hat{\sigma}^2 = \frac{Q_{E_{Fixed}} - (n - M)}{\displaystyle\sum_{m=1}^{M}\left(\sum_{k=1}^{K_m}\hat{w}_{mk} - \frac{\displaystyle\sum_{k=1}^{K_m}\hat{w}_{mk}^2}{\displaystyle\sum_{k=1}^{K_m}\hat{w}_{mk}}\right)}$$

where n is the total number of studies, M is the number of groups, $Q_{E_{Fixed}}$ and $\hat{w}_k$ are estimated using the fixed-effects model as above.

---

heterogeneity will be identical for all fixed-effects models (those which assume no study variation beyond that of additional explanatory variables); for random-effects models (those that assume true variation among the studies, even beyond that modeled by additional explanatory variables), however, the total heterogeneity will vary depending on the specific model. The reason for this should be evident; although the equation for total heterogeneity is constant, the inputs to the equation change in random-effects models but are constant in fixed-effects models. In particular, the weights in random-effects models are dependent on the estimate of pooled variation among studies, which is itself model dependent. Changing the estimate of pooled variance changes the weights, which changes the total heterogeneity. In fixed-effects models this pooled variance is always equal to zero, thus the weights are constant, and the total heterogeneity is also constant.

Instead of splitting our taxa by suborder, what if we had decided to examine a different taxonomic level, such as families? (Note: it is important to emphasize that in a meta-analysis, as in any statistical analysis, one should avoid independently testing for explanatory variable after explanatory variable within the same data set. In this chapter we are doing this for illustrative purposes, but the proper way to perform a meta-analysis is to make an a priori choice of the factors you wish to analyze and to then analyze them in a single meta-analysis rather than a series of small meta-analyses, as has so often been done in the literature.)

Table 9.4 shows the results for individual families. The first thing to note is that the sample sizes per family are quite small. The largest family only has six species, while some families have only two species. Three additional families with just a single species (for which we had data) were eliminated from the analysis. The means differ quite a bit among families, but because of the low sample sizes the confidence intervals tend to be much wider for the family estimates than they were for the suborders. There is no significant variation within any of the families except for the Tortricidae; in this family, $I^2 = 79\%$, indicating that a very large proportion of the observed variation within this family is due to differences among the studies.

Dividing the species into families explains a significant amount of variation; however, a significant amount of variation is still unexplained (Table 9.3C). We could also repeat this analysis with a random-effects model; in this case, little of the variation is explained by dividing the species into families (results not shown).

TABLE 9.4. Family specific results under a fixed-effects model.

Family	K_m	Mean Effect	95% CI	Q_E	P
Noctuidae	6	0.2450	0.0003 to 0.4897	4.9239	0.425
Pieridae	3	0.7304	−0.4282 to 1.8891	3.0609	0.216
Pyralidae	4	0.2441	−0.0635 to 0.5517	3.0375	0.386
Tortricidae	5	0.3228	0.0387 to 0.6069	19.1049	0.001
Papilionidae	2	0.2358	−2.1061 to 2.5778	0.0018	0.966
Gelechiidae	2	0.6515	−0.7009 to 2.0038	2.0062	0.157

A somewhat better approach might be to recognize that the two categorical structures we have independently analyzed are hierarchical; the families are nested within the suborders. We can partition the heterogeneity in this form, just as in a nested ANOVA. The calculations parallel those for the single factor ANOVA. Therefore, the mean of any group at any level is the weighted average of all studies within the group, and the heterogeneity Q is the weighted sum of squares difference between the mean for the group and the mean for the next higher-order group that contains the lower group. Thus, the Q for suborders is the weighted sum of squares for the difference in suborder means and global mean, while the Q for the families (within suborders) is the weighted sum of squares for the difference in family mean and the suborder mean that contains the family.

Although the amount of variation that can be attributed to the suborders is small and not statistically significant (Table 9.3D), accounting for the data structure (families fall within suborders) reveals a marginally significant effect of our model in explaining the variation in the studies, although there is still substantial unexplained variation. Of course, family and suborder are arbitrary taxonomic categories; advanced approaches for dealing with data with phylogenetic structure are described in Chapter 17.

The above analysis was conducted using a fixed-effects model. Unfortunately, there is no easy way to perform a nested analysis of this type with a random-effects model when using the least-squares/moments approach to meta-analytic inference, because nested structures can only be described for a GLM analysis (see below) using an overparameterized model for the design matrix. This then requires a generalized inverse procedure to solve, making the entire approach difficult and unappealing. To solve nested structures with random-effects models one should use either a maximum likelihood or Bayesian inference method (see Chapters 10 and 11).

Simple continuous ("linear regression") model

Another simple structural model for meta-analysis is linear regression, where effect sizes are thought to vary relative to a single continuous covariate, X_k (measured or reported for each study). In this case we are modeling $\theta_k = \beta_0 + \beta_1 X_k + \varepsilon_k$ (Hedges and Olkin 1985, Greenland 1987). Regression models used in meta-analysis are weighted regressions, like the weighted categorical models described above. Although still exact, the formulas for estimating β_0 and β_1 under the moment/least-squares inference approach are quite a bit more complex than those for the global mean or simple categorical model structure.

To test the significance of each β (under the null hypothesis that $\beta = 0$), it can be divided by its standard error (e.g., $\hat{\beta}_0/s_{\hat{\beta}_0}$) and compared to a normal distribution. The square of this value for β_1 is equal to Q_M or

$$Q_M = \frac{\hat{\beta}_1^2}{s_{\hat{\beta}_1}^2}.$$

BOX 9.7.
Regression coefficients and their standard errors.

The slope (β_1) of the regression can be estimated as

$$\hat{\beta}_1 = \frac{\displaystyle\sum_{k=1}^{K} w_k X_k \hat{\theta}_k - \frac{\displaystyle\sum_{k=1}^{K} w_k X_k \sum_{k=1}^{K} w_k \hat{\theta}_k}{\displaystyle\sum_{k=1}^{K} w_k}}{\displaystyle\sum_{k=1}^{K} w_k X_k^2 - \frac{\left(\displaystyle\sum_{k=1}^{K} w_k X_k\right)^2}{\displaystyle\sum_{k=1}^{K} w_k}}$$

with standard error

$$s_{\hat{\beta}_1} = \frac{1}{\displaystyle\sum_{k=1}^{K} w_k X_k^2 - \frac{\left(\displaystyle\sum_{k=1}^{K} w_k X_k\right)^2}{\displaystyle\sum_{k=1}^{K} w_k}},$$

while the intercept (β_0) can be estimated as

$$\hat{\beta}_0 = \frac{\displaystyle\sum_{k=1}^{K} w_k \hat{\theta}_k - \hat{\beta}_1 \sum_{k=1}^{K} w_k X_k}{\displaystyle\sum_{k=1}^{K} w_k},$$

with standard error

$$s_{\hat{\beta}_0} = \frac{1}{\displaystyle\sum_{k=1}^{K} w_k - \frac{\left(\displaystyle\sum_{k=1}^{K} w_k X_k\right)^2}{\displaystyle\sum_{k=1}^{K} w_k X_k^2}}.$$

See Hedges and Olkin (1985) and Greenland (1987) for more information.

TABLE 9.5. Partitioning of heterogeneity for a linear regression model.

Source of Heterogeneity			df
Model	Q_M	$\dfrac{\hat{\beta}_1^2}{s_{\hat{\beta}_1}^2}$	1
Error	Q_E	$Q_T - Q_M$	$K - 2$
Total	Q_T	$\displaystyle\sum_{k=1}^{K} w_k (\hat{\theta}_k - \hat{\mu})^2$	$K - 1$

BOX 9.8.
Random-effects variance estimate when using linear regression.

The estimate of the additional variance component for the random-effects model under linear regression (Rosenberg et al. 2000) is very complicated (although is simply algebra), but is presented here for completeness:

$$\hat{\sigma}^2 = \frac{Q_{E_{Fixed}} - (n-2)}{\displaystyle\sum_{k=1}^{K} \hat{w}_k - \sum_{j=1}^{K} \hat{w}_j^2 \left(\frac{\displaystyle\sum_{k=1}^{K} \hat{w}_k X_k^2 - 2X_j \sum_{k=1}^{K} \hat{w}_k X_k + X_j^2 \sum_{k=1}^{K} \hat{w}_k}{\displaystyle\sum_{k=1}^{K} \hat{w}_k \sum_{k=1}^{K} \hat{w}_k X_k^2 - \left(\sum_{k=1}^{K} \hat{w}_k X_k \right)^2} \right)},$$

where n is the number of studies, $Q_{E_{Fixed}}$ and $\hat{w}_k$ are once again estimated using the fixed-effects model as above. Despite the complexity, most of the terms are actually quite simple when examined individually (e.g., the sum of all of the weights or the sum of the weights × the independent variables, squared or unsquared).

Because a direct formula for Q_E is extremely complicated, it is much simpler to calculate Q_E as the difference between Q_T and Q_M. Again, we can place these heterogeneity estimates in an explanatory table (Table 9.5).

Note that the significance of Q_M determined from a χ^2 distribution with 1 degree of freedom is identical to that found by comparing $\hat{\beta}_1/s_{\hat{\beta}_1}$ to a normal distribution (as described above). As with the simple categorical model, the significance of Q_M can also be examined using a randomization procedure, as described later in this chapter.

Returning once again to the Lepidoptera example, rough estimates exist of the percent of polyandry that occurs naturally for 22 of the 25 species. Treating the percent polyandry as an independent variable, we can use a weighted regression to test whether the effect size of a study (fecundity in females mated with virgin vs. experience males) is influenced by the degree of polyandry in the species. In a fixed-effects framework, our estimate of the slope is 0.0059 and the intercept is 0.1161. Overall, we find that the regression does explain a significant amount of the variation (Table 9.3E), but that there is a significant amount of variation which remains unexplained.

A similar result is found using a random-effects model (Table 9.3F). However, as expected, any remaining variation is accounted for as part of the random-effects model itself.

MORE COMPLEX MODELS AND WEIGHTED GLM

Warning: Contents may include matrices and additional Greek symbols. If these are of concern, you may wish to skip this section.

More complicated explanatory models can also be solved using least-squares approaches to meta-analysis, but as the complexity increases, the advantages of the maximum likelihood and Bayesian approaches (Chapters 10 and 11) become more prevalent. We explain the least-squares approach for more complex modeling here to both illustrate the approach as well as to put the simpler models in a common context. Rather than the algebraic formulas described above, the complex models are best solved using a weighted general linear model (GLM) and matrix algebra (Hedges and Olkin 1985, Rosenberg et al. 2000). Despite their apparent complexity, these formulas are fairly simple and can easily be solved using a matrix-based mathematics program such as MATLAB (MathWorks 2007).

Recall from Chapter 8 that the individual effect sizes can all be represented as a single vector

$$\hat{\Theta} = \begin{bmatrix} \hat{\theta}_1 \\ \vdots \\ \hat{\theta}_K \end{bmatrix},$$

and the explanatory model can be described as a matrix,

$$\mathbf{X} = \begin{bmatrix} 1 & X_{11} & \cdots & X_{p1} \\ \vdots & \vdots & & \vdots \\ 1 & X_{1K} & \cdots & X_{pK} \end{bmatrix}.$$

Alternative construction of matrix $\mathbf{X}$ allows for the solution to any structural model which can be described in a general linear framework. For example, if $\mathbf{X}$ is simply a column of 1's we have the determination of the global mean, while if $\mathbf{X}$ contains a column of 1's and a column of a continuous covariate, we have a linear regression, and so forth. The general model can thus be described as

$$\hat{\Theta} = \mathbf{X}\hat{\mathbf{B}} + \mathbf{E},$$

where $\hat{\mathbf{B}}$ is a vector of regression coefficients. In expanded form, this becomes a general regression model (often called meta-regression), or

$$\theta_k = \beta_0 + \beta_1 X_{1k} + \beta_2 X_{2k} + \cdots + \beta_p X_{pk} + \varepsilon_k.$$

The goal of the meta-analysis becomes the estimation of $\hat{\mathbf{B}}$ and determining the significance of each of its elements (β_0, β_1, etc.). Using a weighted GLM approach, we find that

$$\hat{\mathbf{B}} = (\mathbf{X}^T \mathbf{W} \mathbf{X})^{-1} \mathbf{X}^T \mathbf{W} \hat{\Theta},$$

where $\mathbf{W}$ is a diagonal matrix containing the individual study weights, or

$$\mathbf{W} = \begin{bmatrix} w_1 & 0 & 0 \\ 0 & \ddots & 0 \\ 0 & 0 & w_K \end{bmatrix}.$$

The variance of each of the elements of $\hat{\mathbf{B}}$ can be determined from

$$\Sigma_{\mathbf{B}} = (\mathbf{X}^T \mathbf{W} \mathbf{X})^{-1},$$

where the variance is the corresponding element of the diagonal (i.e., the variance of the first element of $\hat{\mathbf{B}}$ is the value at position $(1,1)$ of $\Sigma_{\mathbf{B}}$, the variance of the second element of $\hat{\mathbf{B}}$ is the value at position $(2,2)$ of $\Sigma_{\mathbf{B}}$, etc.). The significance of the individual regression coefficients can be tested by comparing the ratio of the coefficient and its standard error to a normal distribution (just as we test the significance of the slope and intercept in the simple linear regression model, as above). The overall model heterogeneity can be determined from

$$Q_M = \hat{\mathbf{B}}^{T*} \Sigma_{\hat{\mathbf{B}}}^{-1*} \hat{\mathbf{B}}^*,$$

where the * indicates that each of these matrices has been reduced by rank 1, by removal of the first row and column. The error heterogeneity is found by

$$Q_E = (\hat{\Theta} - \mathbf{X}\hat{\mathbf{B}})^T \mathbf{W} (\hat{\Theta} - \mathbf{X}\hat{\mathbf{B}}),$$

and the total heterogeneity is most simply determined as the sum of Q_M and Q_E. The significance of these is still determinable by comparison to a χ^2 distribution, where Q_M, Q_E, and Q_T have p, $K - p - 1$, and $K - 1$ degrees of freedom respectively (p is the number of columns of $\mathbf{X}$ after the first, or put another way, $\mathbf{X}$ has $p + 1$ columns).

All of these calculations assume a fixed-effects model. As before, the random-effects model uses identical formulas except for the determination of the weight matrix, $\mathbf{W}$. In this most general case, the formula for the random-effects variance component is

$$\hat{\sigma}^2 = \frac{Q_E - (K - p - 1)}{\mathrm{tr}(\mathbf{W}) - \mathrm{tr}\left[\mathbf{W}\mathbf{X}(\mathbf{X}^\mathsf{T}\mathbf{W}\mathbf{X})^{-1}\mathbf{X}^\mathsf{T}\mathbf{W}\right]}$$

where "tr" indicates the trace of the matrix, and Q_E and $\mathbf{W}$ indicate the fixed-effects variants. The random-effects variance estimates shown in Boxes 9.2, 9.6, and 9.8 are derived from special case "simplifications" of this general formula for specific constructs of matrix $\mathbf{X}$. Because a nested hierarchical structure cannot be described using a design matrix $\mathbf{X}$, one cannot use this formula to estimate the random-effects variance for a nested meta-analysis. For this reason one cannot perform a nested random-effects model using least-squares inference, and must instead look to maximum likelihood or Bayesian methods of inference (Chapters 10 and 11).

For illustrative purposes, we analyzed the Lepidoptera data with a model consisting of both the suborders as a categorical variable and the percent polyandry as a continuous variable. This is logically equivalent to an ANCOVA. Matrix $\mathbf{X}$ contained three columns: the first contained all 1's, the second contained the percent polyandry, and the third contained a 1 if the species was in the suborder Heterocera and a -1 if it was in the suborder Rhopalocera. This model could be written as

$$\theta_k = \beta_0 + \beta_1 X_{1k} + \beta_2 X_{2k} + \varepsilon_k.$$

Here β_0 is an intercept, β_1 is the regression coefficient for the percent polyandry, and β_2 is the regression coefficient for the suborder. The estimated values of these coefficients, their standard errors, and their significance are shown in Table 9.6A.

Given the previous independently run analyses it should not be surprising that only the % polyandry is significant. However, we still need to examine the fit of the entire model (Table 9.3G). The model explains a significant amount of the variation, but a significant amount remains unexplained. This was a fixed-effects model where all of the residual variation outside of the % polyandry and the suborders was assumed to be due to the sampling error. If we choose a

TABLE 9.6. Regression coefficients from the GLM models.

Coefficient		Estimate	SE	P
A. Fixed-effects model				
"Intercept"	β_0	0.1235	0.1042	0.2358
% Polyandry	β_1	0.0059	0.0019	0.0021
Suborder	β_2	−0.0106	0.0698	0.8794
B. Random-effects model				
"Intercept"	β_0	0.1471	0.1419	0.2999
% Polyandry	β_1	0.0056	0.0028	0.0444
Suborder	β_2	−0.0552	0.0084	0.5482

random-effects model where we expect real additional variation in the studies beyond sampling error, we estimate the random-effects variance as 0.0447. The regression coefficients are fairly similar to those in the fixed-effects model, although now the percent polyandry is only marginally significant (Table 9.6B). However, when the full model is considered (Table 9.3H), it no longer explains significant variation in the studies; instead, all of the variation is subsumed by the assumption of random-effects among studies.

RESAMPLING METHODS FOR META-ANALYSIS

Resampling methods are a computationally intensive approach to estimating distributions and statistical significance that serves as an alternate to conventional parameteric statistics that generally have more restrictive assumptions, such as normality and homoscedasticity. Resampling tests are often useful when sample sizes are small or the original data do not conform to the distributional assumptions of parametric tests (Manly 1997). They also help when data needed to calculate parametric variance estimates around the mean effect are unavailable in primary studies (Chapter 13). Resampling statistics are generally performed by (1) calculating a metric or statistic from the original data, (2) permuting the original data in some way, (3) recalculating the same metric or statistic, and (4) repeating steps 2 and 3 many times. Finally, the values calculated from step 3 are used to construct a distribution based directly on the data that can then be used for significance testing and the construction of confidence intervals. Because meta-analytic data often have small sample sizes and may violate basic distributional assumptions (e.g., normality), resampling techniques can be extremely useful for accurately determining the significance of meta-analytic metrics (Adams et al. 1997).

Bootstrapping

Bootstrapping is a resampling procedure which can be used to generate confidence intervals around a given statistic. Bootstrapping works by randomly choosing (with replacement) n studies from a sample size of n, and then calculating the desired statistic. For example, if there were fifty studies in total, fifty studies would be chosen for each bootstrap iteration. However, because bootstrapping is sampling with replacement, some of the studies from the original sample would be chosen more than once, while others would not be chosen at all. This procedure is repeated many times to generate a distribution of possible values. The lowest and highest 2.5% values are then chosen to represent the lower and upper 95% bootstrap confidence limits (or whatever percentiles are appropriate for the desired α value).

Confidence intervals generated in this way are called percentile bootstrap confidence intervals, because they are calculated by merely choosing certain percentile values (Dixon 1993). These confidence intervals assume that the distribution of bootstrap values is centered around the original value. When this is the case, the percentile bootstrap is known to produce the correct confidence intervals (Efron 1982, Dixon 1993). However, when more than 50% of the bootstrap replicates are above or below the original value, the bootstrap confidence interval should be corrected for this bias (details can be found in Dixon 1993).

In meta-analysis, bootstrapping is generally used to generate confidence intervals around average effect sizes (either the global average or individual group/category averages, when appropriate). For example, in the Lepidoptera meta-analysis discussed above, the bootstrapped confidence interval for the grand mean under the fixed-effects model was 0.2252 to 0.4978, wider than the parametric confidence interval calculated before, but still excluding the expected null value of zero.

Randomization tests

Randomization tests are most frequently used to determine the significance level of a given test statistic. In meta-analysis, randomization tests can be used to test the significance of the model structure, Q_M, as an alternative to comparison to a χ^2 distribution. This is done by randomizing the relationship between the individual study effect sizes and the corresponding model structure (the $\mathbf{X}$ matrix). For example, for a simple categorical ("ANOVA-like") model, the effect sizes are randomly assigned to each group, such that each group has the same number of effect sizes as in the real data. For a linear regression, the effect sizes are randomly assigned to the independent variables. In the most general GLM framework, this test can be seen as randomly shuffling the rows of the $\mathbf{X}$ matrix. Q_M is recalculated for the randomized data. This represents one possible outcome of Q_M if the relationship between the effect sizes and the model is random. By performing many randomization iterations, a frequency distribution of possible outcomes is generated, and we can then compare the observed Q_M (the model heterogeneity estimated from the original data) to this distribution in order to determine significance. For example, if 42 of 9999 Q_M's generated from a randomization procedure are greater than or equal to the observed value, the significance of the observed value is 43/10000 or 0.043. (You must add one to both the numerator and denominator because the observed data needs to be counted as a potential permutation; see Edgington 1987.)

It should be clear that the fine-scale accuracy of the probability level in a randomization test is a function of the number of iterations performed. Increasing the number of iterations allows one to decrease the smallest detectable probability. For example, a randomization test based on only 49 iterations can detect a probability as low as 0.02 (1/50) and can only return significance values in multiples of 0.02. In contrast, a test based on 4999 iterations can detect a probability as low as 0.0002. The disadvantage of larger numbers of iterations is an increase in computational time. As a general rule of thumb, between 2499 and 4999 iterations are often sufficient for a reliable probability level (Adams et al. 1997). As the number of iterations increases, the variance of replicated tests decreases and there is often little gain in accuracy.

Maximum Likelihood Approaches to Meta-analysis

Kerrie Mengersen and Christopher H. Schmid

IN CHAPTER 9, MOMENT-BASED and least squares approaches to estimating the parameters of the meta-analysis model were described and illustrated. This approach is appropriate if the aim of the analysis is simply to estimate the means and variances (that is, the moments of the model) and if these moments are well defined (and easily computed). For example, if the data are normally distributed, the first two moments (the mean and variance) completely describe the distribution and are easily calculated, as described in Chapter 9. Thus in a simple fixed or random-effects model, the method of moments (MM) is adequate.

However, meta-analysts increasingly have been using more complex models and statistical distributions to better describe the problem at hand. For example, the data may be counts that have a Poisson distribution, or proportions that have a binomial distribution; see Chapter 7 for examples of these types of effect estimates. For example, in the fiddler crab example in Appendix 10.1, the data are counts of crabs. We could take the logarithm of the counts, assume that these are (approximately) normally distributed, and fit a normal fixed- or random-effects model as described in Chapter 9. However, this is approximate; a more appropriate alternative is to directly model the effects based on the Poisson distribution, and we can do this using a maximum likelihood (ML) approach.

As another example, consider the Lepidoptera mating data set, described in Appendix 8.1, which provides 25 estimates of effect (standardized mean differences in fecundity among females mated with virgin or experienced males). These standardized mean difference (Hedges' *d*) estimates can be assumed to have a normal distribution, so we can fit a normal random-effects model using either method of moments or maximum likelihood approaches. Because of the symmetry of the normal distribution, these two approaches will give equivalent results.

However, suppose that now we want to extend the simple random-effects model so that instead of combining all effect estimates within one global mean effect, we wish to group species within suborder, and then combine the suborders within a global effect. We have thus introduced a new "level" to the model, and "nested" species within suborders. This model cannot be easily fitted using the moment-based techniques described in Chapter 9.

Approximate moment-based estimators have been proposed for some of these more complex models, but they are subject to bias (under- or overestimation of the true value). Indeed, even in a simple random-effects model (assuming normal distributions and with no additional nesting), the between-study variance term must be approximated, and the common approximations—such as those proposed by Hedges (Chapter 9) and DerSimonian and Laird (1986)—can be afflicted by bias (Böhning et al. 2002).

BOX 10.1.
DerSimonian and Laird's approximation of the
between-study variance (DerSimonian and Laird 1986).

This approximation of the between-study variance is $X^2 - (k-1)/[\Sigma w_i - \Sigma w_i^2/\Sigma w_i]$, where $w_i = 1/\sigma_i^2$, $X^2 = \Sigma w_i(T_i - \mu')^2$, $\mu' = \Sigma w_i T_i/\Sigma w_i$ and T_i and σ_i^2 are respectively the study-specific effect estimates and associated variances, $i = 1, \ldots, k$. This remains unbiased if the sample variances s_i^2 are used instead of σ_i^2 only in special cases (e.g., if the T_i are normally distributed).

Finally, the meta-analysts may be interested in more than the moments of the model (e.g., the overall mean and variance). For example, they may wish to estimate the probability of exceeding a biological or ecological threshold (as in survival data), or the value corresponding to the most extreme 10% of the distribution (as in ecological risk or climate modeling), or a ranking of the study-specific parameters (taking into account uncertainty in their estimates), and so on. This cannot be easily achieved using a method of moments approach.

In this chapter, we describe an alternative, more general approach based on maximizing the *likelihood* of the data; that is, for a model with unknown parameters, finding the parameter values that are "most likely" to generate the observed data set. As a very simple example, suppose that we are interested in the overall age of a population of people, and we have only two options: 30 years and 100 years. Now suppose that we observe a random sample of people with average age of 20 years. Then this sample is more *likely* to have been observed if the overall age of the population is 30 years as opposed to 100 years. Of course, we have made many assumptions in this example, and in practice there is more information about parameter values and the observed sample. We describe this more general setup in the context of meta-analysis below, and then give some worked examples.

OVERVIEW OF MAXIMUM LIKELIHOOD ESTIMATION (MLE)
The basics of MLE

As in Chapters 8 and 9, we consider a set of independent observations $(T_1, T_2, \ldots, T_k)$. Then the *likelihood* of observing this set is the product of the individual probabilities of observing each of the T_i.

For example, suppose that each of the T_i, $i = 1, \ldots, k$, is normally distributed with unknown mean θ_i and variance σ_i^2. We can then ask: "What is the *likelihood* of observing the set of observations $(T_1, T_2, \ldots, T_k)$ given these unknown parameters?" Since the observations are independent, the likelihood is simply the product of their corresponding probability densities. Thus if $T_i \sim N(\theta_i, \sigma_i^2)$, then the likelihood

$$L(T_1, T_2, \ldots, T_k; \theta_1, \sigma_1^2, \theta_2, \sigma_2^2, \ldots, \theta_k, \sigma_k^2)$$

is given by

$$N(T_1; \theta_1, \sigma_1^2) \times N(T_2; \theta_2, \sigma_2^2) \times \ldots \times N(T_k; \theta_k, \sigma_k^2),$$

which can be written as

$$\prod_{i=1}^{k} N(T_i; \theta_i, \sigma_i^2). \tag{10.1}$$

For example, suppose that we observe two independent observations, $(T_1, T_2) = (1, 3)$ and we assume that they come from a normal distribution with the same unknown mean, μ say, and known variance $\sigma^2 = 1$. Then the likelihood of observing (T_1, T_2) is the probability of observing a value of 1 *and* a value of 3, given specified values of μ. Hence if we hypothesize that $\mu = 0$, then

$$L(T_1, T_2; \mu, \sigma^2) = N(T_1 = 1; \mu = 0, \sigma^2 = 1) \times N(T_2 = 3; \mu = 0, \sigma^2 = 1) = 0.242 \times 0.004 = 0.001.$$

If we now hypothesize that $\mu = 1$, say, then the likelihood of observing $(T_1 = 1, T_2 = 3)$ under these parameter values becomes

$$L(T_1, T_2; \mu, \sigma^2) = N(T_1 = 1; \mu = 1, \sigma^2 = 1) \times N(T_2 = 3; \mu = 1, \sigma^2 = 1) = 0.399 \times 0.054 = 0.022.$$

If we were only choosing between these two options, then the parameter that is *more likely* to have generated the observed data (T_1, T_2) is $\mu = 1$ (with $\sigma^2 = 1$). This value *maximizes the likelihood* and is the *maximum likelihood* estimate of the unknown mean in this simple case.

This likelihood can also be written in a reverse form:

$$L(\mu; T, T_2, \sigma^2).$$

This formulation can be phrased as, "how likely is a given value of μ, given the data (and other parameters)?" This is very close to the Bayesian framework described in Chapter 11. We illustrate this formulation in Case study 1, below.

The set of parameters that we wish to estimate depends on the model. In a meta-analysis context, if we consider a fixed-effects model, the main parameter of interest is the overall mean μ_0. Thus, the likelihood of observing a set of normally distributed effect estimates $(T_1, T_2, \ldots, T_k)$—taking into account their associated variances $(\sigma_1^2, \sigma_2^2, \ldots, \sigma_k^2)$—is given by the equation above, with means θ_i equal to μ_0.

The maximum likelihood estimate (MLE) of μ_0 is the value that maximizes this expression. Of course, since μ_0 can theoretically take any value from $-\infty$ to $+\infty$, this calculation is not as simple as comparing two options for μ_0, as in the above illustration. Instead, the MLE is found by integrating the likelihood over all possible values of μ_0. This is explained in more detail below.

If there are multiple unknown parameters, then the MLEs must be obtained for each of these. For example, referring again to the fixed-effects model, although it is common practice to take the study-specific variances $(\sigma_1^2, \sigma_2^2, \ldots, \sigma_k^2)$ to be fixed at the sample values $(S_1^2, S_2^2, \ldots, S_k^2)$, in reality these are unknown parameters that should be estimated as part of the analysis. Hence the likelihood should be maximized over all of these parameters, which requires joint integration. Moreover, these parameters influence each other—for example, the σ_i^2 values influence the weights of the individual effect estimates and therefore influence the estimate of μ_0—so it is not possible to estimate the parameters individually. Thus some type of iterative approach is required to maximize the combined likelihood.

As another example, in a simple random-effects meta-analysis, the unknown parameters include the study-specific means (and variances), the overall mean, and the between-study variance, so the likelihood must now be maximized over this full set of parameters. Again, since these parameters influence each other (e.g., the study-specific means influence the value of the overall mean and the between-study variance), it is not possible to estimate them individually, and some type of iterative approach is required.

In these cases, it is possible to use a restricted maximum likelihood (REML) approach introduced by Patterson and Thompson (1971), which is closely related to empirical Bayes (EB) estimation (Chapter 11). Here, a MLE for a parameter is found by "plugging in" the current MLEs for the other parameters, then this MLE is "plugged in" to the next parameter, and so on until all of the MLEs achieve stability.

One frequently asked question in the context of a meta-analysis is whether these MLEs are appropriately variance weighted. The answer is yes—this happens through the likelihood, as can be seen in Equation 10.1. Indeed, it is easier to compute appropriate variance-weighted estimates for more general models, often without the need for approximations that are required in a method of moments framework.

We discuss these concepts and methods below for the simple fixed- and random-effects models described in Chapter 8. We follow this with some worked examples using the Lepidoptera data set described in Appendix 8.1, and then discuss some extensions to the model.

Discussion of MLE

Typically, ML, REML, and EB estimates are obtained using computer software. In most cases, the REML model is specified in terms of *variance components* that explain the way in which the variation in the data is apportioned to within-study (fixed), between-study (random), and residual variation (or more components if the model is more complicated).

It may be helpful to think about using REML in situations in which ANOVA would normally be used but the data are unbalanced or correlated, or in which linear regression would normally be used but there is more than one source of variation or correlation in the data. Thus REML can be used to combine information across similar studies conducted at different times or in different places.

The likelihood can be used to determine quantities of interest, such as confidence intervals and the probability that the parameter exceeds a threshold (e.g., zero), often via the *likelihood profile*. This is akin to investigating that part of the likelihood that is of interest for a particular question; for example, if the analyst is interested in the probability that the overall mean is greater than a specified threshold, the "slice" of likelihood corresponding to this area is obtained and its "profile" can be examined. The use of the likelihood profile in meta-analysis is described by Van Houwelingen (1995). Böhning et al. (2008) describe profile likelihood methods in the context of meta-analysis of binary data, and show that this is an effective way of dealing with nuisance parameters in a model. We discuss the likelihood profile a little more below.

Comparisons between models or assumptions, such as whether a fixed- or random-effects model best fits the data, can be performed using a *likelihood ratio* test. In this test, the likelihood of the data under each model is computed and the ratio of these likelihoods is subjected to a chi-square test.

The models described below are all examples of *multilevel* models and can be developed and analyzed within this general framework. An excellent reference for the construction and analysis of multilevel models is the text by Gelman and Hill (2007). See also Chapter 5 of Demidenko (2004), which discusses mixed models for meta-analysis.

Steps in calculating a MLE

In general, the following process applies for finding the MLEs for a given model:

(1) Decide on the model structure, as described in Chapter 8. This primarily includes the description of heterogeneity between studies, specification of relevant covariates, and characterization of the fixed and random components of the model.

(2) Decide on the distribution of the observed estimates; for example, normal, binomial, Poisson, and so forth. A range of effect estimates and associated potential distributions is described in Chapters 6 and 7.

(3) Create the likelihood for the parameters in the model. As described above, the likelihood for a set of independent effects is the product of the individual probability distributions. If the study effects are independent, then their likelihoods can be multiplied.

(4) Find the values of the parameters that maximize this likelihood.

(5) Make inferences about the parameters based on the distributions of the estimates (if possible) and the likelihood profile.

These steps are described in more detail below. Before discussing these, we consider the Lepidoptera mating example described in Appendix 8.1 and analyzed in Chapter 9.

Case study 1: Lepidoptera mating example

In this study, 25 estimates of effect (standardized-means differences in fecundity among females mated with virgin or experienced males) are reported as Hedges' d estimates, with associated variance estimates. In the terminology described above, the study-specific effect estimates are $T_i = d_i$. For this analysis we treat the variances as known, and comment on this assumption later in this chapter.

It is common to assume that a Hedges' d estimate has a normal distribution, so that for the ith study, $T_i \sim N(\theta_i, \sigma_i^2)$. Thus the likelihood for the set of estimates $(T_1, T_2, \ldots, T_{25})$ is given by Equation 10.1.

For the first study, the estimated effect is the Hedges' d estimate, $T_1 = 0.5385$, with corresponding variance of $\sigma_1^2 = 0.0441$. If we consider this study in isolation, the likelihood then becomes $N(T_1; \theta_1, \sigma_1^2)$. As described above, we can calculate the value of this likelihood for given values of θ_1. For example, the likelihood if $\theta_1 = 1.0$ is the value of the normal density at 0.5385, given a mean of 1.0 and a variance of 0.0441—this value is 0.170. Similarly, the likelihood if $\theta_1 = 0$ is the value of the normal density at 0.5385, given a mean of 0 and a variance of 0.0441—this value is 0.071. Note that (considering this study in isolation), the likelihood is maximized at $\theta_1 = T_1 = 0.5385$, with a value of 1.90. As discussed above, we can represent these values of θ_1 by $L(\theta_1; d_1, \sigma_1^2)$, which is depicted in the upper panel of Figure 10.1. The lower panel of Figure 10.1 shows the individual likelihoods for all 25 studies in the data set.

MAXIMUM LIKELIHOOD ESTIMATION FOR SIMPLE MODELS
Fixed-effects model, known variances

Following from Chapter 8, we first consider the basic meta-analysis model for estimating a mean effect μ, based on estimates T_i, $i = 1, \ldots, k$ reported in k studies, using a fixed-effects model:

$$T_i \sim N(\theta_i, \sigma_i^2); \; \theta_i = \mu \tag{10.2}$$

Assume that the T_i are independent and that the σ_i^2 are known. Then the *likelihood* for the T_i, $i = 1, \ldots, k$, is simply the product of these normal distributions, as in Equation (10.1), above.

We now find the value of μ that maximizes the likelihood, that is, the one that gives us the highest probability of observing the data T_i, $i = 1, \ldots, 25$. This likelihood is given by

$$L(T_i; \mu, \sigma_i^2) = \prod_{i=1}^{25} N(T_i; \mu, \sigma_i^2).$$

After a little calculation, we obtain the likelihood for μ, a normal distribution with variance given by $1/(\sum W_i)$, where $W_i = 1/\sigma_i^2$:

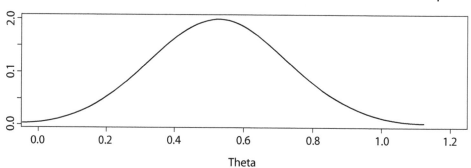

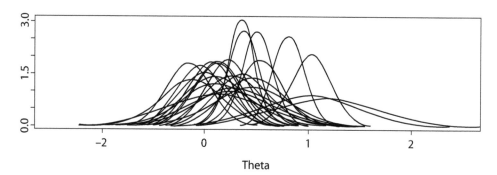

Figure 10.1. Likelihood of observing the estimates T_1 and S_1^2 from the first study given different values of θ (*top panel*), and similarly the likelihood of observing each of the 25 studies (*bottom panel*) for the Lepidoptera data.

$$\mu \,|\, T_i, \sigma_i^2 \sim \mathrm{N}(\mu, 1/(\textstyle\sum W_i)); \; W_i = 1/\sigma_i^2 \tag{10.3}$$

and a maximum likelihood estimate given by

$$\mu_{MLE} = \textstyle\sum W_i T_i \big/ \textstyle\sum W_i. \tag{10.4}$$

Notice that in this case, because the effect estimates are independent and normally distributed, and because the variances are assumed to be known, the MLE is the same as the moment-based estimator described in Chapter 9.

For the Lepidoptera example, recall that T_i are the study-specific effect estimates (standardized differences in fecundity among females mated with virgin or experienced males), θ_i are the corresponding true study-specific effects, and μ is the overall mean effect. In this case, assuming that the study-specific variance estimates are known (i.e., $S_i^2 = \sigma_i^2$), the maximum likelihood estimate of μ is the same as those obtained under the method of moments approach described in Chapter 9, $\mu_{MLE} = 0.3588$.

Random-effects model, known variances

Consider now the random-effects model described in Chapter 8:

$$T_i \sim \mathrm{N}(\theta_i, \sigma_i^2); \; \theta_i \sim \mathrm{N}(\mu, \tau^2). \tag{10.5}$$

If τ^2 and the σ_i^2 are known, then the parameters θ_i and μ are again normally distributed with the variances

$$\text{Var}(\theta_i) = \tau^2 \sigma_i^2/(\sigma_i^2 + \tau^2)$$
$$\text{Var}(\mu) = 1/(\sum (1/(\sigma_i^2 + \tau^2)))' \tag{10.6}$$

and maximum likelihood estimates

$$\theta_{iMLE} = (\tau^2 T_i + \sigma_i^2 \mu)/(\sigma_i^2 + \tau^2)$$
$$\mu_{MLE} = \sum W_i T_i / \sum W_i$$
$$W_i = 1/(\sigma_i^2 + \tau^2)$$
$$\text{Var}(\mu_{MLE}) = 1/\sum W_i. \tag{10.7}$$

Note that the MLE for θ_i is a weighted average of the data T_i and the overall mean μ, where the weights are formed by the inverse variances. Thus for imprecise study estimates, the σ_i^2 is relatively large so that the MLE for that study will be drawn toward the overall mean. Similarly, as the variation τ^2 between studies increases, there is less "shrinkage" toward the overall mean. If there is no between-study variance, so that $\tau^2 = 0$, then the MLE reduces to the fixed-effects estimate.

The following holds for the fixed-effects model described above. Because the T_i's are independent and normally distributed, as are the θ_i's, then if the variances σ_i^2 and τ^2 are assumed to be known, the MLE for μ given by Equation 10.6 is equal to the corresponding moment-based estimator described in Chapter 9 as $\mu_{MLE} = 0.3308$.

As discussed in the Introduction, in more complicated models this equivalence between moment-based and maximum likelihood estimates will not hold and the calculations will not be so straightforward. In these cases we turn to restricted maximum likelihood and empirical Bayes estimation, described in more detail below, or we turn to more generally full Bayesian estimation, described in Chapter 11.

Random-effects model, unknown between-study variance

Suppose that in the random-effects model (Equation 10.4) above, the between-study variance parameter τ^2 is unknown. In this case, we wish to estimate θ_i, μ, and τ^2. The MLEs for θ_i and μ have the same distributions as in Equation 10.6, but since τ^2 is unknown, we use restricted maximum likelihood (REML) estimates:

$$\theta_{iMLE} = (\tau^2 T_i + \sigma_i^2 \mu)/(\sigma_i^2 + \tau^2)$$
$$\tau^2_{MLE} = \sum W_i^2 (k(k-1)^{-1}(T_i - \theta_{iMLE})^2/\sigma^2)/\sum W_i^2$$
$$\mu_{MLE} = \sum W_i T_i / \sum W_i$$
$$W_i = 1/(\sigma_i^2 + \tau^2)$$
$$\text{Var}(\mu_{MLE}) = 1/\sum W_i. \tag{10.8}$$

See Brown and Kempton (1994) for a description of REML estimation and its applications.

If the study-specific estimates of effect, T_i are normally distributed, then confidence intervals (and other probability statements) for the overall mean effect, μ_{MLE}, can be based on the normal distribution. Other inferences are also possible; for example, we can approximate a 95% prediction interval for the θ_i of a new study as $\mu_{MLE} \pm 1.96 \tau^2_{MLE}$.

Because τ^2 is unknown, the study-specific parameters θ_i are not normally distributed. Although approximations to normality can be made if the sample sizes are sufficiently large, confidence intervals and other probability statements based on these approximations can be misleading in terms of both bias and precision for the sample sizes typically seen in the context of meta-analysis.

An alternative is to consider the likelihood profile. Denote the full log-likelihood for the overall mean μ and the between-study variance τ^2 by $LL(\mu, \tau^2)$. Then the profile log-likelihood for μ is obtained as that part of the full log-likelihood corresponding to the maximum value of τ^2. Similarly, the profile log-likelihood for τ^2 is given by maximizing μ. Thus,

$$PLL(\mu) = \max_\tau LL(\mu, \tau^2)$$

$$PLL(\tau^2) = \max_\mu LL(\mu, \tau^2).$$

Then $PLL(\mu_{MLE})$ is the profile of $PLL(\mu)$ at the MLE of μ. Note that this has reduced the multivariate normal likelihood (obtained from the normally distributed data) to a single normal "profile" for μ and a single value for μ_{MLE}. Thus the MLE for μ and τ^2 can be found simultaneously, acknowledging uncertainty in both estimates.

As discussed above, we can use likelihood ratios to build confidence intervals and make inferences. For example, 95% confidence intervals for μ and τ^2 are given by the set of values of μ such that $PLL(\mu) > PLL(\mu_{MLE}) - 1.92$, and similarly by the set of values for τ^2 such that $PLL(\tau^2) > PLL(\tau^2) - 1.92$. The value of 1.92 is based on the fact that twice the difference between two log-likelihoods has a chi-squared distribution with 1 degree of freedom. The value corresponding to the 95% percentile of this distribution is 3.84, so we take $1.92 = 3.84/2$. Notice that this value is not *always* the threshold for a likelihood ratio test, since it is necessary to account for the degrees of freedom that are based on the difference in the number of model parameters.

Random-effects model, all variances unknown

We turn now to the most typical situation in practice. This occurs when not only is the between-study variance τ^2 unknown, but the variances σ_i^2 of the study-level effect estimates are also unknown. Often, we estimate the true variance σ_i^2 (or some surrogate measure) by the sample variance S_i^2 (or similar).

If the sample sizes are sufficiently large, then S_i^2 is a good approximation to σ_i^2. It is common practice to ignore the fact that the S_i^2 are estimates, pretend that $S_i^2 = \sigma_i^2$, and proceed with REML as described earlier.

Alternatively, a Bayesian model can be adopted to take proper account of this variation; a useful prior for σ_i^2 is $\nu S_i^2 / \chi_\nu^2$ where ν is the between-study degrees of freedom (DuMouchel and Harris 1983). Bayesian models are discussed in more detail in Chapter 11.

Case study 1 revisited

We return to the Lepidoptera mating example described above. As discussed, since the Hedges' d estimates are assumed to be normally distributed and the corresponding variances are assumed to be known, the maximum likelihood estimates of the overall effects under the simple fixed- and random-effects models are equivalent to the moment-based estimates reported in Chapter 9.

This equivalence persists when covariates are included in the models. For illustration, we consider a random-effects meta-analysis model with covariates suborder ($0 =$ Heterocera,

1 = Rhopalocera) and % polyandry. The three cases with missing polyandry values are excluded from the analysis.

Table 10.1 provides a comparison of estimates obtained under method of moments using two common estimates of the between-study variance based on Hedges (1983, 1989) and DerSimonian and Laird (1986), and maximum likelihood using REML estimates obtained through the R package metafor (previously called mima); see Appendix 10.2 for details. More details about the Hedges and DerSimonian and Laird (DL) estimators can be found in Raudenbush (1994). As expected, the three sets of estimates are broadly comparable, although there are slight differences between the two methods of estimation under the method of moments approach. Viechtbauer (2006) provides a general comparison of REML, DL, and HE with respect to their bias and efficiency.

TABLE 10.1. Comparison of method of moments (DerSimonian and Laird's [DL] and Hedges' [HE]) and maximum likelihood estimates for the Lepidoptera mating example, using a random-effects meta-analysis with moderators, excluding three studies with missing % polyandry values.

	Method of Moments (DL)	Method of Moments (HE)	Maximum Likelihood (REML)
Residual Heterogeneity			
Estimate	0.045	0.039	0.049
Test: Q-statistic, *df*	26, 19	38, 19	37, 19
Test: P-value	0.01	0.01	0.01
Parameter Estimates:			
Intercept			
Estimate	0.092	0.093	0.091
P-value	0.49	0.47	0.51
95% CI	–0.18, 0.36	–0.16, 0.35	–0.18, 0.36
Suborder			
Estimate	0.11	0.10	0.12
P-value	0.55	0.56	0.54
95% CI	–0.24, 0.49	–0.24, 0.48	–0.23, 0.50
Polyandry			
Estimate	0.0056	0.0056	0.0055
P-value	0.044	0.038	0.049
95% CI	0.0001, 0.011	0.0003, 0.011	0.0001, 0.011
Covariances			
Suborder-Polyandry	2.1E-7	2.0E-7	2.2E-7
Overall Test of all Moderators			
Q-statistic, *df*	4.31, 2	4.56, 2	4.14, 2
P-value	0.12	0.10	0.13

It is seen that there is still substantial between-study heterogeneity (residual heterogeneity $= 0.0487$, $P = 0.0091$) even after accounting for suborder and polyandry, and that these co-variates describe very little of this heterogeneity. (The omnibus test of all moderators is not statistically significant at the 5% level; polyandry is barely significant at this level.) The MLEs for suborder and polyandry are 0.1151 and 0.0055, respectively. These values indicate that the overall mean effect (overall Hedges' d) is larger by an amount of 0.1151 for Rhopalocera compared with Heterocera, and increases by 0.0055 for every 1% increase in polyandry.

Since the covariates considered above do not explain much of the between-study hetero-geneity, we return to the random-effects model with no covariates, as described earlier in this chapter, and illustrate the calculation of the maximum likelihood estimates of the study-specific effects θ_i, based on Equations 10.7. For example, for the first entry in the Lepidoptera mating data set, with $T_1 = 0.54$ and $\sigma_i^2 = 0.044$, we use $\tau_{MLE}^2 = 0.051$ and $\mu_{MLE} = 0.33$ to obtain

$$\theta_{1MLE} = (\tau_{MLE}^2 T_1 + \sigma_1^2 \mu_{MLE})/(\sigma_1^2 + \tau_{MLE}^2) = (0.051 \times 0.54 + 0.044 \times 0.33)/(0.044 + 0.051) = 0.44.$$

As expected, this is "shrunken" from the observed effect (0.54) toward the overall mean (0.33). This shrinkage effect is observed for all 25 studies in the Lepidoptera mating data set.

It is also straightforward to verify the value of $\mu_{MLE} = 0.33$ using Equation 10.7:

$$W_i = 1/(\sigma_i^2 + \tau_{MLE}^2)$$

$$\text{e.g. } W_1 = 1/(0.044 + 0.051) = 10.53$$

$$W_2 = 1/(0.021 + 0.051) = 13.85, \text{ etc.}$$

$$\sum W_i = 219.1$$

$$\mu_{MLE} = \sum W_i T_i / \sum W_i = [(10.5 \times 0.54) + \text{etc}]/219.1 = 0.33.$$

Of course, an alternative to reverting to the simple model with no covariates is to include other covariates in the model. See, for example, the analyses of this data set in Chapter 11.

MAXIMUM LIKELIHOOD ESTIMATION FOR MORE COMPLEX MODELS
Mixed models

As discussed in Chapter 8, more complex meta-analysis models can often be cast in the form of a mixed model in a multilevel or hierarchical framework. Here, the fixed and random effects can be used to describe specific and general sources of heterogeneity, respectively, and the multiple hierarchies can be used to describe clustering or nesting of estimates and studies, or other forms of dependence within and between studies. This is discussed in more detail below and in Appendix 10.2.

General description of a meta-analysis model

The meta-analysis model described above can also be extended to accommodate more complex problems in the following manner. Consider the random-effects model given in Equation 10.6 above. We can recast this model as follows:

$$T_i = \mu + \delta_i + e_i$$

where δ_i is the deviation of the ith true study mean from the overall mean μ, and e_i is the devia-tion of the observed effect T_i from the true effect ($\theta_i = \mu + \delta_i$) for that study.

We can also write this in vector form:

$$\mathbf{T} = \mu\mathbf{I} + \boldsymbol{\delta} + \mathbf{e}$$

$$\boldsymbol{\delta} \sim \text{MVN}(\mathbf{0}, \text{G}); \mathbf{e} \sim \text{MVN}(\mathbf{0}, \text{R})$$

where MVN indicates a multivariate normal distribution. Here, $\mathbf{T}$ is the vector of estimates $(T_1, T_2, \ldots, T_k)$ assumed to be normally distributed. In the "fixed" component of the model, $\mathbf{I}$ is a k-vector of ones and μ is the overall mean. In the "random" component of the model, G is a variance-covariance matrix describing the between-study variation. Finally, $\mathbf{e}$ is the matrix of within-study residuals. If the studies are uncorrelated, then the between-study variance matrix G has τ^2 on each of the diagonals and zero elsewhere, and the within-study variance matrix R has diagonal elements $(\sigma_1^2, \sigma_2^2, \ldots, \sigma_k^2)$ and zeros elsewhere.

In the meta-analysis context, we know $\mathbf{T}$, given by the reported study-specific estimates, and we often assume that we know R, given by the reported variances of these estimates. The aim is to estimate G (and hence τ^2), and the overall mean μ.

A general mixed-effects linear model with covariates can be described as an extension of the above model:

$$\mathbf{T} = \text{X}\boldsymbol{\beta} + \text{Zb} + \mathbf{e}$$

$$b \sim \text{MVN}(\mathbf{0}, \text{D}) ; e \sim \text{MVN}(\mathbf{0}, \text{R})$$

As above, the model includes both (within-study) fixed effects (given by $\text{X}\boldsymbol{\beta}$) and (between-study) random effects (given by Zb). Here, X is a design matrix of dimension $(k \times p)$, which describes covariates that influence the within-study fixed effects, and $\boldsymbol{\beta}$ is the $(p \times 1)$ vector of fixed-effects parameters. Similarly, Z is a design matrix of dimension $(k \times q)$, which describes covariates that influence the between-study random effects.

As above, in this model we know $\mathbf{T}$ and assume we know R. The aim is to estimate the between-study variation represented by D and the effect of the covariates represented by $\boldsymbol{\beta}$. This representation of the model is required by a number of software packages, such as SAS PROC MIXED.

Meta-analysis of nonnormal data

Nonnormal data can often be transformed to normality and analyzed using the above models. Sheu and Suzuki (2001) provide an example from psychology of a meta-analysis of binomial data, using a log-odds transformation.

Turner et al. (2000) provide results of a comparison between method of moments, maximum likelihood, and restricted maximum likelihood estimates for a multilevel meta-analysis of binomial data in a clinical trial setting. Preference is given for the REML estimates in this setting. Likelihood-based methods, Wald methods, and parametric bootstrap methods are compared for constructing confidence intervals for the overall effect and between-study variance, with a preference for bootstrap intervals. The authors also evaluate the use of individual level (primary) data, as discussed in Chapter 18, and describe a bias-corrected bootstrap approach for unbiased estimation of effects and the corresponding confidence intervals.

If the data are nonnormal, the general linear and general linear mixed models described above can be extended further to the generalized linear model (GLM) and generalized linear mixed model (GLMM), respectively. In the generalized models, a *link* function is used to describe the relationship between the response variable and the linear terms in the model. Typical link functions include a log transformation for Poisson data and a logit transformation for binomial data, for example. Bolker et al. (2009) provide a description of the general and generalized linear and mixed models (GLM and GLMM) for meta-analysis of data in the context of ecology and evolutionary biology.

Case study 2: Handedness of fiddler crabs

As discussed above, nonnormal data can often be transformed and analyzed as normal distributions. This is illustrated here through an example of a meta-analysis of the handedness of fiddler crabs across 19 studies, described in Appendix 10.1.

In this example, we define the effect of interest as the probability of left-handedness, which is estimated by the proportion of left-handed crabs in each of the studies. The probability distribution for the effect estimates is thus binomial, and a logit link function is used to transform the effect estimates. For this data, study and species were used as explanatory variables.

The MLE estimates obtained from the analysis of these data are compared in Table 10.2 with corresponding MM estimates obtained using the methods of DerSimonian and Laird (1986) and Hedges (1983, 1989).

The analysis indicates that the overall probability of left-handedness in the crab population is not significantly different from 0.5. However, there is a substantial uncertainty attached to this estimate, as demonstrated by the very wide confidence intervals for the overall mean. This is in part a consequence of the significant heterogeneity between studies ($P < 0.001$). Indeed, a different analysis of these data in Chapter 11 reveals that $P \neq 0.5$ for some crabs.

Goodness of fit of the meta-analysis model can be evaluated through the deviance. This is a measure of the amount of variation in the data that is not described by the model. In this example (not shown in Table 10.2), the deviance divided by its degrees of freedom was less than 1, which indicates adequacy of the model.

MAXIMUM LIKELIHOOD META-ANALYSIS COMPUTATIONS

There is a wealth of programs available for conducting meta-analysis using maximum likelihood, restricted maximum likelihood, and its variants. In addition to built-for-purpose

TABLE 10.2. Comparison of method of moments (DerSimonian and Laird's [DL] and Hedges' [HE]) and maximum likelihood estimates for the handedness of fiddler crabs example, using a random-effects meta-analysis with no moderators.

	Method of Moments (DL)	Method of Moments (HE)	Maximum Likelihood (REML)
Residual Heterogeneity			
Estimate	0.8331	0.9944	1.0628
Test: *Q*-statistic, *df*	486.96, 126	486.96, 126	486.96, 126
Test: *P*-value	<0.001	<0.001	<0.001
Parameter Estimates:			
Overall Mean			
Estimate	0.497	0.4968	0.4966
Variance	0.0102	0.0115	0.0121
P-value	<0.001	<0.001	<0.001
95% CI	0.2992, 0.6947	0.2864, 0.7072	0.2811, 0.7122

software routines, general software packages that are popular for this purpose include SAS, Stata, MLwiN, Splus, and R. We give references to the use of some of these packages below, acknowledging that this list is selective rather than comprehensive. Two of the packages, SAS and R, are described in more detail in Appendix 10.2. The intention is to provide the "flavor" of a range of programs, rather than comprehensive details of their use; the interested practitioner is referred to the respective software package documentation and indicated papers, which provide more complete expositions of their use for meta-analysis. The reader is also referred to Chapter 12 for more general discussion of software for meta-analysis.

A number of papers provide details for performing meta-analysis using SAS; see, for example, Normand (1999), Stijnen (2000), Wang and Bushman (1999), Sheu and Suzuki (2001), and Littell et al. (1996). SAS PROC MIXED has now been used in a wide range of applied settings; for example, see Miyazawa and Lechowicz (2004) and Whitehead (2002). This software routine has also been described in the bivariate setting (Chapter 18) by Van Houwelingen et al. (1993), Paul et al. (2006), and Riley et al. (2008), among others.

As discussed above, nonnormal data can often be transformed to normality and analyzed using standard meta-analysis models. There is a growing body of software and associated "guides" for undertaking such analyses. For example, Kuss and Koch (1996) provide a set of SAS macros for binary outcome data using SAS, including both fixed- and random-effects models. See also SAS PROC NLMIXED for fitting binomial outcome measures.

Riley et al. (2008) promote the use of Stata as a very flexible multilevel modeling platform; they employ Stata to obtain REML estimates for a bivariate meta-analysis because one of their models was reportedly unable to be analyzed using SAS PROC MIXED. The corresponding Stata program used by Riley et al. is also publicly available.

Sterne et al. (2007) provide a review of user-written packages for meta-analysis in Stata. The packages cover a wide range of topics, including meta-analysis of binary data (cell frequencies from 2×2 tables for each study), differences or ratios between parameters (for different subpopulations of data), numerical summaries (mean and standard deviation for each study), reported P-values from each study, and primary data from each study. Fixed-, random-, and mixed-effects models with multiple levels of nesting (hierarchies), meta-regression, and log-linear dose-response regression, among other models, can be fitted. Cumulative meta-analysis plots, forest plots, funnel plots, Galbraith plots, "trim and fill" plots, influence plots, tests of symmetry, and H and I^2 statistics are also available. Likelihood ratios can also be plotted in this package.

Turner et al. (2000) use MLn/MLwiN software to find ML and REML estimates via iteratively generalized least squares (IGLS), and restricted iteratively generalized least squares (RIGLS) algorithms, respectively. They reference Woodhouse (1996) and Goldstein et al. (1998) for further information about the software, and Goldstein (1995) for details of IGLS and RIGLS.

A number of packages are available for conducting meta-analysis in SPlus and R. The reader is referred to the online documents on modeling using R (http://cran.r-project.org/other-docs. html); in particular, see Kuhnert and Venables' document, "An Introduction to R: Software for Statistical Modelling & Computing," and that by Bliese, "Multilevel Modelling in R." An example of a simple package that is available in both platforms is metafor (Viechtbauer 2010; previously called mima in Viechtbauer 2006), which performs fixed- and random-effects models using two method of moments estimators (Hedges 1983, 1989; DerSimonian and Laird 1986), ML and REML, and empirical Bayes. A very popular package is lmer, which fits linear mixed-effects regression models; see Appendix 10.2. Variations of this package fit generalized (for nonnormal data), nonlinear, and multivariate models. An example of a package that allows for more complex analysis is mcmcglmm (Hadfield 2009), which fits multivariate generalized linear mixed models. This package allows for a wide range of distributions, including normal,

Poisson, zero-inflated Poisson, and multinomial. Missing data, certain forms of censoring, and a range of variance structures are also accommodated. Three special types of variance structure are those associated with pedigrees, phylogenies, and measurement error.

The lme and lmer packages in R are used quite widely for meta-analysis in ecology as well as in a wide range of other disciplines. Gelman and Hill (2007) provide a detailed account of the use of lmer for multilevel models, including the advantages and limitations of the method, and comparisons with other approaches (in particular the Bayesian framework). Demidenko (2004, Chapter 5) also describes the use of lmer and lme for meta-analysis in a mixed model framework. Traill et al. (2007) use lmer for a meta-analysis of estimates of minimum viable population size over 30 years, based on 141 sources and 212 species. They fit a series of generalized linear mixed-effects models (GLMM) to $\log_e$MVP using the lmer function, with the random-effects error structure of GLMM used to correct for nonindependence of species due to potential shared evolutionary life history traits.

Nakagawa et al. (2007) also employed lmer in a meta-analysis to assess the function of house sparrow bib size, based on six correlates (fighting ability, parental ability, age, body condition, cuckoldry, and reproductive success). The authors describe the multilevel approach as "a flexible meta-analysis method" due to its ability to accommodate nonindependence in the data, induced by multiple studies from the same research groups, or the same populations or species. The authors concluded that bib size in house sparrows signals dominance and to a lesser extent age, and possibly reflects body condition.

Finally, Milsom et al. (2000) used the Genstat package to fit a generalized linear mixed model of the relationship between presence of ground-nesting birds and assorted grazing marsh habitat variables, based on about 430 marsh sites. Within-site variation was accounted for by the assumption of a binomial distribution for the response (presence/absence of birds); between-site variation was described by specified covariates and by a random term to allow for differences in the proportion of marshes occupied on different landholdings. Although this was not explicitly cast as a meta-analysis, the model structure is the same.

ACKNOWLEDGEMENTS

Ransom Myers was an early collaborator on this chapter. His groundbreaking scientific research based on meta-analysis, as well as his passion for high–quality meta-analysis methodology, underpin this text. He is warmly remembered and greatly missed.

APPENDIX 10.1: HANDEDNESS OF FIDDLER CRABS

The handedness of animals is of interest to a wide range of ecologists. In fiddler crabs, handedness is easily identified by observing whether the dominant large claw is on the left or right side of the body. Table A10.1, kindly compiled and provided by Michael Rosenberg, summarizes the number of fiddler crabs expressing right- and left-handedness. reported by the authors listed below. Note that the number and identity of studies conducted varies between crab species.

The reader is referred to the papers by Rosenberg (2000, 2001) and the website www.fiddler crab.info for full references to the following studies: Rosenberg (2000, 2001), Yamaguchi (1994), Jones and George (1982), Barnwell (1982), Frith and Frith (1977), Ho et al. (1993), Green and Schochet (1972), Takeda and Yamaguchi (1973), Williams and Heng (1981), Frith and Brunenmeister (1983), Ahmed (1976), Spivak et al. (1991), Gibbs (1974), Masunari and Dissenha (2005), Yerkes (1901), Morgan (1923), Yamaguchi and Henmi (2001), Shih et al. (1999), and Masunari et al. (2005).

TABLE A10.1. Handedness of fiddler crabs.

		Counts of handedness reported in primary studies					
Subgenus	Species	Right, Left	Right, Left	Right, Left	Right, Left	Right, Left	Right, Left
Australuca	*bellator*	4,2	15,19				
Australuca	*elegans*	62,46	0,2				
Australuca	*hirsutimanus*	320,273	1,1				
Australuca	*longidigitum*	8,12	1,5				
Australuca	*polita*	69,56	3,4				
Australuca	*seismella*	198,190	2,4				
Australuca	*signata*	180,153	2,1	53,39			
Gelasimus	*borealis*	118,6	47,3				
Gelasimus	*dampieri*	59,4	72,4	2,0			
Gelasimus	*hesperiae*	51,5	8,1				
Gelasimus	*neocultrimana*	812,24	4,0				
Gelasimus	*tetragonon*	272,9	90,3	29,2	12,0	7,0	99,6
Gelasimus	*vocans*	1165,23	32,2	167,12	5,0	554,13	255,8
Gelasimus	*vomeris*	34,0	413,4	8,0	618,26		
Leptuca	*annulipes*	179,179	4,3	37,24	188,161		
Leptuca	*argillicola*	0,1					
Leptuca	*batuenta*	10,11					
Leptuca	*beebei*	22,28					
Leptuca	*bengali*	1,0					
Leptuca	*coloradensis*	3,4					
Leptuca	*crenulata*	2,3					
Leptuca	*cumulanta*	170,183	5,2	14,19			
Leptuca	*deichmanni*	27,25					
Leptuca	*dorotheae*	4,5					
Leptuca	*festae*	18,32					
Leptuca	*inaequalis*	19,14					
Leptuca	*lactea*	2,2	3583,3559				
Leptuca	*latimanus*	2,2					
Leptuca	*leptodactyla*	1,4	13,21				
Leptuca	*limicola*	0,1					
Leptuca	*mjoebergi*	62,52	5,1				
Leptuca	*oerstedi*	5,2					
Leptuca	*panamensis*	4,6					
Leptuca	*perplexa*	56,53	2,0	483,458			
Leptuca	*saltitanta*	23,27					
Leptuca	*speciosa*	1,5	40,57				
Leptuca	*spinicarpa*	2,0					
Leptuca	*stenodactylus*	22,28					
Leptuca	*tallanica*	3,3					
Leptuca	*tenuipedis*	4,7					
Leptuca	*terpsichores*	26,24					
Leptuca	*tomentosa*	1,1					
Leptuca	*triangularis*	17,11	1,1				
Leptuca	*uruguayensis*	1,3	409,411				
Minuca	*brevifrons*	3,3					

(*continued*)

TABLE A10.1. *Continued*

		Counts of handedness reported in primary studies					
Subgenus	Species	Right, Left	Right, Left	Right, Left	Right, Left	Right, Left	Right, Left
Minuca	*burgersi*	193,164	2,3	266,251			
Minuca	*ecuadoriensis*	9,4					
Minuca	*galapagensis*	6,4					
Minuca	*herradurensis*	5,1					
Minuca	*marguerita*	1,0					
Minuca	*minax*	2,4	28,20				
Minuca	*mordax*	169,137	5,3	101,124			
Minuca	*panacea*	1,4					
Minuca	*pugilator*	24,26	36,30	932,869			
Minuca	*pugnax*	25,25	27,33	552,578			
Minuca	*pugnax/pugilator*	241,283					
Minuca	*rapax*	462,479	0,4	290,292			
Minuca	*subcylindrica*	8,5					
Minuca	*thayeri*	4,2					
Minuca	*umbratila*	1,0					
Minuca	*virens*	3,2					
Minuca	*vocator*	4,1	29,34				
Minuca	*zacae*	0,1					
Paraleptuca	*chlorophthalmus*	3,0					
Paraleptuca	*crassipes*	3,1					
Paraleptuca	*inversa*	2,2	64,79				
Paraleptuca	*sindensis*	1,5					
Tubuca	*arcuata*	7,10	17,14	261,257			
Tubuca	*capricornis*	35,32	39,27	0,2			
Tubuca	*coarctata*	87,68	2,1				
Tubuca	*demani*	3,0					
Tubuca	*dussumieri*	47,29	2,2				
Tubuca	*flammula*	156,170	5,5				
Tubuca	*forcipata*	11,17	2,2				
Tubuca	*formosensis*	4,1	0,1	152,120			
Tubuca	*paradussumieri*	1,3					
Tubuca	*rhizophorae*	3,3					
Tubuca	*rosea*	0,2	18,25				
Tubuca	*typhoni*	1,0					
Tubuca	*urvillei*	52,33					
Tubuca	*urvillei*	1,1	29,30				
Uca	*heterpleura*	1,3					
Uca	*insignis*	1,4					
Uca	*intermedia*	1,0					
Uca	*major*	2,1					
Uca	*maracoani*	236,248	3,4	57,63			
Uca	*ornata*	1,0					
Uca	*princeps*	0,1					
Uca	*stylifera*	2,1					
Uca	*tangeri*	0,1					
Uca	*tangeri*	38,32					

APPENDIX 10.2: META-ANALYSIS USING R
WITH METAFOR PACKAGE (PREVIOUSLY MIMA)

As discussed in this chapter, the R software contains a number of packages for maximum likelihood estimation of a meta-analysis model. We describe two such packages here and refer the reader to Chapter 12 for further details.

The MiMa package requires a simple one-line command. For the Lepidoptera mating example, a random-effects model with covariates suborder (0=Heterocera, 1=Rhopalocera) and % polyandry (excluding the three cases with missing % polyandry values for illustration) requires only the commands

```
mods<-cbind(suborder,polyandry)
mima(yi,vi,mods,method="REML")
```

where the data set is constructed as a matrix with 22 rows (corresponding to the different studies) and 4 columns: yi (the study-specific effect estimates), vi (the corresponding study-specific variances of the effects), suborder (coded 0 and 1) and polyandry (coded as a percentage). The methods of analysis available in MiMa include REML (restricted maximum likelihood), ML (regular maximum likelihood), HE (Hedges' method of moments), DL (Dersimonian and Laird method of moments), and EB (empirical Bayes). Other command arguments are available for controlling the algorithm and output. A command is also available for obtaining a fixed-effects analysis if desired. The output from this analysis comprises an estimate of the between-study heterogeneity after accounting for covariates and an associated test for whether this is significantly different from zero, parameter estimates and the associated variance-covariance matrix, an overall test of the influence of all the covariates (moderators) in the model, and individual tests for these moderators.

Meta-analysis using R with lme package

As discussed in this chapter, more complex meta-analysis models can often be cast as a mixed model and estimated using ML or REML. The lme function in R provides a very flexible framework for describing fixed and random effects, ascribing weights (typically study-specific variances in a meta-analysis of mean responses), including covariates and factors, and assessing the relative contribution of within- and between-study heterogeneity.

We can cast the Lepidoptera database in a mixed-model context, and use the software package R to obtain maximum likelihood estimates of the model parameters. We focus on the use of the function "lme." In order to use this function, it is necessary to load the packages MASS, STATS, and NLME into R; in the Windows version of R this can be achieved via the pulldown menu "Packages."

Assume that we have constructed an Excel data set comprising the 25 rows corresponding to the different studies and four columns: No (study-specific indicator), Mating (monandrous or polyandrous), Hd (Hedges' d estimate) and Var (variance of the Hedges' d estimate).

We can import this data set into R in a number of ways. To read it into an R dataframe called "lepi," say, we can do the following:

- Save the Excel spreadsheet as a text file separated by commas—that is, save as a .csv file (say, "lepi.csv")—then read this into R using the "read.csv" command found in the Base package.

```
> lepi=read.csv(file="lepi.csv", sep=",",header=T)
```

This is the most common and robust approach to reading files into R from any other source.

- Copy the required rows and columns in the Excel document and then paste them into R using the command

> lepi=read.table(file="clipboard",sep="\t",header=T)

where "\t" means that the values are separated by tabs.

The file now needs to be "attached" by using the following command:

> attach(lepi)

We first fit a simple normal random-effects model using the command:

> Lepi.1 = lme(Hd~1, weights=~Prec, random=~1|No)
> Lepi.1

Here, the lme statement has three components that define the study-level model: in this case, a single mean response denoted by "~1" with no covariates; study-specific weights; and the random effects (a single random effect (denoted by "~1") across studies (denoted by No), which allows the study means to vary). This is sometimes called a simple random intercepts model in R.

These commands return the log-likelihood of the model (−10.69), the results of the fixed-effects component (in this case, the estimated overall mean, 0.339), and the results of the random-effects component (here this is the estimated standard deviation representing the between-study variation, 0.353).

Other commands can retrieve further information from the Lepi.1 object. These include:

> summary(Lepi.1) # slightly more information
> VarCorr(Lepi.1) # within and between study variances
> fixed.effects(Lepi.1);random.effects(Lepi.1)
> fitted(Lepi.1)

Note that other commands may also be useful; for example, see the lattice plots available using "xyplot."

The addition of a covariate, Mating, is achieved by modifying the lme function as follows:

> Lepi.2=lme(Hd~Mating, weights=~Prec, random=~1|No)
> Lepi.2

The output from this command is the same as above: a goodness of fit term described by the log-likelihood (now −9.56, showing that the additional variance components in this model result in a better fit), a random-effects term described by a between-study standard deviation (0.3286), and a fixed-effects term which now comprises an intercept (corresponding to the estimated effect for the monandrous group, 0.2018) and a coefficient (corresponding with the additional effect for polyandrous group, 0.2861).

We can also nest study effect estimates within family and then combine the family effects to obtain an overall effect using the following commands:

> Lepi.3 = lme(Hd~1, weights=~Prec, random=~1|Family/No)
> Lepi.3

where Hd denotes the Hedges' *d* estimates for each study, Prec denotes the study-specific inverse variances, family denotes the family, and No is an indicator for each study (1, 2, . . .).

The output from these commands will contain a log-restricted-likelihood (−10.69), an over-all mean (fixed intercept, 0.3391), and standard deviations for the two random effects (between family, 6.93E-6, and between study within family, 0.3532). Thus, the three sources of variation described by the VarCorr command will be between families, between studies within families, and within studies (residual). Note that in this case, the variation between families is so small compared with the variation between studies that it has almost no influence on the param-eter estimates of interest; this additional hierarchy therefore appears to be unnecessary in the meta-analysis.

Meta-analysis using R with lmer package

The lmer package in R is used quite widely for meta-analysis in ecology as well as a wide range of other disciplines. Gelman and Hill (2007) provide a detailed account of the use of lmer for multilevel models, including the advantages and limitations of the method and comparisons with other approaches (the Bayesian framework in particular). They introduce the package using an example of estimating the distribution of radon in each of the approximately 3000 counties in the United States, based on a random sample of more than 80,000 houses through-out the country. Although this is not strictly a meta-analysis in the context adopted in the body of this book, the same model structure applies; that is, the data are structured hierarchically as houses within cities and cities within counties, and there is acknowledged variation in radon levels within a house, between houses, and between counties. A simple model with no predic-tors is given by

M0 <- lmer(y ~ 1 + (1 | county).

Here, "1" represents a constant term and "1 | county" allows this term to vary by county. This gives an estimate of the overall mean and corresponding standard error, an estimate of the between-county variation, and estimates for each county with corresponding standard errors. The addition of a covariate x (floor of measurement in the radon example) is achieved by the command

M1 <- lmer(y ~ x + (1 | county)

The results now include estimates of the intercept and the effect of x, as well as components of variance (represented as standard deviations) for the 'error terms" county (between counties) and residual (between houses within counties).

The authors also give detailed commentary on commands for displaying the results of the analysis, including estimates of the intercept and regression coefficients for each county, simi-lar estimates averaged over all counties, county-level errors showing how much the overall intercept is shifted up or down in particular counties, confidence intervals for the overall slope, and so on. A range of models for combining estimates with both individual-level and group-level predictors is also considered.

Note that the lmer package runs much faster than the lme package, but cannot fit correlated level 1 errors.

Meta-analysis using SAS PROC MIXED

We illustrate the use of SAS PROC MIXED by again conducting a meta-analysis of the Lepi-doptera data described in Appendix 8.1, under a simple random-effects model with no covari-ates. We follow the first example of Sheu and Suzuki (2001) to develop the program shown below.

The Lepidoptera data are read in from a text file "Lepidoptera," comprising two columns, *study* (a study identification number, from 1 to 25) and *diff* (the observed study-specific effect estimates). Below we provide PROC MIXED code for meta-analysis of the Lepidoptera mating data under a simple random-effects model.

```
TITLE 'Lepidoptera mating analysis';
DATA Lepidoptera;
      INFILE 'Lepidoptera,asc';
      INPUT study diff;
PROC MIXED DATA = Lepidoptera;
      CLASS study;
      MODEL diff = RESIDUAL SOLUTION CI;
      RANDOM study / SOLUTION;
      REPEATED / GROUP = study;
PARM      (0.0)
      (0.0441) (0.0213) etc (0.0667) / EQCONS = 2 to 25;
      RUN;
```

In the case study of fiddler crab handedness discussed in Chapter 10, the probability distribution for the effect estimates is binomial, and the link function that is used to transform the effect estimates to normality is the logit. For this data, Study and Species are explanatory variables. Since these data are binomial, the events/trials syntax is used to specify the response in the MODEL statement. Profile likelihood confidence intervals for the regression parameters are computed using the LRCI option. PROC GENMOD code for the crab handedness case study under a simple random-effects model is provided below.

```
TITLE "Handedness of Fiddler Crabs":
PROC IMPORT OUT= WORK.HANDED
      DATAFILE= "Handed.xls"
      DBMS=EXCEL REPLACE;
RUN;
PROC GENMOD DATA = HANDED;
      CLASS STUDY SPECIES;
      MODEL RIGHT/TOTAL = STUDY SPECIES / DIST = BIN LINK = LOGIT
        LRCI;
RUN;
```

In the "Criteria For Assessing Goodness Of Fit," the output the value of the deviance divided by its degrees of freedom is less than 1. A *P*-value is not computed for the deviance; however, a deviance that is approximately equal to its degrees of freedom is a possible indication of a good model fit.

Bayesian Meta-analysis

Christopher H. Schmid and Kerrie Mengersen

IN THIS CHAPTER, WE introduce and describe a Bayesian approach to meta-analysis. We discuss the ways in which a Bayesian approach differs from the method of moments and maximum likelihood methods described in Chapters 9 and 10, and summarize the steps required for a Bayesian analysis. We will find that Bayesian methods provide the basis for a rich variety of very flexible models, explicit statements about uncertainty of model parameters, inclusion of other information relevant to an analysis, and direct probabilistic statements about parameters of interest. In a meta-analysis context, this allows for more straightforward accommodation of study-specific differences and similarities, nonnormality and other distributional features of the data, missing data, small studies, and so forth.

As in the previous chapters, we do not dwell on technical details, including mathematical derivations and computation, but provide references to further information for the interested reader. The ecological literature has recently seen a rapid increase in the number of applications of Bayesian methods (Ellison 2004, Kulmatiski et al. 2008).

BAYESIAN INFERENCE

As with the methods described in Chapters 9 and 10, the aim of a Bayesian analysis is to estimate model parameters given the observed data. In a meta-analysis context, the parameters may include study-specific effects, overall mean effects, a between-study variance, or effects from meta-regression. The observed data typically include reported study-specific effect estimates and associated variance estimates, as well as any variables used in meta-regression.

The major distinction between inferences with Bayesian and non-Bayesian models is the interpretation of model parameters. Bayesian inference treats parameters as random variables about which there is uncertainty. We describe our understanding about likely values of these parameters through probability distributions that quantify this uncertainty. For example, a normal distribution with mean 10 and standard deviation 5 indicates that we believe the parameter most likely equals 10 and has a 95% chance of lying between 0 and 20 (two standard deviations around 10). This direct expression of the probability of a parameter enables us to attach probabilities to any set of potential values of the parameters or combinations of parameters. For example, the parameters of a model might be the effects of CO_2 changes on different plant species, and we might want to compare the relative effects on different species.

Knowledge about the likely values of parameters changes as we collect more information. A *prior probability distribution* describes our knowledge about a parameter before collecting and analyzing data about it. This knowledge might come from external sources, such as expert opinion, which have no other way to be incorporated into the analysis. Using Bayes' rule, we can mathematically combine this prior information with that coming from the data via the *likelihood* (Chapter 10) in order to update our knowledge about the parameters, expressed quantitatively through the *posterior probability distribution*. Statements about the probability of scientific hypotheses and parameter values follow directly from the posterior distribution.

This process of combining the likelihood with the prior information to produce a posterior distribution is an evolutionary one. Today's posterior becomes tomorrow's prior when new data become available. Cumulative meta-analysis describing the sequential evolution of the summary estimate is a Bayesian algorithm (Lau et al. 1995; Chapter 15). The current information about the treatment effect is the prior distribution that combines with the next study to update the posterior distribution (Brown and Schmid 1994, Schmid and Brown 2000).

A Bayesian modeling approach requires (1) choice of a model as discussed in Chapter 8, which leads to a likelihood; (2) choice of a prior distribution; (3) computation of the posterior distribution; and (4) checking how well the model fits the data and how sensitive inferences are to choice of the prior and other model features.

In light of the above steps, it is clear that the Bayesian approach builds on much of the same framework as the frequentist approaches described in Chapters 9 and 10. The construction of the meta-analysis model and the corresponding considerations about sources of variation (especially between-study heterogeneity) remain the same for all models. Moreover, good statistical practice, including model checking, should be a feature of all meta-analyses regardless of the statistical method.

To illustrate Bayes' rule in the context of the random-effects model, recall that

$$T_i \sim \mathrm{N}(\theta_i, S_i^2) \tag{11.1}$$

with

$$\theta_i \sim \mathrm{N}(\mu, \tau^2), \tag{11.2}$$

where T_i are the observed study effects (e.g., response ratios) assumed to follow normal distributions with true means θ_i and within-study variances S_i^2. The random effects θ_i are assumed to have mean μ, and between-study variance τ^2. This model can also be written as

$$T_i \sim \mathrm{N}(\mu, S_i^2 + \tau^2), \tag{11.3}$$

which expresses the two variance components (between-study and within-study) associated with the outcomes. Treatment effects depending on covariates in a meta-regression model are easily handled by generalizing the mean to a regression function so that $\mu = \beta_0 + \beta_1 X_{1i} + \beta_2 X_{2i} + \ldots + \beta_p X_{pi}$.

Equation 11.3 supplies the likelihood for μ and τ^2 conditional on the observed T_i and S_i^2, and Equation 11.1 gives the likelihood for θ_i. Bayes' rule states that the posterior distribution is proportional to the product of the likelihood and the prior distribution. In the random-effects model, the posterior probability of the model parameters μ, τ^2 and θ_i, given the observed summary statistics T_i and S_i^2, is

$$\mathrm{P}(\mu, \tau^2 \mid T_i, S_i^2) \propto \prod_{i=1}^{N} \mathrm{P}(T_i \mid \mu, \tau^2, \theta_i, S_i^2) \, \mathrm{P}(\mu, \tau^2, \theta_i)$$

where the first term in the product on the right-hand side, $\prod_{i=1}^{N} \mathrm{P}(T_i \mid \mu, \tau^2, \theta_i, S_i^2)$, is the likelihood, and the second term, $\mathrm{P}(\mu, \tau^2, \theta_i)$, is the prior distribution of the parameters.

Any term in the product forming the likelihood can be written in terms of the probability distribution expressed by Equation 11.3 because knowledge of μ and τ^2 implies knowledge of θ_i. Furthermore, by Equation 11.2 we know the conditional distribution of θ_i given μ and τ^2, so we can re-express the prior as $P(\theta_i | \mu, \tau^2) P(\mu, \tau^2)$ by the laws of conditional probability. In other words, Equation 11.2 gives us the prior for θ_i given μ and τ^2 so we need only specify a prior for μ and τ^2.

CHOOSING THE PRIOR DISTRIBUTION

In many ecology investigations, substantial knowledge about the model parameters is available. Prior distributions should reflect the beliefs of the analyst about the likely values of model parameters. Highest probabilities should be assigned to those values believed most likely, and lowest to those believed least likely. Information about these prior distributions comes from many sources, including previous studies, expert opinion, reviews of the relevant literature, and meta-analyses. Some of these sources are more objective than others, but the final summary distribution is the analyst's choice. Because probability is measured as a degree of belief by the Bayesian (Savage 1954), different analysts may choose different prior distributions and may therefore get different posterior distributions. Methodology and examples in which the prior distributions are based upon formally elicited subjective beliefs are discussed by Chaloner et al. (1993), Low Choy et al. (2009), Kuhnert et al. (2010), and Moala and O'Hagan (2010).

Most frequentist statisticians think of a model as a structural representation of the probabilistic generation of data. Given a set of data and a model for the way these data are generated, there can be only one possible statistical inference. The inference derives completely from the likelihood. Inferences differ only if the likelihoods differ. A Bayesian model, however, consists of two parts: a likelihood and a prior. Therefore, two Bayesians using the same likelihood and different priors may make different inferences because they are using different models, just as they would if they used the same priors and different likelihoods. The frequentist approach suffers because it cannot incorporate external information that might lead two scientists to draw different conclusions with the same data and the same likelihood. Frequentist methods cannot quantify in a single model the scientific method of learning.

As an example, consider how an environmental activist and a coal manufacturer might look at the same data and reach quite different conclusions about the effect of coal production on CO_2 levels and their effects on forests, even if they agreed on the same probabilistic representation of the data as captured in the likelihood. The activist would likely take a dim view of the effect, while the manufacturer would consider mitigating factors. For example, the data may show that CO_2 levels have risen by 5% on average in regions with substantial coal-based energy production compared to areas without. For the environmental activist, this is only a further evidence to bolster a strong prior belief that coal pollutes. The coal mining supporter might point to previous studies that did not find such an association, and based on this, use a prior of no effect that downweights the likelihood. Spiegelhalter et al. (1994) characterized these two approaches as those of the skeptic and the enthusiast. By quantifying the respective "biases" through prior distributions, we can better understand how their posterior inferences differ. While the frequentist might argue that such differences are nonscientific because they are not based on an objective analysis of the data alone, interpretations of all data analyses are to some extent colored by prejudices and so have elements of subjectivity. The interpretations ultimately depend on the experiences that formulate the background to the problem. The Bayesian analysis makes this transparent by quantifying it.

The still skeptical reader might consider the manner in which regression models are developed. A purist not allowing for any "subjective" choices on the part of the analyst should not allow them to choose variables other than by an a priori algorithm. However, we know that a good modeler makes choices that reflect their scientific knowledge, often choosing to include, exclude, or transform variables so as to make the model fit the data better and agree with previous discoveries. Such choices are also often implicitly subjective (Berger and Berry 1988).

In principle, the Bayesian might employ any probability distribution that matches the available scientific information; however, in practice choices have often reflected a desire to simplify computation of the posterior distribution. A favorite is the *conjugate prior* distribution that takes the same mathematical form as the posterior. Examples are the combination of a normal prior with a normal likelihood for continuous data to produce a normal posterior, combination of a beta prior with a binomial likelihood for binary data to form a beta posterior, and combination of a gamma prior with a Poisson likelihood for count data to produce a gamma posterior. While conjugate priors facilitate computation in simple problems, they do not work well in hierarchical structures and for parameters with boundary constraints, such as the requirement of variances to be positive.

Consider the random-effects model given by Equations 11.1 and 11.2. The second equation specifies the prior distribution for the study effect parameters, θ_i in terms of other parameters, μ and τ^2. This prior distribution is a reasonable choice if we believe the study effects are randomly drawn from a distribution of possible study effects that can be well-approximated by a normal distribution. This distribution is conjugate to the normal likelihood implied by Equation 11.1. Little information, however, may be available to construct priors for the population parameters μ and τ^2.

To accommodate this lack of information, Bayesians use *noninformative priors* that spread the probability diffusely so that potential values are equally or almost equally likely. Uniform distributions are an obvious choice because they can be limited to ranges of values that satisfy boundary constraints. For example, a uniform distribution on a variance or standard deviation would allow values greater than zero and less than some value that excludes unreasonably large variances (Gelman 2006). Because a normal distribution with an infinite variance is mathematically equivalent to a uniform distribution, it is often the choice of prior for parameters without boundary constraints, such as means and regression coefficients for which it is also a conjugate prior. For computational purposes, a large value such as one million is often chosen for the prior variance, although any large number that incorporates all potential values of the parameters is acceptable. However, a normal distribution does not make sense for a variance, because a variance must be positive and usually has a skewed distribution. In general, much more care must be given to the form of a noninformative prior for variance parameters because posterior inferences can be sensitive to distributions that allow very large values (Gelman 2006).

When the joint prior distribution of all the parameters is noninformative, the posterior distribution will be the same as the likelihood, and thus the posterior mean and variance will be numerically equal to the maximum likelihood estimate (MLE) and its variance. This provides a Bayesian interpretation of the MLE as the posterior mean under a noninformative prior distribution and permits likelihood-based calculations to be used for probabilistic interpretations of model parameters. Use of fully noninformative priors also provides a benchmark analysis with which to compare results from models with informative prior distributions. Noninformative priors are also therefore sometimes called *reference priors*.

Consider again the random-effects model with known variances (Chapter 10). The MLE given in Equation 10.6 is the posterior mean from a Bayesian model with a noninformative prior for μ. When the between-study variance is unknown, a fully Bayesian analysis requires a prior distribution for τ^2. An alternative is to perform an *empirical Bayes analysis* in which an estimate $\hat{\tau}^2$, based on the data is substituted for the unknown τ^2. Treating τ^2 as known to be equal to this empirical estimate, the remainder of the analysis is carried out as for the Bayesian analysis with known variance. If the estimate of τ^2 given in Box 9.2 (Chapter 9) is used, the posterior mean will also be the same as the method of moments estimate.

The empirical Bayes method has the drawback of ignoring the uncertainty associated with the estimate $\hat{\tau}^2$. It therefore calculates a conditional posterior distribution at $\hat{\tau}^2$. The fully Bayesian analysis correctly estimates the posterior distribution as a weighted average of conditional posteriors with the weights given by the posterior distribution of the between-study variance. This may not only change parameter means, but will increase posterior variances. The empirical Bayes estimate always underestimates the posterior variances of model parameters leading to *credible intervals* (Bayesian confidence intervals) that are too short. The underestimation is especially severe when the variance is poorly estimated, as is often the case in meta-analyses based on few studies.

When the within-study variances are unknown, one can take a fully Bayesian approach and put prior distributions on the unknown variances. Nevertheless, analysts usually take an empirical Bayes approach and assume that the sample variances, S_i^2, are adequate estimates of the true within-study variances, σ_i^2, and ignore the additional uncertainty introduced by their estimation. This approach is justified if the individual studies are large enough, but may lead to difficulties with small studies. The size of the study necessary for a good variance estimate will depend on the type of data. Continuous measurements have much more information than discrete ones and the variance of a proportion is particularly sensitive to its value. Using the general rule of thumb that the product of the sample size and the proportion, p, should be at least 5 implies that the sample size should be at least $5/p$. For continuous data, sample sizes of 10 may be adequate.

In a fully Bayesian analysis, all sources of uncertainty are acknowledged. Thus, if the σ_i^2 are unknown, the model is extended to include prior distributions for these parameters. These priors can be more or less informative based on the amount of prior information available, either expert driven or data driven. Common choices include conjugate inverse gamma priors and uniform distributions. DuMouchel and Harris (1983) suggested an inverse chi-square such that $\sigma_i^2 \sim \nu_i S_i^2 / X_{\nu_i}$ where S_i^2 is the sample variance for the ith study, ν_i is the degrees of freedom (typically one less than the sample size), and X_{ν_i} is a chi-square random variable with n_i degrees of freedom.

The prior distributions may be chosen to apply to either the standard deviation or the variance. A uniform distribution for the standard deviation implies a nonuniform distribution for the variance and vice versa. Current work suggests that a uniform distribution on the standard deviation scale is preferred because it reduces the tendency for the uniform distribution to favor unreasonably large values. The prior on the variance also leads to improper posterior distributions with three or fewer studies in the meta-analysis (Gelman 2006).

Figure 11.1 illustrates two different prior distributions, one an inverse gamma and the other an inverse chi-square. The range of plausible values of the variance under the two priors happens to be similar, indicating that different distributions may give similar prior weights.

BAYESIAN UPDATING

We mentioned earlier that the evolution of the posterior as data are processed corresponds with the process of cumulative meta-analysis. It should come as no surprise then that the posterior mean is a weighted average of the prior mean and the maximum likelihood estimate. The weights correspond to the *precisions* associated with the prior distribution and the likelihood. In the case of normal distributions, this precision is simply the inverse of the variances. Therefore, the posterior mean is weighted toward the component with the smaller variance. If the

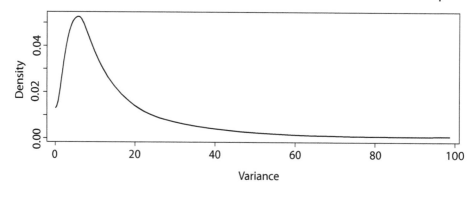

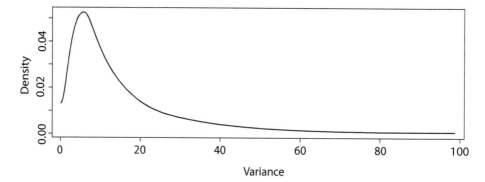

Figure 11.1. Prior distributions for variance parameters. (*Top*) inverse gamma (1,8) distribution; (*bottom*) inverse chi-square with 10 degrees of freedom and $S^2 = 100$.

amount of information in the data is large, then the posterior mean will tend to be like the MLE; if the data provide little information, then the posterior distribution will not move far from the prior distribution.

Figure 11.2 shows the likelihood, prior, and corresponding posterior for two scenarios. In the first scenario, the prior and the data provide equal amounts of information (their probability densities have the same spread), so the posterior mean is an average of the two. The posterior variance is smaller than the prior variance and the likelihood variance, reflected by the tighter distribution for the posterior. This is a consequence of the result that the posterior precision is a sum of the prior precision and the likelihood precision. In the second scenario, the prior provides little information, so the likelihood and posterior distributions are similar.

To quantify these ideas, reconsider the random-effects model with known variances. The mean of the posterior distribution of the random effect μ is a weighted average

$$\hat{\mu} = (\hat{\sigma}_{MLE}^{-2}\hat{\mu}_{MLE} + \sigma_0^{-2}\mu_0)/(\sigma_0^{-2} + \hat{\sigma}_{MLE}^{-2})$$

of its prior mean μ_0 and the maximum likelihood estimate $\hat{\mu}_{MLE}$, with weights given by the inverse of the prior variance and the inverse of the variance of the maximum likelihood estimate. The maximum likelihood estimate and variance were given by Equations 10.5 and 10.6 in Chapter 10.

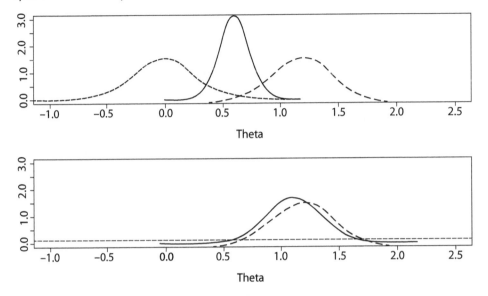

Figure 11.2. Likelihood (*long-dashed line*), prior (*short-dashed line*), and corresponding posterior (*solid line*) when prior and likelihood contributed equal information (*top panel*); the same three quantities with flat prior so that posterior and likelihood are similar (*bottom panel*).

An important special case arises under a noninformative prior on μ when σ_0^{-2} is very small. In that case, the posterior mean is the same as the maximum likelihood estimate and the posterior variance is the variance of the maximum likelihood estimate. Thus, Bayesian estimates agree with maximum likelihood estimates in the absence of prior information.

The posterior mean and variance of the study effects θ_i are also weighted averages of their priors and likelihoods. The likelihood is derived from Equation 11.1 since the data for θ_i is T_i. The prior distribution for θ_i is derived from the posterior distribution of its parameters μ and τ^2 obtained from the information in all studies except the ith study. It can also be shown that the posteriors of the θ_i are equivalent to their maximum likelihood estimate and variance if the priors on μ and τ^2 are noninformative. If the ith study provides only a small amount of information relative to the total of the remaining studies, then θ_i will resemble μ, but if the information in the ith study is large (perhaps because it is a much larger study than all the rest), then θ_i will be quite different from μ (Schmid and Brown 2000).

Although it might seem strange that a better estimate of the effect of a single study may be gained by using data from other studies, this phenomenon, called *shrinkage*, is well known in statistical theory (Efron and Morris 1973, James and Stein 1961). Intuitively, if we can assume that certain studies are estimating the same thing or are exchangeable in statistical terminology, then we can gain more information about each one by using information or borrowing strength from the others (Draper et al. 1993). This phenomenon will be very familiar to ecologists who can use their biological experience dealing with related species or environments to make informed inferences about a new case.

Posterior distributions for the variance terms, τ^2 and σ_i^2, are also generated under the Bayesian model, although with slightly more calculated computations. Again, the posterior means are weighted averages of prior and likelihood, while the posterior precisions are sums of prior and likelihood precision. We shall describe numerical algorithms for carrying out these computations in a bit, but first we consider interpretations of these posterior distributions.

INTERPRETATION OF BAYESIAN ESTIMATES

The two major differences between Bayesian and likelihood inference are the use of prior information and the probabilistic interpretation given to model parameters. For interpretation, the Bayesian takes a "subjective" approach. The modelers' parameters are random variables about which they are uncertain (O'Hagan and Forster 2004). Their beliefs about the likely values of these parameters may differ from those of a colleague, even if both of them examine the same data. Expressed in Bayesian terms, they may have different posterior distributions because their underlying prior distributions differ (assuming of course, that their likelihoods are the same). But the influence of the prior wanes as the amount of data increases, so that with substantial amounts of data, posterior inferences should be similar unless the priors are very strong. However, if study-specific sample sizes are small or the variances of the estimates are large, then priors must be developed with care.

The non-Bayesian modeler adopts a *"frequentist"* interpretation of probability, as the frequency with which an event occurs in the long run. In the frequentist system, parameters are not uncertain random quantities, but fixed and unknown constants. The data are random and estimates of the values of parameters are derived from functions of the data, called sample statistics, like the sample mean. Because the data are random and parameters are fixed, the frequentist calculates the probability of the random data, given hypotheses about the fixed parameters. The Bayesian, on the other hand, calculates the probability of hypotheses about the unknown parameters, given the fixed values of the data observed. Examples are Bayesian statements, such as "there is a 10% chance that the effect size comparing the production of offspring by female Lepidoptera mating with virgin males instead of sexually experienced males is greater than 0.5," or "there is a 70% chance that the effect size is between 0.1 and 0.4."

COMPUTATION OF THE BAYESIAN MODEL

Up to this point, we have discussed the use of Bayes' rule to compute a posterior distribution from a combination of a prior distribution and a likelihood. We mentioned that the computation is simplified when conjugate priors are chosen, but that the Bayesian paradigm is flexible in handling a wide range of problems. In practice, Bayesian methods are used to solve a variety of complex problems involving hierarchical data structures, correlation, and stochastic processes that are not easily handled by frequentist methods. Congdon (2003, 2005, 2007, 2010) provides a voluminous number of applications in all different types of scientific disciplines. What has enabled this explosion of interest in the use of Bayesian methods is a numerical simulation method called *Markov chain Monte Carlo* (*MCMC*) that has vastly simplified Bayesian computations.

Although the Bayesian approach to statistical analysis was developed long before the frequentist approach (Stigler 1990), it was computationally infeasible until the advent of computers and associated numerical algorithms for solving the equations. Bayes' rule states that the posterior probability $P(\Theta|Y)$ of parameters Θ conditioned on data Y can be written as:

$$P(\Theta|Y) = P(Y|\Theta)P(\Theta)/P(Y).$$

The term $P(Y|\Theta)$ is the likelihood, $P(\Theta)$ is the prior, and $P(Y)$ is a normalizing constant so that all the probabilities sum to one. This normalizing constant is the *marginal probability* of the data and can be calculated by weighting the conditional probability of the data given Θ over the prior distribution of Θ. Unfortunately, for continuous distributions this weighting involves high-dimensional multiple integration if there are many parameters. Except in some simple cases, such as multivariate normal distributions, this integration requires numerical analysis that is prohibitively expensive computationally. Various approaches were put forward, but practical methods were restricted mainly to the use of conjugate distributions until the seminal paper by Gelfand and Smith (1990) introducing Markov chain Monte Carlo to the statistical community. MCMC allows these computations to be accomplished by simulation, and as an important byproduct, returns simulated values of the entire joint posterior distribution, thus permitting numerical calculation of any quantities of interest (Besag et al. 1995, Robert and Casella 2010). Most often, we are interested in the marginal distributions of single parameters.

In MCMC, the posterior distribution for all parameters is decomposed into a series of conditional posterior distributions (conditioning on the other parameters and the data) for each parameter or set of parameters. The MCMC algorithm then simulates values of each parameter from its respective conditional distribution, updating the distribution with the other parameter values as it goes.

For example, consider a joint posterior distribution for the overall effect parameters μ and τ^2. (It is acknowledged that there are other parameters, but for exposition we focus only on these two.) Instead of estimating or simulating from the joint distribution $P(\mu, \tau^2 \mid T_1, \ldots, T_k)$ directly, the MCMC algorithm first starts with an arbitrary value of μ, then simulates a new τ^2 from its conditional distribution $P(\tau^2 \mid \mu, X)$ using the current value of μ; then it simulates a new μ from the conditional distribution $P(\mu \mid \tau^2, X)$ using the current τ^2 and so on. After many iterations of this algorithm, the simulation will be drawing from the correct posterior distribution and subsequent draws reflect a random empirical sample from the posterior (Gilks et al. 1996). Each iteration looping through all the parameters produces a single draw from the joint posterior distribution.

From this sample, we can construct any aspect of the joint posterior distribution. For example, the posterior mean of μ is obtained as the mean of the simulated values of μ, a posterior variance as the variance of the simulated μ values, and a posterior percentile as the corresponding empirical percentile of the simulated values. Similarly, a 95% confidence (or credible, in Bayesian terminology) interval is constructed by taking the middle 95% of simulated values of μ, and the interval is interpreted naturally and directly as "there is a 95% probability that μ lies in this interval." The correlation of the sampled μ and τ^2 values provides an estimate of their posterior correlation. If the simulation involves the study effects θ_i as well, we could estimate the probability that ith study effect is greater than the jth as the probability that on any iteration, $\theta_i > \theta_j$.

Because the simulation returns an empirical sample of the posterior distribution and not just estimates and standard errors, it is not restricted to assumptions that these estimates follow normal distributions, as are the likelihood and method of moments approaches. Although these assumptions derive from standard large-sample arguments, many meta-analyses have small numbers of studies, and asymptotic arguments may therefore not hold. For example, 95% posterior credible intervals need not be symmetric around the mean. The empirical posterior distribution also permits inferences to be made about quantities that clearly do not have normal distributions, such as the 95th percentile of the size of fish in a survey of a body of water.

As a simulation-based approach, MCMC requires starting values, determination of convergence of the algorithm to the correct posterior distribution, and sufficient numbers of draws from this posterior to obtain precise and accurate inferences. We give a brief overview here, but the reader interested in more details should consult the MCMC literature. Some good references include Gilks et al. (1996), Gelman et al. (2004), Gelman and Hill (2007), Carlin and Louis (2008), and Robert and Casella (2010).

Starting values for the algorithm

As long as they are realistic, starting values will not affect posterior inferences except in determining the length of time required until convergence. Gelman et al. (2004) recommend that sequences be started at reasonable guesses for the parameters, perhaps computed from approximate solutions. For example, method of moments estimates for the mean and between-study variance, and random draws from their implied distribution of random study effects, might suffice to start an MCMC algorithm for the random-effects model. Often, however, crude starting values of zero for means and one for variance may also be sufficient. Nonpositive starting values for variances will fail, however, as they do not define proper probability distributions.

Convergence of the simulation algorithm

Convergence is a much trickier problem that has spawned a wide literature. The basic idea is that a proper chain must simulate from the entire posterior distribution with appropriate weighting. As the number of parameters increases, the posterior distribution covers an exponentially increasing multi-dimensional space that will be very sparsely covered with simulations. Thus, it will be difficult to determine whether the algorithm has visited all the areas it is supposed to, and whether it has visited them the appropriate number of times. Practitioners have developed two main types of diagnostics to check for convergence.

The first type of convergence diagnostic monitors a single long chain of simulations, evaluating plots of simulated values as well as their autocorrelations. Plots of the simulated sequences should define the domain of potential values and should repeatedly traverse this domain. Frequently, simulated sequences are highly autocorrelated so that the simulation moves slowly through the parameter space. Thus, it may take many thousands of iterations to visit the entire distribution. Several statistics have been suggested to test for convergence of the mean of each parameter, with the main idea being that convergence is achieved when the mean of values in the beginning of the sequence (e.g., the first 10%) is similar to the mean of values late in the sequence (e.g., the last 50%).

The second type of convergence diagnostic is based on simultaneously running multiple chains, often a small number such as three or five, each from a different set of overdispersed starting values (Gelman and Rubin 1992). One set of starting values is usually chosen from the approximate solution, while others are picked so that the chains will traverse different paths toward the region of convergence so as to be sure that the final answers are not sensitive to the choice of starting values. Convergence for any parameter is assessed by the ratio of variance between chains, to that within chains. At the beginning of the simulations, the chains will be much further apart from each other than the distance between successive iterations within a chain, but as the simulation converges, the multiple chains will begin to mix and eventually the variance within-chain will equal that between-chains. When this ratio becomes close to one, convergence is achieved. Brooks and Gelman (1998) recommend that a ratio of 1.1 be used to indicate convergence. The ratio is usually calculated using the second half of each chain so that upon convergence, the second half of each chain may be saved and augmented by additional draws as needed.

In practice, analysts use the above types of diagnostic methods mainly to assess sequences of draws from single parameters. This assures convergence of the marginal distributions, but not necessarily of the joint distributions. Because of the difficulty of multivariate assessment, however, it is usually assumed that univariate convergence implies multivariate convergence. In fact, sometimes if the number of parameters is very large, users may even monitor only those subsets of parameters that are expected to converge the most slowly (e.g., variances).

Simulation accuracy

Upon convergence, the analyst discards the previous simulations as "*burn-in*" and retains the succeeding ones. The number retained will vary with the problem. Simulations, especially in meta-analysis, are inexpensive in both time and computing power, so it is usually advisable to save a large number, such as thousands or tens of thousands. Larger numbers are needed more for multivariate summaries like correlations, than for univariate summaries like means. As highly autocorrelated chains reduce the number of effective draws, some have recommended discarding autocorrelated draws by "thinning" the chains (e.g., saving every tenth draw if auto-correlations up to lag 10 are high); however, this reduces the amount of information available to estimate the joint posterior and, unless computer storage is a major issue, is not recommended. An easy statistic for evaluation of the precision of the simulation is the Monte Carlo standard error. For example, a Monte Carlo error of 0.001 implies that the mean is accurate to the third decimal place with respect to simulation error. The ratio of the Monte Carlo error to the posterior standard deviation assesses the relative error of the simulation. Increasing the number of simulations will reduce the Monte Carlo error. Decreasing the posterior standard deviation requires increasing the amount of data.

Software

A variety of software tools can run MCMC algorithms. The most popular is WinBUGS (Windows version of Bayesian inference Using Gibbs Sampling) (www.mrc-bsu.cam.ac.uk/bugs). An open source version called OpenBUGS is also available (Lunn et al. 2009). Generically, we shall call it BUGS. Given a user-specified model in the form of likelihood and priors, BUGS constructs the necessary posterior distributions, carries out MCMC sampling, provides statistical and graphical summaries of the posterior distributions, and reports some convergence diagnostics. BUGS can also export the simulations to a text file. Online manuals give several worked examples of Bayesian meta-analyses. Congdon (2003, 2005, 2007, 2010) also uses BUGS to illustrate a wide range of Bayesian models and analysis, including meta-analysis. BUGS can also be called from within the R and Stata statistical packages, eliminating the need to export and import saved simulation chains, and extending the possibilities for data exploration and presentation of results.

Other general software for Bayesian meta-analysis and specialized routines are also available for particular approaches; the reader is encouraged to review papers and websites relevant to the specific meta-analysis problem. Chapter 12 provides more information on software for meta-analysis.

CASE STUDY 1: LEPIDOPTERA MATING EXAMPLE

To illustrate the power of the Bayesian approach, we return to the Lepidoptera data described in Appendix 8.1 and analyzed previously in Chapters 9 and 10. Recall that the data consist of effect sizes computed by Hedges' *d* statistic comparing the fecundity of female Lepidoptera when mating with virgin or nonvirgin males. A positive effect size indicates that females are more fertile when mating with virgin males. As the analyses in Chapters 9 and 10 showed, the average effect size is approximately 0.3 showing that mating with virgin males produces more offspring than mating with sexually experienced males. Furthermore, we saw that the mean effect size for polyandrous species was greater than that for monandrous species. We might also be interested in the comparative effects by family and might wish to control for the amount of polyandry among the species.

This last analysis, however, is complicated by the lack of information on polyandry in three of the species. We do have some information, though, because the mating patterns have been defined as either polyandrous or monandrous, indicating that the percentages are either greater than or less than 40%, respectively. We can use this information to help construct prior distributions for the missing percentages and estimate them from the Bayesian analysis. This imputation allows us to use all the data, rather than only the data for those species for which polyandry information is available.

We begin by fitting the simple mean model (Equations 11.1 and 11.2, above) to estimate the mean effect size for comparing the fecundity of females mating with virgin males versus experienced males. We assume noninformative priors for study-specific and overall means (normal with mean 0 and variance 1 million) and for the various standard deviations (uniform over the range 0 to 10,000). Table 11.1 shows the observed mean and variance of Hedges' d, along with posterior estimates: the study-specific mean effects, standard deviations, and quantiles; the

TABLE 11.1. Observed and posterior effect estimates for Lepidoptera example.

Species	Observed Mean	Observed Variance	Posterior Mean	Posterior SD	Lower 2.5%	Median	Upper 97.5%	Posterior Prob >0*
Pectinophora gossypiella	0.81	0.02	0.81	0.02	0.76	0.81	0.85	1.0000
Sitotroga cerealella	0.51	0.02	0.51	0.02	0.46	0.51	0.55	1.0000
Agrotis segetum	0.38	0.02	0.38	0.02	0.33	0.38	0.42	1.0000
Busseola fusca	0.47	0.08	0.46	0.08	0.31	0.46	0.61	1.0000
Helicoverpa armigera	0.30	0.13	0.31	0.12	0.07	0.31	0.54	0.9947
Pseudaletia unipuncta	−0.15	0.05	−0.14	0.05	−0.23	−0.14	−0.05	0.0016
Spodoptera littoralis	0.10	0.11	0.12	0.10	−0.08	0.12	0.32	0.8773
Trichoplusia ni	0.26	0.08	0.26	0.07	0.12	0.27	0.41	0.9998
Chilo partellus	−0.04	0.05	−0.03	0.05	−0.13	−0.03	0.07	0.2654
Diatraea considerata	0.46	0.13	0.44	0.12	0.20	0.44	0.68	0.9997
Ostrinia nubilalis	0.36	0.02	0.36	0.02	0.32	0.36	0.39	1.0000
Plodia interpunctella	0.11	0.05	0.12	0.05	0.03	0.12	0.21	0.9939
Choristoneura fumiferana	0.14	0.05	0.14	0.05	0.05	0.14	0.23	0.9994
Choristoneura rosaceana	1.03	0.04	1.02	0.04	0.94	1.02	1.09	1.0000
Epiphyas postvittana	−0.12	0.09	−0.08	0.09	−0.25	−0.08	0.09	0.1650
Lobesia botrana	0.07	0.05	0.07	0.05	−0.02	0.07	0.16	0.9364
Zeiraphera canadensis	0.02	0.07	0.03	0.07	−0.10	0.03	0.16	0.6659
Acrolepia assectella	0.54	0.04	0.53	0.04	0.45	0.53	0.62	1.0000
Thymelicus lineola	0.11	0.08	0.12	0.08	−0.03	0.12	0.27	0.9391
Jalmenus evagoras	0.37	0.07	0.36	0.07	0.23	0.36	0.50	1.0000
Papilio glaucus	0.23	0.04	0.23	0.04	0.15	0.23	0.32	1.0000
Papilio machaon	0.25	0.16	0.26	0.14	−0.02	0.26	0.53	0.9683
Colias eurytheme	1.01	0.21	0.82	0.18	0.47	0.82	1.17	1.0000
Eurema hecabe	0.11	0.20	0.17	0.17	−0.17	0.17	0.50	0.8379
Pieris napi	1.17	0.26	0.85	0.21	0.45	0.85	1.28	1.0000
Overall Mean			0.32	0.07	0.18	0.32	0.46	1.0000
Overall Variance			0.12	0.04	0.06	0.11	0.22	1.0000

* Since 10000 simulations were done, the number of simulations for which the effect exceeded 0 is just the posterior probability multiplied by 10000.

posterior probability that each effect estimate is greater than 0 for each of the 25 species; and the overall mean effect (μ) and between-study variance (τ^2). The posterior distributions were produced from 10,000 iterations of a single Markov chain after discarding the first 1000 runs as burn-in. Figure 11.3 shows the empirical posterior distributions for species effect sizes, as well as for the overall mean and the between-study variance.

It can be seen that there is some variation in the study-specific estimates, corresponding to the information provided in the data; more precise estimates have higher peaks and more uncertain estimates have lower peaks. The posterior means are generally close to the observed effect sizes (denoted by an x in the empirical posterior distributions of the species effects), although they are shrunken toward the overall mean of 0.33. This shrinkage is greatest for those species for which the Hedges' d statistics are most variable. Observe that *Pieris napi* has a large effect size, but also a large standard deviation. Its posterior mean is therefore downweighted

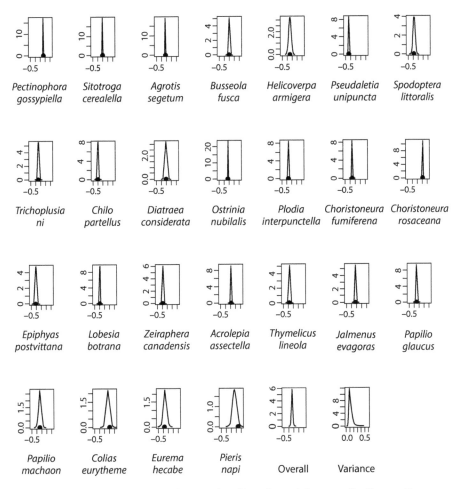

Figure 11.3. Posterior densities of study-specific effects θ_i, and the overall effects μ (*bottom middle*), and τ^2 (*bottom right*). The solid circles indicate the observed mean. The heights of the curves in these figures are proportional to their probabilities so that the areas under the curves sum to 1.

toward the average of all the species because the large effect size is not as believable as some of the other effect sizes. On the other hand, *Pectinophora gossypiella* is precisely estimated and so even though its observed effect size is fairly large, its posterior mean is the same as the observed effect. Shrinkage is much greater for *Pieris napi* because almost 10 times as much data contributed to the estimate for *Pectinophora gossypiella*. Other species with substantial shrinkage toward the mean are *Colias eurytheme* and *Eurema hecabe*. Note that although most distributions are reasonably symmetric, most do not look like normal curves.

The overall mean effect size is 0.32 with 95% credible interval (0.18, 0.46). The posterior probability of an effect size greater than 0 is 1.0000 (since all 10,000 simulations returned positive effect sizes). Thus, we can be almost certain that overall, fecundity of females in these species is greater when they mate with virgin males. Observe that this statement much more directly conveys the information required for testing a hypothesis than the corresponding procedure of a *P*-value calculation with a frequentist procedure.

Among individual species, 16 have a greater than 99% chance of a positive effect size, 1 (*Pseudaletia unipuncta*) has a greater than 99% chance of a negative effect size, and the remaining 7 have less certain probabilities ranging from 16% to 97% for a positive effect. Two of these 7 suggest negative effects, and 5 suggest positive ones. The Bayesian method thus attaches a probability to each species, and using this, we can judge the evidence for the hypotheses. If we were interested in drawing conclusions about the chances of a positive effect, we could state that 16 are strongly indicated (or "significant") and the other 9 are not (or "insignificant"), but the probabilities actually provide us with much more information than a simple test of hypothesis.

Other types of probability statements may be easily constructed from the MCMC simulations. For example, we might be interested in the probability that the effect size for *Papilio machaon* is greater than that for *Epiphyas postvittana*. This is easily calculated by counting the number of simulations for which this occurs. The estimate here is 0.9835 (9835 simulations out of 10,000).

Examination of the posterior distribution of the between-study variance shows that most of the probability falls below 0.3, which corresponds to a standard deviation of about 0.55. However, there is substantial probability, perhaps higher than 90%, that this variance is positive, so it would dangerous to ignore study heterogeneity in this meta-analysis. The posterior estimates of study-specific and population parameters are weighted averages across this posterior distribution of the variance, rather than averages using a single between-study estimate, as in the moment-based estimates described in Chapter 10.

Differentiating mating patterns: Monandrous and polyandrous

The analysis so far has not differentiated among the 25 species in any systematic manner. The summary analysis concerned a single mean. Now, we consider the potential effects of mating patterns. In the data set shown in Appendix 8.1, 13 species were defined as monandrous ($\leq$40% polyandrous females) and 12 polyandrous (>40% polyandrous females). Does the effect of male virginity on female fecundity vary by female sexual preferences? To answer this question, we extend the model to incorporate a binary regression covariate classifying whether the species was monandrous or polyandrous. In effect, this replaces the single mean effect with two effects, one for each level of polyandry.

Analysis with this model (Table 11.2) shows that although in both groups females produced more offspring when mating with virgin males, the effect of male sexual experience was greater on polyandrous females (mean 0.46, 95% CI: 0.27 to 0.65) than on monandrous

TABLE 11.2. Results of Bayesian analysis treating polyandrous and monandrous females as two separate groups.

Species	Observed		Posterior					
	Mean	Variance	Mean	SD	Lower 2.5%	Median	Upper 97.5%	Posterior Prob >0
Monandrous								
Papilio machaon	0.25	0.16	0.24	0.14	−0.04	0.24	0.51	0.96
Eurema hecabe	0.11	0.20	0.14	0.17	−0.20	0.14	0.46	0.79
Jalmenus evagoras	0.37	0.07	0.36	0.07	0.22	0.36	0.49	1.00
Chilo partellus	−0.04	0.05	−0.03	0.05	−0.14	−0.03	0.07	0.25
Diatraea considerata	0.46	0.13	0.41	0.12	0.17	0.41	0.65	1.00
Zeiraphera canadensis	0.02	0.07	0.02	0.06	−0.10	0.03	0.15	0.65
Epiphyas postvittana	−0.12	0.09	−0.09	0.09	−0.27	−0.09	0.08	0.14
Thymelicus lineola	0.11	0.08	0.11	0.08	−0.04	0.11	0.26	0.93
Lobesia botrana	0.07	0.05	0.07	0.05	−0.02	0.07	0.16	0.93
Agrotis segetum	0.38	0.02	0.38	0.02	0.33	0.38	0.42	1.00
Ostrinia nubilalis	0.36	0.02	0.36	0.02	0.32	0.36	0.39	1.00
Choristoneura fumiferana	0.14	0.05	0.14	0.05	0.04	0.14	0.23	1.00
Acrolepia assectella	0.54	0.04	0.53	0.04	0.45	0.53	0.62	1.00
Polyandrous								
Plodia interpunctella	0.11	0.05	0.12	0.05	0.03	0.12	0.21	1.00
Choristoneura rosaceana	1.03	0.04	1.02	0.04	0.95	1.02	1.09	1.00
Colias eurytheme	1.01	0.21	0.84	0.18	0.50	0.84	1.20	1.00
Pseudaletia unipuncta	−0.15	0.05	−0.14	0.05	−0.23	−0.14	−0.05	0.00
Spodoptera littoralis	0.10	0.11	0.14	0.10	−0.06	0.14	0.34	0.91
Papilio glaucus	0.23	0.04	0.24	0.04	0.15	0.24	0.32	1.00
Helicoverpa armigera	0.30	0.13	0.33	0.12	0.09	0.33	0.56	1.00
Sitotroga cerealella	0.51	0.02	0.51	0.02	0.46	0.51	0.55	1.00
Pieris napi	1.17	0.26	0.88	0.21	0.47	0.88	1.31	1.00
Pectinophora gossypiella	0.81	0.02	0.81	0.02	0.76	0.81	0.85	1.00
Trichoplusia ni	0.26	0.08	0.27	0.07	0.13	0.27	0.42	1.00
Busseola fusca	0.47	0.08	0.47	0.08	0.31	0.47	0.62	1.00
Monandrous			0.19	0.09	0.00	0.20	0.38	0.98
Polyandrous			0.46	0.10	0.27	0.46	0.65	1.00
Difference			0.26	0.13	0.00	0.27	0.53	0.97
Overall Variance			0.10	0.04	0.05	0.10	0.20	1.00

females (mean 0.19, 95% CI: 0.00 to 0.38). Density plots of the species posteriors (not shown) differ little from those in the single mean model, although the distributions are a bit tighter (i.e., exhibit less variability) because of the additional information contributed by the grouping into two means. The between-study variability is also reduced slightly as the species distributions are allowed to cluster around two means, rather than just one. The parameter for difference between the polyandrous and monandrous species indicates that on average the polyandrous species have an effect size 0.26 greater than the monandrous species. The posterior probability that the polyandrous species have larger effect sizes than the monandrous species is 0.9748.

Whether this probability is enough to be convincing is an open question, but the weight of evidence favors the hypothesis.

Differentiating mating patterns: Degree of polyandry

Data on the degree of polyandry are also available for 22 of the 25 species (see Appendix 8.1). These data may provide more information than the simple categorization into two groups at a cutoff of 40%. However, the missing data on degree of polyandry in three species pose a problem. We have two options. One is to ignore the information from the three species missing the variable, and the other is to impute values. Ignoring the three species and using information only on those species with complete data may introduce bias if the reason for absence of data is not completely random and is related to some observed or unobserved characteristic of the species. Moreover, restricting analysis to species with complete information only reduces the power of the study since the size of the sample decreases from 25 to 22 (a loss of more than 10% of the information).

Fortunately, as described above, the Bayesian model can treat these missing values as parameters, and after placement of an appropriate prior can accommodate them in analysis. The model fit generalizes Equation 11.2 so that $\theta = \alpha + \beta \times X$, where X is the percentage of polyandry in the species. We use a noninformative prior on both α and β. For the missing covariates, we assume uniform distributions over the range of possible values. These are not completely noninformative because we do know that two of the species were monandrous and one was polyandrous. Making use of the known cutoff of 40%, we can assign uniform priors over the range from 0–40% for the monandrous species, and from 35%–100% for the polyandrous species. (We use 35% as a lower bound rather than 40% because one of the species was declared polyandrous even though its percentage polyandry was 37.5%.)

Table 11.3 displays the resulting posterior estimates and densities. The slope for the relationship between degree of polyandry and effect size is estimated to be 0.005 per 1% added polyandry, and has a 0.9776 probability of being greater than 0. Thus, it is marginally significant. The effect size difference of 0.26 between monandrous and polyandrous species is thus equivalent to a difference of about 52% in the percentage polyandry; this matches closely with the average polyandry percentages of 18% for monandrous and 62% for polyandrous species.

The density plots for the three missing polyandry percentages (Fig. 11.4) indicate that these values were estimated almost uniformly over their potential range, except near the boundaries. This makes sense because we do not have much information about these values other than the bounds imposed by the two-level classification.

Differentiating groups with small numbers: Families

Our final analysis with the Lepitodoptera data deals with differentiating between the nine different families to which these species belong. The results for the families are displayed in Table 11.4 and Figure 11.5; we do not show the species parameters because these are very similar to the results in the previous tables. Because several families are only represented by a single species and no family has more than six species, fitting a fixed-effects analysis of variance model would not be possible. If we treat the families as random effects in a nested structure, however, we can borrow strength from the exchangeability of the effects to obtain estimates. In this case, the regression function consists of the sum of nine terms, one for each family, and each term is assumed to have a noninformative prior.

Species effects with large within-study variance are not shrunk as strongly toward zero, particularly those in the family Pieridae, because they are shrunk toward the family mean rather

TABLE 11.3. Results of Bayesian analysis with percentage polyandry treated as a continuous covariate.

Species	Observed		Posterior					
	Mean	Variance	Mean	SD	Lower 2.5%	Median	Upper 97.5%	Posterior Prob >0
Pectinophora gossypiella	0.81	0.02	0.81	0.02	0.76	0.81	0.85	1.00
Sitotroga cerealella	0.51	0.02	0.51	0.02	0.46	0.51	0.55	1.00
Agrotis segetum	0.38	0.02	0.38	0.02	0.33	0.37	0.42	1.00
Busseola fusca	0.47	0.08	0.47	0.08	0.31	0.47	0.62	1.00
Helicoverpa armigera	0.30	0.13	0.33	0.12	0.09	0.33	0.56	1.00
Pseudaletia unipuncta	−0.15	0.05	−0.14	0.05	−0.23	−0.14	−0.05	0.0014
Spodoptera littoralis	0.10	0.11	0.14	0.10	−0.06	0.14	0.34	0.92
Trichoplusia ni	0.26	0.08	0.28	0.07	0.13	0.28	0.42	1.00
Chilo partellus	−0.04	0.05	-0.04	0.05	−0.14	−0.04	0.06	0.25
Diatraea considerata	0.46	0.13	0.41	0.12	0.17	0.41	0.66	1.00
Ostrinia nubilalis	0.36	0.02	0.36	0.02	0.32	0.36	0.39	1.00
Plodia interpunctella	0.11	0.05	0.12	0.05	0.03	0.12	0.21	1.00
Choristoneura fumiferana	0.14	0.05	0.14	0.05	0.04	0.14	0.23	1.00
Choristoneura rosaceana	1.03	0.04	1.02	0.04	0.94	1.02	1.09	1.00
Epiphyas postvittana	−0.12	0.09	−0.09	0.09	−0.26	−0.09	0.08	0.15
Lobesia botrana	0.07	0.05	0.07	0.05	−0.02	0.07	0.16	0.94
Zeiraphera canadensis	0.02	0.07	0.02	0.07	−0.11	0.02	0.15	0.63
Acrolepia assectella	0.54	0.04	0.53	0.04	0.45	0.53	0.62	1.00
Thymelicus lineola	0.11	0.08	0.11	0.08	−0.04	0.11	0.26	0.94
Jalmenus evagoras	0.37	0.07	0.35	0.07	0.21	0.35	0.49	1.00
Papilio glaucus	0.23	0.04	0.24	0.04	0.15	0.24	0.32	1.00
Papilio machaon	0.25	0.16	0.25	0.14	−0.03	0.24	0.52	0.96
Colias eurytheme	1.01	0.21	0.81	0.18	0.47	0.81	1.15	1.00
Eurema hecabe	0.11	0.20	0.14	0.17	−0.20	0.14	0.48	0.80
Pieris napi	1.17	0.26	0.89	0.21	0.49	0.89	1.30	1.00
Intercept			0.12	0.12	−0.13	0.12	0.36	0.84
Slope			0.005	0.003	0.0001	0.005	0.011	0.98
Overall Variance			0.10	0.04	0.05	0.09	0.20	1.00
Imputed % Polyandry								
Busseola fusca			67.05	18.41	36.87	66.85	98.16	1.00
Papilio machaon			20.28	11.51	1.04	20.39	39.03	1.00
Eurema hecabe			19.52	11.55	0.98	19.20	39.00	1.00

than the overall mean. Each of the families is more likely than not to have a positive effect size, although only Pieridae and Gelechidae are highly likely (probability > 0.95). Again, the simulated joint posterior permits easy comparisons between families. Table 11.5 depicts pairwise comparisons of the probability that one family has a larger effect size than another. Pieridae is more likely to have larger effect sizes on average than any other family.

META-ANALYSIS OF NONNORMAL DATA

All of the methods discussed so far in this chapter have assumed that the observed summary statistics from each study follow normal distributions. In many common applications, however, the outcomes are not normally distributed and may not even be continuous. Often, they are

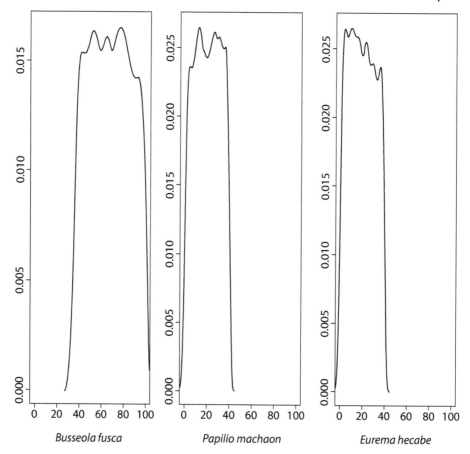

Figure 11.4. Density plots for missing polyandry percentages from regression on percentage of polyandry.

TABLE 11.4. Posterior estimates of family effects.

Family	Mean	SD	95% CI	Prob > 0
Yponomeutidae	0.54	0.37	−0.19, 1.28	0.93
Noctuidae	0.22	0.15	−0.08, 0.52	0.93
Pieridae	0.75	0.25	q0.26, 1.23	1.00
Pyralidae	0.22	0.18	−0.16, 0.59	0.89
Tortricidae	0.23	0.16	−0.09, 0.55	0.92
Lycaenidae	0.37	0.37	−0.36, 1.11	0.85
Papilionidae	0.24	0.27	−0.29, 0.76	0.83
Gelechiidae	0.66	0.26	0.16, 1.17	0.99
Hesperiidae	0.11	0.37	−0.63, 0.84	0.63

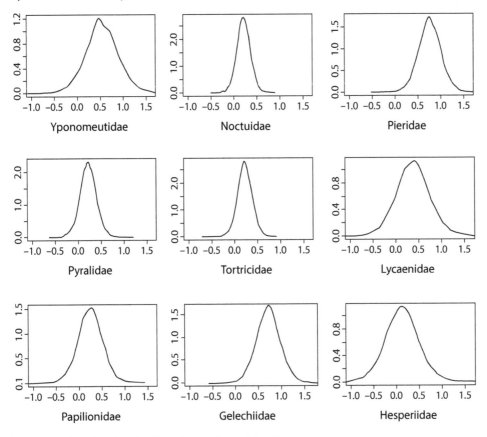

Figure 11.5. Density plots for Bayesian analysis of families.

TABLE 11.5. Posterior probabilities that the effect size in one family is greater than that in another family.

	Y	N	Pi	Py	T	L	Pa	G	H
Y	—	0.81	0.31	0.80	0.79	0.63	0.76	0.38	0.81
N		—	0.03	0.50	0.48	0.35	0.47	0.07	0.61
Pi			—	0.96	0.96	0.81	0.92	0.60	0.93
Py				—	0.48	0.35	0.47	0.08	0.61
T					—	0.36	0.48	0.08	0.63
L						—	0.61	0.25	0.70
Pa							—	0.12	0.62
G								—	0.90
H									—

Y = Yponomeutidae; N = Noctuidae; Pi = Pieridae; Py = Pyralidae; T = Tortricidae; L = Lycaenidae; Pa = Papilionidae; G = Gelechiidae; H = Hesperiidae binary, taking only two possible values, and are summarized in terms of the proportions that fall into each category. For example, an analysis of invasive species might use proportions of individuals observed, or a comparative study of mortality rates between two groups of individuals might use odds ratios to estimate the relative odds of the probabilities in each group. While it is possible to transform the data to achieve normality of distribution (for example, by taking logs of the proportions or undertaking a logistic analysis), this may be misleading or infeasible if the sample sizes are small or if positive (or negative) outcomes are rare. It is much more appealing to model the data directly, without transformation. With the exception of very simple cases, this is difficult using the frequentist approaches described in Chapters 9 and 10, but it is straightforward in the Bayesian framework.

In the general formulation of the Bayesian model, any prior distribution may be combined with any likelihood to give a posterior distribution. Some combinations of prior distributions and likelihoods are easier to compute with than others, but MCMC algorithms can be designed for almost any problem. Uncommon distributions are not nearly as difficult to work with as they are in the likelihood setting, although the algorithm may need to run for a long time before the correct posterior distribution is obtained. When outcomes are binary, the natural likelihood to use is the binomial; when outcomes are counts, the natural likelihood is the Poisson. The hierarchical nature of a random-effects model permits the placement of informative structure on the model parameters, thus providing a structured, informative prior distribution.

CASE STUDY 2: HANDEDNESS OF FIDDLER CRABS

To demonstrate a simple application of Bayesian models to nonnormal data, consider the data on the number of fiddler crabs expressing right- and left-handedness described in Appendix 10.1. In fiddler crabs, handedness is easily identified by observing whether the dominant large claw is on the left or right side of the body. Each report lists the number of right- and left-handed crabs in one or more species. We consider each species detailed in each report as a separate study. In all, there are 157 studies comprising data from 7 subgenera.

In this model, we assume that the observed number X_i of right-handed crabs among a total number n_i in the ith study has an independent binomial distribution with an unknown true probability π_i of right-handedness. Although each study has a different probability, we reduce the number of study parameters by assuming that the studies form a distribution of random effects centered around different means for each subgenus. In order to use a normal distribution for the random effects, we transform the probabilities into logits and place the normal distributions on these logits. The logit transformation takes the natural logarithm of the ratio of the probability to its complement (logit $\pi = \ln[\pi/(1-\pi)]$). Therefore, if we define $\pi_{i(j)}$ as the probability in the ith study with the jth subgenus, we have logit $\pi_{i(j)} \sim N(\mu_j, \tau^2)$, where μ_j is the mean probability of the jth subgenus on the logit scale with between-study variance τ^2. Another way to write this is as a regression with dummy variables Z_j for each subgenus so that the mean of the normal distribution is the expression $\beta_1 Z_1 + \beta_2 Z_2, \ldots, + \beta_7 Z_7$. Thus, the final model has 165 parameters (157 $\pi_{i(j)}$, 7 μ_j, and τ^2).

By modeling the exact binomial distribution, the Bayesian approach avoids one of the major shortcomings of the normal approximation with binary data, the occurrence of zero counts. Note that in the data set being analyzed, quite a few of the studies are very small and contain either no left-handed or no right-handed crabs in their samples. If we had directly modeled the logit observed probabilities, such zeros would have given undefined outcomes. The usual solution to this problem is to add a small constant (often 0.5) to numerator and denominator of the proportion. This makes little difference when the outcome is rare and there

TABLE 11.6. Analysis of fiddler crab data: Percentage right-handed.

Subgenus	Mean	SE	95% CI
Gelasimus	97.0	0.3	96.5 – 97.5
Australuca	53.3	1.3	50.8 – 56.1
Tubuca	52.9	1.3	50.2 – 55.6
Minuca	50.3	0.8	48.8 – 52.0
Leptuca	49.9	0.8	48.2 – 51.4
Uca	48.7	1.9	45.0 – 52.7
Paraleptuca	45.9	4.9	36.3 – 55.7

are few zero cells. In our example, however, the outcome is not rare, so that addition of the constant will have a big effect on the outcome. For example, a study with two right-handed and no left-handed crabs has an observed proportion of right-handed crabs of one (or of left-handed crabs of zero). Adding 0.5 to numerator and denominator changes the observed proportion to 5/6. This is not an issue with the Bayesian approach because the binomial distribution handles zeroes directly.

Applying the random-effects binomial model to the fiddler crab data gives the results in Table 11.6 for the mean proportions of right-handed crabs in each subgenus. The random-effects proportions in each study are displayed in Figure 11.6. We see that the subgenera do differ with respect to the proportion of right-handed crabs. While *Gelasimus* crabs are almost all right-handed, the other subgenera have nearly equal proportions of right- and left-handed crabs. The proportions are well-estimated because we have a substantial number of studies, several of which are quite large. Figure 11.6 shows that species variation within subgenus is small compared to variation between subgenera. If we wished to investigate study-to-study variation within species, we could expand the model to incorporate species as well as subgenus effects.

It is also quite easy to test other hypotheses of interest. For example, the probability that the proportion of right-handed crabs in one subgenus exceeds that in another is easily calculated by the proportion of MCMC simulations in which this event occurred. Table 11.7 shows these posterior probabilities for each pair of subgenera. *Gelasimus* is clearly more right-handed than any other subgenus, but *Australuca* and *Tubuca* are also more right-handed than *Minuca, Leptuca, Uca,* and *Paraleptuca*.

SENSITIVITY ANALYSES AND MODEL CHECKING

Consistent with sound general statistical practice, it is always a good idea to undertake a sensitivity analysis to evaluate the influences of the model setup, choice of priors, distributional assumptions, and any characteristics of the data, such as outliers. This is particularly important when the amount of data is small. None of these ideas is unique to Bayesian analyses except the need to check for the influence of the choice of prior distribution on posterior inference.

The biggest criticism of Bayesian methods has always involved the subjectivity of the choice of prior distributions. We have argued that this subjectivity honestly addresses the differences of opinion that scientists have about evidence. But this also implies that any summary of evidence ought to reflect this difference of opinion, rather than hide it. As a result, Bayesians have recommended that any scientific report ought to include one analysis using a reference prior to summarize evidence from the data alone, as well as others based on different priors that reflect a variety of different perspectives including that of the authors (Spiegelhalter et al. 1994).

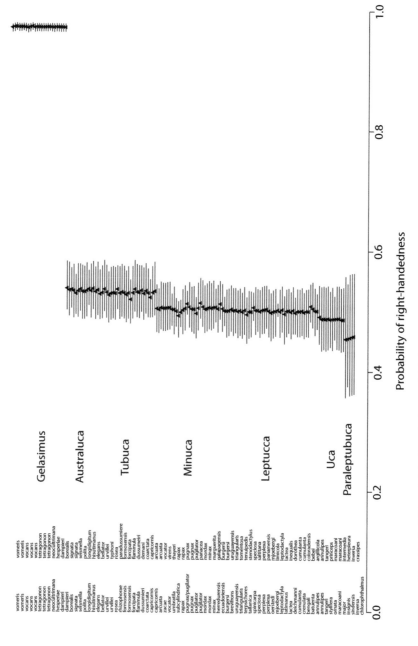

Figure 11.6. Probability of right-handedness of fiddler crabs by subgenus.

TABLE 11.7. Pairwise probability comparisons. Numbers shown are posterior probabilities that subgenus labeled by row is more right-handed than subgenus labeled by column.

	Australuca	Tubuca	Minuca	Leptuca	Uca	Paraleptuca
Gelasimus	1.00	1.00	1.00	1.00	1.00	1.00
Australuca		0.60	0.98	0.99	0.97	0.90
Tubuca			0.93	0.97	0.97	0.89
Minuca				0.66	0.80	0.80
Leptuca					0.75	0.78
Uca						0.72

This sensitivity analysis has another benefit as well. If all of the different priors lead to similar posterior inferences, then consensus may be achievable. Otherwise, such consensus may be unlikely. In some problems, particularly when few data are available to inform an important parameter, inferences may be quite sensitive to small changes in prior assumptions. In such cases, drawing conclusions may be dangerous, and the best advice may be that more research is needed.

Having gone through the exercise of checking for sensitivity to choice of a prior distribution, it is also wise to check how well the models support the data by performing *posterior predictive model checks* (Gelman et al. 2004). These create simulated data sets from the models and the posterior distributions of the model parameters, and compare features of these simulated data sets to the same features in the observed data. For example, if the feature is the largest study effect, we compare the largest effect in the observed data with the largest value in each of the simulated data sets. The location of the observed statistic in the distribution of simulated values provides a measure of the agreement between the model and the data, similar to that provided by a sample statistic relative to the expected distribution under the null hypothesis. Considered this way, the location gives a type of Bayesian *P*-value. Values near zero or one indicate data that do not agree well with the bulk of the data sets simulated from the model, and therefore suggest that the data do not agree with the model. Some have called this a Bayesian bootstrap for its resemblance to the bootstrap resampling procedure (Rubin 1984).

As an example of posterior predictive checks, we will check simulations from the random-effects model for the fiddler crab data. Some potential features that we might wish to check include an overall goodness of fit (summed squared deviation of observed and predicted), the maximal deviation of observed and predicted values, or the identity of the report with the greatest or smallest proportion of right-handed crabs. We want to focus particularly on features of the data not explicitly captured in the model. For example, the mean proportion of right-handed crabs in each subgenus is a model parameter, so we would expect the model to reproduce this data feature well. However, features specific to an individual observation are not reflected directly in the model. One interesting feature of the data is its large number of small samples. In particular, 18 of the reports contain no left-handed crabs. A good model ought to be able to reproduce this feature. Therefore, as a model check we will simulate 3000 data sets from the model, and compare the distribution of numbers of reports with no left-handed crabs in each simulation, with the observed total of 18.

Figure 11.7 shows the distribution of the number of studies that had no left-handed crabs in each simulation. In 3000 simulations, the number of such studies is never less than 7, nor greater than 29. The most likely number was 18, which is the same as the observed number. A two-sided *P*-value for testing the null hypothesis that the observed number is drawn from the simulated distribution is 0.96, thus indicating no evidence of departure from this assumption.

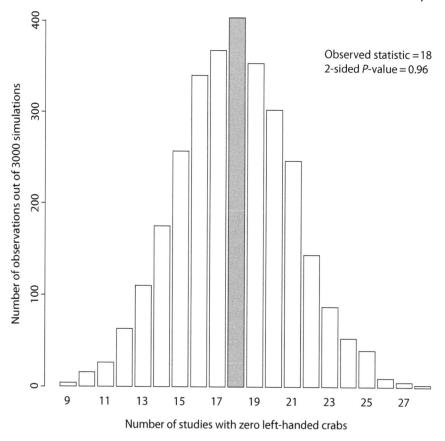

Figure 11.7. Distribution of number of studies with no left-handed crabs in 3000 simulations using posterior distribution of study probabilities with fiddler crab data. Gray bar indicates location of observed number of 18 studies with no crabs in data set.

Of course, many other predictive checks are possible and should be carried out. Given the simplicity of the data structure and our model, and in the interests of moving on to the next chapter, we will not pursue further checking here.

CONCLUSIONS

Bayesian inference is often considered esoteric and extremely mathematical, but modern software makes fitting these models practical. Bayesian analysis offers considerable flexibility in model choice and the resulting conclusions. By requiring explicit descriptions of all sources of uncertainty, it gives the analyst great control over the model, including the incorporation of external knowledge about the biological processes; however, it also imposes the burden of careful checking to ensure that the model can accurately reproduce the observed data. The reward for this effort is the ability to make explicit probabilistic statements about the likelihood of relevant scientific hypotheses.

APPENDIX 11.1:
WINBUGS CODE FOR LEPIDOPTERA MODEL

Note: All of the material in this appendix can also be found on the web site for this book, http://press.princeton.edu/titles/10045.html

```
model
{
for (i in 1:N) {
     d[i] ~ dnorm(mu[i],isigma[i])
     mu[i] ~ dnorm(xb[i], itau)
# Use if fitting single mean                              xb[i] <- meanmu
# Use if comparing monandrous and polyandrous            xb[i] <- alpha + beta*polycat[i]
# Use if fitting continuous variable degree of polyandry  xb[i] <- alpha + beta*polypct[i]
# Use if fitting family effects adjusted for degree of polyandry:
   xb[i] <- beta*polypct[i] + beta1*Yponomeutidae[i] + beta2*Noctuidae[i] +
   beta3*Pieridae[i] + beta4*Pyralidae[i] + beta5*Tortricidae[i] + beta6*Lycaenidae[i] +
   beta7*Papilionidae[i]+beta8*Gelechiidae[i]+beta9*Hesperiidae[i]
     isigma[i] <- 1/(vard[i]*vard[i])
     poly[i] <- step(polycat[i] - 1)
     polypct[i] ~ dunif(a[i],b[i])
     a[i] <- 35*poly[i]
     b[i] <- 40 + 60*poly[i]
}
# Set up prior distributions. Use only those needed depending on form of linear predictor
meanmu ~ dnorm(0,1.0E-06)
alpha ~ dnorm(0,1.0E-06)
beta ~ dnorm(0,1.0E-06)
beta1 ~ dnorm(0,1.0E-06)
beta2 ~ dnorm(0,1.0E-06)
beta3 ~ dnorm(0,1.0E-06)
beta4 ~ dnorm(0,1.0E-06)
beta5 ~ dnorm(0,1.0E-06)
beta6 ~ dnorm(0,1.0E-06)
beta7 ~ dnorm(0,1.0E-06)
beta8 ~ dnorm(0,1.0E-06)
beta9 ~ dnorm(0,1.0E-06)
se ~ dunif(0,100)
tau <- se*se
itau <-1/tau
apoly <- alpha+beta
}
```

WinBUGS data for Lepidoptera example

```
list(N= 25,
vard =
```

```
c(0.0237,0.0217,0.0213,0.0819,0.1283,0.0482,0.1073,0.0750,0.0522,0.1325,0.0171,
    0.0458,0.0479,0.0372,0.0895,0.0463,0.0667,0.0441,0.0781,0.0714,0.0435,0.1551,
    0.2051,0.2003,0.2550),
d = c(0.8090,0.5072,0.3762,0.4688,0.3043,-0.1524,0.1000,0.2631,-
0.0411,0.4577,0.3585,0.1140,0.1371,1.0281,-
0.1176,0.0655,0.0155,0.5385,0.1070,0.3656,0.2317,0.2506,1.0125,0.1105,1.1690)
#,Moth = c(1,1,1,1,1,1,1,1,1,1,1,1,1,1,1,1,1,1,1,0,0,0,0,0,0,0),
,polypct=c(81,68.5,27.5,NA,68,55,61.5,81,5.5,7.5,27.5,37.5,27.5,45,12.25,22.5,8,38,22
    ,0,65,NA,47,NA,70)
,polycat = c(1,1,0,1,1,1,1,1,0,0,0,1,0,1,0,0,0,0,0,0,1,0,1,0,1)
,Gelechiidae = c(1,1,0,0,0,0,0,0,0,0,0,0,0,0,0,0,0,0,0,0,0,0,0,0,0,0),
Noctuidae = c(0,0,1,1,1,1,1,1,0,0,0,0,0,0,0,0,0,0,0,0,0,0,0,0,0,0),
Pyralidae = c(0,0,0,0,0,0,0,0,1,1,1,1,0,0,0,0,0,0,0,0,0,0,0,0,0,0),
Tortricidae = c(0,0,0,0,0,0,0,0,0,0,0,0,1,1,1,1,1,0,0,0,0,0,0,0,0,0),
Yponomeutidae = c(0,0,0,0,0,0,0,0,0,0,0,0,0,0,0,0,0,1,0,0,0,0,0,0,0,0),
Hesperiidae = c(0,0,0,0,0,0,0,0,0,0,0,0,0,0,0,0,0,0,1,0,0,0,0,0,0,0),
Lycaenidae= c(0,0,0,0,0,0,0,0,0,0,0,0,0,0,0,0,0,0,0,1,0,0,0,0,0,0),
Papilionidae = c(0,0,0,0,0,0,0,0,0,0,0,0,0,0,0,0,0,0,0,0,1,1,0,0,0,0),
Pieridae = c(0,0,0,0,0,0,0,0,0,0,0,0,0,0,0,0,0,0,0,0,0,0,1,1,1)
)
```

Lepidoptera example: Initial values

```
list(se = 1, mu = c(0,0,0,0,0,0,0,0,0,0,0,0,0,0,0,0,0,0,0,0,0,0,0,0,0,0),
#Use values needed for linear predictor chosen
meanmu = 0,
alpha = 0,
beta = 0,
beta1 = 0, beta2 = 0, beta3= 0, beta4=0, beta5=0, beta6= 0, beta7=0, beta8=0,
    beta9=0
)
```

Model for Fiddler crabs example

```
model
{
    for (i in 1:N) {
        Right[i] ~ dbin(pi[i],Total[i]);
        logit(pi[i])<- p.right[i];
        p.right[i] ~ dnorm(mu[i],itau);
        mu[i] <- beta1*Paraleptuca[i] + beta2*Uca[i] + beta3*Leptuca[i]
            + beta4*Minuca[i] + beta5*Tubuca[i] + beta6*Australuca[i] +
            beta7*Gelasimus[i]
}
# Following commands set up pairwise comparisons used in Table 11.7
        b12 <- step(beta1 − beta2)
        b13 <- step(beta1 − beta3)
        b14 <- step(beta1 − beta4)
```

```
       b15 <- step(beta1 − beta5)
       b16 <- step(beta1 − beta6)
       b17 <- step(beta1 − beta7)
       b23 <- step(beta2 − beta3)
       b24 <- step(beta2 − beta4)
       b25 <- step(beta2 − beta5)
       b26 <- step(beta2 − beta6)
       b27 <- step(beta2 − beta7)
       b34 <- step(beta3 − beta4)
       b35 <- step(beta3 − beta5)
       b36 <- step(beta3 − beta6)
       b37 <- step(beta3 − beta7)
       b45 <- step(beta4 − beta5)
       b46 <- step(beta4 − beta6)
       b47 <- step(beta4 − beta7)
       b56 <- step(beta5 − beta6)
       b57 <- step(beta5 − beta7)
       b67 <- step(beta6 − beta7)
# Convert logit scale to probability scale
p.Paraleptuca <- exp(beta1)/(1+exp(beta1))
p.Uca <- exp(beta2)/(1+exp(beta2))
p.Leptuca <- exp(beta3)/(1+exp(beta3))
p.Minuca <- exp(beta4)/(1+exp(beta4))
p.Tubuca <- exp(beta5)/(1+exp(beta5))
p.Australuca <- exp(beta6)/(1+exp(beta6))
p.Gelasimus <- exp(beta7)/(1+exp(beta7))
# Noninformative priors
beta1 ~ dnorm(0,.000001);
beta2 ~ dnorm(0,.000001);
beta3 ~ dnorm(0,.000001);
beta4 ~ dnorm(0,.000001);
beta5 ~ dnorm(0,.000001);
beta6 ~ dnorm(0,.000001);
beta7 ~ dnorm(0,.000001);
se ~ dunif(0,100);
tau <- se*se;
itau <- 1/tau;
}
```

Fiddler crab data

```
list(N=157,
Right = c(3, 3, 2, 64, 1, 1, 1, 1, 2, 236, 3,
57, 1, 0, 2, 0, 38, 179, 4, 37, 188, 0, 10, 22, 1, 3, 2, 170,
5, 14, 27, 4, 18, 19, 2, 3583, 2, 1, 13, 0, 62, 5, 5, 4, 56,
2, 483, 23, 1, 40, 2, 22, 3, 4, 26, 1, 17, 1, 1, 409, 3, 193,
2, 266, 9, 6, 5, 1, 2, 28, 169, 5, 101, 1, 24, 36, 932, 25, 27,
552, 241, 462, 0, 290, 8, 4, 1, 3, 4, 29, 0, 7, 17, 261, 35,
```

39, 0, 87, 2, 3, 47, 2, 156, 5, 11, 2, 4, 0, 152, 1, 3, 0, 18,
1, 52, 1, 29, 4, 15, 62, 0, 320, 1, 8, 1, 69, 3, 198, 2, 180,
2, 53, 118, 47, 59, 72, 2, 51, 8, 812, 4, 272, 90, 29, 12, 7,
99, 1165, 32, 167, 5, 554, 255, 34, 413, 8, 618),
Total = c(3, 4, 4, 143, 6, 4, 5, 1, 3, 484, 7, 120, 1, 1, 3, 1, 70, 358, 7,
61, 349, 1, 21, 50, 1, 7, 5, 353, 7, 33, 52, 9, 50, 33, 6, 7142,
4, 5, 34, 1, 114, 6, 7, 10, 109, 2, 941, 50, 6, 97, 2, 50, 6,
11, 50, 2, 28, 2, 4, 820, 6, 357, 5, 517, 13, 10, 6, 1, 6, 48,
306, 8, 225, 5, 50, 66, 1801, 50, 60, 1130, 524, 941, 4, 582,
13, 6, 1, 5, 5, 63, 1, 7, 31, 518, 67, 66, 2, 155, 3, 3, 76,
4, 326, 10, 28, 4, 5, 1, 272, 4, 6, 2, 43, 1, 85, 2, 59, 6, 34,
108, 2, 593, 2, 20, 6, 125, 7, 388, 6, 333, 3, 92, 124, 50, 63,
76, 2, 56, 9, 836, 4, 281, 93, 31, 12, 7, 105, 1188, 34, 179,
5, 567, 263, 34, 417, 8, 644),
Paraleptuca = c(1, 1, 1, 1, 1, 0, 0, 0,
0, 0, 0, 0, 0, 0, 0, 0, 0, 0, 0, 0, 0, 0, 0, 0, 0, 0, 0, 0, 0, 0, 0,
0, 0, 0, 0, 0, 0, 0, 0, 0, 0, 0, 0, 0, 0, 0, 0, 0, 0, 0, 0, 0, 0, 0,
0, 0, 0, 0, 0, 0, 0, 0, 0, 0, 0, 0, 0, 0, 0, 0, 0, 0, 0, 0, 0, 0, 0,
0, 0, 0, 0, 0, 0, 0, 0, 0, 0, 0, 0, 0, 0, 0, 0, 0, 0, 0, 0, 0, 0, 0,
0, 0, 0, 0, 0, 0, 0, 0, 0, 0, 0, 0, 0, 0, 0, 0, 0, 0, 0, 0, 0, 0, 0,
0, 0, 0, 0, 0, 0, 0, 0, 0, 0, 0, 0, 0, 0, 0, 0, 0, 0, 0, 0, 0, 0, 0,
0, 0, 0, 0, 0, 0, 0, 0, 0, 0, 0, 0, 0, 0, 0, 0, 0, 0, 0, 0, 0, 0, 0,
0, 0),
Uca = c(0, 0, 0, 0, 0, 1, 1, 1, 1, 1, 1, 1, 1, 1, 1, 1, 1,
1, 0, 0, 0, 0, 0, 0, 0, 0, 0, 0, 0, 0, 0, 0, 0, 0, 0, 0, 0, 0, 0, 0,
0, 0, 0, 0, 0, 0, 0, 0, 0, 0, 0, 0, 0, 0, 0, 0, 0, 0, 0, 0, 0, 0, 0,
0, 0, 0, 0, 0, 0, 0, 0, 0, 0, 0, 0, 0, 0, 0, 0, 0, 0, 0, 0, 0, 0, 0,
0, 0, 0, 0, 0, 0, 0, 0, 0, 0, 0, 0, 0, 0, 0, 0, 0, 0, 0, 0, 0, 0, 0,
0, 0, 0, 0, 0, 0, 0, 0, 0, 0, 0, 0, 0, 0, 0, 0, 0, 0, 0, 0, 0, 0, 0,
0, 0, 0, 0, 0, 0, 0, 0, 0, 0, 0, 0, 0, 0, 0, 0, 0, 0, 0, 0, 0, 0, 0,
0, 0, 0, 0, 0, 0, 0, 0, 0, 0, 0, 0, 0, 0, 0),
Leptuca = c(0,
0, 0, 0, 0, 0, 0, 0, 0, 0, 0, 0, 0, 0, 0, 0, 1, 1, 1, 1, 1, 1,
1, 1, 1, 1, 1, 1, 1, 1, 1, 1, 1, 1, 1, 1, 1, 1, 1, 1, 1, 1, 1,
1, 1, 1, 1, 1, 1, 1, 1, 1, 1, 1, 1, 1, 1, 1, 0, 0, 0, 0,
0, 0, 0, 0, 0, 0, 0, 0, 0, 0, 0, 0, 0, 0, 0, 0, 0, 0, 0, 0,
0, 0, 0, 0, 0, 0, 0, 0, 0, 0, 0, 0, 0, 0, 0, 0, 0, 0, 0, 0,
0, 0, 0, 0, 0, 0, 0, 0, 0, 0, 0, 0, 0, 0, 0, 0, 0, 0, 0, 0,
0, 0, 0, 0, 0, 0, 0, 0, 0, 0, 0, 0, 0, 0, 0, 0, 0, 0, 0, 0,
0, 0, 0, 0, 0, 0, 0, 0, 0),
Minuca = c(0, 0, 0, 0, 0, 0, 0, 0,
0, 0, 0, 0, 0, 0, 0, 0, 0, 0, 0, 0, 0, 0, 0, 0, 0, 0, 0, 0, 0, 0, 0,
0, 0, 0, 0, 0, 0, 0, 0, 0, 0, 0, 0, 0, 0, 0, 0, 0, 0, 0, 0, 0, 0, 0,
0, 0, 0, 0, 0, 0, 0, 0, 0, 0, 1, 1, 1, 1, 1, 1, 1, 1, 1, 1, 1,
1, 1, 1, 1, 1, 1, 1, 1, 1, 1, 1, 1, 1, 1, 1, 1, 1, 1, 1, 1, 0,
0, 0, 0, 0, 0, 0, 0, 0, 0, 0, 0, 0, 0, 0, 0, 0, 0, 0, 0, 0, 0, 0,
0, 0, 0, 0, 0, 0, 0, 0, 0, 0, 0, 0, 0, 0, 0, 0, 0, 0, 0, 0, 0, 0,
0, 0, 0, 0, 0, 0, 0, 0, 0, 0, 0, 0, 0, 0, 0, 0, 0, 0, 0, 0, 0, 0,
0, 0),

Tubuca = c(0, 0, 0, 0, 0, 0, 0, 0, 0, 0, 0, 0, 0, 0, 0,
0, 0, 0, 0, 0, 0, 0, 0, 0, 0, 0, 0, 0, 0, 0, 0, 0, 0, 0, 0, 0,
0, 0, 0, 0, 0, 0, 0, 0, 0, 0, 0, 0, 0, 0, 0, 0, 0, 0, 0, 0, 0,
0, 0, 0, 0, 0, 0, 0, 0, 0, 0, 0, 0, 0, 0, 0, 0, 0, 0, 0, 0, 0,
0, 0, 0, 0, 0, 0, 0, 0, 0, 0, 0, 0, 0, 1, 1, 1, 1, 1, 1, 1, 1,
1, 1, 1, 1, 1, 1, 1, 1, 1, 1, 1, 1, 1, 1, 1, 1, 1, 1, 0, 0, 0,
0, 0, 0, 0, 0, 0, 0, 0, 0, 0, 0, 0, 0, 0, 0, 0, 0, 0, 0, 0, 0,
0, 0, 0, 0, 0, 0, 0, 0, 0, 0, 0, 0, 0, 0, 0, 0),
Australuca = c(0,
0, 0, 0, 0, 0, 0, 0, 0, 0, 0, 0, 0, 0, 0, 0, 0, 0, 0, 0, 0, 0,
0, 0, 0, 0, 0, 0, 0, 0, 0, 0, 0, 0, 0, 0, 0, 0, 0, 0, 0, 0, 0,
0, 0, 0, 0, 0, 0, 0, 0, 0, 0, 0, 0, 0, 0, 0, 0, 0, 0, 0, 0, 0,
0, 0, 0, 0, 0, 0, 0, 0, 0, 0, 0, 0, 0, 0, 0, 0, 0, 0, 0, 0, 0,
0, 0, 0, 0, 0, 0, 0, 0, 0, 0, 0, 0, 0, 0, 0, 0, 0, 0, 0, 0, 0,
0, 0, 0, 0, 0, 0, 0, 0, 0, 0, 0, 1, 1, 1, 1, 1, 1, 1, 1, 1, 1,
1, 1, 1, 1, 1, 0, 0, 0, 0, 0, 0, 0, 0, 0, 0, 0, 0, 0, 0, 0,
0, 0, 0, 0, 0, 0, 0, 0, 0),
Gelasimus = c(0, 0, 0, 0, 0, 0, 0,
0, 0, 0, 0, 0, 0, 0, 0, 0, 0, 0, 0, 0, 0, 0, 0, 0, 0, 0, 0, 0,
0, 0, 0, 0, 0, 0, 0, 0, 0, 0, 0, 0, 0, 0, 0, 0, 0, 0, 0, 0, 0,
0, 0, 0, 0, 0, 0, 0, 0, 0, 0, 0, 0, 0, 0, 0, 0, 0, 0, 0, 0, 0,
0, 0, 0, 0, 0, 0, 0, 0, 0, 0, 0, 0, 0, 0, 0, 0, 0, 0, 0, 0, 0,
0, 0, 0, 0, 0, 0, 0, 0, 0, 0, 0, 0, 0, 0, 0, 0, 0, 0, 0, 0, 0,
0, 0, 0, 0, 0, 0, 0, 0, 0, 0, 0, 0, 0, 0, 0, 0, 0, 0, 0, 1,
1, 1, 1, 1, 1, 1, 1, 1, 1, 1, 1, 1, 1, 1, 1, 1, 1, 1, 1, 1, 1,
1, 1, 1)
)

Fiddler crab example: Initial values

list(
beta1 = 0,beta2 = 0,beta3 = 0,beta4 = 0,beta5 = 0,beta6 = 0,beta7 = 0,
se=1,
p.right =
c(0,0,0,0,0,0,0,0,0,0,0,0,0,0,0,0,0,0,0,0,0,0,0,0,0,0,0,0,0,0,0,0,0,0,0,0,0,0,0,0,0,0,0,0,
0,0,0,0,
0,0,0,0,0,0,0,0,0,0,0,0,0,0,0,0,0,0,0,0,0,0,0,0,0,0,0,0,0,0,0,0,0,0,0,0,0,0,0,0,0,0,0,0,
0,0,0,0,
0,0,0,0,0,0,0,0,0,0,0,0,0,0,0,0,0,0,0,0,0,0,0,0,0,0,0,0,0,0,0,0,0,0,0,0,0,0,0,0,0,0,0,0,
0,0,0,0,
0,0,0,0,0)

12

Software for Statistical Meta-analysis

Christopher H. Schmid, Gavin B. Stewart, Hannah R. Rothstein, Marc J. Lajeunesse, and Jessica Gurevitch

Disclaimer: Authors of this chapter have participated in authoring several of the software packages discussed in this chapter. CHS is a coauthor of MetaAnalyst, HR is a coauthor of Comprehensive Meta-Analysis, MJL is the author of phyloMeta, and JG is a coauthor of MetaWin.

TO CONDUCT A META-ANALYSIS, a researcher will need to use computer software to perform all but the simplest calculations. There are three types of software that can be used, depending upon the needs of the user. The first option is a spreadsheet, the second is a general purpose statistical package, and the third option is a program developed expressly to carry out meta-analysis.

The most basic meta-analytic tools, such as weighted averages for fixed- and random-effects models, can be programmed by a knowledgeable user in a spreadsheet, such as Microsoft Excel. In the past, spreadsheets have often been used to carry out many of the meta-analyses published in the scientific literature, but we advise that these should no longer be used in research. This is because of the likelihood of programming and transcriptional errors, and because many of the statistical and graphical analyses that have now become standard are not available using spreadsheet analyses, though they are available elsewhere.

General purpose statistical software varies in its ease of use and flexibility for meta-analysis. These packages cannot be used "as is" because their base algorithms are designed primarily for the statistical analysis of primary studies, and thus produce erroneous results applied directly to meta-analysis. Fortunately, a variety of books and journal articles have included code for adapting one or more of these packages for meta-analysis and, for each of the major software packages, meta-analysts have developed publicly available macros and subroutines. We describe some of these in this chapter for SAS, SPSS, Stata, S-Plus, R, and BUGS. Stata, in particular, has become a popular tool because of its relatively intuitive commands and the availability of macros to perform most of the standard and many of the not-so-standard analyses written and promoted by methodological leaders in meta-analysis (Sterne et al. 2009).

Users who are familiar with these packages, and who have the requisite programming and statistical ability, may find it worthwhile to use these programs, or even to create their own routines, functions, or macros; these could be particularly useful for complex analyses such as the multivariate, Bayesian, and longitudinal analyses described in Chapters 11 and 16. These chapters give code for some of the most common types of analyses. However, most nonstatisticians will find this difficult. These statistical packages generally have steep learning curves and require a large investment of time, very little of which is specific to meta-analysis. Several are also quite expensive.

Finally, there are stand-alone packages for meta-analysis that come in many different flavors. We focus on those that are the most flexible and the most suited to the types of analyses carried out by ecologists and evolutionary biologists. Thus, we pass over the many limited programs that are either specialized for other fields, such as psychometric meta-analysis (the Hunter-Schmidt package; Roth 2008), the meta-analysis of diagnostic tests (MetaDiSc; Zamora et al. 2006), or that were written in outdated formats such as MS-DOS in the early years of meta-analysis (e.g., DStat and True Epistat; Normand 1995).

The most well-known and widely available of the general stand-alone programs are Comprehensive Meta-Analysis (CMA), MetaWin, MIX, MetaAnalyst, and RevMan. In general, all these packages are easier to use for meta-analysis than the general purpose packages, but they vary substantially in their ability to import and export data, compute effect sizes and variances for the data from each primary study, handle complex data structures (such as multiple subgroups, outcomes, time-points, or comparisons), use different statistical models, conduct subgroup analyses and meta-regression, perform sensitivity and small study/publication bias analyses, and provide presentation quality graphics. They also differ in user friendliness, technical support, documentation, and price. Thus, their attractiveness depends on which features are important to individual users.

We first review the stand-alone programs (see Table 12.1 for comparison), then discuss the general purpose software (see Table 12.2 for comparison), and finally briefly review two programs that can extract the data underlying a graphical display. Readers will need to keep in mind that software features, cost, and availability all change fairly rapidly over time; while some of the specific information here may be out of date by the time you are reading this chapter, the general issues and principles we discuss in choosing software for meta-analysis will have a longer half-life. Web searches, the Methods sections of recent research syntheses, and professional meetings where research synthesis results and methods are presented, are good resources for keeping up with both software availability and developments in methodology.

STAND-ALONE META-ANALYSIS SOFTWARE
Comprehensive Meta-analysis (CMA)

CMA is a popular package that has been developed under the consultation of a team of meta-analytic experts. It offers an accessible user interface with many options, excellent graphics, and most standard non-Bayesian analyses that do not require iterative computations.

Data entry

Summary data from each study are entered into a spreadsheet. One of the strengths of CMA is the variety of types of summary data it will accept and then convert to the effect size of interest. Additionally, CMA allows the user to select a different format for each study, if needed. For example, the user can provide correlations and sample sizes, correlations and standard errors, or P-values and sample sizes. For binary data, the user can provide the number of events and nonevents, odds ratios and confidence intervals, or chi-square statistics and sample sizes. Altogether, CMA accepts summary data in over 100 different formats. It also allows the user to import data from other Windows-based spreadsheet programs, such as Excel.

Analysis

CMA can perform meta-analysis for the following effect sizes: raw mean difference, standardized mean difference (Cohen's d and Hedges' g—the latter is referred to as Hedges' d in

TABLE 12.1. Comparison of stand-alone software capability for meta-analysis.

	Rev Man	MetaAnalyst	MIX	MetaWin	CMA
Operating System	Windows	Windows	Windows	Windows	Windows
Distributor	Cochrane Collaboration	J Lau, C Schmid	Private	Private	Biostat, Inc
Website	www.cc-ims.net	http://tuftscaes.org/ meta_analyst/	http://www.meta-analysis -made-easy.com/	www.metawinsoft.com	www.metaanalysis.com
Version	5.1	Beta 3.13	2.0	2.1.5	2.2
Price (US $)	FREE	FREE	FREE	60	various
Spreadsheet	∅	✓	✓	✓	✓
Import data	∅	✓	✓	✓	✓
MA interface	✓	✓	✓	✓	✓
Regression	∅	✓	✓	✓	✓
Single group	∅	✓	✓†	✓	✓
Fixed Effects	✓	✓	✓	✓	✓
Random Effects	✓	✓	✓	✓	✓
Multilevel Models	∅	✓	∅	∅	∅
Random Effects Regression	∅	✓	∅	✓	✓
Bayesian Models	∅	✓	∅	∅	∅
Cumulative MA	∅	✓	✓	✓	✓
Small sample/Publication Bias Tests	∅	∅	✓	✓	✓
Binary Data	✓	✓	✓	✓	✓
Continuous Data	✓	✓	✓	✓	✓
Multivariate	∅	✓	∅	∅	∅
Documentation	✓	✓	∅	✓	✓
Forest Plot	✓	✓	✓	✓	✓
Funnel Plot	✓	✓	✓	✓	✓
Data Export options	∅	✓	✓	✓	✓
Technical Support	✓	✓	✓	✓	✓
Programming capabilities	∅	∅	∅	∅	∅
Automated Leave one out sensitivity	∅	✓	∅	✓	✓

†One group analysis available for continuous outcomes only.

ecological literature), odds ratio (and log odds ratio), risk ratio (and log risk ratio), risk difference, correlation (and Fisher's *z*), rate ratio, and hazard ratio. It will also compute point estimates in single group designs, such as the mean, proportion, or rate in a single group. Finally, the program will work with a generic effect size.

Menus and toolbars are used to customize the analyses, and the way they are displayed. The user has the option of choosing either fixed- or random-effects models, or selecting both. The user can also tailor the statistics that will be displayed. Options include any or all of the following: effect size, variance, standard error, *P*-value, confidence interval, study weights, and heterogeneity statistics including τ^2, τ, and I^2.

The program can manage complex data structures, including multiple subgroups, outcomes, time-points, and/or comparisons. Several options are provided for dealing with the various data structures. For example, if some studies include several independent subgroups, the user can restrict an analysis to only some subgroups or combine data across subgroups within studies. Similarly, when some or all studies have more than one time-point at which outcome data were collected, the user can (1) limit the analysis to one or more of the time-points, (2) use the average across time-points, or (3) compare effects in different outcomes at different time-points. However, the program cannot perform an appropriate multivariate analysis across time, as discussed in Chapter 16.

The program can perform analyses using a single categorical (analysis of variance analogue or subgroup analysis) or continuous modifier (weighted regression on continuous data).

Several methods are available to assess small study/publication bias. These include the Begg and Mazumdar (1994) and Egger, Davey Smith, Schneider, et al. (1997) tests, the file drawer method (Orwin 1983), trim and fill (Duval and Tweedie 2000a, 2000b), and cumulative meta-analysis ordered by sample size or precision (Lau et al. 1995).

Sensitivity analysis options include an *omit one* analysis where the meta-analysis is run repeatedly, with one study removed on each pass. Another option is the ability to exclude sets of user-selected studies from the analysis.

Although CMA can do subgroup analyses and fixed-effects meta-regression, it cannot fit Bayesian models.

Graphics

CMA can produce forest plots with many user customizable features. Results can be exported to Microsoft Word or Powerpoint, or can be printed as a high resolution graph. CMA also produces funnel plots and forest plots for cumulative meta-analysis.

Documentation

CMA comes with an online manual, free tutorial examples in PDF form, and free technical support by phone or email. CMA was tested for accuracy against the Stata macros and the RevMan program. The formulas for conducting the synthesis are documented in Borenstein et al. (2009), and the formulas for computing effect sizes are documented in Borenstein et al. (2010a).

Cost and availability

All versions of the software may be downloaded from www.meta-analysis.com, including a fully functional demo program that runs for 10 days, with an on-request extension to one month. Costs vary depending upon whether the purchaser is a student, academic user, or commercial (for-profit company) user, whether a site license or individual use license is needed, and whether the program is obtained through an annual lease that includes free updates, or a perpetual license with optional updates. There are also several versions with different features. As of July 2012, a single perpetual license costs $1295 at the corporate rate, $795 at the academic rate and $395 at the student rate. The price drops to $395 and $265 for corporate and academic institutions if 40 or more copies are purchased. Annual leases for teaching are $95 per copy.

MetaWin 2.0

MetaWin is the only meta-analysis program written by ecologists and specifically designed for ecological data, although it can be used more broadly. MetaWin is easy to use and has options not available in other stand-alone meta-analysis packages, including randomization and bootstrapping procedures. But it can only handle basic analyses, with no facilities for more complex meta-analysis modeling. Its ease of use may tempt novice users to analyze data without necessarily thinking through or fully understanding their choices.

Data entry

Data can be entered directly into a spreadsheet, or imported from a Windows-based spreadsheet such as Excel. The size of the file is limited only by the storage capacity of the user's computer. The spreadsheet can be manipulated with simple commands (save, print, clear cells, cut and paste, insert column, etc.). Effect sizes and their variances can be calculated from means, standard deviations, and sample sizes (Hedges' d, ln response ratio), or from two by two contingency data (odds ratio, rate difference, and relative rate). Statistical functions can be transformed using the MetaWin calculator (e.g., Student's t can be transformed to r, Cohen's d can be transformed to r, F can be transformed to Cohen's d, and so on). Effect sizes of the user's choosing can also be directly input by the user. If correlation coefficients and sample sizes are input, MetaWin will calculate Fisher's z-transform. All commands are carried out by clicking on "radio buttons" to indicate one's choices.

Analysis

Grand mean effect size and its confidence intervals, weighted mean effect sizes for categories of studies (moderators or covariates), and confidence intervals can be calculated using parametric or bootstrap (including bias-corrected bootstrap) techniques. Weighted regressions on continuous moderators can also be computed. Heterogeneity tests (Q statistics) offer the option of parametric or randomization tests. MetaWin does not offer the heterogeneity statistics I^2 and τ^2 for random-effects analyses. Calculations can be based on fixed-effects or random-effects models. Ordinarily, the inverse of the sampling variance is chosen for the weights in the analyses, but the user can choose any other weights desired, or can do unweighted analyses (by inputting a column of ones for the weights). The program will omit any studies with undefined values for means or sampling variances from the analysis, and will indicate in the output which studies were eliminated.

Additional options are the "Refine Categories" and "Refine Studies" tabs. These allow the user to remove studies belonging to particular categories, to remove individual studies (singly or any number of them) and rerun the analyses, or to focus on particular groups; they additionally can be used for sensitivity analysis. One can also perform fail-safe calculations using either Orwin's method (Orwin 1983) or Rosenthal's method (Rosenthal 1979), with options to alter the probability level. Cumulative meta-analyses is available, but not for subgroups. The user can choose any quantitative variable to order the studies for the cumulative analysis.

Graphics

Graphics are directed to data exploration and include normal quantile plots, scatterplots, forest plots, weighted histograms, funnel plots, regression plots, radial plots for odds ratios, and cumulative meta-analysis plots. All can be generated by clicking on the appropriate radio buttons after indicating which data are to be plotted. Most options for detecting small study/publication bias are available except for trim and fill, which is not an automatic option. Graphical output is not of publication quality.

Documentation

MetaWin has an online manual and help buttons. Users can also purchase a printed version of the manual.

Cost and availability

MetaWin 2.0 was published by a commercial scientific textbook publisher, Sinauer Associates, but went out of print before this book was published. Its original cost was $150, discounted for students and for multiple-user licenses (e.g., for teaching). While it is no longer offered or supported by the publisher, it is available for downloading at a reduced price from the original package ($60 as of July 2012). Information about how to purchase it online is available on the MetaWin home page, currently at http://www.metawinsoft.com/.

MIX

MIX is a program for meta-analysis that works as an add-on to Microsoft Excel (Bax et al. 2006). It is easy to use, but is limited in its capabilities.

Data entry

Data are initially input manually through a spreadsheet or by reading in an Excel or text file or a built-in MIX data set. Data may be entered as raw counts for one or two group data, or as summary effect sizes for comparative data. Data can be saved in an internal format.

Analysis

MIX uses computational tools in Excel to do its calculations, but has its own interface that allows the user to choose from different meta-analytic tools. Menus and toolbars are used to customize the analysis. The user can choose to set significance levels, weighting methods, metrics, models, sorting orders, continuity correction for zero cells, graphical output formats, and use of the normal or t-distribution. Graphs are Excel objects and so can be edited with the features of Excel graphs. Data can be saved as text files, .csv files, .xls files, or as MIX data sets. Output can be exported as a graphics object only in .gif or .png format.

MIX can perform basic analyses, but has no sophisticated modeling options and no programming capabilities. MIX fits fixed- and random-effects (DerSimonian-Laird) summary models for both binary and continuous outcomes. Binary metrics include the risk ratio, risk difference, and odds ratio. Continuous metrics include the weighted mean difference, Hedges' g and Cohen's d statistics, and Fisher's z for correlations. For single groups, it supports a generic effect size. The program can do cumulative meta-analysis and fixed-effects regression. It reports only τ^2 as a heterogeneity statistic.

Graphics

MIX has a variety of graphics including the boxplot, normal quantile plot, residual histogram, forest plots, funnel plot, L'Abbé plot (L'Abbé et al. 1987), and radial plot (Galbraith 1994). It also has a variety of quantitative tools for examining small-study/publication bias effects including the fail-safe N, Egger and Begg/Mazumdar tests, and trim and fill. Sensitivity analysis options include an *omit one* analysis where the meta-analysis is run repeatedly, with one study removed on each pass, as well as the ability to exclude sets of user-selected studies from the analysis.

Documentation

Currently, no documentation is available, but this is planned at the next major update.

Cost and availability

MIX is a free add-on to Microsoft Excel and can be downloaded at http://www.meta-analysis -made-easy.com/. A new version 2.0 is now available. This will be distributed in two formats. The professional version will include all features and will require purchase of a license. As of July 2012, fees range from $75 for a student license to $210 for a personal license (for those not at academic or educational institutions). Users in developing countries can get the professional version for $50. A scaled-down "lite" version will be given for free but will not allow the user to save or build data sets. Interested users should check the website for final specifications.

PhyloMeta

PhyloMeta is a simple Windows console program developed for integrating phylogenetic information into all the conventional statistics of ecological and evolutionary meta-analyses (Lajeunesse 2011a). Here the user inputs their meta-analytical data (effect sizes and variances) and a hypothesis on the phylogenetic history connecting these effects (see details in Chapter 17). This phylogenetic hypothesis is then converted into a correlation matrix that is applied to regression models assuming fixed- or random-effects models (described in Lajeunesse 2009). It also calculates model selection criteria (AIC scores) to help distinguish the fit of these different meta-analytical models. It was designed to complement the nonphylogenetic analyses of MetaWin. It is easy to use, but has limited capability for running more elaborate regression models or for testing more sophisticated evolutionary hypotheses.

Data entry

The data entry is simple and requires the input of two text files; one contains the effect size data and moderator groupings, and the other contains the hypothesized phylogeny in NEWICK format. These effect sizes and variances must be computed externally before the analyses. PhyloMeta also cannot analyze multiple effect sizes for single species—users must pool all the effects for a single species prior to phylogenetic meta-analyses. There is also no restriction on the type of effect size metric used to quantify experimental outcomes.

Analysis

The user does not specify what analyses to perform; all are performed simultaneously in a single run of phyloMeta (e.g., fixed- and random-effects, within- and between-study Q-tests, analyses with and without phylogenetic information, or subgroup analyses). This design is meant to facilitate the calling of phyloMeta in R for simulation analyses or for visualizing results. It does not perform meta-regressions or advanced multilevel, multivariate, or Bayesian models. All analyses are outputted to the console screen and to a text file.

Graphics

PhyloMeta is a console program in Windows, and only provides tables of results in text format.

Documentation

Additional details on the implementation of phyloMeta are found in Lajeunesse (2011a). PhyloMeta also has a manual that describes how to perform analyses and includes troubleshooting tips for conducting phylogenetic meta-analyses. This manual is free and currently available on Marc Lajeunesse's website at http://lajeunesse.myweb.usf.edu/publications.html.

Cost and availability

PhyloMeta can be obtained at no cost as a Windows executable file on the web at http://lajeunesse.myweb.usf.edu/publications.

RevMan

RevMan (Review Manager) was developed by the Cochrane Collaboration for use with Cochrane reviews, though its use is not restricted to this application. RevMan is designed to manage all stages of a systematic review, rather than solely to conduct a meta-analysis. Users enter the protocol and the complete review, including text, characteristics of studies, comparison table, and study data. This is different than the other software reviewed in this chapter. Furthermore, since RevMan is designed for health care applications, it is primarily used for analysis of binary data, although it will also analyze mean differences.

The advantage of consistency gained by providing a standardized procedure and software for all reviews also limits its utility. The Cochrane process is quite specific and incorporates many features relevant to research in health fields. Geographically scattered review groups combine to publish reviews intended to be accessible to all interested parties. Thus, it excludes the use of advanced statistical analysis, such as regression, multilevel, multivariate, and Bayesian models, and has very limited capacity to account for random effects. It is not suitable for describing the heterogeneous types of data ecologists and evolutionary biologists frequently encounter.

Data entry

RevMan accepts data in only four formats: events and sample size; means, standard deviations and sample size; observed minus expected frequencies, standardized by the variance;

and a generic effect size. All studies must be entered in the same format. In order to deal with summary data in different formats, the user must compute the effect size and variance for each study externally and then manually input the computed effect size and variance into RevMan. Users can import summary data from spreadsheet programs, such as Excel.

The program can perform meta-analyses using raw mean differences, standardized mean differences (Hedges' *g*), odds ratios, risk ratios, risk differences, and a generic effect size.

Analysis

The user can specify the model (fixed or random effects), statistical method (e.g., Mantel-Haenszel, inverse variance, or one-step), and effect size measure. RevMan provides Q, I^2, and τ^2 as heterogeneity statistics. If data are entered for subgroups, RevMan can conduct a separate analysis for each subgroup, and can also compare effects across subgroups for fixed-effects inverse-variance based methods. It cannot, however, carry out meta-regressions or any advanced multilevel, multivariate, or Bayesian models. Sensitivity analysis in RevMan is limited to the ability to exclude sets of user-selected studies from the analysis.

Graphics

The results are displayed as a forest plot. RevMan does not carry out statistical tests for small-study/publication bias, but it will produce a funnel plot.

Documentation

The Cochrane Collaboration publishes a large handbook which serves as a standard reference for doing systematic reviews, as well as a manual for using RevMan (Higgins and Green 2011). This is available for a free download at http://handbook.cochrane.org.

Cost and availability

RevMan can be downloaded for free at http://ims.cochrane.org/RevMan.

MetaAnalyst

MetaAnalyst is a Windows-based enhancement of an old MS-DOS program distributed for many years as shareware by Joseph Lau. The update implements many advanced analytic capabilities for meta-analysis. Currently, it is undergoing a major revision that will enable it to directly interface with both the R and BUGS languages (Wallace et al. 2012).

Data entry

MetaAnalyst currently supports manual input of data through a spreadsheet format, as well as data imported from Microsoft Excel (.xls) and Comma Separated Value (.csv) formats. Users are prompted as to what type of data (e.g., one or two arm, continuous/binary/diagnostic) they wish to enter and are presented with the appropriate spreadsheet that includes study identifier, study name, and year. Data may be saved in internal format. Summary metrics are automatically calculated in the spreadsheet and displayed in a variety of formats (e.g., odds ratio, risk ratio, or risk difference for binary data), along with confidence intervals. Covariates may be added and removed at any time. Column labels may also be changed.

Analysis

MetaAnalyst supports a large number of analyses. Users can perform analyses of continuous data effects (means, standardized means, Hedges' *g*), binary data effects (risk difference, risk ratio, and odds ratio), and diagnostic accuracy metrics (e.g., sensitivity and specificity). Users may perform fixed-effects analyses with inverse variance, Mantel-Haenszel, and Peto weights; random-effects analyses using the DerSimonian-Laird weights; maximum likelihood fit by the EM algorithm, and Bayesian analyses. Meta-regression is supported for fixed, random, and Bayesian analyses, and allows the use of any number of covariates, including the baseline risk (McIntosh 1996, Schmid et al. 2004). Users may specify the confidence or credible (Bayesian confidence) levels desired. The program supports analyses of subgroups and allows exclusion of any arbitrary set of studies; it will perform sensitivity analyses leaving one study out at a time.

The Bayesian analyses use OpenBUGS (the open source version of WinBUGS, http://www .openbugs.info/w/) to perform Markov chain Monte Carlo (MCMC) analyses. Users may select different inverse gamma prior distributions for variance parameters. Bayesian analyses produce summaries of quantiles of the posterior distribution, as well as posterior probabilities of summary effects. MetaAnalyst currently implements specific OpenBUGS script files, but will shortly have the capacity for users to implement their own script files. The Gelman-Rubin diagnostic (Gelman et al. 2004) is automatically calculated to help diagnose convergence. Users can define the convergence parameter as well as the number of simulations to run and to save. Various other output statistics can be requested, including summaries and plots of the chains run; these encompass the means, medians, and percentiles of each parameter, their autocorrelations, and plots of their empirical probability densities. Numerical output is saved as .rtf files.

Graphics

Graphic types supported include forest plots for single and cumulative meta-analyses, L'Abbé plots, and funnel plots. Plots of parameter traces and autocorrelations, as well as kernel density plots of the parameters are available for Bayesian models. Plots of the observed risk and their shrinkage to posterior estimates are also included. All of these plots can be edited within MetaAnalyst and are saved as .png images.

Documentation

MetaAnalyst has online help and documentation that come with the downloaded program.

Cost and availability

MetaAnalyst is free and available for download from http://tuftscaes.org/meta_analyst/.

MACROS FOR GENERAL PURPOSE STATISTICAL SOFTWARE

Meta-analytic models can be fit with a standard regression or analysis of variance program in any of the major statistical packages, but require modifications to the standard output because of their specific within-study weights. Each of the major packages also has user-written routines that have implemented specific meta-analytic macros. We first describe the manipulations necessary to use standard statistical software for meta-analysis, and then describe meta-analysis macros that can be obtained for each of the four major packages.

TABLE 12.2. Comparison of general purpose statistical software capability for meta-analysis.

	Stata/BUGS	SAS	SPSS	R/BUGS
Operating System	Windows, Mac (only Stata), Linux, Unix	Windows, Linux, Unix	Windows, Mac, Linux	Windows, Mac (only R), Linux, Unix
Distributor	Stata Corp	SAS Corp	SPSS	R Project
Website	www.stata.com	www.sas.com	www-01.ibm .com/software/ analytics/spss/ products/ statistics	www.r-project.org
Version	12	9.3	20	2.15
Price (US $)	From 595 for annual license (Intercooled version)	Various	Various	FREE
Spreadsheet	✓	∅	✓	✓
Import data	✓	✓	✓	✓
MA interface/routines	Macros	Macros	Macros	Macros
Regression	✓	✓	✓	✓
Single group	✓	✓	✓	✓
Fixed Effects	✓	✓	✓	✓
Random Effects	✓	✓	✓	✓
Multilevel Models	✓	✓	✓	✓
Random Effects Regression	✓	✓	✓	✓
Bayesian Models	✓	✓	∅	✓
Cumulative MA	✓	✓	✓	✓
Publication Bias	✓	✓	✓	✓
Binary Data	✓	✓	✓	✓
Continuous	✓	✓	✓	✓
Multivariate	✓	✓	✓	✓
Documentation	✓	✓	✓	✓
Forest Plot	✓	✓	✓	✓
Funnel Plot	✓	✓	✓	✓
Data Export options	✓	✓	✓	✓
Technical Support	✓	✓	✓	✓
Programming capabilities	✓	✓	✓	✓
Automated Leave one out sensitivity	✓	✓	✓	✓

As described in earlier chapters, meta-analytic models are estimated by method of moments, maximum likelihood, and Bayesian methods. With method of moments, the meta-analytic estimate is a weighted average and can thus be calculated by using a regression program supplying weights as an option. When the inverse variance study weights are assumed to be known, however, a slight modification to the regression output is required. This arises because weighted regression programs typically assume that the weights supplied are proportional to the study variances (not equal to them), with the constant of proportionality defined as the residual mean squared error, σ^2. Therefore, the proper standard errors for the meta-analytic estimates require dividing the standard errors output (as provided by the programs) by the estimate of σ. When using a mixed model routine as in Chapter 10 for maximum likelihood, the variance parameters are specified explicitly, so this correction is not needed. Instead, the proper likelihood is calculated from the model specified by the user. Likewise, the Bayesian analysis specifies the within-study variances explicitly, although it is also possible to model a common proportionality constant as well. Whitehead (2002) provides more information and example code for manipulating software packages to do meta-analysis.

Stata

Stata is a general purpose computer package that has considerable flexibility, responds well to user requests, and provides a suite of meta-analytic programs contributed by users (Sterne et al. 2009). Stata has no graphical user interface for meta-analysis and must be used in a command-line format instead. On the other hand, the program syntax is straightforward and most analyses only require one line of code, usually giving the name of the program, the data, and the options required for the analysis. Stata has clean graphics, particularly for the funnel plot, and appears in many publications of meta-analyses. It also has an interface to BUGS (see below) called *winbugsfromstata*, so it is possible to run Bayesian analyses and return the results to Stata. Thus, Stata can be used for data entry, manipulation, and presentation, leaving BUGS to do the calculation.

Data entry

Data are passed into the meta-analytic routines through Stata. Users can input cell frequencies from two by two tables for binary outcomes, means and standard deviations for numerical outcomes, or estimates and standard errors; these are placed in the Stata meta-analysis functions.

Analysis

Stata provides several commands for meta-analysis. *Metan* is the main meta-analysis command; it provides a comprehensive range of methods for meta-analysis using fixed- and random-effects models, and creates new variables containing the treatment effect estimate and its standard error for each study. These variables can then be used as input to other Stata meta-analysis commands. Meta-analyses may be conducted in subgroups by using the by() option. *Metacum* is a command for implementing cumulative meta-analysis and *metap* performs meta-analysis of *P*-values. Meta-regression is implemented in the *metareg* command. Stata also has commands for implementing advanced methods, including *glst* for generalized least squares that is used for trend estimation of summarized dose-response data (Berlin et al. 1993), *mvmeta* for maximum likelihood, restricted maximum likelihood, or method of moments estimation of random-effects multivariate models; and *metamiss* for meta-analysis with binary outcomes when some or all studies have missing data. *Metabias* implements tests for asymmetry in funnel plots and *metatrim* does trim and fill.

Graphics

Stata implements the funnel plot with the *metafunnel* command, and the contour enhanced funnel plot with *confunnel*. Other graphics include the forest plot with *metan*, and the L'Abbé plot with *labbe*.

Documentation

The Stata meta-analysis commands are not part of the official Stata release, but they can be downloaded from the central repository for Stata packages at http://www.repec.org, which is also directly accessible through Stata. Within Stata, you can get help for a description of these commands, as well as their installation instructions. They have been collected and updated by Sterne et al. (2009). Another useful source is the book by Egger et al. (2001). Stata itself comes with documentation and also publishes *Stata Journal* (www.stata-journal.com), a quarterly journal that describes Stata computing resources. The *winbugsfromstata* connection to BUGS can be downloaded at http://www2.le.ac.uk/departments/health-sciences/research/gen-epi/ Progs/winbugs-from-stata/.

Cost and availability

Stata can be ordered from www.stata.com. As of July 2012, the price for the Intercooled version 12 for a government/nonprofit single user is $595 for the annual license and $1195 for the perpetual license.

R

R is the open source version of the S programming language. Users have contributed many packages to its CRAN repository that implement new statistical methods. These can be installed in the local computing environment. The best R package for meta-analysis is *metafor*, developed by Wolfgang Viechtbauer (www.wvbauer.com/downloads.html). It is described below.

Data entry

Data are passed into the meta-analytic routines through R, which means that they can be entered directly or read in through a wide variety of formats using R import functions. The package has a function to calculate common effect size metrics from raw data.

Analysis

The functions available allow users to carry out method of moments analyses for binary and continuous outcomes, including fixed-effects models (Mantel-Haenszel, Peto, and inverse variance weighting methods) and random-effects models (various methods for estimating the between-study variance), meta-regression using generalized least squares, cumulative meta-analysis, tests for small sample/publication bias, and the trim and fill method. Various functions enable users to extract summary statistics, such as estimates and confidence intervals from the fitted model objects, and to work with residuals. Sensitivity analyses leaving one study out at a time are also available.

More sophisticated analyses can be implemented in two ways with R. First, users can write their own functions in the R programming language. Second, users desiring to perform Bayesian analyses can use the R packages *R2WinBUGS*, *R2jags,* and *BRugs* to enable R to call the BUGS environment (see below).

Graphics

The metafor package implements forest plots, funnel plots, residual diagnostic and influence plots, normal probability plots, the radial plot, and plots of the sensitivity of estimates to leaving out one study at a time. Users also have R's powerful graphics facilities available to develop and customize their own presentation quality figures.

Documentation

R packages can be downloaded from http://cran.r-project.org/web/packages/. All R packages have associated help files that are installed when the package is loaded. These explain the various R functions that comprise the package. A variety of books are available for programming in R, including many online documents. The metafor package is described in Viechtbauer (2010).

Cost and availability

R is an open source program available for free download at www.r-project.org. User-contributed packages stored at www.cran.r-project.org/web/packages/ may be downloaded from within R.

SPSS

SPSS is the most widely used statistical package in the social sciences. Mark Lipsey and David Wilson have written a set of macros for doing method of moments calculations for random- and fixed-effects meta-analysis and for meta-regression. These analyses must be performed from the syntax window using the command line interface rather than from the pull-down menus in SPSS.

Data entry

The Lipsey and Wilson macros use the effect size and its standard error as inputs. Therefore, these values must already have been calculated in the SPSS data set including any transformations, such as small sample size bias corrections, the Fisher's z transformation for correlations, or the logarithmic transformation for odds ratios and response ratios that are necessary for analysis.

Analysis

After initialization, each macro is run with the inputted effect size and standard error. *MeanES* calculates basic statistics, such as mean effect size, confidence intervals, z-test, and a test of homogeneity for fixed- and random-effects models. *MetaF* performs the analog to the one-way ANOVA analysis and allows specification of either a fixed-effects model, or three methods of estimating mixed-effects (random-effects) models: method of moments, full-information maximum likelihood, and restricted-information maximum likelihood. It is useful for testing differences across mean effect sizes for a categorical variable, such as treatment. *MetaReg* performs weighted generalized least squares regression for a fixed-effects model, or mixed-effects (random-effects) models using method of moments, full-information maximum likelihood, and restricted-information maximum likelihood.

Graphics

Plots must be created from the output of the macros using SPSS graphics facilities.

Documentation

A "read me" file with instructions for the macros is part of the download (see "cost and availability," below). SPSS itself has extensive documentation that comes with the program.

Cost and availability

The Lipsey and Wilson macros are available for a free download from http://mason.gmu .edu/~dwilsonb/ma.html. SPSS itself has recently been bought by IBM, and the statistics package is now named IBM SPSS and can be purchased from http://www-01.ibm.com/software/ analytics/spss/products/statistics/. Reduced pricing is available for educational purposes and for students. There are also discounts for bulk purchasing.

SAS

SAS is the most widely used statistical package, particularly in the corporate world. In fact, SAS is no longer merely a statistical package but a data environment (which is why you no longer find its original name of Statistical Analysis System anywhere in its documentation). SAS produces many books that show users how to apply SAS coding to statistical problems. Wang and Bushman (1999) have written one for meta-analysis. Kuss and Koch (1996) also provide meta-analysis macros for SAS. Lipsey and Wilson's macros described above for SPSS are also written for SAS. Below, we discuss the use of the macros developed for meta-analysis that are described in Wang and Bushman (1999), and Kuss and Koch (1996). Sophisticated users may also use PROC MIXED (Van Houwelingen et al. 2002) and PROC NLMIXED (Macaskill 2004) to fit all sorts of different meta-analytic models, but these require familiarity with the programming language.

Data entry

Data are input and manipulated using standard SAS programming code. SAS also has a user interface SAS Enterprise Guide that allows input into a spreadsheet type format. SAS stores data in databases that are accessed through programming code in data steps. Point and click options in the guide are automatically translated into SAS programming code.

Analysis

Wang and Bushman (1999) show how to use SAS commands and statistical procedures to fit fixed- and mixed- (random) effects models for continuous and categorical data and correlations. Kuss and Koch (1996) provide several macros, including *metacalc* for fixed- and random-effects meta-analysis of the risk difference, log risk ratio, or log odds ratio (inverse variance, Mantel-Haenzel, or Peto weights).

Graphics

Wang and Bushman (1999) show how to use SAS graphics and macros to make forest plots, funnel plots, quantile plots, and other descriptive plots. Kuss and Koch (1996) provide the macro *metafunn* for producing a funnel plot, *metaci* for a forest plot, and *metagalb* for a radial plot. They also include *metasens* macro for a sensitivity plot that describes the change in the mean estimate as the between-study variance estimate changes.

Documentation

The book by Wang and Bushman (1999) and the paper by Kuss and Koch (1996) provide instructions on using the code. SAS itself has voluminous documentation and a whole library of books on using SAS statistical procedures.

Cost and availability

The Lipsey and Wilson macros can be downloaded at no cost at http://mason.gmu. edu/~dwilsonb/ma.html. The Kuss and Koch and Bushman and Wang macros are available in their publications. SAS itself can be ordered from www.sas.com and is individually priced according to the configurations of the client. Although there is not one standard price, it is very expensive since it requires purchase of a yearly license for at least the SAS/Base and SAS/Stat modules.

BUGS

BUGS (Bayesian inference Under Gibbs Sampling) is the most widely used software for doing Bayesian analysis. It performs Markov chain Monte Carlo (MCMC; Chapter 11) using an intelligent algorithm that determines the appropriate numerical routine from the model structure provided by the user. These include Gibbs sampling, adaptive rejection sampling, and Metropolis-Hastings (in order of increasing complexity). BUGS is implemented in three different versions: WinBUGS, OpenBUGS, and JAGS (Just Another Gibbs Sampler). They use different numerical algorithms so that simulation results can differ slightly. In particular, models may converge at different rates. More importantly, OpenBUGS and JAGS are open source programs that are being maintained, while WinBUGS is no longer updated. Open-BUGS works in the Microsoft Windows environment, whereas JAGS works on Windows, Mac, and Linux/Unix.

All packages can be used in stand-alone form, but interfaces to general purpose software (R and Stata) are also available. These include the R packages *R2WinBUGS*, *R2jags,* and *BRugs,* and the Stata package *winbugsfromstata*. These interfaces offer routines to run the BUGS commands within the R or Stata environment so that MCMC output can be manipulated from within the statistical environment. Because Stata and R have extremely flexible graphic capabilities, quite exceptional plots can be made showing the benefits of using these complex models (see some examples in Chapter 11). We next describe the capabilities of the BUGS environments, independent of their potential inside R or Stata.

Data entry

To run the MCMC algorithm, the user must provide three files: (1) a short program defining the probabilistic structure of the problem including the prior distribution, (2) data for each fixed variable and outcome, and 3) starting values for the parameters of the MCMC algorithm.

Analysis

After compiling the input files, the analyst starts the simulation with a graphical user interface. BUGS decides which type of MCMC algorithm is best and generates simulations from the joint posterior distribution. The user may then investigate these simulations through a variety of statistical summaries and graphical displays. The user supplies the number of simulations to run, and can then check on convergence of the algorithm with a variety of diagnostics available from pull-down menus. Various other output statistics can be requested, including summaries and plots of the chains run; the plots include the means, medians, and percentiles of each parameter. These statistics and plots are useful for diagnosing convergence and for understanding the joint posterior density of the parameters.

Graphics

BUGS implements a variety of graphical summaries that include traces of the parameters simulated, autocorrelations, empirical probability densities (kernel density plots), and Gelman-Rubin-Brooks convergence diagnostics. These displays cannot be directly exported outside of BUGS without an interface to a statistical package. The outputs of the chains can also be saved in text files and read into programs, such as the R package CODA, that make use of these simulations.

Documentation

Online manuals are available and downloaded with the program. These include three volumes of examples that show how to implement various statistical analyses in BUGS. Two of these are written explicitly for meta-analysis. Bayesian meta-analytic methods described in the statistical literature generally use BUGS, and authors often provide code in the papers. Four books by Peter Congdon also provide many examples with code (Congdon 2003, 2005, 2007, 2010).

Cost and availability

WinBUGS is developed and distributed by the Medical Research Council in Great Britain and is available free of charge from www.mrc-bsu.cam.ac.uk; however, a software key is needed to enable the full power for large data sets. This key will be sent to the user upon registration. OpenBUGS is an open source version available at http://www.openbugs.info/w/, and JAGS is open source available at http://mcmc-jags.sourceforge.net.

SOFTWARE FOR DIGITIZING DATA FROM GRAPHS

Before conducting a meta-analysis, the research synthesist must collect the primary data that will go into the meta-analysis. In many papers in the fields of ecology, evolution, and conservation, the data are often presented in graphical form, including bar graphs, points with confidence intervals for means, scatter graphs, and so forth. Although contacting authors to obtain the numerical data is an option, it may be a slow and frustrating process. Many meta-analysts in these fields therefore rely upon using software to convert scanned images of graphs to the underlying numbers, and thus allow inclusion of that study in the meta-analysis.

Many different programs are available to digitize graphs. They are generally quite inexpensive, and include various other functions, depending on the package. They digitize data automatically from scanned files or manually with mouse clicks. The programs generally accept many image file formats and run on common platforms such as Windows, Linux, and less commonly, Macintosh. For most applications any of these will work.

Free programs include DigitizeIt (http://www.digitizeit.de/), Engauge Digitizer (http://digitizer.sourceforge.net/), DataThief (http://www.datathief.org/), ImageJ (http://rsbweb.nih.gov/ij/), and GrabIt! (http://www.datatrendsoftware.com/home.html), which works with Excel. Commercial packages include Dagra (http://www.blueleafsoftware.com/Products/Dagra/) and Ungraph (http://www.biosoft.com/w/ungraph.htm).

Various issues and limitations are involved in accurately digitizing data from graphs. For example, points may be hidden behind other points, accuracy is limited by the resolution of the digitized image, and the translation to numbers may be limited by the accuracy of locating the mouse clicks in some programs. However, we are of the opinion that "pretty good" information is often much better than no information at all, given that users recognize the limitations of the data with which they are working.

CONCLUSIONS

Doing all but the simplest meta-analytic calculations requires a computer and appropriate software. Many different general purpose software packages have statistical and graphical tools that enable users to carry out meta-analyses, but these packages usually require knowledgeable programming and do not include specific tools for meta-analysis. Users without sufficient methodologic and programming background to understand the statistical subtleties may struggle to work out how to correctly implement meta-analyses with these general purpose statistical packages. For nontechnical users, one of the stand-alone packages may be much easier to use, although the modeling options differ considerably among packages. Meta-analysis is becoming a more common scientific enterprise and in the health sciences, it already has a higher citation impact than any other research design (Patsopoulos et al. 2005); as it is increasingly used, the associated software will undoubtedly continue to become more widely available, more flexible, and easier to use.

Statistical Issues and Problems

Recovering Missing or Partial Data from Studies: A Survey of Conversions and Imputations for Meta-analysis

Marc J. Lajeunesse

META-ANALYSIS USES SUMMARY STATISTICS like effect sizes to combine information from multiple studies. Yet a common problem encountered when collecting information for calculating effect sizes is the absence of data from published studies. The incomplete reporting of means, correlations, variances, and sample sizes can bias meta-analysis in many ways: reviews will have smaller sample sizes because studies with missing data are often excluded (Orwin and Cordray 1985, Follmann et al. 1992); effect size metrics like Hedges' d are disfavored because they require too many within-study statistics; approaches to pooling effect sizes will use less restrictive statistical models such as unweighted analyses (Kelley et al. 2004, Furukawa et al. 2006); and meta-analysis may yield spurious results because excluding studies with missing information could further exacerbate publication bias.

In this chapter, I discuss possible solutions for dealing with partial information and missing data from published studies (Box 13.1). These solutions can improve the amount of the information extracted from individual studies, and increase the representation of data for meta-analysis. I rely heavily on advances and observations from the medical literature; this is necessary given that discussion relating to missing information has received limited attention in ecological and evolutionary meta-analysis (Lajeunesse and Forbes 2003). I begin with a description of the mechanisms that generate missing information within studies, followed by a discussion of how gaps of information can influence meta-analysis and the way studies are quantitatively reviewed. I then suggest some practical solutions to recovering missing statistics from published studies. These include statistical acrobatics to convert available information (e.g., t-test) into those that are more useful to compute effect sizes, as well as a few heuristic approaches that impute (fill gaps) missing information when pooling effect sizes (e.g., Follmann et al. 1992, Yuan and Little 2009). Finally, I discuss multiple-imputation methods that account for the uncertainty associated with filling gaps of information when performing meta-analysis.

DEFICIENCIES IN THE LITERATURE

The selective or variable reporting of statistics used to estimate effect sizes, such as means, variances, and sample sizes, can significantly affect meta-analysis and its reliability to synthesize research. A study may report a t-test that evaluates the difference between a control and treatment mean, but may not report information on the standard deviations or sample sizes of

BOX 13.1.
Classification of published studies based on what statistical information they lack, and suggested approaches to filling these gaps.

Usefulness for meta-analysis	Study statistics	What is available		Addressing what's missing
high	Completely reported	Has all the data for inclusion	→	Nothing missing!
	Selectively reported	All the data are available but not in forms that are easily integrated into meta-analysis (e.g., data in figures, sample sizes need to be determined from table, t-tests and means are not reported, etc.)	→	Extract data from figure or tables (see Chapter 5), convert available statistics (e.g., t-test into effect size)
	Partially reported	Has some data (e.g., sample sizes) but is missing information that cannot be estimated directly from what is available (e.g., variance estimates)	→	Recalculation or conversion of available statistics (back calculation from P-values), or within-study imputation methods.
	Qualitatively reported	No useful data except for P-values or discussion regarding the significance or non-significance of analysis	→	Recalculation of statistics, or use within-study imputation methods or multiple-imputation methods
low	Unreported	No statistics or data are available, although may have specified a protocol for the analysis in the Methods section	→	Exclude from meta-analysis or use an alternative approach to meta-analysis (e.g., vote-count methods)

these means. This is a challenge for meta-analysis because in order for this study to contribute any quantitative information to a review, an effect size must be computed to summarize its findings. Thus, extracting this missing information is important to maintain the scope of the review. I discuss below why there is missing information in published studies before outlining methods useful for recovering or imputing missing information.

Mechanisms that cause data to be missing

There are several mechanisms that can generate gaps of information in published studies. One is the perceived lack of importance. Chan, Hróbjartsson, et al. (2004) found that the lack of clinical importance was the primary reason why medical researchers omitted information from publications. Choosing to omit details of study design (e.g., sample sizes) or analyses ancillary to the main topic under study are examples of this type of reasoning. This also applies to excluding summary statistics such as variances and standard deviations—for example, where the statistical test itself (e.g., t-test) is thought to be more noteworthy for describing study outcomes. This issue is further exacerbated by the editorial policies of many journals aiming for brevity and imposing restrictions on the amount of information reported in the main research article (i.e., penalizing studies that overextend pagination with additional publication costs). For example, authors may exclude information when attempting to meet the requirements of editorial policies prior to submitting their research for publication—omitted information might include fully reported and annotated ANOVA tables. These restricted editorial policies often leave authors without any real incentive to fully report results (unless enforced later by referees as a condition of acceptance for publication).

When information is excluded this way, it is assumed to be missing at random, without being related to the outcome of the study. This is because inclusion/exclusion of this information may not affect the interpretation of the study outcome. A more serious nonrandom mechanism that can contribute to missing information is the lack of statistical significance. Chan, Hróbjartsson, et al. (2004) and Chan, Krleza-Jeric, et al. (2004) found that medical studies were half as likely to fully report statistics of nonsignificant outcomes as compared to the significant ones. In ecology, Cassey et al. (2004) also found that studies missing information tended to be nonsignificant or of lower quality. This type of nonrandom reporting is known as dissemination bias. Here, summary information of the data, results, and statistics are partially reported or excluded entirely (e.g., summarized with only a nonsignificant P-value), statistical assumptions are not fully addressed, or exact statistical procedures are unspecified (Hahn et al. 2000, Pigott 2001). Given that statistical significance is an important criterion for publication (or even whether the study is submitted for publication; Chapter 14), the motivations for why null research outcomes get less coverage in publications become apparent. For example, it is known that selective reporting of research findings—emphasizing strong positive or negative effects while understating nonsignificant findings—can significantly improve the chances of publishing (Chan, Hróbjartsson, et al. 2004). Yet, when the hypothesis is not falsified, it is unclear whether this outcome is due to errors in biological or statistical assumptions. For example, here it may be difficult to distinguish between a nonexisting biological effect and an existing effect that remains undetected because of low statistical power (Chapter 14). For meta-analysis, when there is dissemination bias for emphasizing significant results, and null outcomes are underreported in the primary research, then this has potential to generate biased review outcomes. This is because studies lacking information (which are more often null, see Cassey et al. 2004) will likely be excluded from the review, further exacerbating statistical problems associated with publication bias (see Chapter 14).

How does variable reporting of statistics affect meta-analysis?

In the previous section, I described mechanisms that contribute to incomplete reporting of statistics within studies. Here I examine how this lack of information can diminish the power of meta-analysis to detect nonzero research outcomes. One approach to handling studies with incomplete information is to exclude them entirely from the meta-analysis. Taken to this extreme, the variable reporting of statistics will decrease the sample size of meta-analysis. Small review sample size will reduce the power to detect significant research outcomes and the ability to properly evaluate study heterogeneity (Chapter 22). In a Monte Carlo study that simulated studies with incomplete information (e.g., missing means, variances, or sample sizes), Lajeunesse and Forbes (2003) found that a stringent exclusion criterion has the potential to increase the likelihood of making a review level type II error (false negative). This is because meta-analysis, much like a primary study, is sensitive to sampling error when there are too few data for analysis. For example, small review sample sizes (much like small samples within studies) tend to underestimate or overestimate effect sizes, and yield broad confidence intervals (see Figure 22.2 in Chapter 22; Hedges and Olkin 1985).

Rosenthal (1991) refers to this relationship between the meta-analysis sample size and the ability to detect an effect as second-order sampling error; compared to the first-order sampling error of primary studies (Chapter 22). Still, second-order sampling error assumes that studies (irrespective of whether they are included or excluded from analysis) are a *random* sample of a population with common research outcomes. Clearly, publication bias is known to affect the random sampling of studies used in meta-analysis. For example, this occurs when nonsignificant or marginally significant research is less likely to be published and has minor representation in meta-analyses (e.g., file drawer problem, Chapter 14). What is less clear is whether the incomplete reporting of statistics and subsequent exclusion of such studies from meta-analysis can exacerbate this bias. This depends on whether the missing information within studies is omitted completely at random —that is, unrelated to any observed variable, including the missing statistic itself. In this case, the approach of excluding studies with incomplete information would not cause bias, or at least would not exacerbate publication bias, but would simply erode statistical power as predicted by second-order sampling error (Chapter 22).

However, as described earlier (Cassey et al. 2004; Chan, Hróbjartsson, et al. 2004; Chan, Krleza-Jeric, et al. 2004), there is empirical evidence to suggest that studies with partial information are not missing this information at random. It is known that studies with missing information are likely to be nonsignificant or of lower quality, although previous observations implied that they were not (Englund et al. 1999). In terms of meta-analysis, the selective reporting of statistics due to the study's outcome will certainly exacerbate publication bias; low-quality or nonsignificant study outcomes will be vaguely described and only contain partial information of the research. This has the potential to bias conclusions drawn from research syntheses because mostly significant findings with fully reported results are included in the meta-analysis.

A GUIDE TO HANDLING MISSING INFORMATION

I outline below various approaches to handling missing information from published studies. These methods are grouped under three approaches: (a) contacting researchers for missing data, (b) using algebraic recalculations and within-study imputations for estimating effect sizes and variances, and (c) using between-study imputation methods for filling gaps of information when pooling effect sizes. It is important to note that there will always be more uncertainty in the estimation of effect sizes and variances when approximations or imputation techniques are applied, as compared to the case of having a data set with fully reported information (Pigott 1994). However, relative to the alternative of excluding studies with missing information, the need to improve statistical power and issues relating to publication/dissemination bias far outweighs the increased uncertainty associated with imputed analyses.

Contacting researchers for missing data

Before using statistical approximations or imputing data, contacting authors of the publication for the original data is a good start to recovering missing information. Chapters 4 and 5 review some of the aspects regarding this problem; these include approaches to increase reply success, and potential problems for authors retrieving these data (e.g., data are stored on outdated floppy disks). For example, Chan, Hróbjartsson, et al. (2004) found that multiple sequential questionnaires were required to get a reply from 90% of the primary researchers about missing information (though 80% of the researchers that replied to the first questionnaire denied the existence of missing information). Having access to the raw data is ideal for meta-analysis because precise estimates of effects and variances can be calculated, and sources of bias not described in the original publication can be investigated.

Algebraic recalculations, conversions, and approximations

Partial information can often be recovered by recalculating the available summary statistics or by using approximations when information is limited. For example, if a study reports only P-values, these can be calculated directly into t-tests or F-statistics, which then can be converted into effect sizes. Boxes 13.2 through 13.5 provide a roundup of useful equations to recalculate and convert what is available into various effect size metrics. For further information on this material or for additional examples of more complicated situations with incomplete information, see Fern and Monroe (1996), Glass et al. (1981), Gilpin (1993), Chinn (2000), Lipsey and Wilson (2001), Hozo et al. (2005), Pearson (1932), Wiebe et al. (2006), Rosenthal and Rosnow (1991), Terrell (1982), and Walter and Yao (2007).

These conversions and approximations, however, assume that the original data do not violate assumptions of normality (Lipsey and Wilson 2001). These equations are also limited by the numerical precision of the reported summary statistics, and the efficiency of these conversions and approximations become increasingly unreliable when too few digits are reported. Conversions from P-values are particularly sensitive to this problem (Philbrook et al. 2007). Unfortunately, equivalent statistical conversions and approximations for the log response ratio have yet been developed (Gurevitch and Hedges 1999, Hedges et al. 1999, Lajeunesse 2011b). Further, it is always a good practice to test whether studies summarized with conversions or approximations introduce bias to results (especially before pooling results to test ecological hypotheses). This can be evaluated with a sensitivity analysis that contrasts the magnitude and direction of effect sizes from studies with complete and incomplete (but converted) information.

Within-study imputation

When recalculations are impossible, imputation methods can be used to fill gaps of information in order to calculate effect sizes and their variances. To "impute" data means that the missing piece of information is filled with a substitute. For example, without information on the standard deviations (SD) of a study, effect size metrics like Hedges' d cannot be calculated directly (see definition in Box 13.2; Chapter 6). Imputation methods provide a way to filling this missing SD by either using the available data from other studies, or data from previously published meta-analyses. These imputation approaches can be useful given that the standard deviations of the control and treatment means are often not reported in primary studies.

One approach to estimating missing standard deviations is to use the available means ($\overline{X}$) and SD (e.g., from a control or treatment groups) from all the studies with complete information in order to calculate the coefficient of variation (e.g., the SD to mean ratio; Bracken 1992). For example, the missing SD of a given study (denoted with j) can be estimated with

$$\tilde{SD}_j = \overline{X}_j \left(\frac{\sum_i^K SD_i}{\sum_i^K \overline{X}_i} \right), \qquad (13.1)$$

Box 13.2. Hedges' d

key terms

d	effect size
$\bar{X}$	sample mean
T and C	treatment and control groups
SD	standard deviation
n	sample size
J	bias correction factor
s	pooled SD

definition

$$d = \frac{(\bar{X}_T - \bar{X}_C)}{s} J \qquad J = 1 - \frac{3}{4(n_T + n_C) - 9}$$

$$s = \sqrt{\frac{(n_T - 1)SD_T^2 + (n_C - 1)SD_C^2}{n_T + n_C - 2}}$$

approximations

independent t-test

$$d = t\sqrt{\frac{n_T + n_C}{n_T n_C}} \quad d = \frac{2t}{\sqrt{n_{total}}} \quad d = t_R\sqrt{\frac{2(1 - r_R)}{n_{total}}}$$

n_{total} assumes that $n_T = n_C$, t_R = t-test from repeated measures, r_R = corr. between measures

correlation (r)

$$d = \frac{r}{\sqrt{1 - r^2}}\sqrt{\frac{n(n-1)}{n_T n_C}} \quad d^* = \frac{2r}{\sqrt{1 - r^2}}$$

$n = n_T + n_C$, * indicates Cohen's d

F-ratio from one-way ANOVA

$$|d| = \sqrt{\frac{F(n_T + n_C)}{n_T n_C}} \quad |d| = 2\sqrt{\frac{F}{n_{total}}}$$

n_{total} assumes that $n_T = n_C$

Z-score

$$d = \sqrt{\frac{Z\sqrt{n}}{1 - \sqrt{Z^2 n^{-1}}}}$$

$n = n_T + n_C$

Chi-square (2 x 2 χ^2)

$$d = \sqrt{\frac{\chi^2(n_T + n_C)}{n_T n_C}}$$

imputations and conversions

s (pooled SD)

t-test

$$s = \frac{\bar{X}_T - \bar{X}_C}{t\sqrt{\frac{n_T + n_C}{n_T n_C}}}$$

Mann-Whitney P-value

$$s = \frac{\bar{X}_T - \bar{X}_C}{t(P, df^*)\sqrt{\frac{1}{n_T} + \frac{1}{n_C}}}$$

two-way (factorial) ANOVA

$$s = \sqrt{\frac{SS_B + SS_{A \times B} + SS_e}{df_B + df_{A \times B} + df_e}}$$

MS_{error}

$$MS_e = \frac{MS_A}{F_A}$$

A = effect of interest

one-way ANOVA

$$s = \sqrt{MS_e}$$

one-way ANOVA with g groups

$$s = \sqrt{F^{-1}\frac{\sum n_j \bar{X}_j^2 - \frac{(\sum n_j \bar{X}_j)^2}{\sum n_j}}{g - 1}}$$

ANCOVA

$$s = \sqrt{\left(\frac{MS_e}{1 - r^2}\right)\left(\frac{df_e - 1}{df_e - 2}\right)}$$

t-test

$$t = \Phi(P, df)$$
$$t = \sqrt{F}$$

Φ = t-distribution in MS EXCEL as "=TINV(P,df)"
P = P-value, $df = n - 1$,
$n = n_T + n_C$

$t(P, df)$ = t-distribution (Φ), MS = mean squares, e = error, df = degrees of freedom, df^* = n - 2, SS = sums of squares, A and B are the model factors, $n = n_T + n_C$

SD

$$SD = \sqrt{var} \qquad SD = \frac{IQR_U - IQR_L}{1.35}$$

$$SD = \frac{SE}{\sqrt{n}} \qquad SD = \frac{MAX - MIN}{Pearson (1932)}$$

U = upper and L = lower inter-quartile ranges, MAX and MIN ranges, find appropriate values in Pearson (1932) for a given $n = n_T + n_C$

$$\bar{X}_T - \bar{X}_C$$

ANCOVA

$$\bar{X}_T - \bar{X}_C = \bar{X}_T^{LSM} - \bar{X}_C^{LSM} + \beta(\bar{C}_T - \bar{C}_C)$$

LSM = least squares (adjusted) means, C = covariate, β = regression slope with C

Box 13.3. Correlation coefficient (r)

key terms **definition**

r Pearson product-moment correlation

x and y variables under analysis

n total sample size

$$r = \frac{n\sum x_i y_i - \sum x_i \sum y_i}{\sqrt{[n\sum x_i^2 - (\sum x_i)^2][n\sum y_i^2 - (\sum y_i)^2]}}$$

conversions and approximations

linear regression

biserial r (r_b)   point-biserial r (r_{pb})

$$r = \beta\left(\frac{SD_x}{SD_y}\right) \text{ if } y = \alpha + \beta x$$

$r \approx r_b$

$$r_b = \frac{r_{pb}\sqrt{n_T n_C}}{u(n_T + n_C)}$$

SD = standard deviation, α = intercept, β = slope

u = ordinate of unit normal distribution (see Terrell 1982)

independent t-test

Hedges' d

$$r_{pb} = \sqrt{\frac{t^2}{t^2 + n_T + n_C - 2}} \qquad r_{pb} = \sqrt{\frac{t^2}{t^2 + df}} \qquad |r_{pb}| = \frac{P}{\sqrt{P^2 + 4}}$$

$$r = \sqrt{\frac{d^2 n_T n_C}{d^2 n_T n_C + n(n-1)}}$$

df = degrees of freedom, P = P-value

$n = n_T + n_C$

F-ratio of one-way ANOVA F-ratio of ANOVA > 2 groups   Chi-square   Z-score

$$|r_{pb}| = \sqrt{\frac{F}{F + n_T + n_C - 2}} \qquad |r_{pb}| = \sqrt{\frac{SS_{between}}{SS_{between} + SS_{within}}} \qquad |r| = \sqrt{\frac{\chi^2}{n}} \qquad r = \frac{Z}{\sqrt{n}}$$

SS = sums of squares

$n = n_T + n_C$

$n = n_T + n_C$

odds-ratio

coefficient of determination (R^2)

$$r = \frac{n_T^A n_C^B - n_C^A n_T^B}{\sqrt{(n_T^A + n_C^A)(n_C^B + n_T^B)(n_T^A + n_T^B)(n_C^A + n_C^B)}}$$

$$|r| = \sqrt{R^2} \qquad r = \frac{\beta\sqrt{R^2}}{|\beta|}$$

see Box 13.4

β= regression slope

Spearman's rho rank corr. (ρ)

Mann-Whitney U Kendall's tau rank corr. (τ)

$$r = 2\sin\left(\frac{\pi\rho}{6}\right), \text{ if } n < 90; \ r = \rho, \text{ if } n \geq 90$$

$$|r_{pb}| = \frac{1 - 2U}{n_T n_C}$$

$$r = \sin\left(\frac{\pi\tau}{2}\right)$$

where $\overline{X}_j$ is the observed mean of the study with missing information, and K is the number of jth studies with complete information. Hereafter, variables accented with $\sim$ indicate the estimate to be imputed when calculating an effect size (see definitions in Boxes 13.2 through 13.5). This approach assumes that the SD to mean ratio is at the same scale for all studies (Wiebe et al. 2006), and this assumption should be explored for ecological and evolutionary meta-analyses given that experimental scales can differ tremendously between different taxonomic groups or experimental designs.

Box 13.4. Odds-ratio

key terms

		A	B
OR	odds-ratio (2 X 2 contingency table)		
n	cell frequencies		
T and C	treatment and control cell groups		
A and B	cell groups A and B		

	A	B
T	n_T^A	n_T^B
C	n_C^A	n_C^B

definition

$$OR = \frac{n_T^A n_C^B}{n_C^A n_T^B}$$

approximation

group proportions (P)

$$OR = \frac{P_T^A P_C^B}{P_C^A P_T^B}$$

Box 13.5. Z-score

key terms

Z	Z-score
T and C	control and treatment groups
$\overline{X}$	mean
SD	standard deviation

definition

$$Z = \frac{\overline{X}_T - \overline{X}_C}{SD_C}$$

conversions and approximations

Hedges' d

$$Z = \frac{d}{\sqrt{n(d^2 + 4)}}$$

$n = n_T + n_C$

Correlation coefficient (r)

$$Z = \frac{1}{2}\exp\left(\frac{1+r}{1-r}\right)$$

t-test

$$Z = \frac{t\sqrt{n}}{t + \sqrt{n-1}}$$

$n = n_T + n_C$

Mann-Whitney U

$$Z = \frac{U - 0.5(n_T n_C)}{\sqrt{\frac{n_T n_C (n+1)}{12}}}$$

$n = n_T + n_C$

Kendall's tau rank corr. (τ)

$$Z = \frac{\tau}{\sqrt{\frac{2(2n+5)}{9n(n-1)}}}$$

$n = n_T + n_C$

Another approach to imputing missing data uses regression techniques to predict the missing value given the relationship observed among the statistics of studies with complete information (Buck 1960, Pigott 1994). For example, if a study reports sample sizes but is missing information to calculate a pooled standard deviation s (see definition of Hedges' d in Box 13.2), then a prediction of s can be estimated from linear regression between the observed sample size (n), and s from the studies with complete information. This assumes that n is a good predictor of s. Using the regression equation estimated from studies with complete information, the s of a study with missing information is estimated with:

$$\tilde{s}_j = \alpha + \beta(n_j), \tag{13.2}$$

where α is the intercept and β the slope of the linear regression model of n versus s, and n_j is the observed sample size of the study with missing information. Of course, a nonlinear model or any number of covariates can be included in the model in order to improve the efficiency of the regression to predict missing values.

A comparable approach to the regression method is described by Ma et al. (2008), where missing pooled standard deviations are estimated using information from the other studies in the meta-analysis with complete information. Here, the s of the study with incomplete information is estimated as follows:

$$\tilde{s}_j = \frac{\sum_i^K s_i \sqrt{n_i}}{K\sqrt{n_j}}, \tag{13.3}$$

where K is the number of studies with complete information on s and n. This approach uses sampling theory to predict the expected s (see further details in Ma et al. 2008). Alternatively, Follmann et al. (1992) and Furukawa et al. (2006) describe a more impartial estimate of s (independent of the data used in the meta-analysis) that is derived from a previously published meta-analysis based on similar data. This approach can also be used when information on s is not available for any study. Here, the variances (σ^2) and sample sizes of effect size from each study are used to estimate s as follows:

$$\tilde{s}_j = \sqrt{\frac{\sum_i^K [(n_i - 1)\sigma_i^2]}{\sum_i^K (n_i - 1)}}. \tag{13.4}$$

These approaches to generating imputations for $\tilde{s}_j$ when estimating effect sizes are based on several assumptions. For example, they assume some degree of homogeneity among the observed SD and $\overline{X}$ values across studies. Furthermore, unlike effect sizes, imputations are not scaleless estimates; rather, they retain their original units. If there is large variation among estimates, which will be the case when meta-analyses pool research from different species or different measurements of the same ecological of evolutionary construct (e.g., fitness estimated as clutch size or offspring survival), then this may bias imputations. These approaches also assume that information is missing at random and not due to (nonrandom) reporting biases. Unfortunately, it is nearly impossible to test the above assumption in data sets. It is also important to consider that these regression based techniques assume that the missing observations are estimated perfectly by the model. Below, I describe multiple-imputation methods that attempt to account for the error associated with filling gaps of information when observed data are used as the basis for imputation.

Multiple-imputation

Multiple-imputation methods use a random sampling approach to fill gaps of information (Rubin and Schenker 1991). Here, gaps of missing data are filled by sampling the population of observed (available) data, or by sampling a distribution modeled from these available data. These sampling regimes are then repeated and averaged to give an overall "imputed" synthesis. This repetition of sampling is where the "multiple" of multiple-imputation is derived from, because data are filled multiple times to generate complete data sets. These multiple-imputation methods retain the benefits of single-imputation methods where a traditional meta-analysis is performed on imputed data sets. However, they have the advantage that the variability associated with imputing data is explicitly modeled when randomly sampling data; this avoids treating the imputed values as true observations as in single-imputation approaches. For example, the regression approach described with Equation 13.2 does not include the error associated with intercept (α) and slope (β) estimates. Multiple-imputation methods can account for this source of error.

The most intricate aspect of multiple-imputation methods is the way the data are sampled to fill the gaps of missing information. These sampling procedures can apply maximum likelihood or Bayesian models for imputing data (for further details, see Schafer 1997, Little and Rubin 2002), and require specialized software to hypothesize the distributions of missing data and to perform analyses. For illustrative purposes, I describe the simplest sampling model, known as "hot deck" imputation; this involves sampling data to fill gaps of missing information from the observed data derived from studies with complete information. As in the imputation example described earlier, I will explore the situation where a data set is missing several *SD* for calculating the pooled *s*, in order to estimate an effect size. Here a collection of random samples of *s* are first drawn (with replacement) from the total collection of (available) *observed s*. For example, if there are 4 of 30 studies missing *s*, then four *s* will be sampled from the 26 studies with information. These random samples will form a collection of *possible* samples for the missing data. Then a second random sampling (again with replacement) from this collection of *possible s* will generate the data used to fill the gaps of missing information. The imputed data are sampled from the collection of *possible* rather than *observed* values of *s*, because this will create between-imputation variability among the imputed data sets. These random samples of *s* are then imputed to fill the gaps of missing information in order to form a complete data set, and the whole process is repeated to generate *m* number of complete (but randomly filled) data sets.

After *m* complete data sets are generated, a pooled effect size $\bar{\delta}$ and variance $\sigma^2(\bar{\delta}_l)$ is calculated for each data set using traditional meta-analysis (Chapters 8 and 9). The results of each meta-analysis are then averaged into an overall effect size ($\dot{\delta}$) with a variance of $\sigma^2(\dot{\delta})$. Each *l*th result of *m* meta-analyses are pooled using Rubin's average:

$$\dot{\delta} = \frac{\sum_{l=1}^{m} \bar{\delta}_l}{m},$$
(13.5)

which has a variance of

$$\sigma^2(\dot{\delta}) = \frac{\sum_{l=1}^{m} \sigma^2(\bar{\delta}_l)}{m} + \left(1 + \frac{1}{m}\right)\frac{\sum_{l=1}^{m} (\bar{\delta}_l - \dot{\delta})^2}{m - 1}.$$
(13.6)

These results are then treated as the final meta-analysis. Similarly, the total homogeneity test (Chapters 8 and 9) is also averaged across *m* number of data sets as:

$$\dot{Q} = \frac{\sum_{l=1}^{m} Q_l}{m}.$$
(13.7)

There is also a general guideline for how many repetitions (*m*) are necessary to get a good estimate of $\dot{\delta}$ and variance $\sigma^2(\dot{\delta})$ that accounts for the between-imputation variability. Surprisingly, these recommended repetitions are few, and Rubin and Schenker (1991) suggest that if 30% of the data are missing, then an *m* of three is sufficient; whereas when 50% of the data are missing, then at least an *m* of five would be necessary. This guideline assumes that the review sample size is large (e.g., $K > 20$) and that there are more studies with complete information than studies missing information. However, given that this technique applies a random sampling approach, many more repetitions ($m > 100$) should be performed, thereby avoiding the sensitivity of resampling techniques to outliers when few replications are performed.

Nonparametric analyses and bootstrapping

An explicit definition of meta-analysis is (a) quantifying research outcomes using effect sizes, and (b) weighting of these effect sizes based on their relative sensitivity to sampling error. Imputation methods are useful to fill gaps of information when estimates of standard deviations are missing (see above). However, when most studies lack information about *SD*, then an effect size metric that does not require *SD* can be paired with a nonparametric bootstrapping approach that uses a simplified weighting scheme. For example, the log response ratio (ln*R*; Gurevitch and Hedges 1999; Chapter 6) is a less restrictive alternative to Hedges' *d* because it only uses the means to calculate an effect:

$$\ln R = \ln\left(\frac{\overline{X}_T}{\overline{X}_C}\right). \tag{13.8}$$

If the standard deviations are available then calculating an effect with Hedges' *d* is preferred (Lajeunesse and Forbes 2003). When standard deviations are missing, but sample sizes are available, then the inverse of a simplified estimate of the variance can be used to weight studies during the meta-analysis (Hedges and Olkin 1985):

$$\sigma^2(\ln R) = \frac{n_T n_C}{n_T + n_C}. \tag{13.9}$$

Bootstrapping methods are then used to estimate the 95% CI around the pooled mean. See Adams et al. (1997) for further details on this approach. However, it should be cautioned that this is a very crude surrogate for traditional meta-analysis (e.g., using the nonsimplified variances of effect sizes for Equation 13.9), and should never be performed as a shortcut to avoid having to extract *SD* from each study. This approach should only be used as a last resort when *SD*s (or standard errors) are impossible to extract from the majority of studies.

EFFECTS OF IMPUTATIONS ON THE OUTCOME OF REVIEWS

Imputation methods are used to fill gaps of information in meta-analysis by using the data already available from studies that have fully reported statistics. These methods can range from simple to very sophisticated models, but because there is a lack of a standardized protocol for implementing the methods, there is the concern that using some models rather than others will introduce bias or generate misleading results (Riley et al. 2004). However, Rubin and Schenker (1991) argue that for most cases of missing information, time and resources should not be focused toward implementing the most sophisticated models, and these advanced methods are mostly useful when a large number of studies lack information. To put this in perspective, Rubin and Schenker (1991) describe the following hypothetical example about the potential for bias (I have modified this slightly for our theme of meta-analysis). If the imputation method does not introduce bias for 75% of the cases of missing information, and there is a deficiency of information in 20% of the studies, then there is a 25% likelihood that imputations will introduce bias in 20% of the information. In this case, the meta-analysis will then only have a 5% bias due to imputation, leaving the remaining 95% of studies unbiased. If there is continued scepticism of the results obtained using imputation methods, then it has been suggested that studies with imputed information could be further downweighted during meta-analysis (Rief and Hofmann 2009). Alternatively, the appropriateness of imputing data into the overall analysis can be evaluated with a sensitivity analysis where imputed studies are included/excluded to assess overall bias (see Riley et al. 2004, Barzi and Woodward 2004).

Despite the potential for bias, reviews applying imputation methods will have improved variance estimates (e.g., smaller 95% CI) over reviews excluding studies with missing information (Philbrook et al. 2007). These improved variance estimates are due to inclusion of more studies when pooling results compared to a review that simply excludes studies with incomplete information (Chapter 22). Further, imputation methods can also potentially improve the representation of null studies or studies from underrepresented moderator groups. Multiple-imputation methods have an additional benefit of providing more conservative results than approaches based on direct within-study imputations (Riley et al. 2004). This is important given that within-study imputations explicitly treat imputed data as real data, and that not accounting for the uncertainty associated with imputed data can result in an underestimation of the pooled variance (Pigott 2001).

CONCLUSIONS AND FUTURE DIRECTIONS

Many of the challenges associated with a lack of information in the literature can be avoided entirely with thorough reporting of means, correlations, standard deviations, and sample sizes of experiments. To address these gaps of information and to establish a uniform reporting standard for journal publications, the medical sciences launched the CONSORT initiative (CONsolidated Standards Of Reporting Trials; see Moher et al. 2001). This initiative provides guidelines for reporting statistics and data in publications, consisting of a 22-item checklist and a flow diagram to help improve the clarity and transparency of the study. Further, the NIH (National Institutes of Health) has an online database (see www.clinicaltrials.gov) where protocols and results of funded studies must be registered (even when the study is not published). This resource is important to help address publication bias while also allowing for the quick recovery of missing information within publications.

Given that many ecological and evolutionary journals have pagination limits and are increasingly pushing for brevity, a standardized guideline would also serve these fields tremendously. This guideline would not only facilitate data extraction for meta-analysis, but would also increase the reliability and repeatability of primary data analysis. Electronic appendixes have improved the availability of data useful for meta-analysis and have made it easier to publish results and findings tangential to the main article. However, the accessibility of this information is still mostly dependent on the reporting practices of the author and on post-submission editorial/reviewer decisions. With a standardized guideline, authors would submit manuscripts that are fully reported and annotated prior to review. This information can then be moved to electronic appendices when necessary. The prospective registration of data sets and supplementary material of published studies is also an emerging alternative (e.g., see DRYAD at datadryad.org). In fact, many journals are adopting policies that encourage authors to submit raw data to these databases. However, to date, the registration of data in freely accessible databases has had limited success; it may require further work for authors to organize data and the usefulness of these data is limited to how well they are annotated (e.g., description of organization and data manipulations).

ACKNOWLEDGEMENTS

I thank Terri Pigott and Chris Schmid for valuable feedback on this chapter. Work on this chapter was supported by the University of South Florida and a National Science Foundation grant to the National Evolutionary Synthesis Center (EF-0423641).

Publication and Related Biases

Michael D. Jennions, Christopher J. Lortie,
Michael S. Rosenberg, and Hannah R. Rothstein

INCREASED USE OF META-ANALYSIS in ecology and evolution has stimulated greater consideration of the occurrence of publication bias in the scientific literature. A search of Web of Science showed that prior to 1995 no papers in ecology or evolution journals used the term "publication bias" in their title, abstract, or keywords (Fig. 14.1). More frequent occurrence of this term since 1995 seems to coincide with the promotion of meta-analysis in widely read ecology and evolution journals (e.g., Arnqvist and Wooster 1995a) and a subsequent increase in the publication of meta-analyses (Chapter 25). It is noteworthy that of the 84 papers we located using the search term "publication bias," 54 also had "meta-analysis" in the title or as a keyword.

Clearly, publication of meta-analyses has prompted researchers to reevaluate publication practices in ecology and evolution (Møller and Jennions 2001). It is ironic that the quantitative methods associated with meta-analysis have shed the most light on the problem (Song and Gilbody 1998), because the very quantification of potential problems has led to claims that meta-analysis itself is flawed if publication bias is so rife that we cannot estimate the true magnitude, or sources of variance, in effect sizes. Blaming meta-analysis is akin to shooting the messenger (Alex Sutton, pers. comm.). Unfortunately, when publication bias is suspected, this has led some to throw the baby out with bathwater and suggest that meta-analysis is of little value (for an example of this debate, see Kotiaho and Tomkins 2002, Jennions et al. 2004, and Tomkins and Kotiaho 2004).

We offer five short responses to this pessimistic view. First, if the problem of publication bias really is so severe, then we should also avoid narrative reviews. Publication bias affects any attempt to summarize scientific investigation, be it through narrative review or meta-analysis. Second, we should not examine *P*-values in individual studies. It is inappropriate to examine a published study in isolation from knowledge about how many other studies have asked the same question. It is equivalent to using multiple *t*-tests to compare two groups for a suite of characters then, instead of statistically correcting for multiple testing, only focusing on the "interesting" results (i.e., those that are published or reported). Third, there is little hard evidence for publication bias in ecology and evolutionary biology. Most of the available data comes from the health sciences. That said, study areas with similar research protocols in humans and nonhumans suggest that publication bias could be a problem in specific fields. For example, human studies looking for genetic associations (i.e., genes "for" a disease) are subject to considerable publication bias due to data mining (Ioannidis 2003). Given the increased search for candidate genes in biology, similar problems are likely to arise. Fourth, there are methods available to detect and adjust for publication bias. It should be acknowledged forthrightly, however, that none of these methods is entirely satisfactory. The only lasting solution is a change

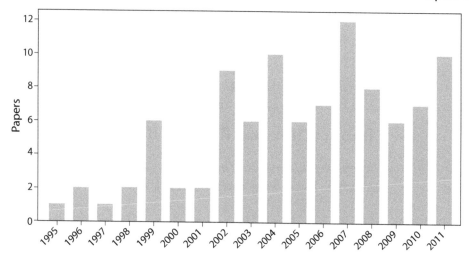

Figure 14.1. Publications in ecological or evolutionary journals indexed in Web of Science (or papers in multidisciplinary journals with a focus on a topic in ecology or evolution) that included the term "publication bias*" in the title, abstract, or as a keyword ($n = 97$). We searched from 1948 to 2011, but no papers prior to 1995 were located.

in scientific publication practices. Fifth, a more positive view is that, unlike narrative reviews, meta-analysis offers explicit methods to detect potential publication bias. This can direct future research and identify limitations on the current scope of inference for specific hypotheses.

Data collection for a meta-analysis is a form of empirical research with many similarities to primary research in field ecology. In both cases the goal is to collect data without introducing biases, and the risk of making mistakes tends to decline with experience and the acquisition of skills (often learned through past failures). Publication bias can influence data collection for a meta-analysis because it reduces the likelihood that the data most readily at hand (published work) are a representative sample of the available data (all studies conducted on the topic). It is analogous to sampling problems encountered when collecting primary data. In Box 14.1 we

---

BOX 14.1.
The similarity between empirical work
and data collection for a meta-analysis.

A seemingly trivial field study of how to measure body size in fiddler crabs illustrates the point that collecting data in an unbiased fashion is a difficult task that requires thought, and is not often perfectly achieved. If the analogy between this example and conducting a meta-analysis is drawn, the similarities with various forms of publication bias should become apparent (effect size = crab size; site on mudflat = place where effect size can be located (i.e., for a meta-analysis, in a journal or elsewhere). The key concern in our hypothetical field study is that the size of crabs living on different parts of a mudflat *might* vary. In the same way, effect sizes collected from different sources *might* vary. Specifically, effect sizes might be greater in published than unpublished studies (see main text).

Consider a study where you want to know the mean and variance in body size for a newly discovered species of mangrove fiddler crab. To avoid the mosquitoes by the

(Box 14.1. continued)

trees you start your study at the sea edge. You collect 100 crabs and find that the mean ± SD for carapace width is 15 ± 4 mm. When you submit the study for publication it is unlikely that a reviewer will ask why you did not measure every crab on the mudflat. This would guarantee that the size distribution is precisely calculated, but everyone accepts principle that a population sample can be used to represent the total population. A reviewer might, however, reject your study by noting that in many mangrove-dwellers, the body size changes with distance from the shore. They suggest that young crabs might live closer to the shore. One solution you could propose is to redo the study and sample 10 crabs every 5 meters as you walk inland from the shore. In so doing you can evenly sample the study area. Upon resubmission, however, another reviewer points out that crab density might also change with distance from the sea. As such, by sampling equal numbers at each site your results might still not reflect the average size of crabs if you systematically undersample sites where crabs are common and oversample those where crabs are rare. You then modify your approach and again sample every five meters. This time, however, you try to collect every crab in 1 m² plots. Even here, however, there is a problem because some areas of the mudflat are so boggy that you often fail to catch crabs there. One irksome reviewer suggests that, based on his own anecdotal evidence, larger crabs prefer muddy spots so that your size estimate is biased downward. You respond by invoking pragmatism and include his concern as a necessary caveat to your conclusions. The editor accepts your case and agrees to publish the study.

The analogy to the problems of compiling a dataset for a meta-analysis should be apparent. When we start a synthetic review we first look in places where it is easiest to collect data, such as performing a keyword search in Web of Science (collecting along the shore). But what if effect sizes differ between studies in journals that are or are not covered by Web of Science, or between published or unpublished studies (crab size differs with distance from the shore)? In our fieldwork we tried to devise a method that allows us to systematically sample the population (we collected equal amounts of size data every five meters). We could do the same when performing a meta-analysis by collecting data from an equal number of published and unpublished studies. The obvious flaw in this approach is that the number of published and unpublished studies differs (crab density varies spatially), and this is a concern if we want to obtain a correct estimate of the mean value *if* effect sizes differ between the two types of studies. We then had to modify our field methods and sample areas in proportion to local crab density so that our size estimate was correctly weighted to obtain a mudflat-wide mean. In the field study, we implicitly assumed that there was no spatial variation in capture rate. The problem when conducting a meta-analysis is that, with extensive effort, we can probably locate almost every published study, but we will only locate a modest proportion of the unpublished ones (capture rate varies with distance from the shore). In the field study, even if capture rate varied, we could correct for this by visually estimating crab density without the need to capture crabs. There is, however, no readily apparent equivalent when collecting data for a meta-analysis because we do not have direct access to unpublished studies (cannot estimate local density). How then do we apportion our efforts so that we collect data from published and unpublished studies in proportion to their true occurrence? Finally, how do we deal with the possibility that an entire class of studies, such as those abandoned prematurely, will almost never come to light (no data from muddy areas)? Is trying to collect all studies (measure every crab) the best solution?

describe a simple field study and use it to draw analogies between factors that bias data collection during fieldwork with those encountered when compiling effect sizes for a meta-analysis.

What exactly is publication bias?

Publication bias is associated with the inaccurate representation of the merit of a hypothesis or idea (Lortie et al. 2007). A strict definition is that it occurs when the published literature reports results that systematically differ from those of all studies and statistical tests conducted; the result is that false conclusions are drawn (Rothstein et al. 2005). Here, we propose an operational definition: *publication bias occurs whenever the dissemination of research is such that the effect sizes included in a meta-analysis generate different conclusions than those obtained if effect sizes for all the appropriate statistical tests that have been correctly conducted were included in the analysis.* This definition takes into account that inclusion of individual studies in a meta-analysis ultimately depends on access to information. For example, the time lag until publication, and the journal, country, and language of publication can all influence the likelihood that a study is located for a meta-analysis. Even if a study is published, failure to report some results can generate a systematic bias in estimates of effect sizes (Chan, Hróbjartsson, et al. 2004; Chan, Krleza-Jeric, et al. 2004; Cassey et al. 2004). Such selective reporting may occur, for example, if several outcomes or predictor variables are examined, but only some of the performed analyses are then reported appropriately. Our definition also takes into account that some published studies are based on fabricated data or use statistical tests inappropriately (e.g., unwarranted exclusion of outliers or a post hoc decision to only report on a subset of the data). The scientific literature is enormous, so whether a study is located is affected by its visibility. This can depend on the relative prominence of the research group that conducted it, or the extent to which their work is cited by others (Leimu and Koricheva 2005a, 2005b). More generally, a distinction can be made between genuine unpublished studies (manuscripts languishing in researchers' file drawers, data sets that have been analyzed but never written up, etc.), and research that is publicly available but difficult to locate or retrieve, such as conference papers, technical and governmental reports, theses, and so on. Both types of studies are less likely to be included in a meta-analysis than those in peer-reviewed journals, but the latter so-called gray literature (Auger 1998, Rothstein and Hopewell 2009) is more likely to be located by meta-analysts. Song et al. (2000) suggest that the phrase "dissemination bias" be used to cover any bias that potentially affects data collection for meta-analysis. The term "publication bias" is, however, so well established that we use it here as a synonym for "narrow sense publication bias and subsequent dissemination bias" (see also Rothstein et al. 2005). When we follow convention and refer to "published and unpublished studies," the context should indicate whether this is shorthand for "located and unlocated studies." A detailed classification of potential sources of dissemination bias is provided in Song et al. (2000), and reviewed in Box 14.2.

Publication bias should not be equated with the failure of a meta-analysis to locate every available study (White 1994). One hypothetical solution to a potential sampling bias in any empirical study is to systematically collect data from the entire population. Meta-analysts sometimes imply that they are pursuing this tactic when they state that their extensive search efforts have located all the relevant studies. It is, however, logically impossible to show that publication bias has been eliminated in this manner because we will never know if we have located every study (Vevea and Woods 2005). The real solution is to minimize biased sampling of the pool of available effect sizes (Box 14.1). The main question is then: "Are the effect sizes used in a meta-analysis a representative sample of those that could potentially be obtained from all the tests that should have been conducted from completed studies?" This can

---

BOX 14.2.
Sources of publication bias relevant to meta-analysis.

Publication bias is driven by differential assessment of merit at any stage in the process of preparing a manuscript (selective reporting), publication of a study (narrow sense publication bias), or the subsequent availability of studies to meta-analysts (dissemination bias) that is influenced by the effect size (significance) of a study.

(A) The significance of results, or the direction or magnitude of an effect size, affects whether a study is published. This could be due to decisions by a researcher to submit, ranking by reviewers, or final decisions by editors (*narrow-sense publication bias*).

(B) The effect size influences the visibility of a study during the data collection phase of meta-analyses (*dissemination bias*). This could be due to the following:

    (1) A narrow sense publication bias
    (2) The impact factor of the journal of publication
    (3) The language of publication
    (4) The frequency with which a paper is cited
    (5) Publication in the gray literature rather than in peer-reviewed journals

(C) Selective reporting such that the effect size (significance) for a measured response variable or potential predictor variable influences whether or not it is included in a manuscript, or survives the review process and stays in the published version of a paper.

---

be split into two key questions. First, do effect sizes of studies that are in a meta-analysis differ from those that are not? In fields such as medicine and the social sciences evidence from both direct tests (which compare published and unpublished studies) and indirect tests (which use the distribution of published effect sizes to infer the effect sizes of unpublished studies) indicates that statistically nonsignificant results are less likely to be published than significant results. Effect sizes therefore tend to be larger in published studies. Indirect tests for publication bias suggest that the same situation occurs in ecology and evolution (Jennions and Møller 2002a, 2002b; Cassey et al. 2004). This conclusion is, however, controversial because indirect tests may attribute to publication bias patterns that could legitimately arise for other reasons (see below). Interestingly, the only two large-scale tests in ecology and evolution that directly compare effect sizes between published and unpublished studies indicate that there is no difference (Møller et al. 2005) or show the opposite trend whereby published studies include fewer significant results than unpublished studies (Koricheva 2003). In terms of selective reporting within a study, there is a clear trend in medicine to more often report significant than nonsignificant results (Chan, Hróbjartsson, et al. 2004; Chan, Krleza-Jeric, et al. 2004), and there is also some evidence for this occurring in ecology (Cassey et al. 2004). Selective reporting is a major concern that has received less attention than narrow sense publication bias. We suggest that in ecology and evolution it could be a greater problem than failure to publish, because many studies involve numerous statistical analyses (e.g., model building and testing the explanatory power of many predictors) and a tradition of then constructing the narrative of a paper around a few key significant results.

The second question is whether publication bias is more common for some types of questions than others (e.g., hypothesis-driven versus descriptive biology), or for some study systems. (For example, are studies of large mammals that take years to complete and have small sample sizes more likely to be published, regardless of the significance of the results, than studies of insects that can be completed in a few weeks with much larger sample sizes?) If publication bias is only prevalent for certain types of studies, this could lead to false conclusions about heterogeneity in the magnitude of effect sizes among groups. This problem is more likely to arise in ecology and evolutionary biology than in the medical or social sciences, due to huge differences in the costs and relative ease of asking the same question in different study systems or species. Indeed, this seems to have led to different criteria for publication. When reviewing papers for journals, greater leeway is often given for small sample sizes when researchers conduct logistically challenging work. For example, there might be a weaker publication bias against nonsignificant results based on small sample sizes for, say, studies of large mammals as opposed to insects because of the widely acknowledged difficulty of conducting large-scale studies on large mammals, This might in turn lead to smaller reported effect sizes for mammals. A subsequent meta-analysis could then promote the potentially false conclusion that the relationship of interest is weaker in mammals than insects.

Statistical significance, sample sizes, hypothesis support, and publication

Fisher (1925) introduced biologists to null hypothesis testing and probability values as an objective measure to assess the likelihood that a result occurs by chance. Unfortunately, current practice has mutated into a false dichotomy between significant and nonsignificant results at a given α level (i.e., $P = 0.05$). For example, Stoehr (1999) notes that identical experiments with outcomes where $P = 0.08$ and 0.80 are treated as being in greater agreement (both nonsignificant) than those where $P = 0.04$ and $P = 0.08$. This overemphasis on significant P-values could lead to nonsignificant results being less often reported. The fact that researchers value significant results more highly than those that fail to reject a null hypothesis is exemplified in use of the terms "positive" and "negative" findings.

In several scientific disciplines there is evidence that studies with significant results are more likely to be published (for major reviews see Song et al. 2000 and Rothstein et al. 2005). It is worth remembering, however, that publication bias is an issue regardless of whether or not rejection of the null hypothesis is the more desirable outcome for an individual researcher. For example, in disciplines where theory is extremely well developed, failure to reject the null hypothesis in novel tests is often seen as healthy confirmation of the strength of the underlying theoretical framework. In these communities, researchers that successfully show how theory can be extended into new areas might be rewarded with high-profile publications. In evolutionary biology, for example, the ability to predict the fig wasp offspring sex ratio under different levels of local mate competition represents a failure to refute a null hypothesis (i.e., there is no significant difference between observed and theoretically predicted sex ratios). The success of sex ratio theory is a major accomplishment of modern evolutionary biology (e.g., Herre 1985, 1987), and this sense of achievement has only grown as increased information on the natural history of fig wasps (e.g., discovery of cryptic species) has improved the fit between observed and expected sex ratios (Molbo et al. 2003).

In a field where theory is generally assumed to explain phenomena successfully (e.g., much of physics) it might be even more difficult to publish results that reject null hypotheses. Such findings are more likely to be attributed to chance (type I error) or methodological flaws than a genuine discrepancy with a predicted outcome (Palmer 2000, 447). Similarly, there might be a disinclination to publish results that are significant in the *opposite* direction to that predicted in hypothesis-driven areas of science, even if the predictive power of the hypothesis is modest (Festa-Bianchet 1996). This might occur if reviewers have personal interests vested in a currently favored hypothesis, or because there is a sensible tendency to be skeptical of results for which there is no readily apparent explanation. This could lead to temporal shifts in effect sizes if certain findings suddenly become "theoretically possible" (Alatalo et al. 1997), or doubts emerge about how a theory is being tested (Simmons et al. 1999) (see Chapter 15).

Despite the above caveats, biologists usually hope that their statistical tests will yield significant results. This means that studies with significant results, if methodologically sound, can be published regardless of sample size. In contrast, there is widespread recognition that nonsignificant results are often inconclusive because they could be due to low statistical power or the genuine absence of a meaningful effect (Cohen 1988). Power is a function of sample size, the true effect size, and its variance, so nonsignificant results are less likely to be published when sample sizes are small. Smaller studies, however, estimate the true effect of a factor or strength of a relationship less accurately than larger studies because of greater sampling error (Ioannidis et al. 1998). They therefore generate a wider range of outcomes than large-scale studies. This is not a problem for publication bias if the true effect size is zero; smaller studies will more often yield significant results, but the estimated effects will be just as likely to be greater or smaller than the true null effect. The null hypothesis is, however, rarely true and publication bias arises when there is a modest true effect size, which is likely to be commonplace. Given the sampling error, small studies will then more often yield results that are significant in the direction of the true effect size than results significant in the opposite direction. These combined results are still symmetric with respect to the true effect size, but if there is a prejudice against publishing nonsignificant results from small studies, then the literature will generate an overestimate of the true effect size. The problem is worse if there is a blanket prejudice against nonsignificant results so that there is no ameliorating effect of large sample size on the likelihood of publishing these results. However, because larger studies yield more accurate estimates of effect sizes (and are given a higher weighting), they reduce the extent to which excess publication of smaller studies with significant results leads to overestimation of the true effect size.

EVIDENCE FOR PUBLICATION BIAS
IN ECOLOGY AND EVOLUTION

Publication bias due to a failure to publish

The first evidence that publication bias was a problem in science came from observations of the abundance of statistically significant results in published studies by Sterling (1959) in psychology, and Smart (1964) in education. Since then similar trends have been noted in many other disciplines (for a review, see Dickersin 2005). In ecology, Csada et al. (1996) claimed that 91.4% of studies reported a significant result for the main hypothesis. The problem with using these data as evidence for publication bias is the lack of a null model. Bauchau (1997) pointed out that biologists are notorious for testing null hypotheses that they expect to reject, rather than designing less readily refutable null hypotheses. For example, they test whether males are the same size as females in a species where this is obviously not the case, rather than formulating a null hypothesis based on prior work that males are 20% larger than females. Inspection of tables of effect sizes and sample sizes in recent biological meta-analyses suggest that a figure of 91.4% is too high. These tables show that many tests do not yield statistically significant results. This discrepancy might have arisen because biological studies are often written up as a narrative to emphasize a particular significant result. The main hypothesis then becomes the one that generated this result. However, most publications contain additional tests of other hypotheses that are less strongly emphasized, and often do not report significant results. So the *main tests* surveyed by Csada et al. (1996) probably overestimate the reporting of significant results in ecology. On the other hand, if hypotheses are labeled main or subsidiary based on *P*-values, there might be a retrieval bias for subsidiary hypotheses and results compared to "main" hypotheses, because most studies have keywords and titles that emphasize the main hypotheses. Thus, the net effect might still be that the studies located for a meta-analysis have smaller *P*-values than those from the average test of the hypothesis in question.

One of the strongest lines of evidence for publication bias is a direct comparison of effect sizes between published and unpublished studies. The trend in most disciplines is that published studies report larger effects. For example, Sterne et al. (2002) used data from 39 medical meta-analyses to show that published studies indicated a far greater beneficial effect of medical invention than their unpublished counterparts. Similarly, McAuley et al. (2000) showed that medical studies published in the gray literature have smaller effect sizes than those in mainstream journals. The only equivalent evolutionary study is by Møller et al. (2005) who found no difference in effect size between published and unpublished studies of fluctuating asymmetry and sexual selection. We should note, however, that a set of unpublished studies often contains recent studies that will be published with the normal time lag after study completion. This decreases the likelihood of detecting publication bias using a "snapshot" comparison. The most compelling direct evidence for publication bias is to track the fate of a cohort of studies registered prior to data collection. At least six cohort studies of registered trials in medicine and one in psychology (Dickersin 2005) have all shown that failure to reject the null hypothesis was the key predictor of whether a study remained unpublished (Dickersin 1997). Most recently, Turner et al. (2008) reported a disquieting analysis of the publication of studies on 12 antidepressants. All these studies were originally registered with the United States Food and Drug Administration (FDA), which conducts its own statistical analyses of the test trials and rules on the efficacy of the drugs. The authors showed that studies with results the FDA considered negative were far less likely to be published (11 of 33) than those with positive results (37 of 38). There was also evidence for selective reporting of results, which we will return to shortly. In the only comparable analysis in ecology, Koricheva (2003) examined unpublished chapters from a cohort of PhD theses and investigated their fate five or more years later. Surprisingly, the proportion of significant results was *smaller* in chapters that were subsequently published.

Failure to publish is the most extreme prejudice against nonsignificant results. A less extreme prejudice could create a time lag to publication so that published effect sizes decrease over time (Dickersin 1997, but see Koricheva 2003). There is evidence for this in the medical and social sciences (Trikalinos and Ioannidis 2005), and the same pattern has been reported in ecology and evolution (Jennions and Møller 2002b). There are, however, plausible explanations other than publication bias for this result (Chapter 15).

Publication depends on assessment of merit. This is composed of a suite of decisions on whether to complete or abandon a study, whether or not to write it up and submit it, how to rate a manuscript under review, and ultimately what an editor finally decides (Lortie et al. 2007). Publication bias can arise due to factors operating at any stage in this process. It is a general finding in medicine and psychology that publication bias is driven by submission bias because authors less often submit nonsignificant results (Dickersin 1997, Cooper et al. 1997). The role of reviewers and editors is less clear (for a mini-review, see Møller and Jennions 2002; for a detailed review, see Song et al. 2000, Neff and Olden 2006). The role of submission bias in ecology and evolution is poorly studied. Koricheva (2003) examined 93 Finnish or Swedish PhD theses, which are presented as a compilation of manuscripts. She then followed the fate of 187 manuscripts that were unpublished at the time of thesis submission. There was no evidence that the statistical significance of their results affected whether they were submitted. Interestingly, however, the studies that were eventually published contained a higher proportion of nonsignificant results than those that remained unpublished (49 versus 41%; $P = 0.02$). It appears that studies with a greater proportion of nonsignificant results were submitted to lower ranked journals where acceptance rates are higher. This raises the broader issue of publication bias arising due to variation in the ability of meta-analysts to locate studies published in different places, or given difference prominence once published.

Publication bias because effect size influences the visibility of published studies

In medicine, studies with significant results are more likely to be published in journals with a higher impact factor (see references in Baker and Jackson 2006), and non-English language journals contain a greater proportion of nonsignificant results (Egger, Zellweger-Zähner, et al. 1997; but see Jüni et al. 2002). These publication patterns create a bias through a dissemination effect because papers in these journals (some of which are not in electronic databases, or are difficult/costly to obtain) are less likely to be located for a meta-analyses. In ecology, Murtaugh (2002) reported a significant relationship between the effect size of the main result and the impact factor of the journal in which it was published. A slightly different pattern occurred in another area of ecology where results that supported a prevailing hypothesis were published in higher impact factor journals than those that rejected it (Leimu and Koricheva 2004). Currently there is no data available to test for a language of publication effect in ecology or evolutionary biology, but it would be interesting to ask nonnative English speaking biologists to compare the effect size or P-values of their English and native language publications (Egger, Zellweger-Zähner, et al. 1997). Do these authors make the extra effort required to write in English when they feel their results are more likely to be published, and do they partly base this sentiment on P-values? Biologists from countries where publication in international English-language journals is rewarded, but still relatively uncommon, would be ideal candidates.

Another source of publication bias is the frequency with which papers are cited. Highly cited papers are more often encountered when compiling studies for a meta-analysis, especially if the search method involves scouring reference lists. This will not lead to publication bias if there is no link between citation frequency and P-values, and/or sample sizes. There is, however, evidence from medicine that studies supporting a medical intervention are more often cited (Ravnskov 1992, Kjaergard and Gloud 2002). In ecology and evolutionary biology this is exemplified by the tendency to cite several confirmatory studies that reject the null hypothesis, followed by one study that failed to do so (the "but see Ref X" phenomenon). Leimu and Koricheva (2005b) examined citation rates for papers testing three different ecological hypotheses (plant stress, carbon-nutrient balance, and costs of antiherbivore defenses). Neither sample size nor P-value affected citation rates, but papers reporting a larger effect in the hypothesis-predicted direction were significantly more often cited for studies of antiherbivore defenses, significantly less often cited for studies of plant stress, and there was no relationship for studies of the carbon-nutrient balance hypothesis. In sum, the available evidence suggests that there is the potential for publication bias in some areas of ecology and evolutionary biology due to the "visibility" of studies being affected by the effect sizes that they report.

Publication bias due to selective reporting
of results within a study

Unfortunately, publication bias is not necessarily eradicated even if we can locate all published and unpublished manuscripts. This is because the way in which the results of a study are written up can itself be biased. Selective reporting of results *within* a study has received far less attention than publication bias *among* studies caused by variation in their publicaton fate. Clearly the worst form of selective reporting is the failure to report that a statistical test was conducted when the outcome is nonsignificant. The main ways in which effect sizes go unreported are the following: (a) only effect sizes for some of the tested predictor variables are presented, (b) only some of the measured response variables are reported, and (c) only data for some subgroups are reported. A lesser crime is poor presentation of results for nonsignificant findings (e.g., simply stating $P > 0.05$), so that it is difficult to estimate the effect size.

In contrast to the objective approach with which data are collected, the ways in which scientists report their findings still involve strong elements of advocacy, censorship, and selective presentation. For example, Ioannidis et al. (1997) noted that in medicine, even when the primary outcome is a negative result, researchers often present their findings in a more positive light by highlighting subgroups for which there were significant benefits, or by focusing on other measured outcomes; see Lortie and Dyer (1999) for an ecological example. Unfortunately, there is relatively little hard evidence that within a study the tests that produce significant results are more often reported than those that do not (Sutton and Pigott 2005). Chan, Hróbjartsson, et al. (2004) and Chan, Krleza-Jeric, et al. (2004) showed, however, that statistically significant outcomes were more likely to be reported than nonsignificant ones in publications from randomized control trials in medicine. One example was mentioned earlier, when Turner et al. (2008) examined the 74 FDA registered studies on the efficacy of 12 antidepressants. For each study, the FDA ruled whether or not the results were positive (i.e., showed that the drug had a beneficial effect, as they found in 38 studies), negative (24 studies), or questionable (12 studies), so that they could then make regulatory decisions. In total, 37 of the 38 studies that the FDA evaluated as having positive results were subsequently published. Of the 36 negative/questionable studies, only 14 were published. More importantly with respect to selective reporting, only 3 of the 14 studies were presented as having negative findings (i.e., in agreement with the FDA ruling on the study), while the other 11 studies were presented in such a way as to create the general impression that the results were positive. This was achieved by selectively reporting on one or more secondary outcomes of the study, rather than the primary outcome that was the main focus when the research was originally registered with the FDA. This selective reporting of results, in combination with the narrow sense "failure to publish" bias already described, resulted in the FDA estimated effect sizes for all 12 medicines being much smaller than those calculated based on the data presented in studies published in journals. On average, the published effect size increased by 32% relative to that calculated by the FDA. Put differently, the FDA only considered 51% of trials to have yielded positive results, while the studies published in journals present evidence that 94% of trials yielded positive results.

There is currently little data on selective reporting in evolution or ecology. One interesting finding comes from a meta-analysis of 40 published studies of correlates of variation in primary sex ratios in birds (Ewen et al. 2004). Cassey et al. (2004) showed that 47% of the 293 original statistical analyses in these studies were presented so that it was impossible to extract an effect size (e.g., direction of effect or the P-value was missing). When Cassey et al. obtained the missing data from the studies' authors, they found that effect sizes for significant predictors of the sex ratio could more often be calculated from publications than those for nonsignificant predictors ($P < 0.01$, $n = 214$). If this is a general trend it means that effect sizes will be over-estimated due to selective reporting.

A second biological study used a novel approach that suggests an even more worrisome trend. In a surprisingly neglected paper, Ridley et al. (2007) surveyed 1000 biological studies in three major journals (*Science, Nature* and *Proceedings of the Royal Society Series B*). They showed that reported *P*-values are more often on or slightly below critical thresholds (such as $P = 0.01$ or 0.001) than slightly above them, compared to the numbers expected based on three different null models. (They specifically excluded a test comparing values slightly above and below 0.05 because of an assumed narrow-sense publication bias against nonsignificant results). Ridley et al. (2007) present several explanations for these findings, but the most plausible are that biologists are either conducting multiple tests ("statistical fishing") and then selectively reporting those that yield significant findings, or they are continuing to collect and analyze data until the desired *P*-value is reached.

A more general concern for ecologists is that stepwise regression, model simplification, and model selection are widely recommended statistical approaches (e.g., Crawley 2002, Johnson and Omland 2004). The problem with these techniques is that nonsignificant terms are removed from the final models. In some cases, authors provide information on the *P*-values associated with excluded terms if each is added individually to the final model. In many cases, however, no *P*-values are presented or, even worse, it is unclear which predictor variables were tested prior to model simplification. This selective reporting based on standard statistical practices has the potential to generate systematic overestimation of the importance of *all* predictor variables. It has even led some authors to recommend that stepwise models be avoided (Whittingham et al. 2006). There is, however, still a trend in ecology and evolutionary biology toward model-building exercises in which parsimonious models are sought and nonsignificant factors are discarded, although model averaging approaches are now becoming more popular in some areas. A similar publication bias arises due to data-mining approaches where multiple statistical tests are conducted. Even if the authors report corrected *P*-values based on Bonferroni or an equivalent approach, this still does not generate an unbiased set of effect sizes for a meta-analysis if only some of the original tests results are reported.

A range of statistical approaches that assess the sensitivity of meta-analyses to selective reporting have been developed (Hutton and Williamson 2000, Hahn et al. 2002), and some *ad hoc* approaches deployed (e.g., Song et al. 2000). The theme that unites these techniques is that when an effect size cannot be calculated due to selective reporting (i.e., information is insufficient or absent but the test is known to have been made), it is better to estimate the likely effect size using information from other sources, or assign a conservative value rather than exclude it from the meta-analysis. Most investigations of publication bias are, however, based on indirect tests, which are primarily designed to detect the absence of studies with certain effect sizes in the meta-analyst's data set. They do not explicitly take into account selective reporting, or try to correct for it. However, because selective reporting effectively leads to the same patterns (e.g., certain effect sizes are missing, while others are overrepresented), these indirect tests are probably still appropriate, although perhaps less powerful, than ones that explicitly take into account selective reporting (Alex Sutton, pers. comm.).

Active data suppression

In some fields, there are strong incentives to suppress studies that produce undesirable results. Halpern and Berlin (2005) review cases in the medicine and social sciences. To date there is little evidence that this is a problem in ecology and evolutionary biology, but it is worth considering. Studies of politically charged environmental issues like the effects of global warming, overfishing, loss of biodiversity, and genetic engineering of crops are often conducted by ecologists working in organizations that have a vested interest in the outcome of the research. It is important to remember that data suppression can be more subtle than simply prohibiting publication. It can be achieved by manipulating the dissemination process. Determining how large-scale reports are presented to the public, ensuring that a result is published in a governmental or NGO report rather than as a peer-reviewed paper, and prohibiting researchers from making public comments on certain studies are seemingly innocuous but effective ways to reduce the likelihood that unpleasant findings become widely available.

HOW TO DEAL WITH PUBLICATION BIAS: INDIRECT TESTS

In this section, we present the five main approaches used to either detect potential narrow sense publication bias or assess how sensitive the results of a meta-analysis are to the possible exclusion of studies (Box 14.3). Where possible we also indicate what evidence these tests have provided about publication bias in ecology and evolutionary biology. Before describing these techniques, however, we must make three points. First, none of the indirect tests is conclusive (e.g., Thornhill et al. 1999). This is an important caveat because some strong statements about the presence of publication bias have been made that are partly based on these indirect tests (for a review, see Palmer 1999, 2000). Second, indirect tests are often applied in ecological meta-analyses to the entire data set, even when heterogeneity in effect sizes attributable to continuous covariates or group differences has been identified. Indirect tests are only applicable to homogenous data sets where an overall mean is being estimated. They should be modified to deal with significant heterogeneity among studies (Terrin et al. 2003). Currently there are no established tests for publication bias designed for meta-analyses that use structural models to explain or explore heterogeneity. We present worked examples of indirect tests in the Appendix to this chapter. Third, we direct the reader to Chapter 21 for other graphical assessments that are useful in meta-analysis. We note that while the recommendations made in the two chapters are not unanimous, the points of difference are mainly due to author preferences.

BOX 14.3.
Indirect tests and sensitivity analyses for publication bias.

(1) Funnel plots. A visual tool to assess "gaps" in the studies used in meta-analysis.
    Only recommended to aid interpretation.
(2) Tests for relationships between effect size and sample size using nonparametric correlation or regression.
    Regression approach recommended if data conform to assumptions.
(3) Trim and fill. Simple method to ascribe values to potentially "missing" studies.
    Recommended as a form of sensitivity analysis.
(4) Fail-safe numbers. The number of studies needed to overturn a result.
    Rosenberg's fail-safe method is recommended over earlier methods.
(5) Model selection. Selection criteria of likelihood models tested (weighted and unweighted).
    Not currently recommended for ecology unless the data set is large and the researcher has access to the necessary statistical expertise.

Funnel plots

Funnel plots are scatterplots of effect sizes against a measure of their variance (e.g., sample size or standard error) (Light and Pillemer 1984) (see Chapter 21 for examples). The variability and range of effect size estimates decrease as sample size increases, due to smaller sampling error. This yields a funnel-shaped plot that is symmetric around the mean effect size (Sterne et al. 2005). However, if there is a modest true effect size (e.g., $r \approx 0.25$; Arnqvist and Wooster 1995a), a publication bias toward significant results creates an asymmetric funnel with small studies with nonsignificant effects missing from the mouth of the funnel on the side opposite to the true effect (Fig 14.2). This occurs because (a) studies are less likely to reach significance in the direction opposite to the true effect by chance, (b) there is a bias against publishing nonsignificant results based on small samples because of low statistical power, and (c) larger studies more accurately estimate the true, significant effect.

When we detect a relationship between sample size and effect size, we cannot assign a causal mechanism to it without further investigation (Sterne and Egger 2005). Effect sizes may be larger in small studies because we have retrieved a biased sample of the available smaller studies, but it is possible that effect sizes *really are* larger in smaller studies (e.g., because smaller studies used different protocols to larger studies). Sterne and Egger suggest use of the term "small-study effect" to describe the relationship between sample size and effect size,

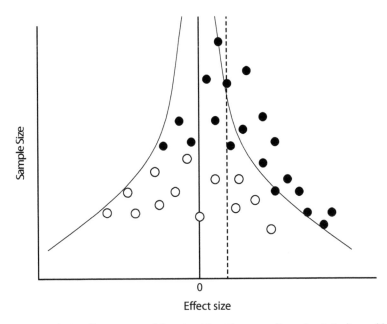

Figure 14.2. A funnel plot illustrating publication bias. The true effect size is indicated by the dotted line. The null value is indicated by the solid line. Data points that lie outside the curves centered around the null value indicate the critical effect size for significance at a given sample size. Solid circles indicate published studies, and open circles indicate unpublished studies. The distribution of all studies around the true effect is symmetric. The unpublished studies are those with smaller sample sizes and nonsignificant results. The net outcome is (i) an asymmetric distribution of published effect sizes, (ii) a relationship between sample size and effect size, and (iii) an overestimate of the true effect size.

rather than labeling it "publication bias." A small-study effect is common in ecology and evolutionary biology (Jennions and Møller 2002a, 2002b). The interpretation of funnel plots based on a visual inspection is discussed more fully in Chapter 21, but we recommend that it is done with caution because (1) funnel plots are not effective when the number of studies is small (<30) because visual asymmetry can often appear by chance, and (2) inspection of funnel plots is unreliable. A study showed that even professional meta-analysts often fail to detect actual publication bias when they examine funnel plots (Terrin et al. 2005). Visual assessment can, however, be facilitated by including contour lines for P-values. This makes it easier to assess whether missing studies are significant or nonsignificant at key threshold values, such as 0.01 and 0.05, and therefore whether or not asymmetry is likely to be attributable to narrow-sense publication bias based on significance criteria (Peters et al. 2008).

The relationship between effect size and sample size

There are two main statistical tests for funnel plot asymmetry. First, one can calculate the nonparametric correlation (Kendall or Spearman) between the standardized effect size and the sampling variance or standard error of a study (these last two terms are equivalent because the tests use ranked data) (Begg and Mazumdar 1994). To standardize an effect size, subtract the mean effect size, and then divide by the square root of the difference between the study variance and the sum of the inverse of all the other studies' variances. This should stabilize the variances so that a plot of standardized effect size against standard error is not funnel-shaped. There will be a correlation if there is a small-study effect. By convention, two-sided P-values are reported because the direction of asymmetry depends on whether the true effect is greater or less than the null hypothesis value.

Second, one can regress standard normal deviates (equivalent to effect size/standard error) against P-values, and weight by the inverse of the study variance (Egger, Davey Smith, Schneider, et al. 1997). The current recommendation, however, is to regress the standard normal deviate of the effect size against precision (equivalent to inverse of the standard error) (Sterne and Egger 2005). If the funnel plot is symmetric, the slope will indicate the size and direction of the effect size and the intercept will be zero. The deviation of the intercept from zero provides an index of the degree of asymmetry. Again, two-sided P-values should be reported. A major advantage of the regression approach is that it can be extended to include other covariates or factors that might affect the relationship (Sterne et al. 2001, Sterne and Egger 2005). Although this extension has not yet been utilized by ecologists and evolutionary biologists, it allows one to test for a small-study effect while controlling for sources of study heterogeneity that might generate an asymmetric funnel plot (e.g., observational versus experimental studies). This approach involves a weighted regression of the effect size against its standard error (square root of variance), where the weighting is the inverse of the variance. In this case, the significance of the test for a nonzero slope provides a measure of funnel plot asymmetry.

To assess the statistical power of the nonparametric correlation approach, Begg and Mazumdar (1994) created data sets assuming publication bias was based on either P-values or absolute effect sizes. These data sets had either 25 or 75 studies. The tests had only modest statistical power with 25 studies (3% to 58%). Begg (1994) therefore suggested that they be used with a liberal significance level (e.g., $P = 0.10$). The regression approach has greater statistical power but, as with any parametric model, there is the potential for a few data points to exert unduly strong leverage (Sterne et al. 2000). There are also concerns that both approaches could yield spurious results if the true effect size is large (Schwarzer et al. 2002). This is, however, unlikely to be a problem in ecology and evolution where effect sizes are generally small (Møller and

Jennions 2002). It should be noted that neither test is informative if variation in study sample sizes is low (Ioannidis and Trikalinos 2007). Another concern is that the regression approach overestimates the significance of a relationship (inflates alpha) for studies with dichotomous outcomes measured using the odds ratio. Some recently modified methods by Peters et al. (2006) and Harbord et al. (2006) partly address this concern.

Finally, it should be noted that new tests are continually being developed. For example, nine tests for funnel plot asymmetry are listed in the *Cochrane Handbook for Systematic Reviews of Interventions* (Table 10.4b in Higgins and Green 2011). When the dependent variable has a continuous outcome, however, only three of these tests can be used (Begg 1994; Egger, Davey Smith, Schneider, et al. 1997; Tang and Liu 2000), and the recommendation is to use the test of Egger, Davey Smith, Schneider, et al. (see Section 10.4.3.1 in Higgins and Green 2011).

Trim and fill

"Trim and fill" is a computationally simple method to adjust for funnel plot asymmetry. It allows one to enter values for "missing" studies to generate a symmetric funnel plot from which a new mean effect size can be estimated (Duval and Tweedie 2000a, 2000b). In general, its findings are consistent with those of the Begg correlation approach and Egger, Davey Smith, Schneider, et al. (1997) regression approach (Sutton et al. 2000, Pham et al. 2001, Jennions and Møller 2002a). That is, when the last two methods show significant funnel plot asymmetry, trim and fill tends to indicate that "missing" studies must be added to create a symmetric funnel plot. Although Duval and Tweedie (2000a) provide a table of *P*-values for the likelihood that a given number of studies would be missing by chance alone, this table is rarely used (but see Jennions and Møller 2002a). Trim and fill was developed as a simple alternative to other methods used to investigate publication bias that are statistically more complex and require skills beyond the reach of many meta-analysts (e.g., on selection models, see Hedges and Vevea 2005).

The technique is readily understood and a fully worked example is provided in the Appendix of this chapter (see also Duval 2005). First, the side of the funnel plot that is assumed to have missing studies (i.e., relative to the observed mean effect, the side closest to the null hypothesis value) is scrutinized and an algorithm is applied to estimate the number of missing studies. By convention, the observed mean effect is always placed to the right of the null value (i.e., if necessary start by reversing the sign of all effect sizes). The trim and fill method assumes that studies with more extreme values on the left side of the funnel plot are those that will be missing due to publication bias. To estimate the number of missing studies, trim off the most extreme values from the right side of the plot until it is symmetric. This is done using an iterative procedure (because the mean changes as studies are removed), which is repeated until the estimate of the total number of studies that need to be trimmed is stable. Next, calculate the new mean effect size based on the symmetric data set. To obtain the variance for this newly estimated mean, replace the original trimmed values and add their missing counterparts. The missing studies are assigned effect sizes and variances that are mirror images of the trimmed studies with respect to the new mean. It should be noted that, even if the plot is strongly skewed to the right, a single outlier on the left side can lead to computational difficulties because the method implicitly assumes that all outliers are on the right side.

The algorithm used to detect asymmetry provides an objective method, or at least one that is consistent among users, to detect small-study effects. For this reason it is arguably preferable to a simple visual inspection of a funnel plot. Duval and Tweedie (2000a) provided three estimators of the number of missing studies, *R*, *L,* and *Q*. Although Duval (2005) suggests that *R* is worth calculating, *L* is currently the only estimator widely used in ecology and evolutionary biology (Jennions and Møller 2002a, Møller and Jennions 2002, Griffin and West 2003, Sheldon and West 2004, Møller et al. 2005, West et al. 2005, Schino 2007). Although trim and fill was not designed to provide a formal statistical test, Duval and Tweedie (2000a) present a table that offers insight into the likelihood that a given value of L will be reported for differently sized databases. These values have been used to informally test whether a funnel plot is more asymmetric than expected by chance (e.g., Sutton et al. 2000, Akçay and Roughgarden 2007).

Conceptually, trim and fill is a form of sensitivity analysis rather than a method to estimate the actual effect sizes of unpublished studies. Several studies have shown that the estimated mean effect size is usually much smaller after trim and fill is used to input putative missing studies. Jennions and Møller (2002a) examined 40 meta-analyses in ecology and evolution and found that 38% of the analyses based on random-effects models had more studies missing than expected by chance. After correcting for this potential publication bias, the estimated mean effect was no longer significantly greater than the null value at the 0.05 level for 21% of the meta-analyses. Trim and fill and Begg correlation estimates of publication bias were positively correlated ($r = 0.40$, $n = 40$, $P = 0.01$). These findings imply that funnel plot asymmetry is widespread in ecology. Although it has not yet been attempted in ecology, it might be worthwhile to use trim and fill in a sensitivity analysis for statements about heterogeneity in effect sizes among groups. That is, when a factor responsible for significant heterogeneity in effect size is detected, separate trim and fill analyses should be run for each group and the heterogeneity test repeated to assess whether group differences persist once putative "missing" studies are included. McDaniel et al. (2006) adopted an approach similar to this when investigating publication bias in employment test validity manuals. (Note that Jennions and Møller [2002a] explicitly limited their analysis to the group with the largest sample size if significant heterogeneity among groups was reported in the original meta-analysis.) It is difficult to justify the use of trim and fill on data sets with known heterogeneity because the underlying assumption of a single symmetric distribution of effect sizes is clearly false (Peters et al. 2007).

Fail-safe or file drawer number

Use of the fail-safe or file drawer number is not based on funnel plot asymmetry. Instead the following question is asked: How many studies (N) averaging an effect size of zero that were not located (e.g., lie in "file drawers") would need to exist to negate the significance of an observed effect size or to reduce it to a specified minimal value? This approach is a form of sensitivity analysis because a large value (especially relative to the number of located studies) suggests that, even if the observed effect is an overestimate, the null hypothesis about the mean effect size can still be rejected. Unfortunately, the fail-safe approach has problems. It has been simplistically applied by ecologists (e.g., Jennions et al. 2001), leading to inflated statements about the robustness of the significance of estimated effect sizes.

Rosenthal (1979) introduced the fail-safe approach by assuming that unlocated studies, on average, provide no evidence against the null hypothesis (i.e., mean value identical to the null value). The overall significance of the set of studies was tested by combining P-values across studies using Stouffer's test. Studies are added to the actual data set until it exceeds the requisite value (e.g., 0.05). Rosenthal suggested that a significant meta-analytic result is robust if N is greater than $5k + 10$, where $k =$ number of studies already in the meta-analysis, although there is no obvious justification for this criterion. Orwin modified this approach to deal with magnitude of effect sizes, rather than significance level. He asked how many additional studies with a mean effect of $d_{missing}$ (usually set at 0) are needed to reduce the observed standardized difference between treatment and control (e.g., Hedges' d) to a scientifically unimportant value ($d = 0.2$ is often used) (Orwin 1983).

The methods of Orwin (1983) and Rosenthal (1979) are explicitly unweighted (i.e., neither method accounts for the weight of the observed and the hypothesized unpublished studies), while modern meta-analyses weight studies by the inverse of their variance (Rosenberg 2005). This difference is problematic given the consensus that publication bias is associated with the absence of studies that have small sample sizes and nonsignificant results. An unsatisfactory solution is to calculate the sample size of a predefined number of studies with the null effect that are needed to make the new mean effect size nonsignificant (L'Abbé et al. 1987). Rosenberg (2005) has, however, provided a solution consistent with modern meta-analyses. If the mean effect size of the additional studies is the same as that of the null hypothesis (e.g., zero if the effect size is r or d), then standard weighted meta-analysis equations can be used to ask how many studies, N, of the same weight (i.e., sample size) as the average of those already being used, are needed to reduce significance to the required level (e.g., 0.05). Alternatively, due to technical problems that arise when using random-effects models, one can interpret N as the relative size of a single additional study (i.e., its weighting relative to the average study).

Rosenberg's approach is readily modified so that additional studies are assigned another weighting (e.g., one could use the mean sample size of the smallest 20% of actual studies). Using the mean weight of the reported studies to calculate how many additional studies are required, yields a conservative estimate of N if, as is widely assumed, unpublished studies have smaller than average sample sizes and consequently lower weighting. Rosenberg (2005) illustrated his approach with a worked example using a meta-analysis of inbreeding from Coltman and Slate (2003). Rosenberg's N was far smaller than Rosenthal's N (e.g., for random-effects models there was up to an eightfold reduction in N).

Becker (2005) recommended that the fail-safe N approach be abandoned because N is often misinterpreted and can be uninformative about actual publication bias. Others have made similar suggestions (e.g., John Ioannidis, pers. comm.). We disagree, at least for ecological and evolutionary studies. The advantage of N is in providing a simple summary value that can be used to compare the relative susceptibility of different meta-analyses to publication bias. We recommend, however, that only Rosenberg's weighted method is used because it is consistent with how mean effect sizes are calculated. Although any weighting value can be assigned to missing studies, we suggest using both the mean of the published studies and the mean of the smallest 20% of studies. The latter figure incorporates the consensus view that unpublished studies have smaller sample sizes. Although the value of 20% is arbitrary, if widely adopted, it will at least allow authors to compare N across meta-analyses. Rosenberg's method can readily be modified to determine the N that will reduce the mean effect size (or the lower limit of the confidence interval) to a specific value, and can thus replace Orwin's unweighted approach.

Finally, it is worth remembering that the fail-safe N approach assumes that the mean effect size of missing studies is that predicted by the null hypothesis. This will inflate N if there is, in fact, a publication bias *against* results in the opposite direction to the current mean effect (i.e. that contradict the prevailing hypothesis).

Selection models

The final approach to correct for publication bias uses two-stage statistical models. A selection model defines the parameters that determine whether a result is published. These parameters can be stated a priori (e.g., Hedges 1984), or can be based on an estimate derived from the distribution of published effect sizes (Dear and Begg 1992). The model provides a weighting reflecting the likelihood that a given result is published. An effect size model then states the expected distribution of effect sizes in the absence of publication bias, which partly depends on whether a fixed- or random-effects model is used. The two models are then combined and the difference between the predicted and observed distribution of effect sizes is attributed to publication bias, which can then be "corrected" for using the selection parameters. These models require a high degree of statistical literacy. They are rarely employed in medical meta-analyses and have not yet been used by ecologists and evolutionary biologists. Given the heavy use of statistics in many areas of ecology and evolution, however, these methods will be within the grasp of some meta-analysts reading this book (or could be conducted in consultation with a colleague in a nearby office). Numerous selection models have been developed so we will only discuss general concepts (these are reviewed in Hedges and Vevea 2005).

Selection models can be divided into those where the selection criteria that weigh the likelihood of publication are based either (a) separately on the effect size and its standard error (Copas and Shi 2001), or (b) solely on P-values (e.g., Iyengar and Greenhouse 1988, Hedges 1992), which is the ratio of the effect size to standard error (i.e., the standard normal deviate, Z). Here we confine ourselves to the latter because it reflects the proximate mechanism thought to generate publication bias, namely studies being primarily vetted based on the significance of their results. In many cases, biological theories make unidirectional predictions so that the direction of an effect influences the likelihood of publication. In such cases, only one-tailed P-values are considered when defining selection criteria, even if the original studies used two-tailed tests. If results in either direction are theoretically possible, then selection models should use two-tailed P-values to define selection criteria.

Selection models that relate P-values to publication can either invoke a threshold function (e.g., only significant studies are published so weighting is 0 or 1) (e.g., Hedges 1984), or treat the likelihood of publication as a continuous or steplike function of decreasing P-values (Hedges 1992, Vevea and Hedges 1995). For any given weighting function it is possible to obtain a maximum likelihood (ML) or Bayesian estimate of the true mean effect size from the set of published effect sizes. The a priori assignment of various weighting functions, especially when a steplike function is assumed so that different weights are assigned to different ranges of P-values, is a form of sensitivity analysis (Vevea and Woods 2005). If the ML estimates of effect sizes are similar for different weighting functions, then one can assume that the reported mean effect is robust against publication bias. In some cases, however, it is impossible to use ML to estimate the mean effect. For example, there is too little information available from the observed effects if the selection criterion is that only significant results in one direction are publishable.

The real advantage of the selection model approach occurs when the selection model is not specified in advance and the weight function is estimated numerically, using specialized software, from the observed distribution of effect sizes. This is only possible, however, when the number of studies is fairly large. To oversimplify, the method takes a given effect size model and adjusts the selection model until it best fits the observed distribution of effect sizes (for technical details, see Hedges and Vevea 2005). Even if the resulting estimate is imprecise, simulation studies suggest that applying the correction for publication bias still led to a better estimate of the mean effect (Hedges and Vevea 1996). The statistical significance of a weighting function model can be tested with a likelihood ratio test to compare it to a simpler nested model where weights are constant across P-values (i.e., no publication bias).

The selection model approach has several pitfalls:

(1) Problems arise if either the effect size model or selection model are incorrectly inferred.

(2) It is difficult to estimate selection model parameters unless the number of studies is large (>100). If parameters are estimated using smaller data sets, then the number of steps in the weighting function has to be reduced (Vevea and Woods 2005).

(3) These methods are computationally intensive and more complex than the "trim and fill" method (the only other approach that "corrects" for publication biases).

The selection model also has some advantages:

(1) When there is effect size heterogeneity due to the presence of subgroups, the method can readily be modified (Terrin et al. 2003). As we suggested earlier, however, trim and fill could also handle this problem if applied to each subset of the data before calculating the grand mean effect.

(2) When there is a known reason for the relationship between effect size and standard error (e.g., pilot studies and power analyses were performed before main studies were conducted), then trim and fill cannot be used, while a selection model can take this into consideration and still test for publication bias (Hedges and Vevea 2005).

(3) It is possible to detect cases where there is a bias against significant results (e.g., because theory is well-developed generating resistance against inexplicable findings; see Palmer 2000).

Finally, Ioannidis and Trikalinos (2007) have developed a simple test that has some similarities with selection models. It determines whether the number of observed significant findings (O) is greater than that expected (E) if we assume a given effect size applies across all studies (i.e., between-study heterogeneity must be low), and is representative of the true effect size. One can then use the power of each study to calculate the probability that it will report a significant effect. Summing these probabilities across all studies gives E, and that can then be compared to O with a chi-square test. They suggest a conservative alpha value (e.g., $P = 0.10$), because of the low power of such a test, especially when O is low. The value used for the effect size could simply be the observed mean effect (although this is likely to be an overestimate given publication bias), the upper and lower values of the 95% confidence interval, or can be drawn from a prior distribution of effect sizes from previously collected data. Ioannidis and Trikalinos (2007) also suggest that one calculates the minimum effect size value for which O is significantly greater than E. If this value falls within the range of estimates of the effect size (based on 95% CI or external evidence), then there is evidence that a bias toward significant results exists.

One advantage of the Ioannidis and Trikalinos method is that it can be used to investigate bias across several meta-analyses. Sample sizes are often too small in any given meta-analysis to detect bias. By pooling meta-analyses across a similar field of research (e.g., studies of interspecific competition in various taxa) one can at least ask whether there is broader evidence for bias. For example, in 7 of 10 meta-analyses of drug efficacy for schizophrenia, O was greater than E.

Statistical tests for publication and reporting bias

In addition to the various recommendations we have provided when using different tests, we encourage researchers to consult the *Cochrane Handbook for Systematic Reviews of Interventions* (Higgins and Green 2011) (http://handbook.cochrane.org/). The above *Handbook* reflects the consensus of experts involved in the meta-analysis of health studies. In the absence of a comparable working group in ecology and evolutionary biology, its recommendations at least allow for a more standardized approach to analyses of publication bias. The *Handbook* provides a more comprehensive summary of the relative value of different tests than we can offer in this chapter. Interested readers should also keep an eye on the output of the Cochrane Bias Methods Group (http://bmg.cochrane.org). Their website provides a regularly updated list of papers by group members, which is a useful starting point to monitor progress on the development of new statistical tests.

What to do if there is publication bias?

Sampling bias is a problem in any empirical investigation, not just in meta-analysis. Our hypothetical field study, where the only goal was to estimate the mean and variance in crab body size, provides a reminder that we always make assumptions about our ability to collect data in an unbiased fashion even when we are asking a very simple question (Box 14.1). In practice, we usually accept that some bias might exist when we collect data. We take this into consideration by treating our results as provisional and assessing their robustness. Sometimes we can test our assumptions and, if needed, modify our conclusions. For example, if muddy sites are hard to reach in our hypothetical study, we might be able to determine crab size and density in a few sites and then estimate what proportion of the habitat is muddy. Or, we could assume a worst case scenario that all the crabs in muddy areas are the maximum size recorded elsewhere and then estimate the mean again to see how much it changes. In an analogous manner, we try to correct for publication bias by using statistical models, such as trim and fill, that make assumptions about which studies are likely to have gone unpublished.

The bottom line is that most progress in science is incremental and conclusions are always provisional. There is no point abandoning a meta-analysis if we suspect there is a publication bias. For now, the best response is to honestly point out the potential problem, suggest solutions, and qualify the conclusions of the meta-analysis accordingly. Future researchers will always be happy to point out when we have erred, and why.

MISIDENTIFICATION OF PUBLICATION BIAS

There are many reasons for an asymmetric funnel plot with a small-study effect (Ioannidis 2005b) that should be considered before concluding that publication bias is responsible for a relationship between sample size and effect size (Thornhill et al. 1999). At some level, they all involve a source of heterogeneity among studies (Box 14.4).

(A) *Study Heterogeneity*. Although an asymmetric funnel plot is likely to arise whenever a heterogeneous set of effect sizes is combined, there is no reason why this by itself will

---

BOX 14.4.
Suggested reasons for funnel plot asymmetry and a small study effect.

Detailed explanations are provided in the main text.

(A) Study heterogeneity can generate asymmetry but cannot, by itself, provide a general explanation for a "small-study" effect
(B) Combining experimental and observational studies
(C) Heterogeneity in baseline values
(D) Continuous sampling and early termination of some studies
(E) Taxon or system related heterogeneity
(F) Study quality
(G) Sample size and measurement error
(H) Incorrect transformation of effect sizes
(I) Actual publication bias against nonsignificant results

generate a small-study effect. Additional factors need to be invoked that generate co-variation between sample sizes and effect sizes among groups.

(B) *Experimental versus Observational Studies.* In ecological meta-analyses, data from observational studies and controlled experiments are often combined. Effect sizes in observational studies might be smaller than those in experiments due to confounding factors that add statistical noise to the relationship, while sample sizes in experiments could be smaller due to the assumption that effect sizes will be larger because of the greater control of confounding variables (as well as logistic constraints). This could generate a negative correlation between sample size and effect size.

(C) *Heterogeneity in Baseline Values.* In ecology, an experimental treatment or natural variation in a causal factor often constrains effects on response variables. There are biophysical or physiological limits on many key parameters, such as growth rates or fecundity. To take the simplest example, survival cannot be greater than 100% or less than 0. So, in a population or experiment where survival is already high, estimates of the effect size for factors that increase survival will, all else being equal, be smaller than those in a population with a lower background rate of survival. This heterogeneity might lead to a small-study effect if researchers are aware of this and adjust their sampling effort accordingly. This phenomenon is well documented in the medical literature (Ioannidis and Lau 1997), and there are modeling techniques available to correct for control/baseline rate variation (Schmid et al. 1998).

(D) *Continuous Sampling and Early Termination.* Sometimes researchers conduct statistical analyses while a study is ongoing, and sample sizes continue to increase as new subjects are included in the study. There is a clear temptation to stop if the desired level of statistical significance is reached (Ridley et al. 2007). In medicine, early termination of a study might be ethical if one would otherwise continue to place patients into a control group after the benefit of a medical intervention becomes clear (Lan et al. 1993). There is far less justification for early termination in ecology and evolution studies, unless one is working with a critically endangered species. If regularly employed, however, one consequence of early termination is that studies of systems or species with smaller actual (and eventually reported) effect sizes will have larger final sample sizes.

(E) *Taxon or System-Dependent Heterogeneity.* It might simply be the case that effect sizes are genuinely larger in some species or systems than other. If this coincides with studies where sample sizes tend to be smaller due to logistical (e.g., remote habitats) or biological constraints (e.g., larger species that are less abundant), a small-study effect will arise. Of course, the reverse pattern could also occur so this does not offer a general explanation for small-study effects.

(F) *Study Quality.* It has been suggested that in some cases in medicine, lower-quality clinical trials with smaller sample sizes are more likely to report strongly beneficial effects of a treatment if the design failure involves a systematic bias toward overestimating effect sizes (e.g., failure to collect data using a double-blind approach may unintentionally inflate measures of treatment effects). In ecology, one would probably predict the reverse trend, because a poorly designed study is less likely to detect a genuine effect due to its failure to take into account confounding variables. In the social sciences, however, some researchers have argued that interventions in smaller studies will produce larger effects because the treatment is more carefully implemented and monitored than in a large study (Lipsey 2003). Similarly, in medical epidemiology, smaller studies might have more accurate data collection methods than larger studies that rely on existing databases.

(G) *Sample Size and Measurement Error*. Although rarely discussed, there is probably a trade-off between the size of a study and measurement accuracy, with less attention paid to each measurement as the numbers increase. Biologists might also be more willing to include aberrant individuals, or statistical outliers, in a large-scale study, arguing that "the results come out in the wash." In addition, larger studies that sample organisms from a wider area or over a longer time frame will collect data on a less homogeneous set of organisms. All these factors should increase statistical noise and reduce effect size estimates in larger studies.

(H) *Incorrect Transformation*. In some cases, an asymmetric funnel plot can occur due to incorrect transformation or failure to transform effect sizes. For example, in many cases the null hypothesis value is 1 for the odds ratio. The expected distribution of studies around an odds ratio of 1 is asymmetric. On the left side you can only have values from 0 to 1, while on the right side they can range to infinity. Log transformation is necessary to make the expected null distribution, now around 0, close to symmetrical. Even then, however, there are problems as there is still a mathematical association between the log odds ratio and its standard error that will affect the outcome of tests for asymmetry; these tests include that of Egger, Davey Smith, Schneider, et al. (1997) when the treatment effect is large. Modified tests for funnel plot asymmetry are therefore required (e.g., Harbord et al. 2006) (see Section 10.4.3.1 of Higgins and Green 2011).

BIASED SAMPLING OF THE NATURAL WORLD

It is impossible to conclude a chapter on publication bias without touching on the biological biases prevalent in the literature. There is a disproportionate amount of information on certain ecosystems, taxa, and species (Gurevitch and Hedges 1999). The effect of this nonrandom sampling of the natural world is worth considering when discussing a meta-analysis. Even if there is no formal publication bias, the fact that some systems are better studied than others means that we should exercise caution when making statements about effect sizes in the natural world. Are the species studied a random sample, at least with respect to the effect of interest, so that the available data can be extrapolated to the wider world? Or should the conclusions of our meta-analysis be restricted to the artificial world of "species that tend to be studied"? Or should we even restrict our conclusions to those species that have been studied?

Some obvious examples of nonrandom data collection that readily spring to mind are the disproportionate number of studies of (a) northern temperate rather than tropical species or Austral temperate species, (b) birds and mammals rather than insects or fish, (c) animals that are easily captured or marked, (d) species that occur at high rather than low densities, and (e) species of economic or agricultural interest. Similarly, there is also a tendency when working on a theoretical question to study species that seem most likely to yield interesting results. For example, a meta-analysis of questions about sexual selection is likely to be biased toward species that are strongly sexually dimorphic rather than monomorphic (Bonduriansky 2007).

In some experimental studies there is a systematic bias toward excluding individuals that cannot be assigned to the desired treatment. For example, in studies testing for a benefit of the number of mates on fecundity, females are randomly assigned different numbers of mates. Females that refuse to complete their allotted mating are, however, discarded. This means that females with more mates are those who are naturally more likely to mate multiply (Torres-Vila et al. 2004). We are sure that researchers can identify similar biases in their own area of specialization (e.g., radio-tracked animals that move further are more likely to disappear from the study area). Perhaps the most general concern is that there is a bias toward studying species or systems that are judged more likely to produce statistically significant results (so-called research bias *sensu* in Gurevitch and Hedges 1999). It is appropriate, indeed essential, for meta-analysts to draw the attention of their readers to these potential biases and gaps in the literature when they discuss the generality of their findings.

CONCLUSIONS

To date, most research on publication bias comes from the medical sciences. Ecological and evolutionary research might not be subject to the same biases, or to biases of the same magnitude. One reason, which we have touched upon, is that biologists often conduct observational studies in which a wide range of parameters are measured and many relationships are then reported. This allows for so many tests that there is little cost to reporting nonsignificant results alongside significant results that are "dressed up" as the main tests of the study. The downside of this, however, is a greater potential for selective reporting of significant findings within a study. Another reason for differing biases in ecology is that a proposed medical intervention has no benefit if it is ineffective, while it is usually still worthwhile knowing whether or not an ecological factor influences a response variable. On the other hand, it could be argued that expensive clinical trials involving the participation of teams of researchers are more visible to the wider research community and are therefore less likely to be buried in a file drawer than ecological studies conducted on species that are only worked on by a single researcher or laboratory (John Ioannidis, pers comm.). One problem in comparing bias between medical and ecological/evolutionary studies is that many different species are tested in the latter, and patterns that are consistent with publication bias might simply indicate genuine heterogeneity among studies. Genuine heterogeneity among studies of different species or systems is likely to be a major impediment to detecting publication bias in ecology and evolutionary biology.

Meta-analysis emphasizes the need to synthesize empirical results rather than treat individual studies as revelations of a wider global truth (Chapter 23). The same approach should be extended to questions about publication bias. At present, the frustrating reality is that insufficient data exist to draw firm conclusions about the extent of publication bias in ecology and evolution. Rather than simply cite the results of medical research on publication bias, we need to collect data on publication patterns in our own fields of research. In our experience, this is an edifying experience that can have a major impact on how one views the world (Hersch and Phillips 2004, Knapczyk and Conner 2007).

Most biologists are familiar with the oft made claim that negative studies are hard to publish because reviewers view them unfavorably. But does anyone really know if this is true in ecology and evolutionary biology? As already noted, the limited evidence suggests otherwise (Koricheva 2003). Conventional wisdom is often wrong because it is driven by a preference for an easy explanation that justifies our own current actions (Levitt and Dubner 2005).

Our final plea is a call to action. It is always worth reminding colleagues that *P*-values are poor criteria by which to decide whether to submit or accept a paper for publication. Reviewing decisions should be based on the question, design, and implementation of a study. When peer-reviewing a "negative study," we need to invoke alternative criteria to assess its merit including logic, generality, and the adequacy of the methods. Finally, we all need to open our file drawers and submit those negative studies. The topic of this chapter will then be one step closer to redundancy.

APPENDIX 14.1: WORKED EXAMPLES
OF ANALYSES FOR PUBLICATION BIAS

We used two data sets. From the Curtis and Wang (1998) data set on the effect of CO_2 levels on plant physiology we extracted cases measuring the parameter PN when there was no confounding treatment (XTRT = NONE). This yielded 50 effect sizes. Here, we treat these as independent data points. We used the reported means and standard deviations to calculate Hedges' d in Metawin 2.1. We also used the 25 effect sizes (Hedges' d) from the species level analysis of the effect on female fecundity of mating with virgin or nonvirgin males; these data are from Torres-Vila and Jennions (2005).

The relationship between effect size and sample size

In Table A14.1 we present the Kendall's rank correlation and Spearman's correlation between standardised effect size (E^*) and the standard error (SE), or between E^* and the sample size (n). The P-values are given in parenthesis. Tests were conducted using Metawin 2.1.

For the data of Curtis and Wang (1998) the use of the standard error (equivalent to using the variance when performing a rank-based test) yields a different conclusion to that reached when using sample size. In some studies the results will be identical; this is the case when the effect size is a correlation and its variance is calculated using the sample size (e.g., the variance of Fisher's z-transformation of r is $1/[n-3]$). In general, the use of the actual variance for a given estimate seems preferable to the use of sample size (which is only an index of the actual variance).

The positive values for the correlations between E^* and SE in Curtis and Wang (1998) suggest that studies with lower precision (greater SE) report larger effect sizes. This is consistent with a publication bias and a right-skewed funnel plot. It should be noted, however, that the marginally nonsignificant correlations ($P = 0.06$) between E^* and sample size lead to the exact opposite conclusion. This contradiction is possible because there is only a weak correlation ($r_s = -0.21$) between sample size and variance or standard error.

The results of Egger's linear regression approach, based on the regression of standard normal deviates against precision (the inverse of the standard error), are given in Table A14.2 (Sterne and Egger 2005). The test for funnel plot asymmetry is based on inspection of the intercept. For

TABLE A14.1. Standardized effect size versus sample size.

Study	Kendall's E^* vs. SE	Spearman's E^* vs. SE	Kendall's E^* vs. n	Spearman's E^* vs. n
Torres-Vila and Jennions (2005)	−0.013 (0.93)	−0.045 (0.83)	−0.06 (0.67)	−0.03 (0.90)
Curtis and Wang (1998)	**0.306 (0.002)**	**0.383 (0.006)**	0.184 (0.06)	0.264 (0.06)

TABLE A14.2. Regression of standardized normal deviate against precision.

Study		B	SE of B	T	P-value
Curtis and Wang (1998)	Intercept	2.057	0.712	2.889	**0.006**
	Slope	0.263	0.444	0.591	0.557
Torres-Vila and Jennions (2005)	Intercept	-0.718	0.829	−0.867	0.395
	Slope	0.511	0.186	2.740	0.012

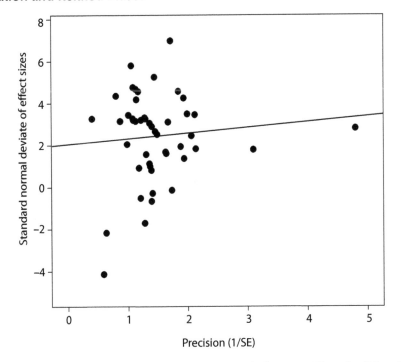

Figure A14.1. Scatterplot of standardized normal deviate of effect size (Effect size/SE) and precision (1/SE) for the Curtis and Wang (1998) data set. The least squares regression line is shown.

Curtis and Wang (1998) the intercept is significantly greater than zero ($P < 0.006$) (Fig A14.1). This is evidence that the funnel plot is asymmetric, such that effect size decreases as precision increases. The results from the regression test and rank correlation test (using SE) are therefore in agreement. For Torres-Vila and Jennions (2005) the intercept is not significantly different from zero, so there is no evidence for funnel plot asymmetry (Fig A14.2). The results from the rank correlation and regression tests are therefore also in agreement.

Trim and fill

We start with Curtis and Wang (1998) and run a fixed model:

(1) The initial mean estimate is $d_1 = 1.43$ (95% CI: 1.26 to 1.61).

(2) Order the effects from largest to smallest. Then calculate the difference between each effect size and d_1. Thirty of the 50 values are positive.

(3) Rank the absolute magnitude of the above differences.

(4) Calculate the sum of the ranks for the positive values (i.e., where the effect size is greater than d_1). This gives $T^+ = 1 + 4 + 5 + 6 \ldots + 47 + 49 = 822$.

(5) Enter this value into the equation to calculate L_0, which is $[4T^+ - n(n-1)]/[2n-1]$. This gives $L_0 = 7.45$, which rounds down to 7.

(6) Recalculate the mean effect size after "trimming" the 7 largest effects (i.e., trim the right-hand side of the funnel plot). Using the 43 cases left $d_2 = 1.21$.

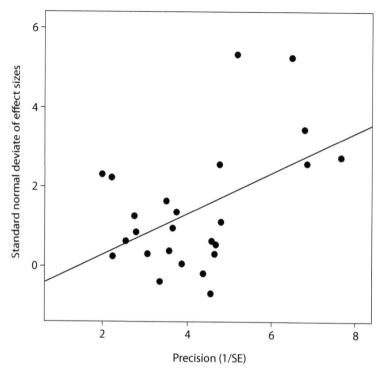

Figure A14.2. Scatterplot of standardized normal deviate of effect size (Effect size/SE) and precision (1/SE) for the Torres-Vila and Jennions (2005) data set. The least squares regression line is shown.

(7) Repeat Steps 2 through 5 using d_2 rather than d_1. The new value of $T^+ = 906$, so $L_0 = 10.85$, which rounds up to 11.

(8) Recalculate the mean effect "trimming" the 11 largest effect sizes (i.e., from the right-hand side of the graph). Using the 39 cases left, $d_3 = 1.11$.

(9) Repeat Steps 2 through 5 using d_3. The new value of $T^+ = 941$, so $L_0 = 12.26$, which rounds down to 12.

(10) Recalculate the mean effect "trimming" the 12 largest effect sizes (i.e., from the right-hand side of the graph). Using the 38 cases left, $d_4 = 1.09$.

(11) Repeat Steps 2 through 5 using d_4. The new value of $T^+ = 944$, so $L_0 = 12.38$, which rounds down to 12.

(12) The estimate of L_0 has therefore stabilized at 12 missing studies.

(13) "Fill" in the 12 putative missing cases so that they are symmetric with respect to d_4 in terms of their value and variance. For example, the two largest effects trimmed were 8.08 (variance: 6.11) and 5.49 (variance: 0.90). Their filled counterparts are -5.90 (variance: 6.11) and -3.31 (variance: 0.90). ($8.08 - 1.09 = 6.99$ and $-5.90 - 1.09 = -6.99$; $5.49 - 1.09 = 4.40$ and $-3.31 - 1.09 = -4.40$)

(14) Finally, use the filled data set of 62 studies to recalculate the mean effect size and its 95% confidence interval. With a fixed model, the value will be identical to d_4 at 1.09. The 95% CI is 0.92 to 1.25. This new estimate is smaller than the initial estimate of d.

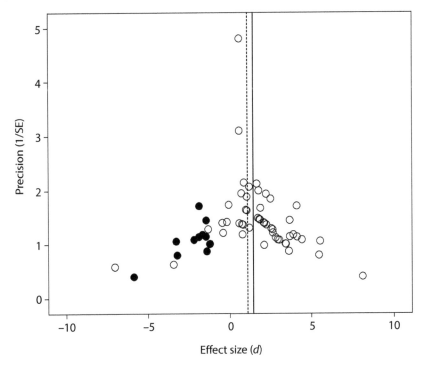

Figure A14.3. Funnel plot showing original (*open circles*) and added "missing" (*solid circles*) effect sizes when using a fixed model to analyze the Curtis and Wang (1998) dataset. The thin vertical line represents the original estimated mean effect size; the thicker, dashed line represents the revised estimate if we assume that there are 12 missing studies.

(15) A funnel plot that includes the "missing" effect sizes is shown in Fig. A14.3. Note that there is one extreme original value on the left-hand side which, as we pointed out in the main text, can lead to difficulties when trimming the right-hand side to create a symmetric funnel plot. We leave it to the interested reader to explore the effect of removing this data point and then reconducting the "trim and fill" on a data set of 49 original effect sizes.

If we used a random-effects model the following occurs:

(1) Initially $d_1 = 1.78$ (95% CI: 1.36 to 2.20).

(2) In this example, many more iterations are required before L_0 stabilizes at 10 missing studies ($L_0 = 2,4,5,6,7,8,9,10$ then 10).

(3) When 10 studies are trimmed $d_9 = 1.27$.

(4) The final estimate of d using the filled data set of 60 cases is not identical to d_9 because the variances of the filled effects influence the mean estimate. The revised estimate of d is 1.25 (95% CI: 0.81 to 1.69). Again, this estimate is smaller than the initial estimate of d.

WHAT IF THE FUNNEL PLOT IS SKEWED TO THE LEFT?   In the data set of Curtis and Wang (1998) the graph is skewed to the right as the putative missing values are on the left-hand side. That is, the sum of the ranks for the positive values (T^+) is greater then the sum of the ranks for the negative values (T^-) (i.e., taking into account distance from the estimated mean there is an excess of effect sizes on the right-hand side). This is the standard expectation given publication bias based on the statistical significance of results.

Now consider the data set of Torres-Vila and Jennions (2005):

(1) Using a fixed-effects model, the initial estimate of d is 0.358 (95% CI: 0.266 to 0.451). The sum of the positive ranks is $T^+ = 132$, that of the negative ranks is $T^- = 192$. The missing studies are therefore on the right-hand side and the graph is skewed to the left.

(2) To trim values from the right-hand side, one simply reverses the sign of all effect sizes, so that the new estimate of d is –0.3584 (95% CI: –0.266 to –0.451) and now $T^+ = 192$.

(3) Having done this, if one follows the protocol described previously for Curtis and Wang (1998), the outcome is that there are five putative missing studies (Fig. A14.4). The final estimate of d using the filled data set of $25 + 5 = 30$ effect sizes is –0.445 (95% CI: –0.360 to –0.560). The signs of these values are then reversed and compared to the initial estimate of d.

(4) In this case, the revised estimate of d is 0.445 which is larger than the initial estimate of 0.358 because effect sizes were missing from the right-hand side.

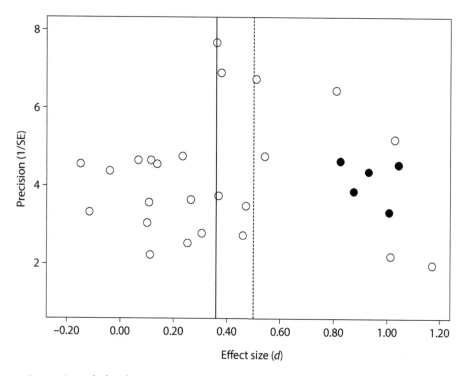

Figure A14.4. Funnel plot showing original (*open circles*) and added "missing" (*solid circles*) effect sizes when using a fixed model to analyze the Torres-Vila and Jennions (2005) dataset. The solid vertical line represents the originally estimated mean effect size; the dashed line represents the revised estimate if we assume that there are 5 missing studies. Note that the "missing studies" are larger than the original effect size, which is not predicted if funnel plot asymmetry is due to a publication bias against nonsignificant results.

This raises an issue. Should we assume there is a potential publication bias whenever a funnel plot is asymmetric, or should we only do so when the asymmetry is such that effect sizes on the side closest to the null hypothesis value are missing? The latter view is more consistent with the general assumption that publication bias is due to a lower rate of publication of studies with nonsignificant values (and the observed effect size is on the same side of the null hypothesis as the true effect size). It should be noted, however, that when Jennions and Møller (2002a) used trim and fill to estimate missing studies, they did this regardless of whether there was funnel plot asymmetry with effect sizes on the side closest to or furthest from the null hypothesis value.

Even if the decision is made to only use trim and fill when the asymmetry is due to missing studies closer to the null hypothesis value it might still be necessary to reverse the sign of the effect sizes to ensure that trimming occurs from the right-hand side of the graph. This usually occurs if the mean effect size estimate is negative, the null hypothesis is an effect size of zero and the missing studies are those closer to zero (i.e., due to the way the original effect sizes are presented the data set is skewed to the left). It is then necessary to first reverse the sign of the effect sizes so that trimmed effect sizes are those furthest from zero (the null value).

Fail-safe or file drawer number

We used the fail-safe calculator developed by Rosenberg, which is freely available at http://www.rosenberglab.net/software.php#failsafe. The fail-safe numbers are shown in Table A14.3.

Fixed-effects models, random-effects models

Rosenthal's fail-safe number is not based on a weighted meta-analysis. It tends to give the largest estimates of a fail-safe N. Rosenberg's N_1 is the relative weight of a single study with the null hypothesis effect size needed to make the estimate of the mean effect not significantly different from the null value. Here, for example, it would need a weighting 349 times larger than that of the average study in Torres-Vila and Jennions (2005) so that the mean estimate of d did not differ from zero. Rosenberg's N_+ indicates that 386 studies with a mean effect of zero would make the estimate of the mean effect nonsignificant. Although Rosenbergs's N_+ can be calculated when significance testing is based on the normal distribution (second row), this is not recommended given the small sample sizes in many meta-analyses. The use of the t-distributions for hypothesis testing is more appropriate.

TABLE A14.3. Fail-safe numbers. (FE=fixed-effects model, RE=random-effects model.)

Fail Safe Number	Torres-Vila and Jennions (2005)	Curtis and Wang (1998)
Rosenthal	440	5452
Rosenberg (based on normal distribution) FE	561	4819
Rosenberg N_1 (based on t-distribution) FE	349	3216
Rosenberg N_+ (based on t-distribution) FE	386	3378
Rosenberg N_1 (based on t-distribution) RE	31	130
Rosenberg N_+ (based on t-distribution) RE	n/a	n/a
Orwin (minimum effect $d = 0.20$)	19.9	308
Sample size of study	25	50
Robust Fail-safe (criterion: $5N + 10$)	135	260

As is often the case, Rosenthal's N is notably larger than Rosenberg's N_+ (t-distribution); it is not possible to calculate Rosenberg's N_+ for a random-effects model, the large decline in the value of N_1 between a fixed- and random-effects model is, however, noteworthy. Finally, Orwin's number shows that far fewer studies with a mean effect of zero are needed to reduce the estimated mean effect size to $d = 0.20$.

15

Temporal Trends in Effect Sizes: Causes, Detection, and Implications

Julia Koricheva, Michael D. Jennions, and Joseph Lau

THE GENERAL AIM OF meta-analysis, as well as of any other form of research synthesis, is to combine scientific evidence scattered through a number of individual studies addressing the same topic. Evidence, however, is not static and tends to evolve over time due to changes in research methods, changes in the characteristics of the subjects being studied, and so forth. New studies might either strengthen or challenge the conclusions of previous reports, resulting in changes in the mean effect size and its variance over time. The magnitude and direction of the mean effect size, and the breadth of its confidence interval, largely determine the conclusions drawn from a meta-analysis. Examples include whether or not a particular treatment, policy, or management strategy works; whether or not factor A has a biologically and/or statistically significant effect on the response variable X; or whether or not a hypothesis is supported by empirical tests. It is important therefore to be aware of the extent of temporal variation in effect sizes and to understand the reasons for this variation.

A number of recent studies in ecology and evolution (Table 15.1) have shown that temporal trends in effect sizes are common and often quite dramatic in these fields. This may perhaps reflect higher heterogeneity of studies included in ecological and evolutionary meta-analyses as compared to those in medicine (Chapter 25). Unlike other sources of heterogeneity, which affect the generality of conclusions drawn in meta-analysis, temporal changes in effect sizes might also jeopardize the stability of those conclusions (i.e., the conclusions of meta-analyses on the same topic conducted in different years might differ). Moreover, some of the methods used for detection of temporal trends in effect sizes, such as cumulative meta-analysis, differ from those used to detect other sources of heterogeneity. For the above reasons, we devote an entire chapter to temporal changes in effect sizes. We first summarize the findings of studies that examined temporal changes in the magnitude and direction of effect sizes in ecology, evolutionary biology, medicine, and the social sciences, and then discuss their possible causes, methods of detection, and implications for the interpretation of the results of the meta-analysis.

EVIDENCE OF TEMPORAL CHANGES IN EFFECT SIZES IN ECOLOGY AND EVOLUTIONARY BIOLOGY

Significant changes in the magnitude and even direction of research findings over time have been reported in many research syntheses from several different areas of ecology and evolutionary biology during the last two decades (Table 15.1). Notably, most of these areas represent

TABLE 15. 1. Outcomes of studies that have tested for temporal changes in reported effect sizes in ecology and evolutionary biology.

Reference	Topic	Pattern revealed
Alatalo et al. 1997	Heritability of secondary sexual characters	Approx. 2-fold increase in the reported estimates of heritability of male ornaments following publication of models supporting the assumptions of good-genes theory
Gontard-Danek & Møller 1999	Relationship between the strength of sexual selection and the expression of secondary sexual characters	Effect size was significantly negatively related to year of publication
Møller & Alatalo 1999	Relationship between male traits and offspring survival	Negative relationship between effect size and the year of publication
Simmons et al. 1999	Relationship between fluctuating asymmetry and sexual selection	Dramatic decrease in effect size and proportion of studies supporting the role of fluctuating asymmetry in sexual selection over <10 years
Poulin 2000	Effects of parasites on host behavior	Significant negative relationship between the effect size and the year of publication over the 30 years of research
Dubois & Cezilly 2002	Relationship between breeding success and mate retention in birds	Weak negative relationship between effect size and year of publication, which became nonsignificant once the effect of clutch size was controlled for
Gardner et al. 2003	Decline in Caribbean corals	Reduction in the rate of coral loss from 1980s to 1990s
Leimu and Koricheva 2004	(1) Effects of N fertilization on phenolics in woody plants (2) costs of plant antiherbivore defenses	Nonlinear approx. 3-fold decrease in magnitude of effect sizes with publication year in both datasets
Nykänen and Koricheva 2004	Damage-induced changes in woody plants	Dramatic nonlinear decrease in effects of damage on phenolic and nutrient concentrations in host plants and on herbivore performance with publication year
Møller et al. 2005	Relationship between fluctuating asymmetry and sexual selection	No difference in mean effect size between studies conducted through 1996, studies conducted in 1997–2001, and the sample of unpublished studies in the present study
Saikkonen et al. 2006	Endophyte-grass interactions	Temporal decrease in effects of endophytes on plant competition, plant performance, and resistance to herbivores
Toth and Pavia 2007	Induced herbivore resistance in seaweeds	Large decrease in magnitude of induced resistance from the late 1980s to early 2000s
Zvereva et al. 2008	Effects of air pollution on species richness of vascular plants	Effect size did not change with the publication year

(*continued*)

TABLE 15. 1. *Continued*

Reference	Topic	Pattern revealed
Kampichler and Bruckner 2009	Role of microarthropods in litter decomposition	Significant decrease in the magnitude of the effect size with the year of publication
Barto and Rillig 2010	Effects of plant herbivory on mycorrhizae	Significant decrease in the magnitude of the effect size with the year of publication
Krist 2010	Egg size and offspring quality in birds	Significant decrease in the magnitude of the effect size with the year of publication
Santos et al. 2011	Dominance and plumage traits	Significant decrease in the magnitude of the effect size with the year of publication
Kelly and Jennions 2011	Sexual selection and sperm quantity	Effect size did not change with the publication year

hypothesis-driven research, and the majority of studies reported a decrease rather than an increase in the magnitude of the effect size with publication year (but see Alatalo et al. 1997). Reported changes in the magnitude of the effect sizes are often quite dramatic (e.g., 2- to 3-fold or more), and sometimes lead to the loss of the statistical significance of the mean effect size, or even to a change in the sign of the effect size (Leimu and Koricheva 2004, Nykänen and Koricheva 2004, Saikkonen et al. 2006). Initially, these temporal trends were treated as isolated occurrences and attributed to paradigm shifts (sensu Kuhn 1970), scientific fads, changes in methodological approaches, or biases in the choice of study systems. However, Jennions and Møller (2002b) have analyzed 44 independent meta-analytic data sets covering a wide range of ecological and evolutionary topics, and found a small but significant decrease in effect size with year of publication across these data sets. It is clear, therefore, that temporal trends in the magnitude of effect sizes represent a general phenomenon in ecology and evolutionary biology, and the most common pattern appears to be a decrease in effect sizes with time.

EVIDENCE OF TEMPORAL TRENDS IN EFFECT SIZES IN OTHER RESEARCH FIELDS

Temporal trends in effect sizes have also been repeatedly reported in clinical medicine. For example, Trikalinos et al. (2004) found that the magnitude of the effect size of therapeutic and preventive interventions in mental health has changed considerably over time; similarly to ecology and evolutionary biology, for three out of four response variables of outcome, a decrease in effect size was more common than an increase. Furthermore, in eight out of 100 meta-analytic data sets examined, the statistical significance of the mean effect size was lost as more trials were published. Similarly, Gehr et al. (2006) demonstrated fading of reported effectiveness over time in three out of four investigated lipid-lowering and anti-glaucoma drugs.

Ioannidis (2005a) showed that the results of 32% of highly cited original clinical research studies that he examined were either contradicted by subsequent research or found stronger effects than subsequent studies. In another study, provocatively titled "Why most published research findings are false," Ioannidis (2005b) argued that the probability of a research finding being false (and thus the probability of this finding being refuted by subsequent research) is particularly high if the study's sample size is small, the effect size in a research field is small, and the scientific field is hot.

In molecular genetics research on genetic associations, a rapid early sequence of extreme, opposite results was observed (Ioannidis and Trikalinos 2005). This phenomenon was named the "Proteus phenomenon" after the mythological god who rapidly metamorphosed himself into different figures. Furthermore, in 21 out of 36 studied meta-analyses of associations in genetic epidemiology, the first study or studies tended to give more impressive results (Ioannidis et al. 2001). Ioannidis and Trikalinos (2005) suggest that the Proteus phenomenon might be characteristic of disciplines where data production is rapid and copious (like "omics" fields), but might be less likely in research where studies take considerable time to perform, as is the case in many ecological field studies and clinical trials in medicine.

Temporal trends in effect sizes have also been observed in the social sciences. For example, the reported magnitude of gender difference in cognitive abilities (Feingold 1988), mathematics performance (Hyde et al. 1990), and sexual attitudes and behaviors (Olivier and Hyde 1993) has declined over time, whereas the effects of the media on body image concerns among American women have increased in the 2000s relative to the 1990s (Grabe et al. 2008).

Recently, Ioannidis (2008) reviewed theoretical work and empirical evidence of decreases in effect sizes over time and suggested that this is a general phenomenon across scientific fields, and that effect sizes in early studies on the topic are often inherently inflated. But why is this the case?

POSSIBLE CAUSES

Early studies are prone to overestimate the magnitude of the effect if they rely on statistical significance testing to establish the existence of an effect, and are based on small sample sizes (Ioannidis 2008). For example, Schmidt (1992) showed that if the actual effect size (standardized mean difference) is equal to 0.5, for a study with a sample size $n = 30$ (equal to statistical power of 0.37) to be significant at $P = 0.05$, the reported effect size must be 0.62 or larger, which is 24% larger than the real effect size (0.50). Furthermore, the average of the significant effect size values in the above example would be 0.89, which is 78% larger than the true value. The lower the sample size and, thus, the lower the statistical power of the early studies, the more likely these are to overestimate the magnitude of the effect (Ioannidis 2008). Ioannidis and Lau (2001) have examined changes in treatment effects over time in two medical fields (pregnancy/perinatal medicine and myocardial infarction). They have shown that the probability that the effect size changes with the accumulation of more data is a function of the cumulative number of patients. It might be expected that stabilization of effect sizes around the mean across studies will take even longer in ecology since unlike medicine where the study subject is a single species (*Homo sapiens*), the diversity of study subjects in biology is much larger and sample sizes are typically small. However, Koricheva et al. (in preparation) have found no evidence of changes in sample sizes over time across 54 meta-analytic data sets on various topics in ecology and evolution. Therefore, earlier studies in ecology and evolutionary biology tend to have the same degree of replication as later studies on the same topic, and thus temporal trends in effect sizes in these fields cannot be explained by lower statistical power of earlier studies.

Jennions and Møller (2002b) suggested that the most general and plausible cause of the observed decrease of effect sizes with time in ecology and evolutionary biology is time-lag bias, the delayed publication of studies reporting small or statistically nonsignificant effects. When time-lag bias operates, the first published studies will report larger effect sizes compared to subsequently published investigations, resulting in a decrease in the mean effect size over time. Such bias has been shown to occur in clinical medicine (Stern and Simes 1997, Ioannidis 1998) as well as in genetic association studies (Ioannidis et al. 2001), but at present there is no direct evidence of time-lag bias against nonsignificant results in ecology (Chapter 14). In contrast, Koricheva (2003) has found that ecological studies with a large proportion of nonsignificant results are more likely to be published than studies with a smaller proportion of nonsignificant results because the latter are submitted to journals with larger impact factors, which have higher rejection rates. She also found no evidence of delayed publication of studies with a large proportion of nonsignificant results.

Another form of selective reporting which might contribute to temporal trends in effect sizes is deliberate withholding or delayed publication of studies that fail to confirm the hypothesis being tested. This type of bias is especially likely to occur in early studies testing a recently suggested hypothesis because of the initial enthusiasm and less critical attitude of scientists toward new and currently popular ideas (Kuhn 1970). Gradually, however, evidence refuting the hypothesis begins to accumulate and, eventually, alternative scenarios and competing theories are suggested; this prompts publication of studies supporting these new theories and leads to temporal changes in magnitude or even sign of the overall effect (Leimu and Koricheva 2004). Evidence that this type of bias is responsible for some of the temporal trends in effect sizes reported in ecology comes from several sources. For example, an increase in reported heritability estimates of secondary sexual characters began after new models that appeared between 1986 and 1988 indicated that such characters can be honest indicators of male viability, providing fitness benefits for choosy females (Alatalo et al. 1997). Similarly, a decrease in the magnitude of reported fitness costs of plant resistance to herbivores began in the early 1990s, after several studies providing theoretical justification for the absence of fitness costs were published (Leimu and Koricheva 2004). The above observations have led to the suggestions that "analyses aimed at assessing the generality of recently advanced paradigms should wait until revolutions have settled" (Simmons et al. 1999). Further evidence for the role that bias against nonconfirmatory evidence plays in temporal changes in effect comes from the study by Poulin (2000). Poulin found that a temporal decrease in effects of parasites on host behavior occurred only among studies that were specific tests of the adaptive manipulation hypothesis, but was not apparent among more descriptive studies that examined the effects of parasites on host behavior in terms of pathology or other consequences for the host. In some fields, however, highly contradictory findings might be more attractive to investigators and editors, resulting in a rapid succession of extreme opposite results; this has been observed in molecular genetics (Ioannidis and Trikalinos 2005).

Temporal changes in effect sizes might also be caused by a bias in the choice of study organisms or systems. Gurevitch and Hedges (1999) suggested that ecologists tend to perform their studies on organisms that are more likely to display statistically significant responses; they called this tendency a "research bias." As the range of study organisms tested increases with time, the magnitude of the cumulative effect sizes diminishes. For example, Tregenza and Wedell (1997) suggested that the increase in published heritability estimates of secondary sexual characters observed by Alatalo et al. (1997) could be due to an increase in the number of studies conducted on birds, in contrast to earlier studies that were conducted largely on insects. Nykänen and Koricheva (2004) have demonstrated that the decrease in magnitude of the reported effects of plant damage on herbivore performance (indicating induced resistance) was partly due to a decrease in the proportion of studies conducted on mountain birch, which manifested stronger induced resistance than other tree species. Similarly, Saikkonen et al. (2006) have shown that most of the conceptual framework for endophyte-plant interactions has been based upon studies of two economically important grass species (tall fescue and perennial ryegrass), particularly tall fescue cultivar Kentucky 31. This cultivar is, however, a misleading model system for endophyte-grass interactions because it performs better than other cultivars and plants collected from nature. As the diversity of study systems increases over the years, the cumulative effect of endophytes on plant competition, performance, and resistance to herbivores decreased; it eventually became nonsignificant in the case of effects on plant competition and performance.

Another potential cause of temporal changes in effect sizes are changes in research or statistical methods over time. For example, Simmons et al. (1999) suggested that the temporal decline in the proportion of studies supporting the role of fluctuating asymmetry in sexual selection was due to an increase in the proportion of studies that used repeatability analysis to distinguish fluctuating asymmetry from measurement error. Changes in statistical analyses, such as better control for confounding variables, could potentially have the same effect. Similarly, as another explanation for the observed temporal increase in heritability estimates of secondary sexual traits observed by Alatalo et al. (1997), Tregenza and Wedell (1997) pointed out that before 1988 the majority of studies used artificial selection; however, after 1998 there was a marked increase in studies using parent-offspring or sib-sib regression in the wild. More recently, Timi and Poulin (2007) demonstrated how changes in the analytical methods used to study patterns of species composition in parasite communities resulted in increases in the likelihood of finding nestedness over time. Such changes in research or statistical methods are common in ecology as well in many scientific fields. This could potentially account for temporal changes in effect sizes observed in some studies, and might results in either increases (Timi and Poulin 2007, Barto and Rillig 2010) or decreases (Simmons et al. 1999) in effect sizes. (See also Worked example 1.)

Since the majority of effect size metrics represent comparisons of frequencies or means between a control and treatment groups (e.g., odds ratios, response ratios, standardized mean differences), temporal changes in event rate or magnitude of the mean in the control group could also account for temporal changes in effect sizes. In medicine, effects of the control rate (the proportion of patients in the control group with the event of interest) on treatment efficacy are well known (Schmid et al. 1998). Temporal changes in the control rate could be due to many factors, such as improved standards of medical care and diagnostic tools, public awareness of the need to get care sooner, and so forth. For example, Antman and Berlin (1992) reported a decrease in the incidence of ventricular fibrillation (VF, cardiac muscle arrhythmia) in patients with acute myocardial infarction (AMI), presumably as a result of dramatic improvements in the general care of AMI patients since the 1960s; the authors pointed out the need to reassess the risk-benefit ratio of using lidocaine prophylaxis treatment for VF, especially in view of a previously reported trend toward excess mortality in lidocaine-treated patients. Similarly, Gehr et al. (2006) also demonstrated that a fading reported effectiveness of several pharmaceuticals could be explained to a large extent by the decrease in the baseline values of the parameter of interest (i.e., patients who had been included in the earlier trials were sicker than patients in later trials).

In ecology, temporal changes in control rates and resulting changes in effect sizes might be expected. Examples include studies assessing losses of biodiversity, habitat fragmentation, and global climate change; as the causes of these responses continue to change in frequency or extent, we might see corresponding changes in the magnitude of the effect sizes assessing these responses over time. Gardner et al. (2003) showed that the rate of coral loss (as measured by the annual rate of change in percent coral cover) in the Caribbean basin decreased in most areas during the 1990s, compared to the 1980s. This decrease in the effect size might suggest alleviation of some of the pressures causing coral mortality. However, it could also indicate that the remaining types of corals are hardier and less sensitive to human-caused disturbance than the corals that disappeared first. More pessimistically, the decrease in rate of coral loss could be a "control rate" type effect, and simply reflect the fact that there is relatively little coral left to lose and thus, as coral cover approaches zero, the rate of coral loss is expected to slow down.

Finally, in some cases temporal changes in effect size might reflect real biological phenomena and be due to rapid adaptation to changes in the strength or direction of the selection pressure. A well known example in medicine is the development of resistance to drugs by bacteria and viruses, which might decrease treatment efficacy for the same treatment over time (Fischbach et al. 2002). Similar adaptive responses may occur in ecological and evolutionary studies; examples include the response to selection pressure imposed by herbicide and pesticide application, or overharvesting by humans (Strauss et al. 2008). Furthermore, Strauss et al. (2008) suggest that the evolutionary history of the study population prior to the experimental manipulation may also strongly influence both the initial magnitude of the treatment effect and the trajectory of subsequent evolution.

To summarize, various factors might explain temporal changes in effect sizes across studies, and several of them might be operating in each particular case. Temporal changes in effect sizes are often indicative of other sources of heterogeneity that might have been missed initially if temporal trends were not examined; exploration of temporal trends in effect sizes is thus a useful diagnostic tool to reveal those causes of heterogeneity. In order to demonstrate that the observed temporal changes reflect real changes in the magnitude of the biological effect over time, the researchers have to rule out other possible explanations, such as publication bias (Chapter 14), heterogeneity between study organisms (Chapter 17), and changes in research methods (see Worked example 1).

METHODS OF DETECTION
Graphical methods

The simplest way to visualize a potential temporal trend in a meta-analytic data set is to produce a scatterplot of effect sizes versus publication year (Figs. 15.1A and 15.2A). Another graphical technique, called cumulative meta-analysis (CMA), was introduced to examine temporal trends in effect sizes in medicine (Lau et al. 1992). In order to conduct CMA, one has to sort individual studies in chronological order, and the earliest available study is then entered into the analysis first. At each step of the CMA, one more study is added to the analysis and the new mean effect size and 95% confidence interval are recalculated. This allows estimation of the contribution of individual studies and assessment of temporal change in the magnitude and direction of research findings. In the absence of biases and heterogeneity, the CMA plot should exhibit a fairly constant estimate of treatment effects over time, with some fluctuations due to chance only in the early steps. As more studies are added to the analysis, the cumulative effect size stabilizes around the mean, and the width of the confidence intervals decreases.

Originally, CMA was proposed as a tool in medicine to detect the earliest year at which a treatment effect became statistically significant, and thus a conclusion about its clinical efficiency could be drawn and a decision about its use made. For example, by using CMA, Lau et al. (1992) demonstrated that the evidence in favor of streptokinase drug therapy for patients with myocardial infarction became significant 13 years before the experts recommended its widespread use. Thus, patients continued to receive inferior treatment long after the evidence was available to demonstrate that other treatments were more effective. More recently, Fergusson et al. (2005) reviewed the results of 64 clinical trials of aprotinin, a serine protease inhibitor used to reduce bleeding during cardiac surgery; the trials were conducted between 1987 and 2002. They showed that the use of CMA would have allowed establishing a clinically significant effect of the drug after 12th trial published in 1992, making the subsequent

42 trials redundant, unethical in regard to the patients, and wasteful of time and resources. Interestingly, Fergusson et al. (2005) revealed that a large number of redundant trials evaluating efficacy of aprotinin were conducted because researchers were not adequately citing previous research.

Mullen et al. (2001) suggested that CMA could be used to assess sufficiency and stability of cumulative knowledge. Consideration of sufficiency addresses the question: "Are additional studies needed to establish the existence of the phenomenon?" If the answer to the above question is no, then collecting additional evidence for an already established effect might waste time and resources, and delay implementation of effective treatments (in medicine) or conservation management policies (in conservation biology). The consideration of stability addresses the question: "Will additional studies change the evidence of the phenomenon's existence and strength?" This aspect directly relates to temporal changes in effect sizes; if the cumulative mean effect on the CMA plot keeps changing with each new study added to the analysis, the results of the meta-analysis should be interpreted with caution because new evidence might change those conclusions. Moreover, instability of the CMA plot might suggest that an important source of heterogeneity exists among the studies, and thus calculation of the mean effect across the whole data set is less meaningful.

CMA is usually applied to a collection of studies retrospectively, to check whether and when the evidence for phenomena under consideration has achieved sufficiency and stability (e.g., Lau et al. 1992). However, Mullen et al. (2001) also recommended CMA as a prospective tool for newly emerging topics with relatively few studies available. They argued that such a prospective approach would help inform researchers about the necessity for investing additional resources in conducting studies on the topic when sufficiency and stability remain uncertain. Mullen et al. pointed out that this prospective approach to CMA removes a commonly raised objection to meta-analysis, namely that the compiled database is too small for a meta-analysis; the prospective approach therefore makes the application of meta-analysis to new research fields imperative rather than suspect.

While the application of CMA in medicine is now widespread, this method has been introduced into ecology only recently (Leimu and Koricheva 2004), even though it is available in MetaWin (Rosenberg et al. 2000), the meta-analysis software package most widely used by ecologists. Somewhat alarmingly, the first applications of CMA in ecology detected severalfold changes in the magnitude of the effect (Leimu and Koricheva 2004, Toth and Pavia 2007), losses of statistical significance of the effects reported in earlier studies (Saikkonen et al. 2006), and even changes in the sign of the cumulative effect size over time (Nykänen and Koricheva 2004). Future applications of CMA to a larger number of ecological data sets will reveal whether these are extreme cases or standard patterns as in studies of genetic associations, where every possible temporal pattern for cumulative effect trajectories has been observed (Ioannidis et al. 2001).

Note that in CMA, data can be arranged not only in chronological order but also in order of any other continuous variable or covariate of interest to the reviewer— for example, by study sample size or control rate (Lau et al. 1995), or by impact factor of the journal in which the study is published (Leimu and Koricheva 2004). In addition, CMA can be conducted on subgroups of studies to take into account heterogeneity in study organisms or in research methods so that one can compare temporal patterns across these subgroups. This is important, because heterogeneity in the data might cause spurious temporal patterns in effect sizes (see Worked example 1).

Several meta-analytic statistical software packages, such as MetaWin and Comprehensive Meta-Analysis, include an option to conduct CMA and to produce a CMA plot (see Chapter 12 and worked examples). However, note that in MetaWin, if a random-effects model is chosen for CMA, it will automatically switch to calculate a fixed-effects model if the between-study variance estimate is less than or equal to zero. Because the estimates of between-study variance might vary at each step of CMA when a new study is added, the resulting CMA plot in MetaWin will often represent a mixture of mean effects and 95% confidence intervals calculated on the basis of fixed- and random-effects models. Because random-effects models usually produce broader confidence intervals, it might be difficult to draw conclusions about the convergence of effect size from such plots. CMA plots produced by the Comprehensive Meta-Analysis software are free from this problem.

Ioannidis and Lau (2001) suggested an extension of CMA, a recursive cumulative meta-analysis (RCMA), which shows the relative change in the magnitude of the treatment effect as a new study is added to the meta-analysis. The relative change at each step of the cumulative analysis is calculated as E_{t+1}/E_t, where E_t and E_{t+1} are the cumulative mean effect sizes at steps t and $t + 1$, respectively. The benefit of RCMA is that results from several cumulative meta-analyses using different metrics of effect sizes and reporting different magnitudes of effects can be plotted on the same graph to compare the patterns (Ioannidis and Lau 2001). Observed relative changes in the magnitude of the cumulative effect size reflect the uncertainty of the treatment effect. Moreover, Ioannidis and Lau (2001) showed that early fluctuations in the magnitude of the treatment effect might sometimes signal further major changes in the magnitude of the effect sizes. Note, however, that recursive meta-analysis based on relative change cannot be used in situations where the sign of the effect size varies between different information steps (t and $t + 1$) of CMA, which is often the case for Fisher's z, Hedges' d, and the log response ratio. In these cases, one can use the absolute difference ($E_{t+1} - E_t$) instead. Furthermore, RCMA works best for symmetric effect sizes such as odds ratios, where the result of the analysis would be the same if E_t/E_{t+1} were used instead of E_{t+1}/E_t (Trikalinos and Ioannidis 2005); this is not true for nonsymmetric effect sizes, such as the relative risk. In view of the above limitations and the prevalence of metrics other than odds ratios in ecology and evolutionary biology, RCMA may prove to be of limited use in these fields.

Statistical methods

Graphical tools like scatterplots or CMA plots are useful for initial inspections of data but, as all visual methods, they might be subject to misinterpretation (compare with funnel plots, see Chapter 14) and should be supplemented by formal statistical methods. In addition, from a frequentist perspective CMA suffers from the problem of multiple testing of the same hypothesis and an inflated type I error (Bender et al. 2008). An infinitely updated CMA would eventually yield a statistically significant finding even when the true effect size is 0 (Berkey, Mosteller et al. 1996). Several techniques have been proposed to adjust P-values and test statistics to multiple testing in CMA (Pogue and Yusuf 1997, Lan et al. 2003). Some authors argue, however, that accumulating meta-analyses are best interpreted in a Bayesian framework and that there is no need to adjust for multiple testing in CMA (Lau et al. 1995). Bender et al. (2008) suggest that the relevance of adjusting for multiple testing in CMA depends on whether the review is intended for descriptive or decision-making purposes. If the former, the adjustment is not required, but if the latter, an adjustment might be advisable. The recently proposed sequential approaches (Brok et al. 2008, Wetterslev et al. 2008, Higgins et al. 2011) may reduce the risk of false positives in cumulative meta-analysis by using the approach analogous to sequential monitoring boundaries. This method deserves attention in future research and we refer interested readers to the above publications, which describe the method in more detail.

Various statistical tests can be conducted to assess the significance of the temporal changes. For example, one can compare whether the results of the first study (studies) differ more than expected by chance from those of the subsequent studies on the topic by using the formula:

$$z = \frac{T_1 - \overline{T}_2}{\sqrt{v_1 + v_2}}, \tag{15.1}$$

where T_1 is the effect size of the first study (studies), T_2 is the mean effect size of all subsequent studies, and v_1 and v_2 are their corresponding variances (Trikalinos and Ioannidis 2005). Absolute values of $z > 1.96$ indicate statistically significant differences at the 5% significance level. The above formula works for effect sizes that follow an approximate normal distribution. It should therefore be used on Fisher's z rather than Pearson's correlation coefficients, and on log response ratios and log odds ratios rather than response ratios and odds ratios, respectively.

One can also use a nonparametric test (e.g., the sign test) to compare the number of steps in cumulative meta-analysis where the effect size increases rather than decreases (Trikalinos and Ioannidis 2005). The assumption is that in the absence of bias, the number of steps in which effect size is increasing and decreasing should be equal. However, in our experience such tests are very conservative because they take into account only the direction, but not the magnitude, of change in the effect size at each step. Even if each step where a decrease in effect size occurs results in a much larger change in effect size than each step where an increase occurs (resulting in a visible decrease in effect size on the CMA or regression plot), no significant difference will be detected by the test; however, this is only the case if the number of steps where decreases and increases occur are similar. Therefore, we do not recommend the use of this test.

Mullen et al. (2001) have suggested quantitative indicators of sufficiency and stability that can be derived from CMA. As an indicator of sufficiency, they recommend calculation of the fail-safe ratio, which simply indicates whether the fail-safe number (Chapter 14) for the current step of CMA exceeds the benchmark of $5N + 10$. If and when the fail-safe ratio exceeds 1, it indicates that cumulative weight of evidence is sufficiently tolerant for future null results. This approach suffers from the same shortcomings as calculation of a fail-safe number (Chapter 14). Some of these problems can be alleviated by using the weighted method of fail-safe sample size (N_{fs}) calculation (Rosenberg 2005); however, the N_{fs} method still severely overestimates the number of studies needed to make the magnitude of the effect nonsignificant if missing studies report an effect with an opposite sign rather than null results, as is often the case in ecology. In addition, calculation of N_{fs} makes sense only if the magnitude of the effect size is significantly different from zero, and thus it cannot be applied to steps in CMA where the effect size is not significant. This limits its usability as sufficiency can never be reached in meta-analysis where the mean effect size is not significant.

As an indicator of stability, Mullen et al. (2001) suggest calculation of the "cumulative slope," which is the slope of the regression of cumulative effect sizes from all the previous and current steps of CMA, along with each new step in CMA. Stability is achieved when the slope of the regression approaches 0, indicating that adding another study causes little change in the cumulative effect size. Note that the significance of the slope cannot be formally estimated because of the problem of multiple testing and the fact that meta-analysis data violate the assumptions of the general linear model for statistical inference. Therefore, even though the estimate of the slope itself is not biased, the decision as to when the slope becomes negligibly small remains somewhat arbitrary.

One can also conduct linear weighted regression analysis for the relationship between effect sizes and publication year (Chapters 8 and 9; Gehr et al. 2006). This method captures temporal trends well when the magnitude of the effect size exhibits a uniform and monotonous decrease or increase with time. However, this is not always the case, and uneven, irregular shifts in effect size in opposite directions have been observed both in ecology (e.g., Nykänen and Koricheva 2004) and in molecular genetics research on genetic associations (Ioannidis and Trikalinos 2005). Two alternative curve-fitting methods, fractional polynomial regression and spline regression, have been suggested to quantify nonlinear associations in meta-analysis (Bagnardi et al. 2004).

Finally, Kulinskaya and Koricheva (2010) have recently proposed the use of statistical quality control (QC) charts, in particular CUSUM charts (Hawkins and Olwell 1997), to assess significance of effects and detect trends over time in cumulative meta-analysis. Methods of statistical quality control were initially developed in industrial applications of statistics to assess whether the variability of a production process was due to chance or to assignable causes. When there is no temporal shift, the process is in control and all effect estimates are normally distributed with the same mean. If a shift happens at some point in time, the mean of the process deviates from the mean, and the process can be considered out of control. Nowadays, quality control charts are commonly used in medicine, epidemiology, and public health to detect a start of an epidemic or to control quality within the United Kingdom's National Health Service. The QC procedures are available in most major statistical packages, including R. Kulinskaya and Koricheva (2010) illustrate the use of QC charts for detection of temporal trends by using several examples, including the meta-analysis by Torres-Vila and Jennions (2005), which is used as one of the data set examples in this book.

If evidence of temporal changes in effect sizes is obtained, the next step is to examine underlying causes of this trend. Barto and Rillig (2010) used a simple approach to find out which moderators changed with publication year; this could then explain the observed temporal patterns in studies of herbivory effects on mycorrhizae. For each potential moderator (e.g., research method used or taxonomic group studied), they have sorted the studies in chronological order and established which level of moderator was used in the earliest study. This level was assigned the value of 1, the next level of moderator was assigned the value of 2, and so on. The authors then performed correlation analyses between the levels of each moderator and the year of publication; this analysis revealed that temporal changes in effect size were most likely to be due to changes in the type of plants used in experiments and the treatment methods.

WORKED EXAMPLES
1. Effects of elevated CO_2 on net CO_2 assimilation in woody plants

As the first example, we have selected a subset of studies from the database compiled by Peter Curtis (Curtis 1996, Curtis and Wang 1998), and reporting the effects of elevated CO_2 on net CO_2 assimilation in woody plants (response variable PN) from 39 studies published over a 10-year period (1987–1996). The effect size metric used in this meta-analysis is the log response ratio. Positive values of the effect size indicate an increase in net CO_2 assimilation in plants under elevated CO_2 as compared to ambient CO_2 levels. In the original database, several individual studies contribute more than one data point because they report results for different plant species. In order to begin the analysis of temporal trends, we have to first average effect sizes and their variances by study because results within the same study, even if considered relatively independent as in the case of different plant species, have been published simultaneously; it does not make sense to add them sequentially in cumulative meta-analysis.

Once we have calculated mean effect sizes and variances per study, we can subject the data to a formal meta-analysis. The overall mean effect size is 0.369 and it is significantly different from 0, as indicated by the 95% confidence intervals. The heterogeneity analysis using the fixed-effects model yields a total heterogeneity estimate of $Q_t = 1018$, which is much larger than 70.703, the critical chi-square value for $df = 38$ ($P = 0.001$). This suggests that between-study variation is significantly larger than would be expected from sampling error alone. We can therefore proceed to search for moderators.

Even though the studies included in this meta-analysis have been conducted over a relatively short time period (10 years), important methodological developments in terms of exposure facilities have taken place during this time (Curtis 1996, Curtis and Wang 1998). It might be interesting to examine, therefore, whether any temporal trends in effect size are apparent. In order to examine whether the magnitude of the effect size changes with time, we first produced a scatterplot of effect size against publication year (Fig. 15.1A). The plot appears to indicate an increase in the magnitude of the effect with publication year. Figure 15.1B shows the CMA plot for the same data set based on the random-effects model. Visual inspection of the CMA plot reveals that the cumulative effect size is increasing with time from being weak (and not significantly different from 0) in early studies, to being significantly positive by the end of the meta-analysis. This indicates an increase in CO_2 assimilation in response to elevated CO_2 concentrations. Overall, the magnitude of the effect size changed more than 2-fold over 10 years, but the increase in effect size was nonlinear and most of the changes took place during the first 10 experiments. However, the magnitude of the effect changed little in the last 20 studies, although the confidence interval kept getting smaller.

In this data set, more than one study on the topic was published each year, except in 1988 and 1996 (Fig. 15.1A). It is difficult to determine the exact chronological order in which such studies were conducted and published (especially given the practice of some journals to provide early online access to articles which will be included in future issues). Yet, the order in which studies published in the same year are entered in the cumulative meta-analysis might affect the shape of the CMA plot (on the CMA plot in Fig 15.1B the order is alphabetic). Therefore,

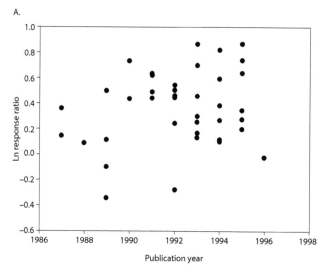

Figure 15.1. Scatterplot (*A*), cumulative meta-analysis plot (*B*), and cumulative meta-analysis trajectory (*C*) of studies examining effects of elevated CO_2 on net CO_2 assimilation in woody plants (data from Curtis and Wang 1998). The CMA plot was plotted using Comprehensive Meta-Analysis software.

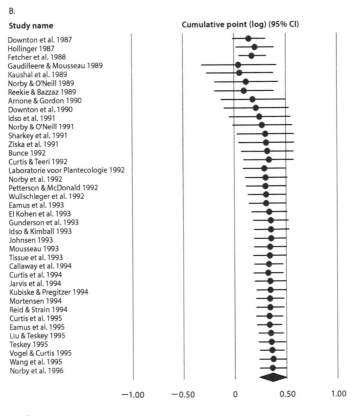

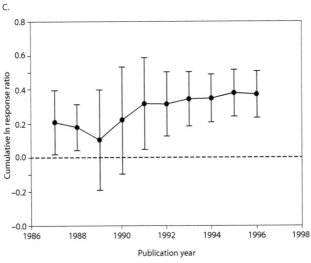

one might prefer to draw the CMA plot based not on study-specific effect sizes, but on year of publication. We can do this by using the values of cumulative means and 95% CI at the end of each calendar year regardless of how many studies have been published in that year. This gives the trajectory of change in cumulative effect size and confidence intervals over the years (Fig. 15.1C). The CMA trajectory plot shows that there is a clear difference in magnitude of effect sizes before and after 1991, thus supporting the conclusion that a nonlinear temporal change in effect sizes took place.

We then performed a weighted least square regression with publication year as an explanatory variable. Since between-study variation is significant, as indicated by heterogeneity analysis, using the random-effects model is more appropriate. This analysis can be conducted by using either the Comprehensive Meta-Analysis software (by selecting method of moments computational option) or the MetaWin software (by selecting continuous random model option in the summary analysis menu). A separate column containing publication year has to be created in the data file and selected as a moderator/predictor. Both of the software packages produced identical results. Variation in effect sizes explained by the model was not significant ($Q_M = 1.86$, $df = 1$, $P = 0.172$), neither was the slope (0.031) nor the intercept (-63), suggesting that publication year is a poor predictor of elevated CO_2 effects on net CO_2 assimilation in woody plants. This conclusion appears to disagree with the results of the CMA discussed above. Recall, however, that the linear regression model that we have applied assumes a linear relationship between the dependent and independent variable, and both the scatterplots and CMA plots suggest that the relationship between publication year and effect size is nonlinear. Therefore, the use of alternative regression methods which allow quantifying non-linear associations in meta-analysis (e.g., Bagnardi et al. 2004) would be preferable in this case.

We could also estimate whether the effect size of the first study in this analysis differed more than would be expected by chance from the results of all subsequent research. The effect size of the first study (Downtown et al. 1987) is 0.148 and variance is 0.007 (Fig. 15.1A). Mean effect size of all other studies excluding Downton et al. could be easily calculated with the Comprehensive Meta-Analysis software by selecting the "one study removed" option (or by removing the row containing the results of the first study from the data file and repeating the meta-analysis). Based on random-effects models, it gives an effect size of 0.375 and a variance of 0.005. By using Equation 15.1, $z = \frac{0.148 - 0.375}{\sqrt{0.007 + 0.005}} = -2.072$, which is larger than the critical value of 1.96 for $P = 0.05$, suggesting that the results of the first study by Downton et al. (1987) differed beyond chance from the results of the subsequent studies.

What could be the cause of the observed increase in the magnitude of the effect with publication year? It cannot be explained by publication bias against nonsignificant results, and publication bias against studies reporting significant negative effects of elevated CO_2 on net assimilation appears unlikely. Neither the ambient and elevated levels of CO_2 nor duration of the exposure, changed significantly over the examined period of 1987–1996. However, previous meta-analyses of the same database by Curtis (1996) and Curtis and Wang (1998) revealed that exposure facility and pot size (rooting volume) significantly affect plant responses to elevated CO_2 in terms of net CO_2 assimilation. Studies conducted indoor in controlled-environment growth chambers (GC) reported smaller effects on net assimilation than studies conducted in greenhouses (GH) and in the field-based open top chambers (OTC), presumably because of the lower light levels in GCs. The use of GCs decreased with time from 35% of all studies conducted between 1987 and 1991, to 21% of studies conducted in 1992–1996, and the use of GH/OTC facilities increased from 65% to 79% for the same time periods. In addition, Curtis and Wang (1998) have shown that net CO_2 assimilation was significantly lower in experiments where plants have been grown in smaller pots (< 2.5 L) than in plants grown in larger pots or in the ground. Only 1 experiment published in 1987–1991 was conducted in the ground as compared to 14 experiments in the ground in 1992–1996. Among the experiments conducted in the pots, average pot size increased from 3 liters in 1987–1991 to 6 liters in 1992–1996. Therefore, methodological changes in exposure facilities and pot size are likely to contribute to the observed temporal changes in net CO_2 assimilation. Although the above sources of heterogeneity have been revealed in previous meta-analyses by Curtis (1996) and Curtis and Wang (1998), it has not been shown before that these methodological changes may lead to temporal trends in effect sizes. Note that if one conducted a meta-analysis on effects of elevated CO_2 on net CO_2 assimilation in 1990, after the first 9 studies on the topic were published, the magnitude of the mean effect size obtained (0.219) would be only 60% of that obtained by 1996 (0.369). Therefore, in the presence of temporal changes in effect sizes, early meta-analyses might considerably over- or underestimate the magnitude of the effect size.

2. Effects of male mating history on female reproductive output in Lepidoptera

The second example is based on a meta-analysis by Torres-Villa and Jennions (2005) that compared reproductive output of female Lepidoptera that mated with virgin males, with those that mated with experienced males. The data set includes 29 studies published in 1971–2003. The metric of effect size used is Hedges' d and positive effects indicate higher reproductive potential of females that mated with virgin males. In this data set, each study contributes only one effect size to the analysis, so there is no need to average the effect by study. The overall mean effect size for 29 studies is $d = 0.335$ and is significantly different from 0, suggesting that females had higher reproductive output when mated with virgin rather than with experienced males. The heterogeneity analysis using a fixed-effects model yields a total heterogeneity estimate of $Q_t = 57.48$, which is significant at $P = 0.001$, suggesting that between-study variation is significantly larger than would be expected from sampling error alone; we thus can proceed to search for moderators.

We have no a priori reasons to suspect that the magnitude of reproductive benefits derived by females mating with virgin males is different in studies conducted in the 1970s than those reported in later studies. However, because the time span of the studies included in this analysis is quite extensive (over 30 years), it might be prudent to examine whether the magnitude of the effect size changes with time. We first produced a scatterplot of effect size against publication year (Fig. 15.2A), which revealed no obvious temporal trend apart from the very last study reporting a larger estimate than all the previous ones. The CMA plot and trajectory (Fig. 15.2B–C) do not reveal considerable temporal variation in the effect size either. There is fluctuation in the cumulative effect size from the first 7 studies, but then the cumulative effect quickly converges to the mean value and remains stable in the last 20 studies.

Note that CMA is much less sensitive to changes in effect sizes that occur in later studies. For example, the last study in this meta-analysis reported an effect size which is two times larger than any effect reported in previous studies (Fig. 15.2A), but the cumulative mean effect barely changes (Fig. 15.2B–C). This is because to change the pooled effect size at later stages of CMA, new studies have to overcome an increasing amount of inertia (accumulating evidence). This is not a big problem if the majority of temporal changes take place early (as is often the case, compare with Worked example 1). However, if the scatterplot suggests that temporal changes take place mainly at the most recent time interval, one might want to conduct CMA in reverse chronological order (i.e., entering the most recent studies first).

To statistically test for the presence of a temporal trend, we have also performed a weighted least square regression, with publication year as an explanatory variable. A random-effects model was used; variation in effect sizes explained by the model was not significant ($Q_M = 0.14$, $df = 1$, $P = 0.709$), nor was the slope (-0.003) or the intercept (5.49). The z test shows that the results of the first study do not differ from the subsequent research more than would be

A.

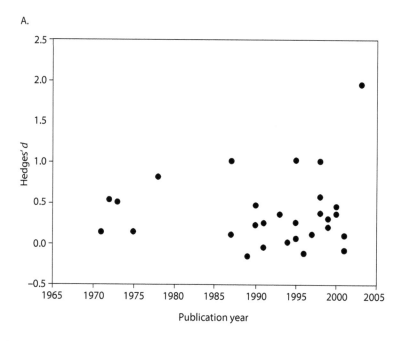

B.

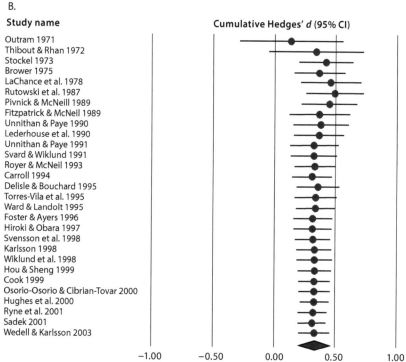

Figure 15.2. Scatterplot (*A*), cumulative meta-analysis plot (*B*), and cumulative meta-analysis trajectory (*C*) of studies examining effects of male mating history on female reproductive output (data from Torres-Vila and Jennions 2005). CMA plot was plotted using Comprehensive Meta-Analysis software.

C.

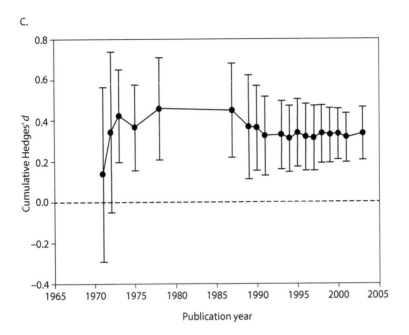

expected by chance ($z = -0.89$, $P < 0.05$). Therefore, effects of male mating history on female reproductive output do not exhibit pronounced temporal changes, indicating that the conclusions of the meta-analysis are stable. Nearly the same results would be obtained if the meta-analysis was conducted in the early 1990s instead of 2003.

CONCLUSIONS AND BEST PRACTICE RECOMMENDATIONS

Significant temporal changes in the magnitude of effect sizes appear to be common in ecology and evolutionary biology as well as in other scientific disciplines. They often lead to changes in statistical significance of the mean effect over time and present a fundamental problem for research synthesis if meta-analyses conducted at different points in time provide different conclusions. If the effects decrease with time (as often appears to be the case in ecology), then early meta-analyses are likely to overestimate the effect. This is of particular concern in conservation biology and applied ecology because it is hoped that the use of meta-analysis in these fields will facilitate communication of research findings to policy makers and lead to the development of evidence-based management policies (Chapter 26).

We recommend that ecologists and evolutionary biologists always explore temporal patterns in effect sizes when conducting a meta-analysis; particularly when the total heterogeneity of effect sizes is significant and the data set offers a sufficient temporal span (at least 10 years). If graphical methods (scatterplots or CMA plots) indicate temporal trends, the extent of the temporal changes in effect sizes should be tested by using the statistical methods (e.g., regression analysis) described above.

Examination of temporal trends in effect sizes should become a routine procedure in ecological and evolutionary meta-analyses, just as tests for publication bias currently are (Chapter 14). Scrutinizing temporal trends is useful for several reasons. First, exploration of temporal trends in meta-analytic data sets is crucial for assessment of stability of the results and the sufficiency of the data. Second, temporal changes in effect size are often indicative of other sources of heterogeneity, such as publication bias or heterogeneity among studies with respect to research methods or the study organisms. Examination of temporal trends in effect sizes is thus a good diagnostic tool for detection of sources of heterogeneity. It might also allow early detection of the point in time when the evidence is sufficient and stable enough to provide the basis for management recommendations. Timely detection of temporal changes in effect sizes that might indicate the need for changes in previously accepted management policies is also possible. This could ultimately result in a significant savings of time and resources in the development of management strategies, and would result in more effective conservation actions. The lack of stability in effect sizes over time could also be used as justification for the need for more research on the topic.

We also recommend that, as meta-analysis becomes more common in ecology and evolutionary biology, biologists adopt the practice of updating meta-analyses on the same topic when a sufficient number of new studies becomes available. In medicine, regular, biennial updating of research evidence by meta-analyses is a standard practice in initiatives like the Cochrane Collaboration (Higgins and Green 2011). In contrast, there is no similar procedure yet in ecology and evolutionary biology and researchers are often discouraged from repeating a review, largely because the majority of journals strongly emphasize novelty as a major criterion for publication (Palmer 2000).

Cumulative meta-analysis represents a useful tool for updating summary results as evidence accumulates. In terms of sufficiency and stability of data, the CMA plot would serve as an indicator of the robustness of the conclusions drawn from the analysis (Mullen et al. 2001). If the cumulative effect size and the confidence interval show no evidence of stabilization, the results of the analysis should be interpreted with caution and potential causes of the temporal changes in effect sizes should then be examined. However, because of the limitations of the CMA as a statistical procedure, we recommend that it is always complemented by the formal statistical analysis of the temporal trend.

Statistical Models for the Meta-analysis of Nonindependent Data

Kerrie Mengersen, Michael D. Jennions,
and Christopher H. Schmid

IN PREVIOUS CHAPTERS WE considered meta-analysis in which each primary study contributes one estimate of an effect size to the analysis. Each study and its corresponding estimate were treated as statistically independent. In many meta-analyses, however, independence is questionable because there are several effect estimates per study and/or some of the individual studies included in the meta-analysis might not provide independent estimates of the effect (e.g., if the studies have been conducted at the same site). Within-study nonindependence can arise due to multiple measures of the same effect on the same experimental units being made over time, multiple treatments being compared to the same set of control individuals, or different measures being taken (e.g., plant height, dry weight, and photosynthesis rate) from the same experimental units to generate several different effect size estimates. Reasons for likely nonindependence among studies include multiple studies being made at the same location or on the same animals or plants. Sometimes there is a judgment call as to whether studies will be non-independent. For example, a research team might differ slightly in their approach to measurement, the available equipment, or the study system being examined. Different studies by the same team are then likely to produce results that are more similar to each other than to those from another team (i.e., the team is a thus moderator that is a source of heterogeneity in effect sizes). Effect size estimates from the same team are then nonindependent. On the other hand, different research groups might use very standardized techniques on generic study systems so that there is little reason to anticipate team identity influencing effect size estimates. It is also important to remember that several different studies are often published from a single experimental setup in ecology (e.g., FACE, or free air carbon enrichment) experiments are expensive, long term, and complex to set up) so that separate publications are not necessarily independent of each other. All of these examples are concerned with nonindependence among response variables, including effect size estimates. Another common type of nonindependence encountered in meta-analyses is that among explanatory moderator variables; this is dealt with separately in Chapter 20. Finally, a special, but ubiquitous, case of nonindependence among response variables (and effect sizes) in biology occurs due to phylogenetic relationships among study species. This topic is covered in Chapter 17, which we recommend to the reader, even if they are not specifically interested in accounting for phylogeny. It provides an excellent worked example of a general approach to dealing with nonindependence that can be adapted to contexts beyond a shared phylogenetic history (e.g., spatial or temporal correlation).

TABLE 16.1. Examples of situations in which nonindependence arises in meta-analysis.

Type of nonindependence	Aim of the meta-analysis
Studies are independent but can be grouped by site, year, research group, and so on, which might be factors that affect the effect size estimates.	Learn about the group and overall mean effects (and sources of heterogeneity).
A study reports repeated estimates of an effect. Examples include repeated measures, estimates over time, or experimental studies at different doses (i.e., in a dose-response study).	Learn about the mean effect at individual time points or set dosages, compare effects at different time points or doses, and/or assess the overall effect or trend.
A study reports estimates of more than one effect relating to a common ecological/evolutionary phenomenon (e.g., offspring size at birth, survival, immune function as measures of evolutionary fitness or performance).	Create a composite effect for each study, and combine these composite effects to estimate the overall mean effect.
A study reports estimates of more than one effect (e.g., different estimates of plant or animal performance such as biomass, mortality, etc.).	Learn about each effect, while taking the other estimated effects into account.

Here, we discuss nonindependence among effect sizes both within and among studies. We confine ourselves to describing four commonplace situations where nonindependence can occur in ecology and evolution meta-analyses (Table 16.1). Each of these four situations is illustrated with a single case study (data sets in Appendix 16.1). In the real world, a data set will often contain several types of nonindependence. For example, each study may provide several different effect sizes obtained from the same individuals, and there may be some sites that contribute several studies to the data set. In such cases the hierarchical approaches developed in this chapter might need to be extended.

The effects of not accounting for nonindependence in meta-analysis are now well documented in the literature. For example, Riley (2009) argues against the practice of meta-analyzing each of a set of outcomes independently. He shows that ignoring within-study correlation in this way can, on average, increase the mean squared error and standard deviation of pooled estimates. Similarly, Jones et al. (2009) argue against the practice of ignoring correlation between time periods and analyzing a set of time points independently. These authors also show that this practice can result in different pooled estimates and corresponding standard errors. Note that these problems are not resolved by another common practice of conducting the independent analyses and performing some form of post hoc adjustment for nonindependent tests, such as the Bonferroni test.

As a simple example of the effect of nonindependence, consider the problem of computing the difference between two effects, say T_a and T_b. The estimated difference is $D_{ab} = T_a - T_b$. If the effects are independent the variance of D_{ab} is simply the sum of the variances, say $\sigma_a^2 + \sigma_b^2$. If the outcomes are dependent, however, the variance is:

$$\sigma_a^2 + \sigma_b^2 - 2\rho_{ab}\sigma_a\sigma_b \tag{16.1}$$

where ρ_{ab} is the correlation between the effects. If the effects are positively correlated, then the variance of the difference is smaller after taking this correlation into account. For example, if $\sigma_a^2 = 0.02$, $\sigma_b^2 = 0.018$, and $\rho_{ab} = 0.2$, then the variance of D_{ab} is equal to 0.0304, compared to 0.038 if the effects were assumed to be independent.

How can nonindependence be addressed in a meta-analysis? Three options are available. First, we could exclude multiple nonindependent estimates and/or only focus on a single response variable. This reduces the problem to a simple meta-analysis that can be modeled using the approaches described in Chapters 8 to 11. However, in doing this we are ignoring relevant data and losing potential information. Second, we can (wrongly) assume that all the effect size estimates are independent and press ahead with the approaches described in Chapters 8 to 11. Unfortunately, ignoring dependence between effects might increase the likelihood of a type I error (false positive), bias the parameter estimates, and generate incorrect estimates of the corresponding variances. This is because correlated observations contribute a different amount of information about an effect than independent observations. The key idea is that correlated data are not as informative as independent data; the true sample size is smaller if the observations are positively correlated. This affects not only the weighting, but also the variances, and is clearly shown in the example above. There, the incorrect weighting (inverse of variance) will be given to the effect D_{ab} if we erroneously assume that ρ_{ab} is always 0 in all the studies we wish to combine to estimate the mean value of D_{ab}. Third, we can extend the models introduced in Chapter 8 to accommodate nonindependence. We will now describe how to do this for each of the four situations described in Table 16.1. For each situation, we discuss the relevant multivariate hierarchical (sometimes called a multilevel or nested) model. The multivariate component captures the nature of the correlation between effects, and the hierarchical component captures how these effects are nested and combined. We also consider some extensions and alternatives to the general multivariate approaches.

In practice, it is not always possible to fit the desired model due to a lack of information or incomplete reporting within and across studies compiled for a meta-analysis. The problems this creates, and how to deal with them, are considered in the final section of the chapter.

SITUATION 1: GROUPED ESTIMATES OF A SINGLE EFFECT

Effect estimates can often be grouped based on an external factor that might lead to their being nonindependent (e.g., when effect sizes are calculated from data collected at the same study site or on the same species). Data set 1 (Table A16.1) contains 25 effect size estimates and corresponding variances (i.e., 25 studies) that can be assigned to five groups (labeled A to E). Each group contains 3 to 8 studies. We have kept this example deliberately vague since it is intended to represent a wide range of common scenarios. Readers would do well, however, to visualize a plausible example relevant to their own field of study.

The first question to ask is whether effect estimates are statistically independent *within* a grouping, so that they each provide unique data about the effect. This can require some biological insights. For example, if the grouping is by population but each study collected data from different animals, we might be happy to treat each study within a given population as independent. If, however, we have the additional information that studies were collected at different times of the year, we might suspect that there is greater similarity between those studies conducted in autumn than those conducted in spring; this would lead to the correlation among studies varying within each population. Second, we need to decide whether the groups themselves are independent. Again, this can require biological insights, and will often depend on the study question. For example, consider a study measuring growth rates of island-dwelling birds, where the grouping variable is "island." We might be willing to treat data collected on islands two km apart as independent if we are asking a question about growth rates on islands in a Finnish lake. We would be less happy to do so if the data set included islands from lakes across Europe, because the similarity in food availability among islands on a Finnish lake is now high relative to the variation in food availability among islands in the full data set. The correlation among effect sizes is thus likely to be far stronger for some pairs of islands than for others in the European data set.

The general meta-analysis model assumes that each study reports an estimate of a true effect for that study, the true effects within groups are distributed around an overall group effect, and the group effects are distributed around a global effect. In general, suppose that there are I groups, and that the ith group reports P_i estimates of the effect of interest. In the ith group, let T_{ij} and S_{ij}^2 denote the jth estimate and associated variance of that estimate. Each T_{ij} is an estimate of a true effect θ_{ij}. Within the ith group, the θ_{ij} are assumed to be distributed around an overall group effect μ_i, with variance γ_i^2. These overall effects, μ_i, are then assumed to be distributed around a global effect μ_0 with variance ω^2 representing the global effect for all I groups. As stated above, independence is assumed among effects within groups, and among group effects.

In Data set 1 there are five groups. Hence $I = 5$, $P_1 = 4$, $P_2 = 3, \ldots, P_5 = 8$, $T_{11} = 0.2$, $T_{12} = 0.6$, $T_{21} = 0.8$; the estimated variance of the first effect in the first group is given by $S_{11}^2 = 0.10$, likewise $S_{23}^2 = 0.04$, and so on. This hierarchical model is depicted in Figure 16.1.

Assume that all of the groups are independent, and that the studies within groups are independent. Further, assume that the effect estimates (T_{ij}) and true effects (θ_{ij}) are normally distributed with true variances σ_{ij}^2 (which are estimated by S_{ij}^2 described above). Then the meta-analysis model can be written:

$$T_{ij} \sim N(\theta_{ij}, \sigma_{ij}^2)$$
$$\theta_{ij} \sim N(\mu_i, \gamma_i^2) \qquad\qquad (16.2)$$
$$\mu_i \sim N(\mu_0, \omega^2).$$

Note that it is not necessary to stipulate normal distributions in Equation 16.2. Other distributions can be used if these are more applicable. This is discussed elsewhere in this book.

A fixed-effects model within groups is obtained by setting $\gamma_i^2 = 0$. A fixed-effects model among groups is obtained by assuming that for all studies, $\mu = \mu_0$ and hence, $\omega^2 = 0$. If necessary, covariates can be added at any level of the hierarchy of the model (Chapter 7). A variety of software is available to fit this model (e.g., SAS [PROC MIXED], Stata, MLwiN, WinBUGS,

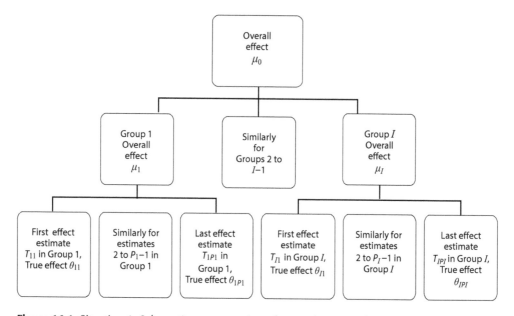

Figure 16.1. Situation 1: Schematic representation of general meta-analysis structure.

TABLE 16.2. The estimates of the mean effect size (and its variance) for meta-analyses of Data set 1 (Table A16.1) that make different assumptions about the independence of effect size estimates.

Research Group:	A	B	C	D	E	Mean
Analysis 1: Ignoring grouping information and assuming all estimates are independent						
FE estimate (SE)	-	-	-	-	-	0.65 (0.040)
RE estimate (SE)	-	-	-	-	-	0.63 (0.076)
Analysis 2: Include group effect: MM approach						
FE estimate (SE)	0.331 (0.137)	0.704 (0.132)	0.647 (0.098)	0.578 (0.067)	0.806 (0.072)	0.65 (0.040)
RE between-study variance	0	0.025	0.162	0.285	0.141	
RE estimate (SE)	0.331 (0.137)	0.701 (0.162)	0.679 (0.194)	0.785 (0.277)	0.855 (0.154)	0.63 (0.113)

Notes:

FE = fixed-effects analysis; RE = random-effects analysis; MM = method of moments.

Analysis 1: In the RE analysis, the between-study variance was approximated as $\hat{\gamma}^2 = \max(0, (X^2 - (k-1))/(\sum W_i - \sum W_i^2 / \sum W_i))$ where $X^2 = \sum W_i (T_i - \hat{\mu}_{0FE})^2$ and k = number of estimates. Hence $W_i' = 1/(S_i^2 + \hat{\gamma}^2)$. See DerSimonian and Laird (1986). Note that a similar equation can be used to estimate variance between-study effects within-groups or between-group effects.

Analysis 2: Method of moments estimates were obtained using a two-step approach, with separate meta-analyses of studies for each group to obtain group effect estimates, and the group effect estimates were then combined in a further (separate) meta-analysis. Between-group variance = 0.026.

R [lme, lmer, mcmcglmm]) (Chapter 12). Alternatively, a simple two-step procedure can be used. A meta-analysis is fitted for each group separately by direct calculation using, say, an Excel spreadsheet or statistical software. The calculated group means and variances are then used in a second meta-analysis to calculate the global mean effect.

Returning to Data set 1, we first ran a meta-analysis in which group effects were ignored. We fitted both a simple fixed- and random-effects model (Chapter 9). We then consider a meta-analysis in which groups were included, as in Equation 16.2. The results based on method of moments estimates (Chapter 9) are given in Table 16.2. As you would expect, the inclusion of dependence via the group effect does not change the estimated global mean and its variance when looking at fixed-effects models (since there is no allowance for between-study or between-group variation). In contrast, there was a difference between the random-effects models. The 50% increase in the variance of the global mean effect estimated when fitting a model that includes groups reflects the fact that there is substantial variation in effect sizes among groups. Maximum likelihood and Bayesian estimates of the global mean effect size can also be obtained that take group into account. The WinBUGS code to do so is provided in Appendix 16.2.

An important modification

If there is dependence between studies within groups, Equation 16.2 can be modified so that within each group the study effects θ_{ij} have a multivariate normal distribution. Thus the second line of Equation 16.2 is replaced by $\theta_i \sim \mathrm{MVN}(\mu_i, \Gamma_i)$, where θ_i is a vector of the true study effects θ_{ij} in the ith group, and Γ_i is a variance-covariance matrix that contains the variances, γ_i^2, of these study effects along the diagonal and their covariances (describing the relationship between the true study effects) in the off-diagonal elements.

Similarly, if there is dependence among groups, the third line of Equation 16.2 can again be modified so that the group effects (μ_i) have a multivariate normal distribution with a variance-covariance matrix containing the (conditional) variances of the group effects (γ_i^2) along the diagonal and covariances among the group effects in the off-diagonal elements.

In short, Situation 1 is a restricted example (all covariances = 0) of more general situations.

SITUATION 2: REPEATED MEASURES ESTIMATES OF A SINGLE EFFECT

Repeated measures studies, including time series and dosage response studies, are common in ecology. The repeated measurements might be taken on the same individual, or they might represent unique sets of individuals measured at different times or at different dosages within a study. In the former case, there is obvious dependence between measurements for each individual. In the latter case, while there is no dependence within individuals, there is still dependence over time within studies. If the aim is simply to obtain an overall estimate of the effect of interest, then the meta-analysis can be collapsed to Situation 1. For example, each time period or dosage can be treated as a separate "study" and the original studies as "groups." We can then modify Equation 16.2 by replacing γ_i^2 with a variance-covariance matrix where the strength of covariance either decreases as the time interval between "studies" increases, or is only dependent on the previous study (autocorrelation). In many cases, however, our aim is to evaluate time or dose trends within the original studies and estimate the overall trend based on the combined information.

We illustrate models that describe trends by considering Data set 2 (Table A16.2), which contains information on annual estimates of shrimp biomass from nine North Atlantic sites; this is part of a large database analyzed by MacKenzie et al. (2003). The data, consisting of biomass (log scale) each year for each study, are provided in Appendix 16.1. The least squares (LS) regression slope coefficient (and associated standard error of this estimate) for biomass per year, fitted for each study independently, is also provided in Appendix 16.1.

When we have access to the primary data (in our example this is the estimated biomass at each time point for each study), our meta-analysis model can have a two level hierarchy. First, we can fit a linear regression over time for each study, and then we can combine the study-specific intercepts and slope coefficients. If, on the other hand, we only have access to the estimated regression coefficients and corresponding standard errors from each study, these estimates can be combined directly across studies using a fixed- or random-effects model.

We will first consider the case in which we have access to the primary data from each study. At the first level, we fit a linear trend assuming that the vector of responses in the ith study has a linear trend over time and that the residuals are normally distributed. Thus

$$
\begin{aligned}
T_{ij} &= \beta_{0i} + \beta_{1i} X_{ij} + e_{ij} \\
e_{ij} &\sim \mathrm{N}(0, \sigma_T^2)
\end{aligned}
\tag{16.3}
$$

where T_{ij} is the observed response in the ith study at the jth time (X_{ij}), β_{0i} and β_{1i} are the study-specific intercept and linear trend parameters, respectively, and e_{ij} is the residual between the observed and fitted responses in the ith study, $i = 1, \ldots, I$.

Note that Equation 16.3 assumes a common variance for residuals. The model does not necessarily have to be this restrictive; a different variance could be assumed for each study by replacing σ_T^2 with σ_i^2 in the second line of the equation.

At the second level, a random-effects model used to combine these study-specific regression estimates is:

$$\begin{pmatrix} \beta_{0i} \\ \beta_{1i} \end{pmatrix} \sim \text{MVN}\left(\begin{pmatrix} \mu_0 \\ \mu_1 \end{pmatrix}, \Gamma\right)$$

$$\Gamma = \begin{pmatrix} \gamma_0^2 & \gamma_{01} \\ \gamma_{10} & \gamma_1^2 \end{pmatrix}$$

(16.4)

where $(\mu_0, \mu_1)'$ is the vector of global regression parameters and Γ is the between-study variance-covariance matrix for these parameters (i.e., to take into account that intercept and regression coefficient estimates might covary across studies, rather than assuming that they do not). Thus, μ_0 and μ_1 are the overall intercept and trend values obtained when we synthesize the data from all I studies.

Second, we consider the case in which we only have estimates b_{0i} and b_{1i} of the intercept and slope parameters β_{0i} and β_{1i} for each study, and estimates S_{0i}^2 and S_{1i}^2 of the corresponding variances σ_{0i}^2 and σ_{1i}^2. (Note that although we have used the same notation S and σ, here they are not directly comparable to the terms in Equation 16.3, since we do not have the primary data.) In an ideal world, we would also have estimates of the corresponding covariances, but since these are rarely provided, we will ignore them here. In this case, we can again use Equation 16.4 after first defining

$$b_{0i} \sim \text{N}(\beta_{0i}, \sigma_{0i}^2); \qquad b_{1i} \sim \text{N}(\beta_{1i}, \sigma_{1i}^2).$$

(16.5)

If the focus of the meta-analysis is on a single regression parameter, such as the linear trend, then the above model reduces to a simple univariate fixed- or random-effects model involving only b_{1i}, β_{1i} and μ_1.

Sometimes, we may want to relate the regression slope or intercept to other variables with a regression model called a meta-regression; this is used to explain their heterogeneity. It is tempting to do separate analyses in each primary study and use the resulting estimates in the meta-regression. This type of two-stage analysis must account for the uncertainty in the estimates from the primary studies in the first stage. Without such weighting, the meta-regression estimates will have standard errors that are too small because they will have assumed the first-stage estimates to be true values.

The above models explicitly accommodate both within-study variation (i.e., whenever the regression in the first line of Equation 16.3 has $R^2 < 1$ or the variances in Equation 16.5 are not equal to zero), and between-study heterogeneity. Note, however, that if there is high between-study heterogeneity in slopes, it might be misleading to combine them and discuss a global trend. Again, if necessary, other potential sources of heterogeneity can be included as covariates in the model by directly extending it to a meta-regression or, for categorical moderators, through a multilevel model in which different subsets of studies are first combined within, and then across, the next level. In our example we have tested for a linear trend, but other forms of temporal changes could readily be investigated. Nonlinear trends can be represented by quadratic or higher order polynomial functions or nonparametric splines. The main point is that if each study generates one or more estimated parameters (and associated variances) that describe the trend, these can be subject to meta-analysis using the above model. If normality cannot be assumed, it is sometimes possible to use transformations (e.g., log transformation for count data or logit transformation for binary data), or generalized linear modeling.

If residuals are assumed to be normally distributed, models can be estimated using ordinary least squares (OLS) or a variant of this approach; an example of a variant is the iterative least squares or generalized least squares (GLS). The OLS estimator assumes that the residual term (e_i) has a constant variance of σ^2. In contrast, the GLS estimator assumes that the variance is not constant, which is more typical in ecology. A transformation is then applied by dividing all terms in the regression by the square root of the nonconstant term in the variance of e_i, resulting in a constant variance for the error term that satisfies OLS assumptions. Multivariate meta-analysis methods that utilize GLS have been proposed by Hedges and Olkin (1985), Raudenbush et al. (1988), and Becker (1992). The GLS approaches to meta-analysis are discussed by Dear (1994), Kalaian and Raudenbush (1996), and Timm (1999). The use of GLS models to correct for phylogeny is also covered in Chapter 17. The models can also be estimated using maximum likelihood (ML) or Bayesian approaches (Chapters 10 to 11).

The general model we have described is extensively discussed by Becker and Wu (2007) in a more general statistical context, and by Arends (2006) as one of many methods for multivariate meta-analysis. McIntosh (1996) describes both ML via the EM (expectation-maximization) algorithm, and Bayesian estimation via Markov chain Monte Carlo (MCMC); these descriptions are for analysis of a bivariate two-level hierarchical regression model for treatment effects, when using the control group event rates as covariates. A worked example of its application in epidemiology is provided by Bagnardi et al. (2004). Gelman and Hill (2007) give an excellent description of these models in the wider context of general multilevel models and ML estimation using the software package R.

The choice of a statistical model for meta-analysis of repeated measures data depends on the available data and the aim of the analysis. We briefly consider some examples in the context of time series data, noting that they also apply to dosage-response and general repeated measures studies.

As a first example, a single time point might be of particular biological interest in a time series experiment (e.g., when animals mature, or the endpoint of the study). In these cases, we can extract a single effect estimate from each primary study, and apply a standard meta-analysis model (Chapter 8). The major limitation of this approach is the tremendous loss of information. Moreover, while this approach might make studies with differences in sampling dates more consistent, it is important to critically consider the comparability of the estimates being combined. (For example, is it appropriate to use the end point of studies that differ widely in duration?) It might be possible to include the time interval since the application of a treatment (or some key biological date such as time since hatching) as a covariate in a meta-regression. It is also important to recognize that it is easy to introduce bias into an analysis by only defining the relevant time point after the data have been assessed (Higgins and Green 2011).

As a second example, a common aim in many ecological meta-analyses is to compare effect sizes at two time points. For example (case A), if each study reports effect estimates at the two time points, and/or reports (or we can calculate) a measure of the difference between the two effect estimates, the aim is then usually to estimate the overall mean difference in effect across studies. Alternatively (case B), if independent sets of studies report effect estimates at only one of the two relevant time points, the aim might be first to combine the reported effect estimates for studies looking at each time point, and then to estimate the difference in the mean effect estimates between the two types of studies. (Case A and case B can be considered to be equivalent to a paired and unpaired experimental design, where time periods are paired by study in the first example). The way in which time points are defined can influence the analysis. For example, they might be relative to an event (e.g., weights before and after a treatment, or treatment-control differences before and after an intervention), depict an epoch (e.g., different stages of development), or be chronological (e.g., effect of hunting on population growth in 1970 and 2000).

Cases A and B can both be addressed through a hierarchical meta-analysis model. For case A, at the first level, a single estimate of the measure of interest, such as a standardized difference in the effect estimates between the two time points is obtained for each study (taking into account the covariance between the data at these time points by using Equation 16.1. At the second level, these study-specific measures are combined using a random- or fixed-effects model. (Of course, if the studies directly report the measure of interest, then only the second level of this hierarchy is required.) In case B, at the first level, a random- or fixed-effects model is used to combine the estimated effects within each time point; at the second level, the mean effects are compared by calculating the difference in the means and using Equation 16.1 to calculate the 95% CI for this difference. In both cases, the two levels can be performed as separate steps or, more preferably, a mixed-effects model can be fit to perform both steps in a single analysis. Depending on the nature of the two time points being considered, it might be prudent to include the interval between the two measures in each study as a covariate. Worked examples of assorted approaches to the problem of variation in the interval between estimates are provided by Guo and Gifford (2002), Côté et al. (2005), and Stewart, Pullin, and Coles. (2007). An obvious extension of the previous aim is to compare multiple time points. Again it is possible that either case A or B applies. When comparing, however, it is necessary to correct for multiple testing using, say the Bonferroni correction (Quinn and Keogh 2002, 49) or the false discovery rate (Benjamini and Hochberg 1995).

As a third example, there is the option of combining effect estimates from all time periods across all studies. This approach is useful if there is no well-defined way to choose a particular time point from each study, and if temporal trends are of minor interest. Depending on the data and research question, "time" can be treated as a replicate or as a continuous or categorical covariate. For example, effect estimates can be averaged, ideally taking into account the temporal dependence in the data (Equation 16.1), since observations at close time points are likely to be more correlated than those that are further apart. Alternatively, time can also be treated as a fixed factor in a mixed model setting if the variation among time points is of specific interest to the experimenter, or as a random effect if the time points simply represent a sample of times in a time frame. Rustad et al. (2001) provide a useful worked example from a meta-analysis of the effect of experimental warming. They estimated effects for each site, year, and outcome of interest for three ecosystem soil variables (soil respiration, net nitrogen mineralization, and above-ground plant growth), and then combined the effect sizes across experimental sites at each time period.

We now use Data set 2 to illustrate some of the above approaches. Here, we treat the problem as a meta-analysis of nine individual time series studies (Fig. 16.2).

To illustrate the full hierarchical model, we fit a linear regression for each site at the first level of the hierarchy, and then combined the site-specific intercept and slope parameters (weighted by their respective inverse variances and covariance) in a bivariate model at the second level of the hierarchy. We did this in a one-step process, taking a Bayesian approach using WinBUGS (Chapter 11; code is provided in Appendix 16.2), with normal priors for the trend parameters, a uniform prior for the standard deviation of the residual of the within-site regressions, and Wishart priors for the precision matrices (the inverse of the variance-covariance matrices). The key results are presented in Table 16.3 (posterior estimates for intercepts were obtained but for brevity are excluded).

The estimated overall linear slope is 0.0199 with a 95% credible interval of (−0.046, 0.50). We might therefore conclude there is no evidence for a directional change in shrimp biomass over the study period. Note that, as expected, the posterior estimates of the slope at each site have "shrunk" from the observed LS estimates to the posterior mean. This is particularly evident for Site 2J, where the observed strongly positive trend (LS = 0.16) was an evident outlier.

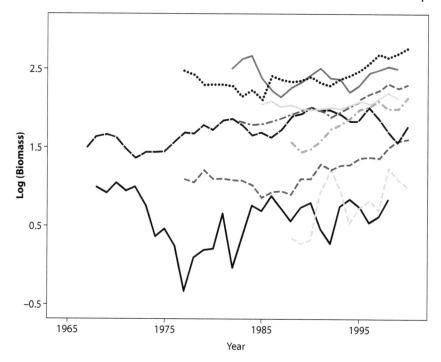

Figure 16.2. Plot of trends in shrimp biomass at 9 study sites between 1967 and 2000 (Data set 2, Table A16.2). Each line represents a different site.

TABLE 16.3. Posterior estimates of the trend for each site in the Bayesian analysis of Data Set 2 (Table A16.2) and the corresponding 95% credible intervals for these values. Also shown are the mean intercept and slope and the corresponding precision matrix for the mean parameter estimates. The precision is equal to the inverse of the variance (i.e., smaller variance equals greater precision). The ordinary least squares (LS) estimates of slopes are included for comparison. The WinBUGS code for this model is in Appendix 16.2.

Site	Parameter	LS Estimate	Posterior Estimate	95% CI lower	95% CI upper
2J	Slope	0.16	0.02	0.006	0.03
3K	Slope	0.06	0.05	0.03	0.08
3M	Slope	0.05	0.05	0.02	0.08
NGSL	Slope	0.03	0.03	0.01	0.04
ESS	Slope	0.02	0.02	0.01	0.03
5Y2	Slope	0.000	0.001	−0.02	0.01
ICE	Slope	0.01	0.01	0.01	0.02
BS	Slope	−0.002	0.00	−0.02	0.02
SKAG	Slope	0.008	0.01	−0.01	0.03
Mean	Intercept		1.03	0.36	1.71
	Slope		0.02	−0.46	0.50
$1/\gamma_1^2$	1/Variance		1.56	0.102	5.35
$1/\gamma_{12}$	1/Variance		−0.025	−2.75	2.65
$1/\gamma_2^2$	1/Covariance		2.92	0.211	9.03

There was, however, substantial heterogeneity in the overall estimates, as indicated by the precision matrix. This cautions us about how we should interpret the mean slope.

Data set 2 was also analyzed using a linear mixed model estimated by restricted ML (REML) in the statistical software package R. We arranged the shrimp biomass data into a file "shrimp" with three columns: Study (1, 2, . . . , 9), Year (rescaled by "Year-1972," so that 1 = 1973), and Biomass (log). The meta-analysis model was then fit using the command

lme(Biomass~Year,random=~1|Study/Year,data=shrimp, method=REML)

This command describes the fixed-effects (within-study) component of the model (i.e., the linear regression of Biomass on Year) and the random-effects (between-study) component (an overall effect, denoted by "~1," which allows the mean to vary within each study and year). If the variation between studies is considered to be a fixed effect, the above equation is changed to "Study*Year." The output includes the log-restricted likelihood for the model, the overall regression estimates (intercept and slope), and the standard deviation corresponding to the random effects (the between-study variation in trend). Note that in this meta-analysis the studies are automatically weighted by the uncertainty of their respective estimates, since the regression analyses, and consequently the variation in the regression estimates, are included as part of the model.

For additional worked examples of meta-analyses of repeated measures data we refer the reader to Marsh (2001), Harley et al. (2001), Harley and Myers (2001), Barrowman et al. (2003), and Blenckner et al. (2007).

SITUATION 3: COMPOSITE EFFECTS

We next consider the problem of combining effect sizes for different response variables obtained from the same individuals. One option is to perform a separate meta-analysis for each response variable, which means that we must take into account that we will then need to correct for multiple testing, as there is nonindependence between effect sizes estimated from the same study. Sometimes, however, we are interested in obtaining an overall estimate of a general phenomenon (a "latent" or "composite" measure) that is described by a synthesis of these variables (Borenstein et al. 2009, 225).

The hypothetical example we use is inspired by experimental studies that assess the benefits of polyandry by comparing the offspring of females assigned a single male they can mate with x times (monogamous), with the offspring of females assigned x males that they can each mate with once (polyandrous) (Simmons 2005). Data set 3 (Table A16.3) contains data from six studies, each of which reported an effect size (Fisher's z transformed correlation between number of mates and outcome) for up to three different traits: offspring size at birth, survival, and immune function. Each outcome can be considered an indirect measure of "offspring fitness" or "offspring performance." Two points should be noted. First, the variances are very similar for the different traits in a given study. This is because the effect is a correlation, so sample size determines the variance (i.e., $1/[n-3]$ for z-transformed Pearson's correlation coefficient) and sample sizes tend to be similar within studies. There is, however, no general expectation for variance to be similar for different response traits within a study. Different traits can have different variances, and differences in sample sizes among traits will also influence the variance of the estimates. This is a potential problem (see below). Second, certain traits were not measured in some studies; in studies C and F, the researcher did not measure immune function. Our analysis does not require that every study measures every response trait.

Our first step is to combine the outcomes for a given study in order to calculate a composite mean effect size for the effect of polyandry on offspring performance. There is no information

about which response variables (size, survival, immune function) are more important indicators of performance, so we given them equal status. A composite effect estimate for a study can
then be calculated as the variance-weighted average of the available estimated effects, X_i say,
for that study (i.e., $\hat{\mu}_{OFE} = (\sum W_i X_i)/(\sum W_i)$, with $W_i = 1/V_i$, assuming a fixed-effects model,
and m = number of estimated effects) . If the variances are equal for each response, as in study
E, this simplifies to the arithmetic mean:

$$[0.07/0.01+0.08/0.01+0.08/0.01]/[1/0.01+1/0.01+1/0.01] =$$
$$[0.07+0.08+0.08]/3=0.0766.$$

The corresponding variance estimate for the mean effect must take into account the nonindependence (i.e., covariances) between the response traits. In general, if variables are correlated,
the variance of their sum is the sum of their covariances. Thus the variance of the arithmetic
(unweighted) mean of $X_1, \ldots, X_m$, say, is obtained by dividing this value by m^2, so that (depart-
ing from the above notation for a little while):

$$V_{mean} = (\sum_{i=1}^{m} \sum_{j=1}^{m} \text{Cov}(X_i X_j))/m^2. \tag{16.6}$$

Note that the term in the outside brackets includes all combinations of i and j, (i.e., both $X_1 X_2$
and $X_2 X_1$) and the variances, since $\text{Cov}(X_1 X_1) = \text{Var}(X_1)$. Let V_i represent the variance of X_i, V_{ij}
represent the covariance between X_i and X_j, and r_{ij} represent the correlation between X_i and X_j.
Then the above equation can be written as follows:

$$V_{mean} = \left(\sum_{i=1}^{m} V_i + 2\sum_{i,j} V_{ij}\right)\bigg/ m^2 = \left(\sum_{i=1}^{m} V_i + 2\sum_{i,j} (r_{ij} \sqrt{V_i}\sqrt{V_j})\right)\bigg/ m^2 \tag{16.7}$$

where r_{ij} is the correlation between outcomes i and j, the summation over i, j is for distinct pairs
of i and j with $i \neq j$ (i.e., only one combination of a pairing is required, hence the multiplication
by 2 of the second term).

As an illustration, for study B the variance of the composite mean is given by

$$V_{mean} = [0.01 + 0.095 + 0.09 + (2 \times r_{size\ survival} \times \sqrt{0.1} \times \sqrt{0.09}) +$$
$$(2 \times r_{size\ immune} \times \sqrt{0.1} \times \sqrt{0.095}) + (2 \times r_{survival\ immune} \times \sqrt{0.09} \times \sqrt{0.095})]/3^2.$$

Note that the above variance equations are appropriate for an unweighted mean (or, equivalently, when the weighting is identical for all effects so that it can be effectively ignored when
calculating the mean). If all the study variances are equal (to V, say) and all the correlations are
also equal (to r, say), these equations reduce to

$$V_{mean} = (V/m)[1 + (m-1)r] \tag{16.8}$$

Hence, when the m effect sizes are uncorrelated ($r = 0$), then the variance of the composite effect is V/m so the variance halves each time the number of estimates is doubled. When $r = 1$,
however, there is no extra information provided so that variance of the composite effect is the
same as that of the individual effects. The multiplier term $(1 + [m - 1]r)$ is known as a variance
inflation factor (Borenstein et al. 2009).

Once the composite mean and variance for each study are calculated, they can be combined
using a study level meta-analysis, based on either a fixed- or random-effects model. Note that
both the estimated mean effect and its variance will be affected by correlations between traits,
and although the mean effect may not change much if the variances are very similar within
studies, its associated variance may still change considerably. Strictly speaking, we cannot use
Equation 16.7 with Data set 3 because the variances are not identical within each study. For

TABLE 16.4. The effect of different levels of correlation between traits when calculating the composite mean for Data set 3 (Table A16.3). Note that the composite mean for each study was calculated giving each response trait equal weighting (i.e., ignoring slight differences in variance within studies). Similarly, so that the original response traits all had the same variance, we used the arithmetic mean variance in Equation 16.7. For simplicity, we assume the same correlation between all three original traits. For ease of interpretation we present the 95% CI for the overall mean. The variance can be calculated as: Upper value − mean = $2.57 \times$ sqrt(Variance) $[t_{5,\ 0.05(2\text{-tailed})}\ 2.57]$.

Study	No traits	Composite Mean	Variance of original traits	Variance of Composite mean when $r =$			
				0	0.3	0.6	1
A	3	0.3	0.019	0.0063	0.0101	0.0139	0.019
B	3	0.4	0.095	0.0317	0.0507	0.0697	0.095
C	2	0.2	0.05	0.025	0.0325	0.04	0.05
D	3	0.5167	0.03	0.01	0.016	0.022	0.03
E	3	0.7667	0.01	0.0033	0.0053	0.0073	0.01
F	2	0.325	0.065	0.0325	0.0423	0.052	0.065
			FE Mean	0.545	0.538	0.534	0.530
			95% CI	0.445–0.645	0.413–0.663	0.388–0.679	0.361–0.699
			RE Mean	0.436	0.442	0.448	0.456
			95% CI	0.154–0.719	0.158–0.726	0.163–0.733	0.172–0.741

illustrative purposes, however, given that they differ only slightly, we will use the mean variance to illustrate the effect of changes in the correlation between traits on our overall estimate of the mean composite effect and its variance. The results are shown in Table 16.4.

If there had been several independent estimates of effect sizes for any of the response traits in a study, then we would need to take a hierarchical approach. First, for each study we would separately calculate the weighted mean and its variance using the standard fixed- or random-effects model for each trait. Once we had one effect size per trait per study, we could then calculate the weighted mean composite effect size and its associated variance using Equation 16.7.

Composite based on differences in effect sizes

If the researcher is interested in comparing the difference between two nonindependent effect sizes, such as offspring size and survival function, then Equation 16.1 can be used to obtain the variance of the estimated difference for each study. The estimated differences can then be combined using a fixed- or random-effects model. For example, if $V_{immune} = 0.02$, $V_{size} = 0.018$, and $r_{immune\text{-}size} = 0.2$, then $V_{immune\text{-}size} = 0.0304$. As the correlation becomes larger, the shared variation increases and $V_{immune\text{-}size}$ decreases. For example, if $r = 0.9$ in the above calculation, $V_{immune\text{-}size}$ declines to 0.004. Combining the estimated differences between offspring size and survival for the six studies under a random-effects model when $r = 0.2$, the overall method of moments estimate given by MetaWin is −0.097 (95% CI; −0.301 to 0.117)—which in this case is equivalent to a fixed-effects model since the between-study variance collapses to zero. When $r = 0.9$, the overall estimate is −0.072 (95% CI: −0.214 to 0.069). There is therefore insufficient evidence that the effect size for polyandry differs for offspring size and survival (although there is a trend for it to be smaller for size).

It is apparent from closer inspection of Equations 16.1 and 16.7 that a larger positive correlation induces a *larger* variance for the sum (or mean) of two effect sizes (see Table 16.4), but a *smaller* variance for the difference between two effect sizes, compared with independence

($r = 0$). This makes sense biologically. Consider a case where we want to measure offspring performance, using a number of proxy measures. Imagine that the breeding season is long and that environmental conditions vary widely. When conditions are good then offspring growth, survival, and escape speed are high; when conditions are poor they are low. Consequently, the more measures we take, the more we will over- or underestimate the offspring performance relative to their average performance during the breeding season. A combined effect therefore increases variance among individuals. In contrast, when we calculate the difference between two measures (say, growth and survival), we are removing differences that arise due to environmental noise (i.e., $r > 0$). Hence, the mean difference is estimated with greater precision and the variance is lower when we account for environmental noise (Borenstein et al. 2009).

In the same way that we can compare the effect sizes of two original traits, we can also compare two composite effect sizes based on different outcomes, or even one composite effect with that for a specific trait. Of course, if multiple comparisons are made using subsets of the same data, we have to adjust for multiple testing. One issue that obviously arises when calculating composite effects is how to obtain estimates of the correlation between effect sizes (r_{ij}). We will return to this problem at greater length after discussing Situation 4. There is, however, one situation that is a special case, but that is fairly commonplace, where it is easy to calculate correlations. It arises when we want to combine or compare the effect of several treatments, when the treatment subjects are independent, but the control always involves the same set of subjects. For example, many studies investigating the benefit of polyandry, mate females with either a single male four times (= control, monandry), two males twice each (treatment 2, polyandry), and four males once each (treatment 4, polyandry) (Tregenza and Wedell 1998). We can then ask two questions. First, is there a positive effect of polyandry (regardless of whether it involves two or four males) on offspring performance? Second, is the effect greater for four males than two males?

To answer the first question, before calculating the mean effect across studies we have to calculate the combined effect size for control versus treatment 2, and control versus treatment 4 for each study using Equation 16.7. In this case, we know the correlations between the effect sizes: treatment 2 and treatment 4 females have no relationship ($r = 0$) while control females do ($r = 1$). If the treatment and the control have equal sample sizes, then the correlation between the effect size for treatment 2 and that for treatment 4 is $r = 0.50$ (the average of the two). If the sample sizes differ, this needs to be weighted accordingly. For example, if the same sizes are $n_{control} = 20$, $n_{treatment\ 2} = 10$, and $n_{treatment\ 4} = 5$, then $r = 1 \times (20/35) + 0 \times (15/35) = 0.57$. To answer the second question, one option is to calculate the difference between the treatment 2 versus control effect, and the treatment 4 versus control effect (using Equation 16.1). It is, however, better to take advantage of the fact that the control effectively "cancels out" when we do so. In short, if the original study provides the data (e.g., means and a measure of variance for each treatment), we simply calculate the effect size for the difference between the original outcomes in the two treatments (Borenstein et al. 2009, 241).

SITUATION 4: MULTIPLE EFFECTS

Suppose that I studies each report estimates of P effects, and our interest is in the analysis of each of these effects, allowing for the possibility that they could be correlated. Note that this is different from Situation 3, in which we had the same type of data, but the aim was to create a composite effect. Here, we wish to estimate each effect.

For example, experimental studies of the effects of elevated CO_2 report a number of effects on plants such as CO_2 assimilation rate, growth rate, and stomatal conductance. We might be interested in finding the overall estimates of these effects. To do this, we could conduct separate meta-analyses for each response variable, but this would ignore any possible correlation between these effects so that any broader inferences or estimates might be misleading. Consequently, the use of a single meta-analysis is preferable because it can take into account correlations between different outcome measures. We pursue this example below.

We can denote the effect estimates for the ith study by the vector $\mathbf{T}_i = (T_{i1}, \ldots, T_{iP})$, and the true effects for this study by the vector $\theta_i = (\theta_{i1}, \ldots, \theta_{iP})$. The overall effects (combined across all studies) will be denoted by $\mu = (\mu_1, \ldots, \mu_P)$. As noted in the Introduction, the models for the various situations described in this chapter are all variations of the same theme, introduced in Equation 16.2. It is no surprise then, that the following model is analogous to that described at the end of the discussion of Situation 1, where a single outcome is replaced by multiple outcomes, and the estimate of variance at a given level of a hierarchy is replaced with a variance-covariance matrix that contains the variances of the effects along the diagonal, and covariances between the effects in the off-diagonal elements. The differences in the models are simply due to the nature of the correlation structure in the data and between effects, and to the questions we wish to answer.

For the ith study, then, let C_i be the (observed) variance-covariance matrix for the effect estimates $\mathbf{T}_i$ and let Σ_i be the (unknown) variance-covariance matrix for the true effects θ_i. Thus the observed and "true" variance-covariance matrices, C_i and Σ_i, are given by

$$
C_i = \begin{bmatrix} S_{i1}^2 & S_{i12} & \cdots & S_{i1P} \\ \vdots & S_{i2}^2 & & \vdots \\ & & \cdots & \\ S_{iP1} & \cdots & & S_{iP}^2 \end{bmatrix} \qquad \Sigma_i = \begin{bmatrix} \sigma_{i1}^2 & \sigma_{i12} & \cdots & \sigma_{i1P} \\ \vdots & \sigma_{i2}^2 & & \vdots \\ & & \cdots & \\ \sigma_{iP1} & \cdots & & \sigma_{iP}^2 \end{bmatrix}.
$$

Here, for example, for the first study, S_{11}^2 denotes the estimated variance of the effect estimate T_{11} (the first effect in the first study), $S_{11}^2 S_{123}$ denotes the estimated covariance between effect estimates T_2 and T_3 (in this study, where σ_{11}^2 is the variance of the first (true) effect θ_1), and σ_{123} is the covariance between the true effects θ_2 and θ_3 in the first study.

As before (Situation 2), the variance-covariance matrix Γ describes the relationship between the estimated overall parameters (here, $\mu_1, \ldots, \mu_P$):

$$
\Gamma = \begin{bmatrix} \gamma_1^2 & \gamma_{12} & \cdots & \gamma_{1P} \\ \vdots & \gamma_2^2 & & \vdots \\ & & \cdots & \\ \gamma_{P1} & \cdots & & \gamma_P^2 \end{bmatrix}.
$$

Commonly (but not necessarily) the study-specific effect estimates are assumed to be multivariate normal. This is often justified based on the assumed distribution of the corresponding population effect (or an appropriate transformation), or on the basis of a large sample size and appeal to the central limit theorem, so that the sample means (standardized differences, mean log response ratios, etc.) are approximately normally distributed. Under this assumption, the model can be written as

$$
\mathbf{T}_i \sim \text{MVN}(\theta_i, \Sigma_i)
$$

$$
\theta_i \sim \text{MVN}(\mu, \Gamma).
$$

Note that we do not attempt to combine these to obtain a single effect, μ_0, as we did in Equation 16.2, since here the effects are measuring phenomena that we wish to examine separately (e.g., CO_2 assimilation and stomatal conductance).

If the meta-analysis contains two effects, for example, the model expands as follows:

$$\begin{pmatrix} T_{i1} \\ T_{i2} \end{pmatrix} \sim \text{MVN}\left(\begin{pmatrix} \theta_{i1} \\ \theta_{i2} \end{pmatrix}, \Sigma_i\right), \qquad \Sigma_i = \begin{pmatrix} \sigma_{i1}^2 & \sigma_{i12} \\ \sigma_{i12} & \sigma_{i2}^2 \end{pmatrix}$$

$$\begin{pmatrix} \theta_{i1} \\ \theta_{i2} \end{pmatrix} \sim \text{MVN}\left(\begin{pmatrix} \mu_1 \\ \mu_2 \end{pmatrix}, \Gamma\right), \qquad \Gamma = \begin{pmatrix} \gamma_1^2 & \gamma_{12} \\ \gamma_{12} & \gamma_2^2 \end{pmatrix}$$

and Σ_i is estimated by C_i.

Note also that it is sometimes easier to describe the model in terms of correlations. For example, if r_{i12} is the observed correlation between the first and second effect estimates, then $S_{i12} = r_{i12}S_{i1}S_{i2}$ and similarly, if ρ_{12} is the correlation between the overall effects μ_1 and μ_2, then $\gamma_{12} = \rho_{12}\gamma_1\gamma_2$.

By merging the levels of the model, the observed effects are related to the overall effects as follows:

$$\begin{pmatrix} T_{i1} \\ T_{i2} \end{pmatrix} \sim \text{MVN}\left(\begin{pmatrix} \mu_1 \\ \mu_2 \end{pmatrix}, V_i\right), \qquad V_i = \begin{pmatrix} \sigma_{i1}^2 + \gamma_1^2 & \sigma_{i12} + \gamma_{12} \\ \sigma_{i12} + \gamma_{12} & \sigma_{i2}^2 + \gamma_2^2 \end{pmatrix}.$$

Thus the observed effects are centered on the overall effects (μ_1, μ_2) and, as for a simple random-effects model (Chapters 7 and 9), the combined variance is the sum of the within-study variances (given by the S^2's as our best estimates of σ^2) and the between-study variances (given by the γ's).

The addition of covariates to the above model has been described by Van Houwelingen et al. (2002), and has been developed further by Arends et al. (2003), Riley, Abrams, et al. (2007), Riley, Simmonds, et al. (2007), and Riley et al. (2008).

Sometimes it is of interest to conduct univariate meta-analyses even if multiple effects are reported in each study. For example, if one of the effects measured in the primary studies is of key interest, the other effects can either be ignored, in which case the model becomes a simple univariate random- or fixed-effects meta-analysis; or, if the primary effect is highly correlated with the other effects, one might use the latter as covariates predicting the primary effect, in which case the model reduces to a univariate regression meta-analysis. To illustrate the latter case, if $T_1, \ldots, T_P$ effects are reported in each study and T_P is the principal effect of interest, the meta-analysis model becomes

$$T_{iP} = \beta_{0i} + \beta_{1i}T_{i1} + \ldots + \beta_{P-1,i}T_{i,p-1} + \varepsilon_i$$

where ε_i is the residual. Other covariates can be added as appropriate.

The meta-regression approach has the advantage of accounting for all of the effects while retaining the simplicity of a univariate meta-analysis. Moreover, the relationship between T_{iP} and another effect T_{il}, say, given the presence of all the effects, is indicated by the regression coefficient β_{li}. Although the effects can be included as independent variables in the above model, dependency between them can also be included via a study-level variance-covariance matrix for both β and ε. The disadvantage of this approach is that relationships between other effects are not evaluated, so the inferential capacity of the meta-analysis under this formulation is constrained.

An alternative to the above approaches, where effects are considered separately, is to calculate a composite measure of the effects for each study and then combine these using standard univariate meta-analysis models (Situation 3). The most simplistic way of dealing with multiple effects is to assume that the effects are independent. The meta-analysis then reduces to a series of separate univariate meta-analyses for each effect (Rosenthal and Rubin, 1978), using the models and methods described in Chapters 8 to 11. Rosenthal and Rubin (1982) and Hedges and Olkin (1985) argue that this approach is useful if outcomes are independent or weakly correlated. If, however, outcomes are highly related this approach can lead to biased parameter estimates, inaccurate variance estimates, and misleading inferences (Walsh 1947; Holmes and Matthews 1984; Hedges and Olkin 1985; Sohn, 2000; Ades 2003). The type and extent of this bias will depend on the type and degree of dependency between outcomes. In general, most evolutionary and ecological meta-analyses to date have taken a univariate approach when dealing with multiple effect sizes per study.

We illustrate the meta-analysis of multiple effects using Data set 4 (Table A16.4). This is a real data set showing the response of plants to increased CO_2 concentration. Peter Curtis compiled and meta-analyzed a comprehensive database of the effect of increased CO_2 concentration on plants (Curtis and Wang 1998).

In Data set 4 we only consider two response variables: net CO_2 assimilation (PN) and stomatal conductance (GS) under "No Stress" condition. The reported effect size estimates are the standardized mean difference (Hedges' d) and corresponding variance based on the mean response of plants grown under ambient and elevated CO_2 conditions. The data consist of 34 "studies" published in 17 papers. The scatterplot of the pairs of effect estimates (GS and PN) across studies suggests there is a positive correlation between the pairs of effects (Fig. 16.3). This motivates a meta-analysis that takes into account potential nonindependence.

The multivariate model described above is extended to include an additional hierarchy that first combines study estimates derived from the same paper (i.e., "Paper" is a grouping factor; Situation 1). Let T_{ijk} denote the estimated effect for the kth outcome (GS, PN) from the jth study in the ith paper. The meta-analysis model proceeds as follows: (i) estimate the true effect θ_{ijk} for each T_{ijk}—since we don't have any information about the correlations between estimates we use a univariate distribution; (ii) based on the vectors $(\theta_{ij1}, \theta_{ij2})$, estimate a combined paper effect (μ_{i1}, μ_{i2}), and here we use a bivariate distribution to allow for correlation between the effects; and (iii) based on the vectors (μ_{i1}, μ_{i2}), estimate an overall effect (μ_{01}, μ_{02}) for GS and

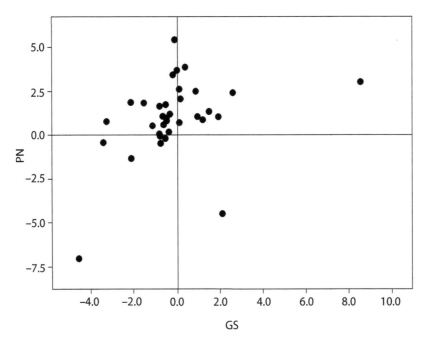

Figure 16.3. The relationship between effect sizes for stomatal conductance (GS) and net CO_2 assimilation (PN) from 34 studies in Data set 4 (Table A16.4).

PN—again we use a bivariate distribution to allow for correlated effects. Formally, the model is as follows:

$$T_{ijk} \sim N(\theta_{ijk}, \sigma_{ijk}^2); \quad \theta_{ij} = (\theta_{ij1}, \theta_{ij2})'$$

$$\theta_{ij} \sim MVN(\mu_i, \Gamma_i); \quad \mu_i = (\mu_{i1}, \mu_{i2})' \qquad (16.9)$$

$$\mu_i \sim MVN(\mu_o, \Omega); \quad \mu_0 = (\mu_{01}, \mu_{02})'$$

where σ_{ijk}^2 is estimated by the reported variance S_{ijk}^2, Γ_i is the variance-covariance matrix for the effects within papers, and Ω is the variance-covariance matrix of the overall effect.

For publications with a small number of reported estimates ("studies"), the terms of the matrix Γ_i may be poorly estimated, particularly the covariance. An alternative model is to maintain a univariate representation for θ_{i1} and θ_{i2} for such a publication, then combine these at the global level, at the last line of Equation 16.8. The method of Riley et al. (2008) for cases in which covariances are unreported and are not easily estimated is also applicable here.

We adopted a Bayesian approach to this meta-analysis, with a normal prior for μ_0 and Wishart priors for Γ_i and Ω. An equivalent univariate analysis, in which each response is analyzed separately, can be fit by changing this model slightly. The models were fit in WinBUGS (the code is in Appendix 16.2 as are the changes to the DATA list needed for the univariate analysis).

Table 16.5 shows the posterior estimates of the overall mean for GS and PN, and the three terms in the overall variance-covariance matrix for the multivariate and univariate models. It can be seen that there is some difference in posterior global variance–covariance terms, and the univariate model undergoes much greater shrinkage of the effects θ_{ijk}, θ_{ij}, which in turn drives the slight differences between multivariate and univariate estimates of the overall means for each response.

DEALING WITH MISSING CORRELATIONS

As noted earlier, a problem that often arises in meta-analysis of dependent outcomes is that the correlations between study effect estimates within groups, and/or between group effect estimates are not reported. The question that then arises is whether to ignore or estimate these correlations. The extreme positions are to assume that the outcomes are independent or that they are completely related (i.e., measuring both outcomes is an exercise in redundancy, aside from the fact that doing so reduces sampling error because the mean of two estimates is closer to the true mean). The same assumptions can be made at the group level. If independence is

TABLE 16.5. The estimates of the mean effect size (and its standard deviation) for net CO_2 assimilation (PN) and stomatal conductance (GC) from Bayesian meta-analyses of Data set 4 (Table A16.4), based on two different modeling assumptions. The WinBUGS code for these models is provided in Appendix 16.2.

Global	Multivariate		Univariate	
Parameter	Mean	SD	Mean	SD
Mean 1 (GS)	−0.3509	0.6104	−0.2336	0.5078
Mean 2 (PN)	1.236	0.6005	1.069	0.4164
Precision 1	0.3703	0.1917	0.2593	0.2037
Precision 2	0.5372	0.254	0.8079	2.931
Precision12	−0.179	0.1547		

assumed, covariance terms in matrices are equal to zero and the full model reduces to a simpler model—that is, of the basic form shown in Equation 16.2. Assuming independence will overestimate, and assuming complete dependence will underestimate, the precision with which the mean effect is known. It is strongly recommended that the researcher undertake a sensitivity analysis to assess the effect of these extreme assumptions by generating study level means and variances using both $r = 0$ and $r = 1$ (Table 16.4).

Alternatively, the researcher can estimate the missing correlations. Consider the problem of combining multiple outcomes within each study (Situation 3). If the outcomes are different measures of the same effect, then they will almost certainly be positively correlated. As a starting point, the researcher should ensure that there is a common predicted direction of the effects. For example, if improved offspring performance is associated with being larger but having a shorter development time, then the sign of the treatment versus control difference for size should be opposite that for development time (hence signs should be reversed for one of the measures). In the case of repeated measures over time, however, it is possible that there is a negative correlation. For example, a large effect of a treatment at time x might lead to a trade-off with a crash at time $x + 1$, so that the difference in outcome between the treatment and control reverses between the two time periods. As an illustration, sexual selection models show that it is possible for males in good condition to "overinvest" in a courtship signal in the present, because the short-term gains more than compensate for any detrimental effect on signaling in the future (Kokko 1997, Hunt et al. 2004).

The researcher might be able to make an assumption about the magnitude of missing correlations by a comparison with available estimates from other studies of the same or closely related outcomes (Berkey, Anderson, et al. 1996). A number of imputation schemes are available (Little and Rubin 2000). If a correlation matrix is estimated or imputed, it is important to check that it is positive definite. Missing within-study correlations can sometimes be estimated from the observed data; Gleser and Olkin (1994) provide relevant formulas for multiple treatment and multiple endpoint studies respectively. For example, if in a polyandry study (Situation 3) the correlation between offspring size and offspring immune function is given separately for the monandry and polyandry treatment, the mean of these correlations may be used as an estimate of the correlation between the effect size for polyandry on size and immunity. A correlation based on the whole data set from a study would be trickier to interpret because it includes any treatment effect of polyandry that changes the values of both the traits of interest. Although it might seem risky to assume correlation from a single study will be similar to those in other studies, Berkey, Anderson, et al. (1996) has noted that correlations from one study are often closer to unreported correlations for the remaining studies than is zero; this last is the assumed correlation when treating outcomes as independent.

When imputing, it is most important to bear in mind that the imputed value is only an estimate of the unknown value. The uncertainty introduced by the imputation must be accounted for in the model to get an unbiased estimate of the truth with a proper estimate of the standard error. Pretending as if the imputed value were the real missing value will lead to underestimation of the variability of the parameter estimate.

A final issue that has to be considered is that when true correlations differ among studies, this should affect the relative weight assigned to each study because of its effect on the variance estimate for that study (e.g., Equation 16.7). This could lead to very different estimates of the overall mean effect (Borenstein et al. 2009, 233). When correlations are unknown and a hypothetical value is assigned, it might be prudent to also conduct a sensitivity analysis to calculate the mean effect after judiciously assigning different extreme values ($r = 0$ and 1) to different studies in order to estimate the mean effect.

DEALING WITH MISSING STUDIES

A problem arises in the analysis of multiple effects if studies report on different subsets of the effects. If there are only a few missing studies it might be possible to estimate them using techniques such as randomization or multiple imputation. Alternatively, if there is sufficient overlap in the subsets, enabling reasonable estimates of the covariances between effects to be obtained, then the multivariate model described above can be altered to allow for different effects in different studies (for an example, see Nam et al. 2003). Difficulties arise, however, if there is insufficient overlap of effects reported by the studies. These include increased variation in effect estimates that are positively correlated due to the inclusion of uncertainty about covariance estimates. This can lead to substantially biased estimates (Riley, Abrams, et al. 2007). Analysts must also beware of reporting bias when encountering studies that report different subsets of effects. Reporting bias arises when authors selectively report some effects and not others, perhaps because of statistical significance or other reasons related to the message they are trying to communicate. Analysis of available effects, even if appropriately done, will lead to bias if the effects are not missing at random (Little and Rubin 2002)

DISCUSSION

We have focused on several possible approaches to the meta-analysis of dependent data, where dependence arises for a number of different reasons. We have shown that the meta-analysis model that is adopted partly depends on the question being answered, the type of data available (i.e., primary or aggregate), the format of these data (e.g., means or regression coefficients), the hierarchical or multilevel nature of the data, the correlation structure in the data and between the effects, and the extent to which we are willing to estimate or ignore covariance. Our intention has been to consider a restricted number of situations in which dependence is commonly exhibited in ecology and evolution meta-analyses, and to describe the meta-analysis models in terms of a general multivariate hierarchical structure that describes the dependencies relevant to that situation. We hope that the reader will now see that the models are simple variations on the same structure, and realize that this structure can be generalized to other situations of interest. We have not gone into details about other, more complex, approaches to the meta-analysis of dependent data, but we provide a brief resume in Appendix 16.3 for the interested reader.

Many of the models we have described can be applied using either primary or aggregated data from the relevant studies. Access to primary data has the strong advantage of ensuring that the same model is fit for each study. It has the disadvantages of necessarily excluding studies for which appropriate data cannot be obtained, and requires that the meta-analyst, who has typically not been involved in the primary studies, has sufficient background knowledge to adequately undertake the study-specific analyses.

One question that often arises is how to decide between a univariate and a multivariate approach. One consideration is on the basis of the available data. If the study-specific effect estimates, variances and within-study variances, *and* covariances are known (the latter usually obtained from the correlation between traits) or estimated (at least partially), then a multivariate analysis can be a profitable approach. However, if the within-study covariances are unknown or cannot be estimated, or if the covariance between outcomes across studies can only be poorly estimated, then a univariate approach is more defensible. Of course, there is no reason not to present analyses based on both approaches as a form of sensitivity analysis to differing model assumptions.

We end by noting that in some published meta-analyses, the authors use several types of meta-analysis approaches to the question being answered (e.g., Rustad et al. 2001, McCarthy et al. 2006). There is rarely a single, perfect solution. Issues of data availability and format influence the choice of approach. Ultimately, however, there is little use running a meta-analysis if it does not address the question it was designed to answer. Ensuring this is the case should always be the priority. Put slightly differently, a sophisticated model that is poorly estimated is often no better than a simple approach in which missing information is acknowledged (e.g., lack of information on correlations between effect sizes at various levels) and then simplifying assumptions are made (e.g., test when $r = 0$ and $r = 1$, or include and ignore grouping factors that might be a source of nonindependence). If these weaknesses are clearly stated the reader can then, at least, interpret the results in a straightforward manner. Equally, it is important to acknowledge the possible biases that can be introduced when using simpler but less appropriate models, so care should be taken in any inferences that are made based on these analyses.

APPENDIX 16.1: DATA SETS

Note: All of the material in section 16.1 and 16.2 of this appendix can also be found on the web site for this book, http://press.princeton.edu/titles/10045.html

TABLE A16.1. Data set 1 ("grouped data set"): A hypothetical data set used to illustrate the analysis of Situation 1. There are 25 studies that can be assigned to five groups (a group could be a study site, research time, year, the technique used to collect the data, and so on). The mean effect size and estimate of its variance from each study are shown.

Group	Effect	Variance
A	0.2	0.10
A	0.6	0.15
A	0.5	0.05
A	0.1	0.06
B	0.8	0.08
B	0.4	0.05
B	0.9	0.04
C	0.2	0.09
C	0.7	0.11
C	0.5	0.03
C	0.2	0.05
C	1.1	0.06
C	1.4	0.07
D	0.1	0.01
D	0.8	0.02
D	0.9	0.02
D	1.4	0.04
E	0.2	0.03
E	0.3	0.05
E	0.8	0.06
E	0.9	0.02
E	1.4	0.07
E	1.2	0.08
E	1.3	0.05
E	0.9	0.04

TABLE A16.2. Data set 2 ("shrimp data set"): Shrimp biomass at nine sites over successive years (ranging from 13 to 34 per site). These data are used to illustrate the analysis of Situation 2.

Year	2J	3K	3M	NGSL	ESS	5Y2	ICE	BS	SKAG
1967							1.50		
1968						0.99	1.64		
1969						0.92	1.66		
1970						1.05	1.63		
1971						0.96	1.48		
1972						1.01	1.37		
1973						0.76	1.43		
1974						0.36	1.44		
1975						0.46	1.45		
1976						0.25	1.56		
1977	2.47				1.09	−0.34	1.68		
1978	2.43				1.06	0.10	1.68		
1979	2.30				1.22	0.19	1.79		
1980	2.30				1.10	0.21	1.73		
1981	2.31				1.10	0.66	1.84		
1982	2.27			1.86	1.09	−0.04	1.86	2.51	
1983	2.16			1.83	1.08	0.38	1.80	2.63	
1984	2.22			1.80	1.02	0.76	1.65	2.67	
1985	2.11			1.81	0.86	0.69	1.69	2.39	2.04
1986	2.40			1.83	0.92	0.88	1.62	2.22	2.09
1987	2.37			1.87	0.94	0.70	1.73	2.16	2.02
1988	2.33	1.56	0.34	1.92	0.91	0.56	1.90	2.26	2.04
1989	2.35	1.44	0.28	1.98	1.10	0.74	1.92	2.33	1.98
1990	2.39	1.48	0.33	2.02	1.11	0.80	2.04	2.42	1.98
1991	2.32	1.58	0.91	2.01	1.28	0.46	1.96	2.51	1.97
1992	2.29	1.73	1.22	1.87	1.22	0.27	1.96	2.38	2.01
1993	2.36	1.79	0.97	1.95	1.26	0.75	1.92	2.37	2.00
1994	2.41	1.86	0.52	2.02	1.28	0.84	1.83	2.21	2.03
1995	2.48	1.98	0.73	2.11	1.35	0.75	1.84	2.29	2.08
1996	2.57	2.01	0.81	2.18	1.37	0.54	2.00	2.44	2.06
1997	2.68	2.11	0.71	2.23	1.37	0.62	1.86	2.48	2.13
1998	2.66	1.99	1.23	2.30	1.51	0.84	1.68	2.53	2.20
1999	2.69	2.00	1.09	2.26	1.59		1.56	2.50	2.11
2000	2.77	2.13	0.99	2.30	1.61		1.77		
No. Yrs	24	13	13	19	24	31	34	18	15
Trend T	0.160	0.058	0.054	0.028	0.022	0.000	0.012	−0.002	0.008
SE(T)	0.004	0.006	0.020	0.003	0.004	0.007	0.002	0.007	0.003

TABLE A16.3. Data set 3 ("polyandry data set"): A hypothetical data set used to illustrate the analysis of Situation 3. There are 16 effect sizes from six separate studies. Each study provides an estimate of the effect of polyandry on two or three offspring traits (offspring size, survival rate, and immune function).

Study	Response	Effect Estimate	Variance
A	size	0.3	0.02
A	survival	0.4	0.018
A	immune	0.2	0.019
B	size	0.5	0.1
B	survival	0.3	0.09
B	immune	0.4	0.095
C	size	0.2	0.05
C	survival	0.2	0.05
D	size	0.3	0.03
D	survival	0.6	0.03
D	immune	0.65	0.03
E	size	0.7	0.01
E	survival	0.8	0.01
E	immune	0.8	0.01
F	size	0.35	0.07
F	survival	0.3	0.06

TABLE A16.4. Data set 4 ("CO_2 data set"): The effect of elevated CO_2 levels on two plant traits (GS and PN) that is used to illustrate the analysis of Situation 4 (Curtis and Wang 1998). There are 68 effect sizes from 17 studies. Each study has at least one pair of effect sizes for PN and GS (range: 1–4).

Paper No[1]	GS Hedges' d	GS Var (d)	PN Hedges' d	PN Var (d)
121	−0.4862	0.3432	0.9817	0.3735
121	0.1535	0.3343	2.0668	0.5113
121	0.0965	0.3337	2.6101	0.6172
121	0	0.2857	3.6586	0.7638
168	−1.5271	0.4305	1.8355	0.4737
222	−3.2447	1.544	0.7747	0.7167
222	−3.4039	1.6322	−0.4267	0.6818
233	−4.57	1.4442	−7.0561	2.8895
290	−2.1284	0.3916	1.8644	0.3586
290	−0.6377	0.2627	1.0313	0.2832
290	0.1082	0.2504	0.7069	0.2656
340	0.9597	0.3717	1.0391	0.3783
340	−0.7788	0.3586	-0.0805	0.3336
506	1.9111	0.5826	1.0516	0.4553
506	2.6041	0.7391	2.39	0.6856
550	8.5458	2.0258	3.0211	0.8563
582	−0.3432	0.5074	1.1915	0.5887
582	−0.6008	0.5226	0.5762	0.5208
582	−0.5389	0.5182	−0.1952	0.5024
582	−2.1085	0.7779	−1.3417	0.6125
582	−0.7445	0.5346	−0.476	0.5142
745	−0.7991	0.18	1.6405	0.2227
745	0.8741	0.1826	2.4785	0.2946
2035	0.3736	0.2261	3.8708	0.7222
2035	−0.5231	0.188	1.7512	0.2515
2045	−0.3751	0.0848	0.1703	0.0836
2045	−0.8026	0.09	0.0557	0.0834
2047	2.1102	0.5189	−4.4809	2.3398
2065	−0.1068	0.3338	5.4279	1.5609
2121	1.4902	0.1661	1.3346	0.8151
2121	1.201	0.1432	0.878	0.7309
2122	−0.1738	0.4015	3.4233	0.986
2129	−0.4823	0.5145	0.8128	0.5413
2131	−1.1225	0.4797	0.532	0.4308

[1] Refers to paper identification numbers in Curtis and Wang (1998)

APPENDIX 16.2: CODE FOR ANALYSES
1. *WinBUGS* code for Bayesian analysis of Data set 1

```
model{
    for (i in 1:No.Ests){                    # No.Ests is number of estimates
        prec.Y[i] <- 1/Var[i]                # prec.Y is precision of Y = 1/var(Y)
        Y[i] ~ dnorm(th[i], prec.Y[i])       # th[i] is true effect theta[i]
        th[i] ~ dnorm(mu[T[i]], prec.th[T[i]])   # mu[T[i]] is true effect for group T[i]
        }                                    # T[i] is the group for the ith estimate
    for (j in 1:No.Groups){                  # No.Groups is the number of groups
        prec.th[j] <- 1/(sig.th[j]*sig.th[j])   # sig.th[i] is the SE of the theta[i]
        sig.th[j] ~ dunif(0,100)             # noninformative uniform prior
        mu[j] ~ dnorm(mu.0, prec.mu)         # mu.0 is the overall effect
        }
    mu.0 ~ dnorm(0.0, 1.0E-2)                # noninformative normal prior
    prec.mu <- 1/(sig.mu*sig.mu)             # sig.mu is the SE of the mu[i]
    sig.mu ~ dunif(0,100)                    # noninformative uniform prior
    }
```

Data
list(No.Ests=25, No.Groups=5,
Y=c(0.2, 0.6, 0.5, 0.1, 0.8, 0.4, 0.9, 0.2, 0.7, 0.5, 0.2, 1.1, 1.4, 0.1, 0.8, 0.9, 1.4, 0.2, 0.3, 0.8, 0.9, 1.4, 1.2, 1.3, 0.9),
Var=c(0.10, 0.15, 0.05, 0.06, 0.08, 0.05, 0.04, 0.09, 0.11, 0.03, 0.05, 0.06, 0.07, 0.01, 0.02, 0.02, 0.04, 0.03, 0.05, 0.06, 0.02, 0.07, 0.08, 0.05, 0.04),
T=c(1, 1, 1, 1, 2, 2, 2, 3, 3, 3, 3, 3, 3, 4, 4, 4, 4, 5, 5, 5, 5, 5, 5, 5, 5)
)

Initial values
list(sig.mu=1, sig.th=c(1,1,1,1,1))

2. *WinBUGS* code for Bayesian analysis of Data set 2

```
# full hierarchical model

model {
    for( i in 1 : N ) {
        beta[i , 1:2] ~ dmnorm(mu.beta[], R[i, , ])
        R[i,1:2 , 1:2] ~ dwish(Omega[ , ], 2)
        for( j in 1 : T[i] ) {
            Y[i , j] ~ dnorm(mu[i , j], tauC)
            mu[i , j] <- beta[i , 1] + beta[i , 2] * x[i,j]}}
    mu.beta[1:2] ~ dmnorm(mean[],prec[ , ])
    tauC <- 1/pow(sigma,2)
    sigma ~ dunif(0.0,10.0)
}
```

```
# DATA
list( N = 9,
T=c(24,13,13,19,24,31,34,18,15),
     Omega = structure(.Data = c(1.0, 0.0, 0.0, 1.0), .Dim = c(2, 2)),
     mean = c(0,0),
     prec = structure(.Data = c(1.0E-6, 0, 0, 1.0E-6), .Dim = c(2, 2)),
Y=structure(.Data=c(
2.47,2.43,2.3,2.3,2.31,2.27,2.16,2.22,2.11,2.4,2.37,2.33,2.35,2.39,2.32,2.29,2.36,2.41,
2.48,2.57,2.68,2.66,2.69,2.77,0,0,0,0,0,0,0,0,0,0,1.56,1.44,1.48,1.58,1.73,1.79,1.86,
1.98,2.01,2.11,1.99,2,2.13,0,0,0,0,0,0,0,0,0,0,0,0,0,0,0,0,0,0,0.34,0.28,0.33,
0.91,1.22,0.97,0.52,0.73,0.81,0.71,1.23,1.09,0.99,0,0,0,0,0,0,0,0,0,0,0,0,0,0,0,0,0,
0,0,0,1.86,1.83,1.8,1.81,1.83,1.87,1.92,1.98,2.02,2.01,1.87,1.95,2.02,2.11,2.18,2.23,
2.3,2.26,2.3,0,0,0,0,0,0,0,0,0,0,0,0,0,0,1.09,1.06,1.22,1.1,1.1,1.09,1.08,1.02,0.86,
0.92,0.94,0.91,1.1,1.11,1.28,1.22,1.26,1.28,1.35,1.37,1.37,1.51,1.59,1.61,0,0,0,0,0,
0,0,0,0,0,0.99,0.92,1.05,0.96,1.01,0.76,0.36,0.46,0.25,-0.34,0.1,0.19,0.21,0.66,-0.04,
0.38,0.76,0.69,0.88,0.7,0.56,0.74,0.8,0.46,0.27,0.75,0.84,0.75,0.54,0.62,0.84,0,0,0,
1.5,1.64,1.66,1.63,1.48,1.37,1.43,1.44,1.45,1.56,1.68,1.68,1.79,1.73,1.84,1.86,1.8,
1.65,1.69,1.62,1.73,1.9,1.92,2.04,1.96,1.96,1.92,1.83,1.84,2,1.86,1.68,1.56,1.77,2.51,
2.63,2.67,2.39,2.22,2.16,2.26,2.33,2.42,2.51,2.38,2.37,2.21,2.29,2.44,2.48,2.53,2.5,0,
0,0,0,0,0,0,0,0,0,0,0,0,0,2.04,2.09,2.02,2.04,1.98,1.98,1.97,2.01,2,2.03,2.08,2.06,
2.13,2.2,2.11,0,0,0,0,0,0,0,0,0,0,0,0,0,0,0,0,0,0,0),.Dim=c(9,34)),
x=structure(.Data=c(
11,12,13,14,15,16,17,18,19,20,21,22,23,24,25,26,27,28,29,30,31,32,33,34,0,0,0,0,0,0,0,0,0,0,
22,23,24,25,26,27,28,29,30,31,32,33,34,0,0,0,0,0,0,0,0,0,0,0,0,0,0,0,0,0,0,0,0,0,
22,23,24,25,26,27,28,29,30,31,32,33,34,0,0,0,0,0,0,0,0,0,0,0,0,0,0,0,0,0,0,0,0,0,
16,17,18,19,20,21,22,23,24,25,26,27,28,29,30,31,32,33,34,0,0,0,0,0,0,0,0,0,0,0,0,0,0,0,
11,12,13,14,15,16,17,18,19,20,21,22,23,24,25,26,27,28,29,30,31,32,33,34,0,0,0,0,0,0,0,0,0,0,
2,3,4,5,6,7,8,9,10,11,12,13,14,15,16,17,18,19,20,21,22,23,24,25,26,27,28,29,30,31,32,0,0,0,
1,2,3,4,5,6,7,8,9,10,11,12,13,14,15,16,17,18,19,20,21,22,23,24,25,26,27,28,29,30,31,32,33,34,
16,17,18,19,20,21,22,23,24,25,26,27,28,29,30,31,32,33,0,0,0,0,0,0,0,0,0,0,0,0,0,0,0,0,
19,20,21,22,23,24,25,26,27,28,29,30,31,32,33,0,0,0,0,0,0,0,0,0,0,0,0,0,0,0,0,0,0,0),
.Dim=c(9,34)))
```

3. *WinBUGS* code for Bayesian analysis of Data set 4

```
model {
     for(n in 1:Nobs) {
          for (j in 1:2){
          tau.d[n,j] <- 1/vard[n,j]
          d[n,j] ~ dnorm(mu.d[n,j], tau.d[n,j])}
          mu.d[n,1:2] ~ dmnorm(mu.i[StudyNo[n],],tau.i[StudyNo[n],,])}
     for (i in 1:NStudy) {
          mu.i[i,1:2] ~ dmnorm(mu.o[], tau.o[,])
          tau.i[i,1:2,1:2] ~ dwish(Gamma.i[,],2)}
     mu.o [1:2] ~ dmnorm(mu.c[], tau.c[,])
     tau.o[1:2,1:2] ~ dwish(Omega[,],2)}
```

```
# Data
list(Nobs=34, NStudy=18,
StudyNo=c(1, 1, 1, 1, 2, 3, 3, 4, 5, 5, 5,6, 6,7, 7,8, 9, 9, 9, 9, 9,
10, 10, 11, 11, 12, 12, 13, 14, 15, 15, 16, 17, 18),
d=structure(.Data=c(-0.4862, 0.9817, 0.1535, 2.0668, 0.0965, 2.6101,
0, 3.6586, -1.5271, 1.8355, -3.2447, 0.7747, -3.4039, -0.4267,
-4.57, -7.0561, -2.1284, 1.8644, -0.6377, 1.0313, 0.1082, 0.7069,
0.9597, 1.0391, -0.7788, -0.0805, 1.9111, 1.0516, 2.6041, 2.39,
8.5458, 3.0211, -0.3432, 1.1915, -0.6008, 0.5762, -0.5389, -0.1952,
-2.1085, -1.3417, -0.7445, -0.476, -0.7991, 1.6405, 0.8741, 2.4785,
0.3736, 3.8708, -0.5231, 1.7512, -0.3751, 0.1703, -0.8026, 0.0557,
-4.4809, 2.1102, -0.1068, 5.4279, 1.4902, 1.3346, 1.201, 0.878,
-0.1738, 3.4233, -0.4823, 0.8128, -1.1225, 0.532),.Dim=c(34,2)),
vard=structure(.Data=c(0.3432, 0.3735, 0.3343, 0.5113, 0.3337,
0.6172, 0.2857, 0.7638, 0.4305, 0.4737, 1.544, 0.7167, 1.6322,
0.6818, 1.4442, 2.8895, 0.3916, 0.3586, 0.2627, 0.2832, 0.2504,
0.2656, 0.3717, 0.3783, 0.3586, 0.3336, 0.5826, 0.4553, 0.7391,
0.6856, 2.0258, 0.8563, 0.5074, 0.5887, 0.5226, 0.5208, 0.5182,
0.5024, 0.7779, 0.6125, 0.5346, 0.5142, 0.18, 0.2227, 0.1826,
0.2946, 0.2261, 0.7222, 0.188, 0.2515, 0.0848, 0.0836, 0.09,
0.0834, 2.3398, 0.5189, 0.3338, 1.5609, 0.1661, 0.8151, 0.1432,
0.7309, 0.4015, 0.986, 0.5145, 0.5413, 0.4797, 0.4308),.Dim=c(34,2)),
mu.c = c(0,0),
tau.c = structure(.Data=c(1.0E-1,1.0E-6,1.0E-6,1.0E-1), .Dim=c(2,2)),
Gamma.i = structure(.Data=c(10,0.1,0.1,10.0), .Dim=c(2,2)),
Omega = structure(.Data=c(10,0.1,0.1,10.0), .Dim=c(2,2))
)

# Univariate model − different variances for each study
model {
    for(n in 1:Nobs) {
            mu.d[n,1] ~ dnorm(mu.i[StudyNo[n],1],tau.i[StudyNo[n],1])
            mu.d[n,2] ~ dnorm(mu.i[StudyNo[n],2],tau.i[StudyNo[n],2])
        for (j in 1:2){
            tau.d[n,j] <- 1/vard[n,j]
            d[n,j] ~ dnorm(mu.d[n,j], tau.d[n,j]) }}
    for (i in 1:NStudy) {
        mu.i[i,1] ~ dnorm(mu1.o, tau1.o)
        mu.i[i,2] ~ dnorm(mu2.o, tau2.o)
            tau.i[i,1] <- 1/pow(sigma1.i[i],2)
            tau.i[i,2] <- 1/pow(sigma2.i[i],2)
            sigma1.i[i] ~ dunif(0,100)
            sigma2.i[i] ~ dunif(0,100) }
    mu1.o ~ dnorm(0.0,1.0E-4)
    mu2.o ~ dnorm(0.0,1.0E-4)
    tau1.o <- 1/pow(sigma1.o,2)
    tau2.o <- 1/pow(sigma2.o,2)
    sigma1.o ~ dunif(0,100)
    sigma2.o ~ dunif(0,100) }
```

```
# Data – univariate model
list(Nobs=34, NStudy=18,
StudyNo=c(1, 1, 1, 1, 2, 3, 3, 4, 5, 5, 5,6, 6,7, 7,8, 9, 9, 9, 9, 9,
10, 10, 11, 11, 12, 12, 13, 14, 15, 15, 16, 17, 18),
d=structure(.Data=c(-0.4862, 0.9817, 0.1535, 2.0668, 0.0965, 2.6101,
0, 3.6586, -1.5271, 1.8355, -3.2447, 0.7747, -3.4039, -0.4267,
-4.57, -7.0561, -2.1284, 1.8644, -0.6377, 1.0313, 0.1082, 0.7069,
0.9597, 1.0391, -0.7788, -0.0805, 1.9111, 1.0516, 2.6041, 2.39,
8.5458, 3.0211, -0.3432, 1.1915, -0.6008, 0.5762, -0.5389, -0.1952,
-2.1085, -1.3417, -0.7445, -0.476, -0.7991, 1.6405, 0.8741, 2.4785,
0.3736, 3.8708, -0.5231, 1.7512, -0.3751, 0.1703, -0.8026, 0.0557,
-4.4809, 2.1102, -0.1068, 5.4279, 1.4902, 1.3346, 1.201, 0.878,
-0.1738, 3.4233, -0.4823, 0.8128, -1.1225, 0.532),.Dim=c(34,2)),
vard=structure(.Data=c(0.3432, 0.3735, 0.3343, 0.5113, 0.3337,
0.6172, 0.2857, 0.7638, 0.4305, 0.4737, 1.544, 0.7167, 1.6322,
0.6818, 1.4442, 2.8895, 0.3916, 0.3586, 0.2627, 0.2832, 0.2504,
0.2656, 0.3717, 0.3783, 0.3586, 0.3336, 0.5826, 0.4553, 0.7391,
0.6856, 2.0258, 0.8563, 0.5074, 0.5887, 0.5226, 0.5208, 0.5182,
0.5024, 0.7779, 0.6125, 0.5346, 0.5142, 0.18, 0.2227, 0.1826,
0.2946, 0.2261, 0.7222, 0.188, 0.2515, 0.0848, 0.0836, 0.09,
0.0834, 2.3398, 0.5189, 0.3338, 1.5609, 0.1661, 0.8151, 0.1432,
0.7309, 0.4015, 0.986, 0.5145, 0.5413, 0.4797, 0.4308),.Dim=c(34,2)))
```

APPENDIX 16.3: LITERATURE GUIDE TO ADVANCED MODELS
Mixed models

A general mixed model was developed by DuMouchel (1995) for meta-analysis of dose response estimation. This paper and later ones by DuMouchel allow for the combination of studies representing multiple outcomes or multiple treatments, accommodating both binary and
continuous outcomes (although a common scale is required for all the outcomes in any particular model). This method does not require that correlations between outcomes within studies
are known. Instead they are estimated by the model. This model extends the standard random-
effects model of DerSimonian and Laird (1986), allowing heterogeneous study designs to be
accounted for and combined. Unlike the methods of Gleser and Olkin (1994) for multiple
treatment designs, no common treatment or control group is required to be present in all studies. Different treatment groups can consist of results from separate (sub)groups of subjects, or
from groups that cross over and are subject to multiple treatments (e.g., repeated measures of
same individuals). A key difference from univariate random-effects models is that each group
of subjects is modeled separately, so the model does not treat them as truly multivariate. Van
Houwelingen et al. (2002) also adopted a mixed model extension to his 1993 approach to
investigate if covariates at the study level are associated with baseline risk via multivariate
regression. Finally, Von Ende (2001) discusses how to analyze a repeated measures response
over levels of a treatment.

Composite models

There are many variations in the calculation of composite effect sizes. For example, Gleser
and Olkin (1994) illustrate the calculation of composite measures for combining effect sizes in
multiple treatment and multiple endpoint situations. Marín-Martínez and Sánchez-Meca (1999)

compare three methods of estimating standardized mean differences, each with different assumptions about the correlation between estimates within studies; the authors conclude that the methods can yield different overall estimates, depending on the degree of variation and covariation between effects. An alternative that has received increasing attention is the use of factor analysis to construct a composite summary measure. This approach is being developed and applied in such diverse disciplines as psychology (e.g., Bushman et al. 1991 and later publications) and marketing (e.g., Peterson 2000); see also Hunter and Schmidt (2004) for a brief discussion of this and related approaches. Of course, the interpretation of a summary effect is crucial. If such an effect is adopted, the ability to examine differences in the individual effects is largely lost. On the other hand, if the effects are very highly correlated, the use of a summary effect is only slightly superior to the selection of any single estimate (Hedges and Olkin 1985, 221).

Full multivariate models

A number of Bayesian approaches to the full multivariate model (Situation 4) have been proposed. General Bayesian mixed model approaches for meta-analysis have been promoted and illustrated by DuMouchel and coauthors (DuMouchel and Harris 1983; DuMouchel 1990, 1995), with particular focus on combining studies of interest to the pharmaceutical sciences (DuMouchel 1990). Dominici et al. (1999) proposed a hierarchical Bayesian group random-effects model and presented a complex meta-analysis of both multiple treatments and multiple outcomes. This analysis evaluates 18 different treatments for the relief of migraine headaches that were grouped into three classes. Multiple heterogeneous reported outcomes (including continuous effect sizes, differences between pairs of continuous and dichotomous outcomes) and multiple treatments were incorporated via relationships between different classes of treatments.

Three Bayesian multivariate models were proposed and compared by Nam et al. (2003). The preferred model was the full hierarchical representation that we described. Nam et al. (2003) also detail a Bayesian approach to the same problem considered by Raudenbush et al. (1988) and Berkey, Anderson, et al. (1996, 1998), focusing on multiple outcomes after adjusting for study level covariates. Turner et al. (2006) provide details of a Bayesian hierarchical model for multivariate outcome data that is applicable to randomized cluster trials. These are studies in which entire clusters of subjects are randomized across treatments, but outcomes are measured on individuals (e.g., fields are given different fertilizer treatments and then growth of individual plants is then measured). Here, correlation between observations in the same cluster, as well as correlation between outcomes, creates dependence between the data that must be accounted for. The authors also consider a parameterization of the model that allows for the description of a common intervention effect using data measured on different scales. The models proposed by the authors are applicable to a multivariate meta-analysis with clustering among studies.

Other examples

There are many other examples of meta-analysis related to dependent variables in the literature. Two examples are given here, as encouragement for the reader to search for situations that are closer to their needs. First, Becker (1992, 1995) describes methods for combining correlations. Inter-relationships between variables in the correlation matrix can be analyzed using fixed- or random-effects models. A set of linear models is derived from the correlation matrix that describes both direct and indirect relationships between outcomes of interest and explanatory variables. Inferences are made on the coefficients of the resultant standardized regression equations, including tests for partial relationships.

Second, a fixed-effects meta-regression model was proposed by Berkey, Mosteller, et al. (1996) to combine studies that report multiple continuous outcomes and/or multiple treatments groups, allowing for study and treatment level covariates corresponding to one or more outcomes. As with the above methods, each study can report all or only a subset of outcomes, and consider all or a subset of treatments on the outcome of interest. This model also allows comparisons without a common control group and, similar to Dear (1994), inclusion of trials that consider only one treatment or control group (single-arm trials), hence allowing better use of the available data. It differs from the models previously described in that outcomes and differences between treatments are measured in the original units and not in terms of effect sizes. This can simplify interpretation and allow for a wider range of applications.

Missing correlations

Missing correlations can be estimated in a Bayesian framework by giving them prior distributions and including them as variables to be estimated in the analysis. An advantage of this approach is that the uncertainty induced by estimating missing correlations is both quantified (in terms of variances of the estimates) and is propagated through the remainder of the analyses. As is intuitively reasonable, this results in larger variances for other estimates. This is arguably better than inserting a point estimate for missing data and pretending that it is "known" (i.e., equivalent in value to observed data).

Riley et al. (2008) address the problem of estimating unknown within-study correlations in the context of a bivariate random-effects meta-analysis. By assuming a single overall correlation parameter ρ, and using the fact that a covariance σ_{ij} is equal to $\rho\sigma_i\sigma_j$, they develop a model that uses only the data required for univariate random-effects meta-analysis of each effect/outcome. They show that, unless the estimate ρ is very close to 1 or -1, the pooled estimates derived from their model have little bias, and better statistical properties than those obtained from separate univariate meta-analyses of each effect.

Models with mixed summary and primary data

Riley et al. (2010) discuss models for situations when primary data are available for some studies and only summary data for others. Such models may have two types of risk factors, those that apply to the individual and those that apply to the study. For example, survival of fish populations may depend on the weights of individual fish captured, but also on the average temperature of the water measured in the study. While the weights will differ by fish within a study, the temperature is the same for each fish in the study. Riley's models permit the primary data to inform the estimates of the individual factors, while both primary and summary data inform estimation of the study-level factors.

Phylogenetic Nonindependence and Meta-analysis

Marc J. Lajeunesse, Michael S. Rosenberg, and Michael D. Jennions

AN IMPORTANT STATISTICAL ASSUMPTION of meta-analysis is that effect sizes are independent (Landman and Dawes 1982, Hedges and Olkin 1985, Gleser and Olkin 1994). This statistical independence means that the collection of effect sizes pooled in a meta-analysis does not have a correlated structure, and that each effect size (or sample) represents an independent piece of information. There are several reasons why a data set might have a correlated structure—for example, when multiple effect sizes are extracted from a single experiment or from different time points throughout a study. These forms of nonindependence are reviewed in Chapter 16. Here we focus on the statistical and conceptual issues arising from the nonindependence that emerges when meta-analysis is based on research from multiple species, which is nearly always the case in ecology and evolutionary biology.

Pooling research from multiple species can be a problem for ecological and evolutionary meta-analysis because species form a nested hierarchy of phylogenetic relationships. This shared phylogenetic history can introduce a correlated structure to effect size data because studies on closely related species may yield similar outcomes, and therefore similar estimates of effect sizes. This similarity is the product of shared (i.e., phylogenetically conserved) morphological, physiological, or behavioral characteristics (Harvey and Pagel 1991). Given that an explicit goal of many ecological and evolutionary meta-analyses is to include research from a diversity of taxa and to generalize across a broad range of species, there is the potential for bias should the literature synthesized be affected by the shared ancestry of taxa (Møller and Thornhill 1998; Verdú and Traveset 2004, 2005; Adams 2008; Lajeunesse 2009).

In this chapter, we describe statistical methods to account for phylogenetic nonindependence of species when pooling and testing for homogeneity of effect sizes. We also describe a method that compares the results of a traditional and a phylogenetically independent meta-analysis to evaluate which approach was more effective at explaining variation in research outcomes. Using these methods, we provide a worked example of a meta-analysis on trade-offs among plant antiherbivore defenses. Finally, we end the chapter with a discussion on approaches for collating phylogenetic information for meta-analysis.

The "apples and oranges problem": It's literal for ecology

The "apples and oranges problem" has a long history in meta-analysis and is a criticism of mixing studies with different conceptual and operational definitions; or more generally, a criticism of mixing studies belonging to different groups (e.g., different experimental designs, different measures of outcome; Wolf 1986, Lynn 1989). The analogy here identifies that apples differ from oranges, and thus aggregating them is not appropriate or meaningful. The "apples and oranges problem" is not a valid criticism because the rationale for aggregating particular studies is relative to the focus of the review; if the focus is on apples then including oranges is inappropriate, but if the focus is on fruit, then grouping apples and oranges is meaningful. This distinction is important because it recognizes that, conceptually, objects form a hierarchy; apples and oranges are fruit because they share similar characteristics (e.g., in an informal sense they are edible, sweet, and have seeds). This concept of hierarchy should also resonate for all ecologists and evolutionary biologists because all species form a nested hierarchy of phylogenetic relationships. Apples and oranges have similar characteristics because of their shared ancestry among flowering plants (angiosperms). Considering the phylogenetic history of taxa is thus important for ecological and evolutionary meta-analyses because quantitative reviews often seek to generalize as broadly as possible by pooling or "mixing" all studies on a given conceptual topic.

But how should meta-analysts approach the problem of mixing research from different taxa? We believe they should first collate all the studies relevant to the conceptual topic under study—irrespective of whether they are based on birds, beetles, or nematodes. With these data, reviewers are then able to assess publication or research biases of taxa (e.g., if there is significant overrepresentation of beetle research over that of other taxa; see Chapter 14), conduct sensitivity analyses to determine whether the inclusion of certain taxa can bias results, and determine whether the phylogenetic history of these taxa is an issue for meta-analysis. An important benefit of collating all studies is the increased sample size of the review—large samples sizes decrease the likelihood of review type I and type II errors (Lajeunesse and Forbes 2003; Chapter 22) and allow for greater representation among moderator groupings essential for hypothesis testing.

Why is phylogenetic history an issue for meta-analysis?

Similarity of traits due to shared ancestry is a problem for meta-analysis because it violates two assumptions. First, that data are drawn from independent samples; this assumption is violated because effect sizes derived from closely related species may be similar in magnitude or direction due to their shared evolutionary history. Second, that effect sizes are sampled from a population that has a normal distribution with an expected variance. However, when effect sizes have a phylogenetic (correlated) structure, different variances are expected because lineages within phylogenies may have evolved at different rates (Harvey and Pagel 1991). Violating these assumptions increases the likelihood of a type I error (false positive) (Harvey and Pagel 1991). Below we review both assumptions in detail because they form the basis for the statistical methods we outline in this chapter.

First statistical violation: Lack of independence

Shared ancestry among species violates the assumption of nonindependence in meta-analysis because it can result in effect size data with a correlated structure—that is, some groupings of data points are more likely to have values that exceed (or are less than) the average value due to a shared error structure. Here these shared correlations are due to the nonindependent origins of traits and characteristics of related taxa. This phylogenetic conservatism (sometimes described as phylogenetic inertia; Wilson 1975) can emerge in effect size data in many ways. For example, effect size data quantifying the magnitude of difference or the correlation between characteristics (Table 17.1) can be nonindependent if the characters themselves are phylogenetically

TABLE 17.1. Variance estimates $\sigma^2(\delta)$ of effect size metrics (δ) that can be used with phylogenetically independent meta-analysis. Effect sizes and their variances are used to define the variance-covariance matrix (**V**). This matrix is then applied to GLS models that estimate the phylogenetically independent weighted mean of a collection of studies. Here, effect sizes are estimated from a study that has a sample size (n), and control (C) and treatment (T) means ($\overline{X}$). Finally, the log odds ratio is based on count data of members belonging to groups A or B. Examples of these effect sizes are found in Van Zandt and Mopper (1998) for Hedges' d and $\ln R$, Koricheva et al. (2004) for correlation coefficients, and Beirinckx et al. (2006) for $\ln OR$.

effect size metric	effect size	effect size variance
Hedges' d[†]	$d = \dfrac{\overline{X}_T - \overline{X}_C}{s}\left(1 - \dfrac{3}{4(n_T + n_C) - 9}\right)$	$\dfrac{n_T + n_C}{n_T n_C} + \dfrac{d^2}{2(n_T + n_C)}$
log response ratio ($\ln R$)	$\ln R = \ln\left(\dfrac{\overline{X}_T}{\overline{X}_C}\right)$	$\dfrac{s_T^2}{n_T \overline{X}_T^2} + \dfrac{s_C^2}{n_C \overline{X}_C^2}$
Pearson's product-moment correlation coefficient (r)	$r = r$	$\dfrac{(1 - r^2)^2}{n - 2}$
Fisher's z transformation of r	$z = \dfrac{1}{2}\ln\left(\dfrac{1+r}{1-r}\right)$	$\dfrac{1}{n - 3}$
log odds ratio ($\ln OR$)	$\ln OR = \ln\left(\dfrac{n_{A_C} n_{B_T}}{n_{A_T} n_{B_C}}\right)$	$\dfrac{1}{n_{A_C}} + \dfrac{1}{n_{A_T}} + \dfrac{1}{n_{B_C}} + \dfrac{1}{n_{B_T}}$

[†]Hedges' d uses the pooled standard deviation (s) between the control and treatment; this is calculated as $s = ([(n_T - 1)s_T^2 + (n_C - 1)s_C^2]/[n_T + n_C - 2])^{1/2}$.

conserved. Body size is an example of a trait that is often used in experimentation as a surrogate for fitness; but for many animals, closely related species or whole lineages may share similar sizes. Another way effect size data could be phylogenetically correlated is when the effect sizes themselves are phylogenetically conserved. An example is when the magnitude of difference or correlations between traits in body size occurs among animals that are sexually dimorphic.

Phylogenetic conservatism can generate effect size data with phylogenetic structure where studies based on related species yield very similar experimental outcomes (e.g., effect sizes). Typically, to statistically account for this source of dependence, researchers use Felsenstein's (1985) phylogenetically independent contrasts (PIC), which is a statistical method in comparative biology used for cross-species generalizations across traits. It transforms trait data into a set of contrasts (pairwise differences between trait values) with zero correlations. Here, although the absolute values of these data may be conserved, the pairwise differences between traits are not (Harvey and Pagel 1994). These contrasts are then analyzed using linear regression to evaluate whether traits share a correlated evolutionary history (for further details, see Martins and Garland 1991). Unfortunately, PICs are not practical for meta-analysis data because each effect size has a different variance and this violates the significance testing and variance estimating procedures of traditional regression models (Hedges 1994). This difference in variances among effect sizes is due to studies having different sample sizes (Hedges and Olkin 1985). Fortunately, PICs are special case of generalized least squares (GLS) models that can include phylogenetic information (Garland and Ives 2000, Rohlf 2001). Here a GLS model can be modified to fit the specific significance testing and variance estimation procedures for meta-analysis.

Second statistical violation: Heterogeneity of variances

Effect sizes with a phylogenetic structure violate the second statistical assumption that data are sampled from a normal distribution with an expected variance (Hedges and Olkin 1985). Variance here is defined as the rate of evolutionary change, and effect size data based on different taxa could have different expected variances because they belong to lineages evolving at different rates with varying divergence times (Harvey and Pagel 1991). Typically, comparative analyses account for this form of heterogeneity in variances by making assumptions on how lineages evolve (Felsenstein 1985, Martins and Garland 1991, Pagel 1997). For example, PICs assume that evolution proceeds as a Brownian motion (BM) process (e.g., random drift), where the expected variance of change is constant throughout the phylogeny; fitting this model of evolution is achieved by standardizing all the contrasts with the square root of the sum of all phylogenetic branch lengths (Martins and Garland 1991). In terms of GLS modeling, fitting a BM model of evolution is straightforward. Simply assume that the phylogenetic correlations used to define the phylogenetic structure of effect sizes are linearly related with the relative time since the divergence of each taxa. The following sections outline in more detail how to fit this evolutionary model into meta-analysis. Note also that although many of the statistics below are described as being "phylogenetically independent," this is only the case if the assumption of Brownian motion evolution holds (e.g., the phylogenetic branch lengths are proportional to the expected evolutionary change). Several diagnostics have been developed to estimate whether data fit this evolutionary model (e.g., Pagel's lambda; Pagel 1997), and these should be used with many of the statistics presented below to assess violations of this evolutionary assumption (Lajeunesse 2009).

STATISTICAL METHODS FOR PHYLOGENETICALLY INDEPENDENT META-ANALYSIS

Several comparative phylogenetic methods have been developed to account for phylogenetic nonindependence of species (Felsenstein 1985, Cheverud et al. 1985, Maddison 1990, Grafen 1989, Martins and Garland 1991, Pagel 1997). Felsenstein's (1985) phylogenetically independent contrasts (PIC) remains the most widespread; however, alternative methods applying the generalized least squares (GLS) theory are increasing in popularity (Pagel 1994). In this chapter, we apply this GLS framework to account for phylogenetic history because meta-analysis can also be modeled under the GLS family of statistics (for further details, see Lajeunesse 2009). The advantage of using a combined statistical framework is the full flexibility to apply different statistical models to effect size data. The following methods can also be applied to any effect size metric (Table 17.1), but require that the variance of each effect size and a hypothesized phylogeny (how each effect size is correlated) are known.

Pooling effect sizes with phylogenetic correlations

Using matrix notation, we can pool K number of effect sizes (δ) into a weighted-mean (pooled) effect size ($\bar{\mu}_+$) with this GLS regression equation:

$$\bar{\mu}_+ = (\mathbf{X}^\mathrm{T}\mathbf{W}\mathbf{X})^{-1}\mathbf{X}^\mathrm{T}\mathbf{W}\mathbf{Y}. \tag{17.1}$$

The 95% confidence intervals (CI) for $\bar{\mu}_+$ are calculated as follows:

$$95\% \, \mathrm{CI} \, [\, \bar{\mu}_+ \pm 1.96\sqrt{(\mathbf{X}^\mathrm{T}\mathbf{W}\mathbf{X})^{-1}}\,]. \tag{17.2}$$

The components of Equations 17.1 and 17.2 are as follows. The superscripts ^{T} and $^{-1}$ indicate the transposition and inverse of a matrix, respectively. The effect sizes are contained in the $K \times 1$ column vector $\mathbf{Y}$. The averaging behavior of Equation 17.1 (i.e., pooling all effect sizes) is defined by $\mathbf{X}$, which is a $K \times 1$ column vector of ones and is known in GLS terms as the design matrix (Groß 2003). The most important feature of Equation 17.1 is the weighting matrix ($\mathbf{W}$), which is the inverse of the variance-covariance matrix ($\mathbf{V}$), such that $\mathbf{W} = \mathbf{V}^{-1}$. In traditional meta-analysis, $\mathbf{V}$ is defined as a $K \times K$ matrix containing the effect size variances $\sigma^2(\delta)$ on its main diagonal, and zeros in all the remaining (off-diagonal) elements. When $\mathbf{V}$ is defined this way, and applied to Equation 17.1, it will yield the same pooled effect size as the one generated by the traditional weighting method of meta-analysis described in Chapter 9.

To calculate a "phylogenetically independent" mean effect size, we need to integrate information in $\mathbf{V}$ on how effect sizes vary together based on their correlated phylogenetic history. This is achieved by defining all the off-diagonal elements of $\mathbf{V}$ with the phylogenetic covariance for each pair of effect sizes (for further details on the statistical background of this approach, see Hedges and Olkin 1985; Becker 1992; Gleser and Olkin 1994; Marín-Martínez and Sánchez-Meca 1999; Cheung and Chan 2004; and appendix of Lajeunesse 2009), such that the $i = 1, \ldots, K$ rows and $j = 1, \ldots, K$ columns of $\mathbf{W}$ become

$$\mathbf{V}_{i,j} = \begin{cases} \sigma^2(\delta_i) & \text{when } i = j \text{ (on diagnoal)}, \\ \text{cov}(\delta_i, \delta_j) & \text{when } i \neq j \text{ (off diagnoal)}, \end{cases} \tag{17.3}$$

where each covariance is calculated as

$$\text{cov}(\delta_i, \delta_j) = P_{i,j} \sqrt{\sigma^2(\delta_i)} \sqrt{\sigma^2(\delta_j)}. \tag{17.4}$$

Here, information on shared ancestry between each ith and jth effect size is found in the phylogenetic correlation matrix $\mathbf{P}$. The elements of $\mathbf{P}$ are defined by the shared internode branch-length distance between species on a phylogenetic tree. Figure 17.1 describes how this branch-length information in converted into a phylogenetic correlation, and Figure 17.2 illustrates how $\mathbf{P}$ is then integrated into $\mathbf{V}$ using Equation 17.4. This approach to defining $\mathbf{P}$ forms the basis for many comparative methods based on the GLS framework (Rohlf 2001). For clarity, we now refer to $\bar{\mu}_+$ as the traditionally weighted pooled effect size, and $\bar{\rho}_+$ as the pooled effect size weighted by variance *and* phylogenetic correlations.

Finally, for all the phylogenetically independent analyses outlined in this chapter, only ultrametric trees (i.e., where all the tips of the phylogeny are aligned or contemporaneous; Figure 17.1) should be used to define $\mathbf{P}$. This is because Equation 17.4 assumes that the elements of $\mathbf{P}$ are statistical correlations (i.e., with values ranging from zero to one) and not raw branch-length distances (Hedges and Olkin 1985). More precisely, $\mathbf{P}$ should only contain elements on the main diagonal (i.e., species that are 100% correlated with themselves), and no off-diagonals should exceed one. When the tree is not ultrametric, such that each species has a different BL_{max} (Fig. 17.2), then it is possible to generate a $\mathbf{W}$ matrix that violates the statistical assumptions of GLS modeling (e.g., $\mathbf{W}$ must be symmetric and positive definite; see Groß 2003). Using an ultrametric tree also meets the evolutionary assumption that phylogenetic branch lengths are proportional to the expected evolutionary change (see Rohlf 2001).

Phylogenetically independent homogeneity tests

Homogeneity statistics (Q) in meta-analysis test whether a collection of effect sizes share a common effect (i.e., $\delta_1 = \delta_2 = \ldots = \delta_K$). A more thorough explanation and interpretation of Q is found in Chapter 9. The following GLS equation allows testing for homogeneity of effect sizes while accounting for phylogenetic correlations:

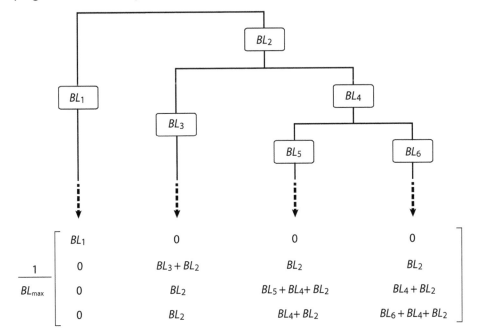

Figure 17.1. The phylogenetic correlation matrix **P** and how it is defined by the phylogenetic re-lationships of four species. Here, the internode branch lengths (*BL*) of the phylogeny are used to calculate the strength of correlations between species, and BL^{max} is the sum of all the internode branch-length distances from the root to tip of the phylogeny. The values within the matrix are the sum of the branch lengths of shared phylogenetic history between two taxa, divided by the depth of the tree (the maximum sum of branch lengths from the root to the tips). Thus, entry *ij* within the matrix represents the proportion of evolutionary history reflected in the tree that is shared by species *i* and *j*. By definition, the amount of history shared by a species with itself is 100%, and since the tree is ultrametric, the values on the diagonal will equal 1.

$$Q = \mathbf{Y}^{\mathrm{T}}\mathbf{W}\mathbf{Y} - \bar{\rho}_{+}^{\mathrm{T}}(\mathbf{X}^{\mathrm{T}}\mathbf{W}\mathbf{X})\bar{\rho}_{+}, \qquad (17.5)$$

where again **W** is the inverse of **V** described in Equation 17.3. The significance of Q is tested against the critical value of a chi-square distribution (χ^2) with $K-1$ degrees of freedom, such that if $Q \le \chi^2_{K-1}$ then $\delta_1 = \delta_2 = \ldots = \delta_K$. Microsoft Excel provides a function that calculates the P-value of this Q-test: =CHIDST(Q,K-1).

Polytomies in phylogenetic trees: A bias for homogeneity tests

Polytomies may influence type I error rates of homogeneity tests because they influence the degrees of freedom of analyses (Purvis and Garland 1993). The degrees of freedom are a mea-sure of the number of independent pieces of information ($K-1$ for a completely resolved tree); polytomies in phylogenetic trees cancel some of this independence because multiple lineages are specified with shared divergences. Polytomies are tree nodes that have more than two im-mediate descendents, and Madisson (1989) divides these into two types, one biological and the other statistical. The first type is referred to as a "hard" polytomy and depicts a true biological event where a group of sister taxa have diverged (speciated) simultaneously. Hard polytomies are most likely to arise due to explosive (rapid) speciation events, simultaneous fragmentation of populations, or introgressive hybridization and recombination events (Meinick and Hoelzer

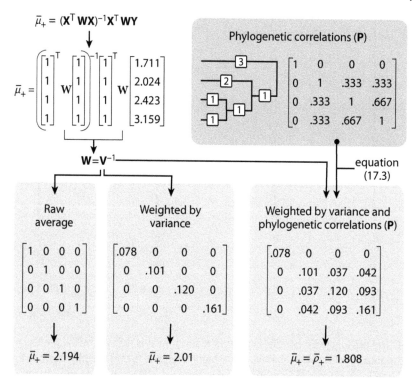

Figure 17.2. An illustration of applying three models of covariance matrices **W** (*light gray*) to the GLS regression Equation 17.1 to generate a raw, weighted by variance (traditional meta-analysis or $\bar{\mu}_+$), and phylogenetically independent mean effect size ($\bar{\rho}_+$). Here $\mathbf{X} = [1 \quad 1 \quad 1 \quad 1]^T$ is the design matrix, and $\mathbf{Y} = [1.711 \quad 2.024 \quad 2.423 \quad 3.159]^T$ is the column vector of effect sizes. The phylogeny depicts how effect sizes are related (*dark gray block*) (effect sizes are aligned with the tips of the phylogeny), and **P** is the phylogenetic correlation matrix of the shared internode branch-length distances from this phylogeny. These distances are in small font on the branches of the phylogeny.

1994). Walsh et al. (1999) describe an example of a hard polytomy, but more generally they appear to be uncommon. Polytomies more often occur because of statistical uncertainty when resolving relationships among taxa; these are known as "soft" polytomies. Statistical approaches to building phylogenetic trees (e.g., maximum likelihood) will generate soft polytomies when they are unable to statistically break up true speciation or bifurcation events among sister taxa.

The correct degrees of freedom for Q-tests should reflect this statistical imprecision in the phylogenetic tree. One approach is to conservatively bind the degrees of freedom to minimize type I errors. Purvis and Garland (1993) suggest a conservative df to be $K - p$. Here, p is the total number of polytomies in the phylogeny. Should the significance of Q be lost after correcting the degrees of freedom for polytomies, then all conclusions drawn from homogeneity tests should be considered as inconclusive given the current resolution of the phylogeny. Alternatively, Purvis and Garland (1993) suggest arbitrarily resolving each polytomy until the tree is fully bifurcated (no more polytomies), and then repeating this method to create a collection of phylogenies with alterative solutions. The results of applying each tree solution are then averaged to provide an overall synthesis. Unfortunately, GLS modeling summarizes phylogenetic relationships using the **P** correlation matrix. Here, arbitrary resolution of polytomies may not significantly change the shared branch-length distances between species because polytomies are resolved by setting the phylogenetic branch-length distance between intervening nodes as a very small distance (Rohlf 2001, Diniz-Filho 2001). Thus, GLS analyses may be insensitive to these resolutions if there are too few polytomies to resolve. However, when there are many polytomies, it is important to generate a broad sampling of tree solutions to avoid this issue. This approach is also useful when the internode branch-length distances are themselves arbitrarily set. This occurs when information on the topology of the phylogeny (i.e., the nodes on tree) is available, but intervening branch-length distances are not. Purvis and Garland (1993) outline multiple procedures for arbitrarily setting internode branch-length distances. Alternatively, missing internode branch-length distances can be simulated with different models of evolution (e.g., Brownian motion).

Integrating multiple studies of a single species

Including multiple studies of the same species requires only a slight modification to phylogenetically independent meta-analysis. These studies are important because they serve as species-level replicates, and their inclusion in meta-analyses can improve variance estimates. The challenge is to modify the GLS model such that these replicates retain their within-species study-level independence, while still accounting for the phylogenetic nonindependence that occurs when effect sizes are combined with other taxa. Figure 17.3 shows an example of how to modify the **P** phylogenetic correlation for three species where data from one species are derived from two studies. Note that using the **P** matrix in Figure 17.3 requires that the **Y** column vector contains all the effect sizes of studies a, b, c_1 and c_2 (i.e., here the column vector rank becomes the number of effect sizes, not the number of species). This also applies for the design matrix **X** with a column rank of K number of effect sizes. The degrees of freedom for this analysis are K number of taxa. Finally, this approach to integrating multiple studies is identical to pooling all the effect sizes and their variances of a single species before a phylogenetically independent meta-analysis (e.g., using Equation 17.1 and 17.2; Gurevitch and Hedges 1999); then, each species only has a single effect size estimate and variance for the final meta-analysis. These pooled effect sizes and variances are based on either a fixed- or random-effects meta-analysis.

Finally, we do not advocate treating these replicate studies as polytomies on the stem of a phylogeny for a given species. When treated this way, the whole group of replications will be

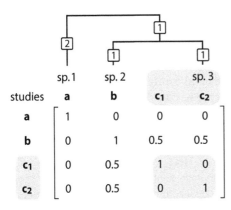

Figure 17.3. Example of a phylogenetic correlation matrix **P**, including multiple studies (effect sizes) from a single species. Here, species 1 and 2 only have one effect size (**a** and **b**, respectively), whereas species 3 has two (**c1** and **c2**). The gray block emphasizes the correlation submatrix of species 3, which is parameterized to account for the between-study independence of studies **c** and **d**. These two studies share the same relative phylogenetic correlation with species 2. Finally, the small font numbers on the branches of the phylogeny indicate their length.

significantly underweighted in the final meta-analysis. This underweighting occurs because polytomies are often set to have a very short branch-length distance, and this small branch length will yield a very strong phylogenetic correlation ($P_{i,j} \rightarrow 1$). Here, the individual variation for each replication is lost because these strong phylogenetic correlations penalize the weighting of each effect size in the final analysis. Thus, in effect, the meta-analysis would generate the same result as if these replications were assumed to have equal variances.

Testing for a phylogenetic signal among effect sizes

Any meta-analysis for which the phylogenetic history is considered a source of bias should evaluate the degree to which effect size data show a phylogenetic signal. Detecting a phylogenetic signal is a test for phylogenetic conservatism. This test quantifies the degree to which related species tend to be more similar than distantly related species. When data show little phylogenetic conservatism, they are considered evolutionarily labile, and analysis of these data is assumed to be the least biased by phylogenetic nonindependence. Unfortunately, the application of information regarding phylogenetic signals is contentious, and it has been advised that if a phylogentic signal is not detected, then researchers can skip phylogenetically controlled analyses (Westoby et al. 1995a, 1995b, 1995c, Ricklefs and Starck 1996, Björklund 1997). However, we believe testing for a signal should be used to test hypotheses rather than as a guide for directing (potentially informative) analyses; remember, the absence of evidence is not evidence of absence. The potential for error in these statistics is due to the continued statistical difficulty in detecting phylogenetic signals from phylogenies with few species (Martins 2000, Freckleton et al. 2002).

Several statistical tests for phylogenetic signals have already been formalized in the literature; these are Pagel's (1994, 1997) λ, and Blomberg et al.'s (2003) K (not to be confused with the K used as the review sample size of meta-analysis). These tests can be applied to effect sizes (Lajeunesse 2009), and are useful because they include calculations of 95% confidence intervals (CI) around λ or K. The interpretation of these statistics is straightforward; should λ or K be nonzero (i.e., the 95% CI does not overlap with zero), then closely related species tend to share similar effect sizes. This information can be used to assess the relative degree of phylogenetic bias in a meta-analysis, and to help evaluate whether "phylogenetically independent" analyses meet the important assumption of Brownian motion evolution.

Testing the fit of phylogenetic correlations

An alternative to evaluating a phylogenetic signal is to evaluate which statistical model, the traditional ($\bar{\mu}_+$) or a phylogenetically independent meta-analysis ($\bar{\rho}_+$), is better at explaining variation among effect size. What is needed to test model fit is the error sums of squares (SSE) of each meta-analysis; this is simply the homogeneity statistic Q (Equation 17.5; see also Hedges and Olkin 1985). With the homogeneity statistic forming the basis for model selection, Akaike's information criterion (AIC) is then used to compare each model. AIC scores are useful for model selection because they penalize models that (a) poorly fit effect sizes, and (b) use additional parameters to explain variation. AIC scores can be calculated as follows:

$$\mathrm{AIC} = 2m - K\left[\ln\left(2\pi\frac{Q}{K}\right) + 1\right] \tag{17.6}$$

where m is the number of parameters in the meta-analysis. Here, traditional meta-analysis will have an $m = 1$, whereas the phylogenetically independent meta-analysis will have $m = 2$. We penalize the phylogenetically independent meta-analysis because it uses an additional parameter (phylogenetic correlations) to explain variation among effect sizes. Finally, the model that best fits the effect size data is the one with the lowest AIC score; for example, if $AIC_{\bar{\mu}_+} < AIC_{\bar{\rho}_+}$ then the traditional meta-analysis is the better fit. It should finally be noted that this approach to estimating AIC scores with Equation 17.6 may only be appropriate for comparing evolutionary models with equal K. Lajeunesse (2009) describes an alternative to calculating AIC using likelihood functions that may be more justifiable for complex model comparisons.

Alternative statistical approaches

Phylogenetically independent meta-analysis is a rapidly growing field of statistics and several alternative approaches to the one presented in this chapter have been proposed. One statistical approach was the use of phylogenetically informed simulations to evaluate the validity of meta-analysis (Verdú and Traveset 2004, 2005). Here, simulations are used to estimate phylogenetically independent critical values for the statistical tests of a weighted regression model (e.g., F-test; Garland et al. 1993). A limitation of this approach is that analyses use the significance tests (F-tests) and variance estimating procedures of weighted regression analyses. This is because effect size data do not meet the assumptions of conventional regression or ANOVA style analyses, given that individual effect sizes will have different variances due to their inverse relationship with sample size (Hedges and Olkin 1985, Hedges 1992). This difference in variance is the reason why the tools of statistical inference for meta-analysis differ from these conventional statistics and rely heavily on chi-square tests. Another limitation is that although this approach is only useful to evaluate bias in meta-analysis, it does not provide a solution to integrate phylogenetic information when pooling results or when testing for homogeneity.

Adams (2008) was the first to propose a method to integrate phylogenetic information when pooling effect size data. Here the effect size data (vector $\mathbf{Y}$) and the design matrix (vector $\mathbf{X}$) are first transformed to have zero phylogenetic correlations before analysis using a traditional weighted regression. This transformation is achieved by multiplying $\mathbf{X}$ and $\mathbf{Y}$ with the inverse square root of the phylogenetic correlation matrix $\mathbf{P}$ (Garland and Ives 2000; Groß 2003). These transformed vectors are then integrated in a traditional weighted regression model as follows:

$$\bar{\mu}_+^{Adams} = [(\mathbf{P}^{-1/2}\mathbf{X})^T\mathbf{W}(\mathbf{P}^{-1/2}\mathbf{X})]^{-1}(\mathbf{P}^{-1/2}\mathbf{X})^T\mathbf{W}(\mathbf{P}^{-1/2}\mathbf{Y}), \qquad (17.7)$$

where $\mathbf{W}$ contains the variances of each effect size on its main diagonal. However, Lajeunesse (2009) found that this approach can introduce a small bias when pooling effect sizes because the phylogenetic transformation has an effect of converting effect sizes into evolutionary units. This decreases the efficiency of the weights used to penalize studies with a large sampling error (Lajeunesse 2009). A more serious issue of this approach is that it applies traditional significance tests of comparative methods or weighted regression models to analyze effect size data. For example, Adams (2008) uses the t-test of a weighted regression to evaluate the significance of the pooled (phylogenetically independent) effect size. However, effect size data are not the same as the quantitative trait data typically analyzed with comparative methods, because each effect size has a different variance, and this violates the assumption of homogeneous variances of traditional significance tests. Using the incorrect statistical tests can significantly bias the outcome and interpretation of results.

Hadfield and Nakagawa (2010) also put forward a general mixed model framework for phylogenetic comparative methods and meta-analysis. This framework is by far the most flexible and sophisticated of all the approaches described in this chapter, and has the advantages of assuming nonnormal distributions and implementing meta-regression or multivariate models (see

Chapters 11 and 16). We predict that this framework and similar mixed modeling approaches will become essential to testing elaborate statistical models with phylogenetic meta-analysis.

Applying the GLS framework for other sources of nonindependence

Correlations used to estimate distances among effect sizes need not be limited to phylogenetic information; experimental, genetic, temporal, and spatial/geographic correlations could also be applied to the GLS framework presented in this chapter. One application could be to control for the overrepresentation of single studies due to extracting multiple effect sizes from a single experimental design (Rosenthal 1991). Here effect sizes are not independent because many of the experimental comparisons are based on multiple traits measured from the same individuals. Should a study report information on how these traits are correlated (i.e., Pearson product moment correlations), then a within-study pooled effect size (controlling for nonindependence among treatment effects) can be estimated using Equation 17.3 with a **P** matrix containing the correlations among traits (see also Gleser and Olkin 1994).

A similar approach can be applied to calculate a within-study pooled effect size from studies based on repeated measure designs. For example, the duration of measurement intervals in a time series analysis will dictate the independence of data, assuming that data measured at short intervals are more likely to be similar. The duration among intervals (as a linear distance) can be treated as correlations in the **P** matrix, where each effect size (drawn from each repeated measure) can be pooled based on the assumed temporal correlations. Simpler correlation structures could also be assumed, such as when measurements are only dependent on the previous measurement (autocorrelation). This avoids treating each repeated measure as a separate effect size or having to apply strict selection criteria that could lose information (e.g., extracting only the final endpoint as the study's effect size; Chapter 16).

AN ILLUSTRATIVE EXAMPLE:
TRADE-OFFS IN ANTIHERBIVORE DEFENSES

Plants that defend themselves from herbivores using multiple (chemical or structural) defenses are predicted to show a negative association among these defenses (Bergelson and Purrington 1996). Plant defense theory predicts this negative association if defenses come at a cost to growth and reproduction so that, given limitations to resource investment, a trade-off in resource allocation is expected among different defenses. However, a meta-analysis of 31 independent studies testing this theory could not detect a trade-off in multiple defenses found among grasses, herbaceous plants, and trees (Koricheva et al. 2004). Although this meta-analysis provided a powerful framework to test this theory by synthesizing and exploring heterogeneity of results from multiple studies, while giving less weight to studies with poor sample precision, it did not provide an evolutionarily robust test of these findings. What was missing was an analysis that integrated the shared phylogenetic history of plants; this would account for any bias due to phylogenetic conservatism of antiherbivore defenses among related plant species (Agrawal 2007). Below, we compare models with and without phylogenetic information to evaluate the degree of phylogenetic bias in the findings of Koricheva et al.'s (2004) meta-analysis.

Methods

A hypothesized phylogenetic tree (Fig. 17.4) of the 22 plant species included in the original meta-analysis was generated using Phylomatic (Webb and Donoghue 2004). Phylomatic is a

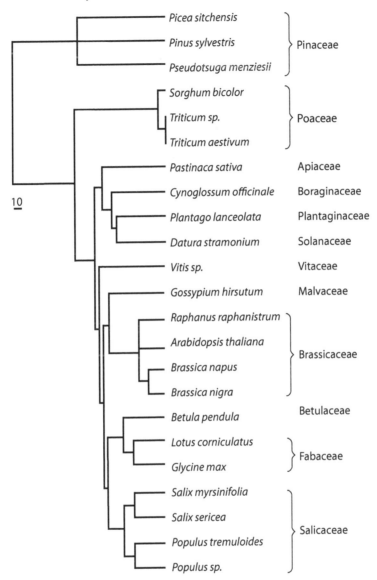

Figure 17.4. The hypothesized phylogenetic relationship of 23 plant species originally synthe-sized in a meta-analysis on tradeoffs in antiherbivore plant defenses; see Koricheva et al. (2004). See Table A17.1 for the NEWICK version of this tree. The plant families of these species are found to the far right. The small bar at the far left indicates the scale of branch-length distance within the phylogenetic tree (10 = 10 million years).

useful online program that estimates plant phylogenies from a modified (and updated) super-tree containing many of the major plant groups (see www.phylodiversity.net/phylomatic). Our phylogeny also includes information on the temporal ordering of nodes. This information was based on estimates of the age of major divergence events among plant groups by Wikström et al. (2001), and was integrated in our tree using BLADJ (also on www.phylodiversity.net). Unfortunately, Phylomatic is only useful for connecting broad phylogenetic relationships, so our tree is not entirely resolved (i.e., not completely bifurcated) among closely related species. For example, species belonging to Brassicaceae and Pinaceae families are described using polytomies.

The branch-length data of this tree are used to define the elements of the phylogenetic correlation matrix **P** (Fig. 17.2), and **P** is then applied to Equation 17.4 to estimate the covariance matrix **V** for phylogenetically independent meta-analysis. However, in Koricheva et al. (2004), meta-analysis results are pooled among different categories of plant defenses such that each category contains only a subset of the species found in Figure 17.1. To resolve this issue, we used a subset tree for each category containing only the relevant species for which effect size data were available (Appendix Table 17.1). These subset trees (used to estimate **P**) conserve all the phylogenetic branch-length information found in Figure 17.1.

One final note is that the original meta-analysis had more studies (experiments) than species. This is due to multiple studies testing different categories of defenses for a single species. To simplify analyses, we pooled the effect sizes of these replicate studies prior to our phylogenetically independent meta-analysis (Appendix Table 17.1). Thus, only one representative (pooled) effect size was used for each species within a given category of antiherbivore defenses. These pooled within-species effect sizes and their variances were estimated using traditional (fixed-effects) meta-analysis (Chapter 9). Finally, all the results reported in Koricheva et al. (2004) assume a random-effects model for pooling effect sizes across species. Here, we also apply a random-effects model, but for simplicity we assume that the between-study variance (τ) is not phylogenetically correlated among taxa. See Chapter 9 for details on how τ is estimated and Lajeunesse (2009) for information on how to calculate a phylogenetically independent version of τ. Thus before our phylogenetically independent analyses, each effect size variance was modified as follows to fit a random-effects model: $\hat{\sigma}^2(\delta_i) = \sigma^2(\delta_i) + \tau$ (details in Chapter 9).

Results and discussion

Although fewer effect sizes were used for each category of plant defenses (here K equaled the number of species in the analysis), our traditional meta-analysis yields the same conclusions reported by Koricheva et al. (2004) (Table 17.2). Studies contrasting individual defense compounds within a group of chemical defenses tended to show positive correlations among these compounds; studies contrasting constitutive (always present) versus induced plant defenses were more likely to find a negative correlation. However, phylogenetically independent meta-analysis revealed that the latter result (contrasting constitutive vs. induced defenses) was significantly influenced by the phylogenetic history of plants (Table 17.2). Here, the AIC scores indicated that the model including phylogenetic correlations (which also found no overall trade-off between these categories of defenses) was more useful in explaining variation among effect sizes. This effect of including phylogenetic correlations in the traditional meta-analysis may have occurred because it contained two species pairs that were very closely related: *Triticum* sp. and *T. aestivum*; and *Brassica napus* and *B. nigra* (Fig. 17.1). The traditional meta-analysis weighted these pairs equally in the overall analyses, even though the data derived from these taxa were not evolutionary independent.

DISCUSSION AND CONCLUSIONS

Integrating phylogenetic information into meta-analysis is challenging. Meta-analysts will have to collate data from published studies *and* gather information to generate a phylogeny for the taxa included in their meta-analysis. Below, we discuss issues that emerge when assembling

TABLE 17.2. Traditional and phylogenetically independent meta-analysis (MA) of studies testing for trade-offs in antiherbivore defenses among 21 plant species (see Figure 17.4). These studies are divided among different categories of plant defenses, and results are based on a random-effects model for pooling effect sizes. Lower (L) and Upper (U) bounds of 95% confidence intervals (CI) are also provided for the traditional ($\bar{\mu}_+$) and phylogenetically independent ($\bar{\rho}_+$) pooled effect size. Pooled effect sizes in bold indicate significant nonzero effects, whereas bolded AIC scores indicate the statistical model that best fit these pooled results. See Table 1 of Koricheva et al. (2004) to compare these results with the original findings.

Category of defenses among plant species	K	Traditional MA				Phylogenetically independent MA			
		$\bar{\mu}_+$	LCI	UCI	AIC	$\bar{\rho}_+$	LCI	UCI	AIC
Chemical defenses	13	−0.012	−0.392	0.368	**46.0**	0.033	−0.129	0.585	54.7
Individual compounds within a group	7	**0.707**	**0.395**	**1.019**	**21.4**	**0.804**	**0.463**	**1.145**	25.0
Groups of compounds within a class	4	−0.119	−0.325	0.086	**12.0**	−0.109	−0.448	0.230	14.1
Classes of compounds	4	−0.589	−1.710	0.532	**11.0**	−0.562	−1.652	0.528	13.4
Mechanical vs. chemical	2	0.158	−0.323	0.640	**6.3**	0.158	−0.323	0.640	8.3
Constitutive vs. induced	11	**−0.499**	**−0.639**	**−0.359**	29.1	−0.445	−1.243	0.353	**16.8**

phylogenetic information. We finally end with a discussion on how to test evolutionary hypotheses using the methods outlined in this chapter.

Phylogenetically independent meta-analysis requires a hypothesis about the evolutionary relationships of taxa. This information on phylogenetic history and divergence events is explicitly used to define the correlations among effect sizes with the intention of penalizing effect sizes from closely related taxa. However, a lack of information can influence the validity of these correlations. In our example, we benefited from an online resource (Phylomatic) that generated a phylogenetic tree using a working hypothesis on the evolutionary history of all plants. Resources like this do not yet exist for other groups of taxa, and reviewers will have to mine published information or make use of the available genetic data to statistically construct phylogenies. However, each approach has its limitations. Cobbling a tree together using published phylogenies will provide a topology that will likely connect most of the taxa reviewed, but branch-length information on relative divergence times will be missing. Using Linnaean rankings would be the simplest way to sketch the topology of a tree. Constructing a tree using genetic information will provide these divergence times, but will be limited to species with genetic information. For example, a brief survey of GenBank revealed that although genetic data were available for 95% of the species found in Figure 17.4, these data were unevenly parsed among four genes useful for phylogenetic construction. A significant technical understanding of phylogenetic construction would be needed to make use of all these data, but even simplifying construction to a single gene would result in a subsampling of effect size data (Lajeunesse 2009). The difficulty of managing molecular data is further exacerbated when meta-analyses review a collection of species spread across broad taxonomic classes (e.g., a mix of plant and animals with different genes used for phylogenetic analyses).

Given these limitations, we suggest using a composite of these approaches to generate a phylogenetic tree that (a) is inclusive in that it contains many of the species under review (even if some species are specified as polytomies), and (b) contains some information about the relative timing of divergence events. Remember that this tree is a hypothesis on the phylogenetic relationships among taxa, and that in comparative analysis, even trees specified with coarse (albeit correct) phylogenetic information can improve the inference of analyses (Freckleton et al. 2002). Of course, the closer the tree is to designating the "true" phylogenetic relationships, the better it will serve as a hypothesis for explaining variation in research outcomes. If there are, however, multiple solutions to designating phylogenetic relationships, then averaging meta-analytical results across these hypotheses is one approach to reconcile these differences.

Finally, the above approach to phylogenetically independent meta-analysis can be used to test different models of evolution. This approach extends the scope of meta-analysis by emphasizing a conceptual (evolutionary) interpretation of analyses rather than treating phylogenetic correlations as a statistical nuisance that needs correction (Westoby et al. 1995b, 1995c). For example, in our phylogenetically independent meta-analysis we assume only the purely neutral BM (Brownian motion) model of evolution. BM evolution assumes that drift is the major force acting to erase the strength of phylogenetic correlations (i.e., the strength of a phylogenetic correlation is linearly related to the time since divergence; Felsenstein 1985, Martins 1994). Figure 17.1 shows how a more recent divergence between two taxa will yield a stronger phylogenetic correlation than one derived from an ancient divergence. However, hypothesizing alternatives to BM can offer a powerful approach to providing more biologically realistic explanations for research outcomes. For example, it is also possible to assume that stabilizing selection is a force that influences the strength and maintenance of phylogenetic correlations; this model is known as an Ornstein-Uhlenbeck process (OU). Here selection is treated as a separate evolutionary parameter (in addition to drift) that acts to erase the strength of phylogenetic correlations (Hansen 1997). Hypothesizing OU models with meta-analyses can be useful when contrasted with results from a BM model, because it can help distinguish between neutral versus adaptive hypotheses in explaining variation among effect sizes (Butler and King 2004). Thus, by contrasting the fit of different evolutionary models, meta-analysis can become a powerful tool for exploring the historical processes responsible for the diversity and distribution of experimental outcomes across taxa (Lajeunesse 2009).

ACKNOWLEDGEMENTS

Work on this chapter by M.J.L. was supported by the University of South Florida and a National Science Foundation grant to the National Evolutionary Synthesis Center (EF-0423641).

APPENDIX 17

TABLE A17.1. The raw effect size data for each species grouped by each category of antiherbivore defenses. Effect sizes are the within-species pooled z-transformed Pearson's correlation coefficients ($\bar{Z}_r$), and their variances $\sigma^2(\bar{Z}_r)$ are shown in parentheses. For further details of these data, see Koricheva et al. (2004). Abbreviations for each category are as follows: chemical defences (CD), individual compounds within a group (CDWG), groups of compounds within a class (CDWC), classes of compounds (CDCC), mechanical vs. chemical (MC), and constitutive vs. induced (CI).

Species *	CD	CDWG	CDWC	CDCC	MC	CI
		Subgroups of CD				
Arabidopsis thaliana	-	-	-	-	0.401 (0.007)	-
Betula pendula	−0.793 (0.143)	-	-	−0.793 (0.143)	-	−0.237 (0.071)

(*continued*)

TABLE A17.1. *Continued*

Species *	CD	Subgroups of CD			MC	CI
		CDWG	CDWC	CDCC		
Brassica napus	-	-	-	-	-	-0.698 (0.2)
Brassica nigra	-	-	-	-	-	-0.56 (0.008)
Cynoglossum officinale	-	-	-	-	-	-0.327 (0.071)
Datura stramonium	0.5 (0.01)	0.5 (0.01)	-	-	-	-
Glycine max	-	-	-	-	-	0.006 (0.2)
Gossypium hirsutum	-	-	-	-	-	-0.148 (0.143)
Lotus corniculatus	-1.673 (0.333)	-	-	-1.673 (0.333)	-	-
Pastinaca sativa	0.617 (0.02)	0.617 (0.02)	-	-	-	-
Picea sitchensis	1.306 (0.012)	1.306 (0.012)	-	-	-	-
Pinus sylvestris	-2.092 (0.083)	-	-	-	-	-
Plantago lanceolata	0.788 (0.012)	1.002 (0.035)	-	0.673 (0.019)	-	-
Populus sp.	-	-	-	-	-	-0.695 (0.111)
Populus tremuloides	0.441 (0.012)	0.971 (0.02)	-0.243 (0.026)	-	-	-
Pseudotsuga menziesii	-0.104 (0.143)	-	-0.104 (0.143)	-	-	-
Raphanus raphanistrum	0.161 (0.04)	-	0.161 (0.04)	-	-0.09 (0.01)	-
Salix myrsinifolia	-0.166 (0.03)	-0.045 (0.083)	-0.235 (0.048)	-	-	-0.789 (0.333)
Salix sericea	0.355 (0.044)	0.355 (0.044)	-	-	-	-
Sorghum bicolor	-0.829 (0.125)	-	-	-0.829 (0.125)	-	-
Triticum aestivum	-	-	-	-	-	-1.875 (1)
Triticum sp.	-	-	-	-	-	0.123 (0.5)
Vitis sp.	-	-	-	-	-	-0.321 (0.25)

* phylogenetic relationships in NEWICK format (as described in Fig. 17.4): ((Picea_sitchensis:154.761902, Pseudotsuga_menziesii:154.761917,Pinus_sylvestris:154.761917):170.238083,((Sorghum_bicolour:12,(Triticum_sp: 0.545457,Triticum_aestivum:0.545457):11.454545):149,((Pastinaca_sativa:112.000003,((Plantago_lanceolata: 87.666664,Datura_stramonium:87.666664):9.666672,Cynoglossum_officinal:97.333336):14.666667):12,(Vitis_sp: 113.333336,((Gossypium_hirsutum:95,(Raphanus_raphanistrum:33.166668,Arabidopsis_thaliana:33.166668, (Brassica_napus:16.583334,Brassica_nigra:16.583334):16.583334):61.833336):10.666664,((Betula_pendula:88.5, (Lotus_corniculatus:56,Glycine_max:56):32.5):9.5,((Salix_myrsinifolia:40.333328,Salix_sericea: 40.333328): 20.333334,(Populus_tremuloides:30.333332,Populus_sp:30.333332):30.333332):37.333336):7.666664):7.666672): 10.666667):37):164):75;

Meta-analysis of Primary Data

Kerrie Mengersen, Jessica Gurevitch,
and Christopher H. Schmid

THE STATISTICAL METHODS THAT receive most focus in this book are those that allow the combination of summary data from each study. This chapter addresses the situation in which the primary data from each study are available for inclusion in the meta-analysis. This type of meta-analysis is appealing for a number of reasons. Most importantly, each study can be analyzed in a consistent manner, thus producing directly comparable effect estimates that have been similarly controlled for potential biases and other study-specific issues. This avoids, for example, the problem of attempting to combine published effect estimates that have been adjusted for different covariates (Chapter 7). Moreover, this type of analysis affords an opportunity to address other questions that typically arise in meta-analysis, such as the role of covariates, potential confounders, and alternative models that can be investigated using the primary data. This may not be possible when the analysis is only based on reported summary data.

In medicine, meta-analysis based on primary data, or individual patient/participant data (IPD) as it is known, is considered the "gold standard" (Simmonds et al. 2005). Even in medicine, however, this approach can be controversial if there is concern about the comparability of studies or the ability of the meta-analyst (who generally was not involved in the individual studies) to undertake consistent analyses that appropriately account for the specific characteristics of each study. Obtaining the IPD (or original ecological data) and associated information relevant to the meta-analysis can also be difficult, costly, and time-consuming (Schmid et al. 2003). Finally, the ability to undertake multiple analyses with these data raises the possibility of selection bias or data dredging (Higgins and Thompson 2002).

In ecology and evolutionary biology, these concerns are magnified; the studies are almost always heterogeneous and need careful study-specific analyses to account for their particular characteristics. However, for the same reason, a meta-analysis based on the primary data might provide more comparable effect estimates across studies; this is because a consistent method of analysis might resolve concerns regarding differences in study-specific designs, analytic methods, covariate adjustment, control of confounders and biases, and so on. As in the medical context, this type of analysis may also provide effect estimates of interest that are not available in the published papers.

As an example of such an analysis, Barrowman et al. (2003) obtained complete time series data from 14 studies in order to investigate a fundamental ecological problem of population dynamics at low population sizes. Their focus was on estimating the maximum reproductive carrying rate capacity and depensation (decreased survival or reproduction due to a decrease in density of adults) among populations of coho salmon on the west coast of North America. Here, the available time series data were typically of short duration and of variable quality, so a meta-analysis model provided a framework for "borrowing strength" among all of the available data sets.

Another example of a meta-analysis of primary data is given by Richards and Bass (2005), who analyzed and then combined sequence data from 13 small subunit ribosomal rDNA surveys in order to investigate patterns of ecological and geographical structuring of eukaryote microbial diversity and associated evolutionary relationships. The authors describe a consistent approach to analyzing the 49 discrete environmental samples contained in the 13 studies in order to construct an alignment of all relevant GenBank nr database sequences.

In a similar manner, Krasnov et al. (2009) analyzed census data from 28 regional surveys of gamasid mites parasitic on small mammals throughout the Palaearctic; the goal was to assess how the abundance of individual mite species is influenced by the abundance and diversity of other mite species on the same host. The data set comprised data on 310,098 individual mites collected from 248,031 individual mammals. The authors calculated a number of indices of abundance and diversity for each mite species on each host species. They used a regression to take account of important covariates, among-host sampling effort, and body size. The residuals of these regressions were then used in subsequent analyses. A combined estimate of the effect size across studies was calculated using both fixed- and random-effects algorithms, and using the number of host species on which a mite was recorded across regions as sample size for each observation. The effect of adjusting for multiple testing using a Bonferroni correction was also evaluated.

Cardinale et al. (2006) describe an analysis and meta-analysis of 111 field, greenhouse, and laboratory experiments that manipulated species diversity in order to examine the effect on functioning of numerous trophic groups in multiple types of ecosystems. Two metrics were used to estimate species diversity. The authors fit a Michaelis-Menton function to data from each study, and used the maximum likelihood estimates of function values to compare key features of the diversity-function curves across systems. In this meta-analysis, it is unclear how or whether between-experiment variation was included.

Despite their appeal and potential for more informed inference, full meta-analyses of primary data are relatively rare in ecology and evolutionary biology; surprisingly, they are even unusual in medicine. After a very careful search of five electronic databases from 1990 to 2001, Simmonds et al. (2005) estimated that less than 5% of the total meta-analysis literature involves primary analyses. The difficulty of obtaining all relevant data for such analyses was also highlighted by these authors.

In this chapter, we discuss statistical approaches to the meta-analysis of primary data in ecology and evolutionary biology, and expand on the particular issues that differentiate this type of analysis from the more common meta-analysis of published summary data. Before doing this, however, we address the important issue of collecting primary data in preparation for a meta-analysis. Note that in much of the following discussion we deliberately use cross-disciplinary references, particularly drawing from the meta-analysis literature in medicine, in order to demonstrate progressive approaches to commonly encountered problems.

COLLECTING PRIMARY DATA FOR A META-ANALYSIS

As in any meta-analysis, there are other important issues associated with the extraction, management, and reconciliation of data for a meta-analysis of primary data. In the ecological arena, there has been little formal discussion of these issues. Stewart and Pullin (2008) illustrate some of the difficulties in their assessment of grazing and conservation of mesotrophic pasture. The issues are also acknowledged in a review of meta-analysis by Stewart (2010).

In the context of medicine, Schmid et al. (2003) discuss the construction of a database of individual studies for clinical trials. The observations and recommendations made in this paper are relevant to ecology as well. The authors requested individual patient data from 11 randomized controlled trials and were rewarded with databases "constructed in several different languages using different software packages with unique file formats and variable names" (Schmid et al. 2003, 324). Conversion to a standardized database of more than 60,000 records took a period of four years and required overcoming "a variety of problems including inconsistent protocols for measurement of key variables; varying definitions of the baseline time; varying follow-up times and intervals; differing medication-reporting protocols; missing variables; incomplete, missing and implausible data values; and concealment of key data in text fields" (324). Despite this, the authors concluded, "analyses based on individual patient data are extremely informative" (325). This example highlights the observation that a closer collaboration in fields of common interest, such as meta-analysis, can be very fruitful, as perhaps the similarities of the endeavors outweigh the discipline-specific differences.

GENERAL HIERARCHICAL MODELING APPROACH

As described in Chapter 8, a meta-analysis of reported effect estimates can be considered as a hierarchical model; in a random-effects framework, for example, each study estimate is considered to be distributed around a study-specific (true) mean. These study-specific estimates form the lowest level of the model. The second level of the model is comprised of the study-specific true means distributed around an overall global mean, which is at the highest level of the model. Depending on the aim and context of the meta-analysis, the true means are estimates of the study-specific and global effects of interest (Fig. 18.1).

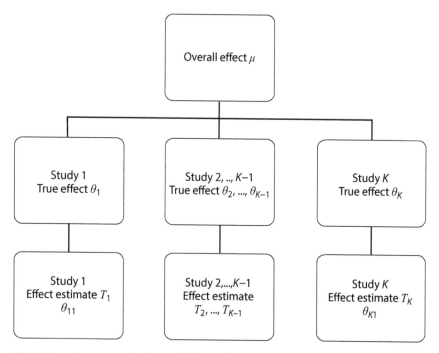

Figure 18.1. General meta-analysis model for combining summary estimates.

A meta-analysis of primary data can be considered as a simple extension of this model, in which the primary data are modeled at the study-specific level, and provide a study-specific effect estimate; these are then combined to provide an overall effect in a hierarchical or multi-level model (Fig. 18.2). The estimates are combined via a fixed- or random-effects model, with covariates included if relevant (see below), to account for within-study sampling variation and between-study heterogeneity.

Simmonds et al. (2005) describe two types of analysis of primary data that were found in their review, discussed above. The first is a "one-stage" or "megatrial" method in which all the primary data from all studies are analyzed as if they belong to a single trial. The second is a "two-stage" approach in which the studies are analyzed separately and the effect estimates of interest are then combined using the meta-analysis techniques for summary data, as described in Chapter 8.

A more complete hierarchical model for the analysis of primary data involves the simulta-neous modeling of data within each study at one level of the model and the combination of the study-specific effect estimates at a second level of the same model. This is obviously an extension of the "megatrial" approach, since it allows for study-specific analyses and a gener-alization of the above "two-stage" approach. Here, analyses are conducted within studies at one level of the model, and the study-level effect estimates are combined within the same model. This type of model encompasses mixed-effects and meta-regression approaches, as described in Chapter 8.

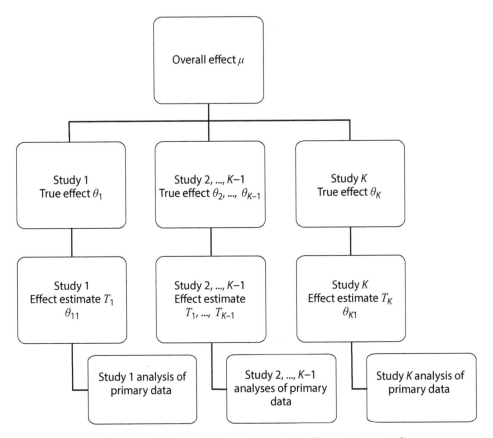

Figure 18.2. General meta-analysis model for combining data from primary studies.

Thompson et al. (2001) provide details of such multilevel meta-analysis models for combining either summary data or individual level data from each study, and for including either study level or individual level covariates from each study. We now describe their approach, first for summary data and then for individual data obtained from studies that comprise two groups (e.g., treatment and control).

As in Chapter 8, consider summary effect estimates T_i, $i = 1, \ldots, I$ from the I studies included in the meta-analysis, with a random-effects model given by

$$T_i = \mu + \eta_i + \varepsilon_i \qquad (18.1)$$

where μ is the overall effect of interest, $\eta_i \sim (0, \tau^2)$ represents the between study heterogeneity, $\varepsilon_i \sim N(0, v_i)$ represents the within study heterogeneity, and the within- and between-study variances are uncorrelated so that $\mathrm{Cov}(\eta_i, \varepsilon_i) = 0$. Following the exposition in Chapter 7 and in Thompson et al. (2001), the model in Equation 18.1 can be written as a multilevel model:

$$
\begin{aligned}
T_i &= \mu_i + \varepsilon_i & \text{at level 1 (within studies)} \\
\mu_i &= \mu + \eta_i & \text{at level 2 (between studies)}
\end{aligned}
\qquad (18.2)
$$

Consider now the case in which the primary data, y_{ij} for the jth individual in the ith study, are available, with corresponding indicators x_{ij} of group membership (e.g., 0 for control, 1 for treatment). Then the above random-effects meta-analysis model (18.1) can be extended as follows:

$$y_{ij} = \phi_i + (\mu + \eta_i)x_{ij} + \varepsilon_{ij} \text{ at level 0 (within studies)} \qquad (18.3)$$

where now ϕ_i are fixed study effects, $\eta_i \sim N(0, \tau^2)$ and $\varepsilon_{ij} \sim N(0, \sigma_i^2)$. Thompson et al. (2001) acknowledge that the ϕ_i can also be modeled as random effects, but caution that while this might be appealing if the number of studies is large and there is little information in each study, it also necessitates a distribution for these parameters which may be unrealistic and can alter the conclusions.

The model may be estimated in two ways using a one-step approach in which Equation 18.3 is fit as a mixed model. Alternatively, it may be fit in a two-step approach in which

$$y_{ij} = \phi_i + \mu_i x_{ij} + \varepsilon_{ij} \text{ at level 0 (within studies)}$$

is fit to each study separately, and the resulting estimates $\hat{\mu}_i$ are fit with a meta-analysis model weighted by the inverse of the residual variance from each separate regression.

For binary data, a suitably transformed effect estimate (e.g., log odds ratio) might be assumed to be normally distributed with known variance, or alternatively the original data can be used directly. Thus, if $y_{ij} = 0$ or 1 for the jth individual in the ith study, then

$$
\begin{aligned}
y_{ij} &\sim \text{Binomial}(1, \pi_{ij}) \\
\text{logit}(\pi_{ij}) &= \phi_i + (\mu + \eta_i)x_{ij}
\end{aligned}
\qquad (18.4)
$$

where the overall effect μ is now a log odds ratio.

COMBINING INDIVIDUAL AND AGGREGATE DATA

When some, but not all, studies provide primary data, it is wasteful to throw away either the primary data or the secondary data to try to fit the data structures into either a primary or secondary data format. Fortunately, Riley et al. (2008, 2010) have developed models that can incorporate both types of data in a single model. These models rely on the observation that the full data in primary studies provide outcomes y_{ij}, whereas the full data in secondary studies provide outcomes $\hat{\mu}_i$. The variances differ as well. By using a dummy variable D_i to indicate whether the data are IPD, $(D_i = 1)$ or not $(D_i = 0)$, one can model both types of data simultaneously:

$$y_{ij}^* = D_i\phi_i + \mu_i x_{ij} + \varepsilon_{ij}^*$$
$$\mu_i = \mu + \eta_i$$
$$\eta_i \sim N(0,\tau^2)$$
$$e_{ij}^* \sim N(0,V_i^*)$$

where $y_{ij}^* = y_{ij}$, $V_i^* = \sigma_i^2$ for IPD data $(D_i = 1)$, and $y_{i1}^* = \hat{\mu}_i$, $V_i^* = V(\hat{\mu}_i)$, $x_{i1} = 1$ for the single response $(j = 1)$ with aggregate data $(D_i = 0)$. The resulting data set is a combination of these two types of data and can now be fit in the usual way.

DEALING WITH COVARIATES

An important aim of many meta-analyses is not only to obtain an overall estimate of an effect of interest, but to explain the heterogeneity observed between published effect estimates. This involves the identification of important covariates that influence the effect, and may allow for a better "model" for estimation or prediction of the effect for particular populations, locations, and so forth.

As described in Chapter 8, covariates can be included in a meta-analysis based on summary statistics through a hierarchical mixed-effects or meta-regression model. This involves the formulation of a linear (or other) model with the study-specific effect estimates as the response, study-specific covariates as explanatory variables in the model, and weights provided by the standard errors of these effects.

Using the notation of Thompson et al. (2001), the random-effects meta-analysis model (18.1) for combining summary data from each study becomes

$$T_i = \beta_0 + \beta z_i + \eta_i + \varepsilon_i \tag{18.5}$$

where β is the regression coefficient indicating the magnitude of the influence of the covariate z_i on the effect estimate T_i. As noted above, this model can be written in various forms; for example, Simmonds and Higgins (2007) express it as

$$\begin{aligned} T_i &= \beta_0 + \beta z_i + \varepsilon_i \\ \varepsilon_i &\sim N(0,\sigma_i^2 + \tau^2) \end{aligned} \tag{18.6}$$

or in multilevel format as

$$T_i = \mu_i + \varepsilon_i \quad \text{at level 1 (within studies)}$$
$$\mu_i = \beta_0 + \beta z_i + \eta_i \quad \text{at level 2 (between studies)}$$

As discussed in more detail in Chapter 8, if τ^2 is equal to zero, this model reduces to a fixed-effects meta-regression.

While a conventional meta-regression is generally the only choice of approach if just summary data are available, the results must be interpreted with some caution. First, if the meta-analysis is observational rather than experimental in nature, covariates may only be asserted to have an association, rather than a causal relationship, with the effect of interest. Second, even if a covariate is apparently influential at the study level, it may not be influential at the level of the units within the study, especially if these units display heterogeneity. This is known as "ecological bias" in medical literature (e.g., Berlin et al. 2002); it can occur when, for example, associations observed at a regional scale may not be the same as those operating at the scale of individuals at a local scale within the region. Finally, it is rarely the case that all studies publish all of the information required to deal with all of the covariates and confounders that are suspected to be important; failure to do this may lead to biased estimates and spurious inferences.

It is sometimes possible to extend the above meta-regression model to address these problems. For example, ecological bias can often be avoided by extending Equation 18.6 to a multilevel model that describes different ecological or geographical scales of measurement. Similarly, Equation 18.6 can be extended to allow for estimation of missing covariates or at least acknowledge the additional bias or uncertainty that their absence may induce in the results. These models are discussed more fully in Chapters 8 and 11.

As discussed above, an alternative and often more complete method of addressing these concerns is through a meta-analysis of the primary data. This allows for consistent, explicit inclusion of covariates within the study-specific analyses as well as between studies. Compared with a conventional meta-regression approach described above, a meta-analysis of primary data is less prone to biases (Stewart and Parmar 1993, Jeng et al. 1995) and has more power to detect the effect of a covariate, particularly when there is little variation in the covariate at the study-level (Berlin et al. 2002, Lambert et al. 2002, Schmid et al. 2004).

Continuing with the notation of Thompson et al. (2001), the meta-analysis model for primary data (18.3) is extended to include covariates as follows:

$$y_{ij} = \phi_i + (\beta_0 + \beta z_i + \eta_i)x_{ij} + \varepsilon_{ij}. \tag{18.7}$$

Simmonds et al. (2005) describe one-stage and two-stage approaches to the inclusion of covariates in the meta-analysis of primary data. In their setup, one-stage methods include covariate adjustment and interaction terms, whereas two-stage methods include separate subgroup analyses or meta-regression within studies (see Chapter 8). The authors found that the former was by far the most common approach in the medical papers that they reviewed, with only one meta-analysis reporting a meta-regression.

In a subsequent paper by Simmonds and Higgins (2007), the issue of accommodating covariates in a meta-analysis is discussed in more detail. They consider a model that is analogous to (18.7) but includes individual level covariates and more explicitly describes the treatment effect and the interaction between the covariate and the treatment. Using their notation, they express their model as

$$y_{ij} = \phi_i + \mu_i z_{ij} + \theta_i x_{ij} + \gamma_i z_{ij} x_{ij} + \varepsilon_{ij}$$
$$\varepsilon_{ij} \sim N(0, \sigma_{yi}^2) \tag{18.8}$$

where x_{ij} again denotes treatment status (coded 1 for treated and 0 for control), ϕ_i is the study effect, θ_i is the treatment effect, and μ_i is the effect of the covariate before treatment. The treatment-covariate interaction, γ_i, is the parameter that indicates the influence of the covariate on the effect of interest. Simmonds and Higgins take ϕ_i to be fixed effects (estimated separately in each study) and the other parameters to be either common effects (e.g., $\theta_i = \theta$ for all studies) or random effects (e.g., $\theta_i \sim N(\theta, \tau^2)$ if the covariates are common across all studies, or $\theta_i \sim N(\theta, \tau_i^2)$ for study-level covariates).

A model considered by Simmonds and Higgins (2007) uses a linear model similar to that described above and then combines the separate estimates using a standard fixed-effects or random-effects meta-analysis. The authors refer to this two-stage approach as a meta-analysis of interaction estimates. Under this model, the overall effect estimate, T, is given by the familiar expression

$$T = \sum \frac{T_i}{\text{var}(T_i) + \tau^2} \bigg/ \sum \frac{1}{\text{var}(T_i) + \tau^2}.$$

Here T_i is the effect estimate from the ith study and the summation is overall all studies. Taking $\tau^2 = 0$ gives a common fixed-effects model. This model makes no assumptions about the relationships among treatment effects across studies. Simmonds and Higgins observe that unlike the previous model, it does not involve simultaneous analysis of the full primary data and can use a study-specific estimate of the interaction and its standard error if full primary data are unavailable.

Based on a number of simplifying assumptions, Simmonds and Higgins (2007) compared the above three approaches (method 1, meta-regression; method 2, full primary meta-analysis; method 3, meta-analysis of interaction estimates) with respect to the power to detect the treatment-covariate interactions of interest. Recall that these are different ideas because they use either individual-level data or study-level data. The authors conclude that the full primary meta-analysis model was always at least as powerful as the other two methods. However, they caution that preference for this model depends on the model itself being correct; choosing an appropriate model for ecology data and meeting the assumption of common treatment, covariate, and interaction effects across trials may be questionable in practice. Formulas are provided for the comparison of a meta-analysis of interactions and a meta-regression. In essence, the choice between these two approaches depends on the nature of the variation of the covariate around the mean within and between studies.

ANALYSIS OF REPEATED MEASURES AND TIME SERIES DATA

Methods for meta-analysis of repeated measures and time series data are discussed in some detail in Chapter 16. These approaches typically require that each study has analyzed the data using the same trend or repeated measures model in a consistent manner, and that all of the necessary estimated variances and covariances are reported. Since this is rarely the case, it is often important in these situations to obtain access to the primary data from the relevant studies. Examples of repeated measures meta-analysis using raw data are available in the ecology literature, including work by Marsh (2001), Barrowman et al. (2003), Harley et al. (2004) and Blenckner et al. (2007).

Marsh (2001) use raw data to investigate factors associated with amphibian population fluctuation. The raw data were from the United States Geological Survey's amphibian time-series database; this contains data from published studies in the United States, Canada, and Europe. Sixty-nine time series were extracted by Marsh; using generalized linear models, variables were identified as being important for prediction of population fluctuation. There are few details on how the meta-analysis was actually carried out, but since each time series is independent, it is assumed that the generalized linear model included a term for study site, thus using the one-stage approach.

Barrowman et al. (2003) describe maximum likelihood and Bayesian estimation of a meta-analysis of 14 time series in order to estimate a fundamental ecological problem of population dynamics at low population sizes. Their focus was on estimating the maximum reproductive carrying rate capacity and depensation (e.g., allele effects) among populations of coho salmon on the west coast of North America. Here, the available time-series data were typically of short and different durations, obtained under different sampling designs and of variable quality. In order to "borrow strength" among these data sets, a hierarchical (mixed-effects) framework was adopted, whereby the ecological parameters of interest among populations are assumed to be drawn from a common population. Thus, the procedure adopted by these authors involves two steps: (i) fit the species dynamics model to the individual data sets and obtain maximum likelihood estimates (and corresponding standard errors) of parameters of interest for each population, and (ii) combine the population-specific parameters in a random-effects meta-analysis to obtain an overall estimate of the parameter of interest.

In more detail, the steps that were taken in this paper were as follows:

(1) All populations were standardized so that they were in the same units.

(2) Attention was focused on particular estimates and portions of the data set that were common across the populations and that allowed meaningful interpretation and comparison with other external data sources.

(3) A small number of candidate nonlinear mixed models for species dynamics were constructed in light of the literature, biological reasoning, and the capacity of the data.

(4) The data were log-transformed so that the residuals under the models were normally distributed.

(5) The models were fit to each data set individually and maximum likelihood estimates of parameters of interest were obtained using a standard statistical software package, SPlus.

(6) The goodness of fit of the candidate models was compared and a "best" model was chosen on the basis of the Akaike information criterion (AIC).

(7) The (log-transformed) population-specific parameter estimates of interest are then assumed to be normally distributed and are combined through a simple random-effects meta-analysis model.

It was acknowledged by the authors that this approach to combining data sets assumes that there is no correlation between the populations included in the meta-analysis.

Harley et al. (2004) employ a similar hierarchical meta-analysis model for combining time series, this time in order to improve abundance estimates for spawning biomass of hoki in Cook Strait, New Zealand. Three approaches to dealing with the time-series nature of the data were considered in which the estimates of interest were assumed (i) to be independent from year to year, (ii) to be fixed (constant) across years, and (iii) to be a random variable from an overall distribution. Again, the "borrowing of strength" or "sharing of information" across years obtained under the third approach was found to provide better estimates in terms of accuracy and stability. This was also the conclusion reportedly reached by Adkison and Su (2001).

Blenckner et al. (2007) conduct a meta-analysis of 23 years worth of data on 18 lakes in Europe that were available from a database. Their interest was in whether the lakes respond coherently to climate forcing even though they are situated in different parts of Europe (northern, western, and central). Access to the raw data allowed the investigators to carry out the same statistical analysis for each lake, permitting assessment of the same target variables in each. The winter North Atlantic Oscillation (NAO) index was correlated with a number of target variables for each of the 18 lakes. Blenckner et al. used a two-stage approach for the meta-analysis of raw data. The repeated measures within each lake were accounted for by calculating correlation coefficients, these were then meta-analyzed across lakes using a fixed-effects model, with estimates of the effect size (here, correlations) weighted by the inverse of the corresponding variances of the estimates.

MODEL CONSIDERATIONS
Other effect estimates

Meta-analysis of primary data necessarily involves careful attention to the model adopted for analysis of the within-study data. Simple summary estimates and regression approaches have been discussed above. Other approaches are also applicable depending on the type of data available within studies.

This is well illustrated by the meta-analyses of Barrowman et al. (2003) who take a multilevel approach in their analysis of population dynamics time-series data within studies. The authors adopt a mixed-effects framework in which a fixed-effects component described withinstudy covariates and a random-effects component described variation between studies. Both maximum likelihood and Bayesian approaches were considered.

Mittelbach et al. (2001) provide an example of a meta-analysis of primary data extracted from 171 studies reporting on species richness patterns along gradients of productivity. The aim of the analysis was to classify patterns from individual studies by the shape of the responses (humped, U shaped, positive or negative monotonic curves). Although this paper focuses predominantly on vote counting (the largely discredited practice of basing inferences on the proportion of studies that reported "statistically significant" estimates; see Chapter 1), a formal meta-analysis is also embedded in the paper. Here, the authors fit a quadratic regression model to the data obtained from each study and then combine the coefficients of the regression (intercept, linear term, quadratic term) in a multivariate analogue of the model described in Equation 18.2 above. Thus, assume that the fitted model for the jth study is:

$$\hat{y}_j = b_{0j} + b_{Lj}X_j + b_{Qj}X_j^2$$

where $\hat{y}_j$ is the vector of expected responses (e.g., species richness measure), and X_j is the vector of values of the independent variable (e.g., productivity measure). This model can be fit to the original or transformed outcome (e.g., a log transformation for count data such as species richness). Alternatively, a generalized linear model (GLM) can be used. Moreover, the model may be represented and estimated using ordinary least squares (OLS) or a variant of this approach (iterative least squares, generalized least squares, etc.). A simple random-effects model that then combines these study-specific regression estimates is as follows:

$$\begin{pmatrix} b_0 \\ b_L \\ b_Q \end{pmatrix}_j \sim N(\mu, \Sigma + C_j)$$

$$\mu = \begin{pmatrix} \beta_0 \\ \beta_L \\ \beta_Q \end{pmatrix}, \quad \Sigma = \begin{pmatrix} \sigma_0^2 & \sigma_{0L} & \sigma_{0Q} \\ \sigma_{0L} & \sigma_L^2 & \sigma_{LQ} \\ \sigma_{0Q} & \sigma_{LQ} & \sigma_Q^2 \end{pmatrix}$$

where μ is the global mean vector of regression estimates, Σ is the between-study variancecovariance matrix, and C_j is the study-specific variance-covariance matrix for the parameter estimates. In the context of Mittelbach et al. (2001), the estimate of the overall nonlinear term, the quadratic regression estimate, β_Q, is of primary interest.

This approach to meta-analysis of regression coefficients is well established in the ecological literature. Jones et al. (1994) describe an early example of its use for the meta-analysis of 42 published experiments on mitochondrial electron transport; they used nonlinear regression to estimate the relationship of interest in each study, and the results of the regression analyses were synthesized by a random-effects model. Van Houwelingen et al. (2002) and Becker and Wu (2007) provide a general description of the bivariate version of this model (i.e., intercept and linear term only); Paul et al. (2005) applied this in an ecological context, with the aim of analyzing the relationship between fusarium head blight and deoxynivalenol content of wheat among 126 field studies. A Bayesian analogue of the model has also been described by Novick et al. (1972) and Riley, Simmonds, et al. (2007). The same model framework can be used to combine other parameters of interest, such as correlations; see Paul et al. (2005) for an example and discussion. Note that the above model explicitly accommodates both within-study variation and between-study heterogeneity. Specific sources of heterogeneity can be included as covariates by directly extending the above model to a meta-regression, or through a multilevel model in which different subsets of the studies are combined within and across the different levels.

Another illustration is given by Whitehead et al. (2001) who describe meta-analysis of ordinal outcomes using individual patient data in a medical context. These authors propose a hierarchical model, with a proportional odds model at the within-study level, and a fixed- or random-effects model to combine the resultant study-specific log-odds ratios. Both frequentist and Bayesian approaches were considered, and details of implementation in software packages are provided.

Generating Primary Data

Sometimes the meta-analyst requires primary data but has only summary statistics available. If these statistics are "sufficient," in that they describe all of the features of the statistical distribution of the data (e.g., the mean and variance of a normal distribution), then it is possible to generate data sets that match these statistics for each study.

A simple example of this approach is if a mean, variance, and sample size are provided for a response that is assumed to be normally distributed. In this case the statistics are sufficient, and one can simulate a set of data of the required sample size from a normal distribution with the estimated mean and variance. The simulated data set can then be included in a meta-analysis of primary data. Of course, the simulated data will not replicate the observed data exactly, but they will retain the statistical characteristics of the observed data which will then be translated into the meta-analysis. For small sample sizes, it is useful to undertake repeated simulations and corresponding meta-analyses, and then average the results.

Note that if the aim of the analysis is to find an overall estimate of the measure described by the summary statistics, then the combination of sufficient statistics is equivalent to combining the original data. Thus, the above approach would only be applicable if some other interesting feature of the data is missing from or not directly reported in the publication. As discussed above, Riley's combined approach to synthesizing primary and summary data improves on this model.

Random effects assumptions

The above random-effects models make the common assumption that the random effects representing between-study variation are normally distributed. Thompson et al. (2001) sound a serious note of caution about this assumption, claiming that there is little empirical evidence to support it. They argue: "In a random-effects meta-analysis, we should be interested in the true effects in all the studies, as represented by their distribution and not just their mean. By assuming normality, we might summarize this distribution by the range $\mu \pm 2\tau$. However, if distributions other than normality are more appropriate, such a range could be quite misleading" (Thompson et al. 2001, 383–384). They observe that while this assumption is required for maximum likelihood estimation, other more flexible distributions can be used in a Bayesian model with estimation via MCMC (Markov chain Monte Carlo), as in WinBUGS; see the discussion below. Suggestions include a Gamma distribution for skewed random effects and a t-distribution if the effects are heavy tailed (e.g., if some studies appear to be "outliers").

ESTIMATION

As discussed in Chapters 9, 10, and 11 respectively, estimation of meta-analysis models based on summary data is typically undertaken using moment-based, maximum likelihood (ML), or Bayesian approaches. The ML approach is usually based on restricted maximum likelihood (REML), which corrects ML for the downward bias in estimation of the between-study variance component (τ^2); however, these approaches often still fit τ^2 as a "plug-in" estimator and ignore the uncertainty associated with its estimation. This results in estimates that are overly precise (Thompson et al. 2001). The problem can be circumvented by using either a parametric bootstrap procedure or a Bayesian approach. The Bayesian framework also accommodates nonstandard (e.g., binary, nonnormal) effect estimates in a straightforward manner; see Thompson et al. (2001) for further discussion.

Thompson et al. (2001) used MLwiN (Goldstein et al. 1998) to obtain REML estimates, and WinBUGS (Spiegelhalter et al. 2003) to obtain Bayesian estimates. The meta-analyses conducted by Barrowman et al. (2003) employed both REML and Bayesian estimation, using SAS PROC MIXED and WinBUGS.

For fixed-effects models, Whitehead et al. (2001) obtained maximum likelihood estimates of the study-level and overall effects using PROC LOGISTIC in SAS, OLOGIT in Stata, and macros by Yang et al. (1996) in MLn (Woodhouse 1996). A generalized estimation equations (GEE) approach was suggested for more general fixed-effects models (e.g., where trials have different classification schemes) and was implemented using PROC LOGISTIC and PROC IML in SAS. Mixed-effects models, which included both fixed and random effects (see Chapter 8) were fitted using specialist Fortran routines; a simpler approach for approximate inference (approximate maximum likelihood and restricted maximum likelihood; see Chapter 10) was suggested using MLn. Their Bayesian model was implemented using WinBUGS.

Concrete examples of analyzing primary data are given in Chapter 16. The first example refers to a shrimp biomass study comprising biomass estimates over different temporal periods at a number of locations. The meta-analysis proceeds by fitting linear trends to the biomass data for each location separately, and then combining the trend estimates across locations. These two steps are combined in a hierarchical mixed model. A maximum likelihood solution is obtained using R, and a Bayesian analysis is performed using WinBUGS. The second example is based on the fish growth meta-analysis described by Helser and Lai (2004). Here, a nonlinear growth model is fit at each of a number of sites, and the site-level parameters are combined at a population level, taking into account the correlation between the parameters in the model.

DISCUSSION

As discussed in this chapter, the meta-analysis of primary data has strong benefits and substantial drawbacks. As meta-analysis grows in popularity and rigor, the interest in primary data meta-analyses will also grow. This will motivate further discussion about appropriate protocols for these types of analyses and additional research into appropriate statistical methods.

The importance of this discussion is highlighted by the numerous published meta-analyses of primary data that do a thorough job of obtaining comparable within-study estimates, but then take a limited approach to combining these estimates across studies by ignoring within- or between-study variation. This is not confined to ecology. For example, Simmonds et al. (2005) found that 28 out of the 44 meta-analyses of primary data published in 1999–2001 adopted only the latter approach, 6 adopted only the former approach, and 8 performed both. Only 7 of the 36 that performed a two-stage approach used random-effects methods, and although 29 reported testing for between-study heterogeneity, none quantified the impact of this. The authors reported that no paper considered detailed model-fit criteria.

Simmonds et al. (2005) argue that in common with systematic reviews of published literature, meta-analyses involving primary data should follow a specific protocol. In ecology, such a protocol could also inform about the type of analyses that should be taken at both the within-study and between-study levels, the use of random effects, the accommodation of covariates, dealing with missing data, and so on. The use of random-effects models in both one-stage and two-stage meta-analysis models of primary data is recommended in general. Exceptions to this may be made if there are biological or statistical reasons for assuming sufficient homogeneity between studies.

In ecological research, systematic assembly and analysis of large databases put together from primary data from many separate studies are likely to become more available and much more important in the future. Funding agencies and journals are increasingly requiring that authors make data available when papers are published, and the large-scale synthesis of ecological responses has become increasingly urgent and of high general interest. Consortia such as the Collaboration for Environmental Evidence (http://www.environmentalevidence.org) will no doubt play an increasingly important role in these endeavors.

ACKNOWLEDGEMENTS

Ransom Myers played an early instrumental role in the construction and content of this chapter. He is fondly remembered as a passionate scientist and an inspiring collaborator.

Meta-analysis of Results
from Multisite Studies

Jessica Gurevitch

RESEARCH SYNTHESIS IN ECOLOGY has typically been based on literature reviews, as is also common in other fields. That is, a search is conducted for relevant data addressing a particular research question, the utility of published and unpublished data is assessed, and the results are synthesized to address questions based on all of the available evidence. However, another not uncommon scenario in ecological research occurs when a group of researchers wishes to address the same or similar questions, using similar methodology, and is subsequently interested in synthesizing their results both to determine overall answers to those questions and to determine patterns of variation in response among studies. Sometimes those researchers have coordinated the design of their experiments from the start, and may have met many times to discuss the implementation of the experiments, as well as the ongoing results; often the experiments may involve long-term field manipulations. In recent years, people involved in such research networks are increasingly interested in using meta-analysis to carry out their research syntheses. Some of the issues discussed in this chapter share similarities with the case of individual researchers who wish to combine the results of distinct experiments that they have conducted themselves or within a single research group, sometimes over the course of many years.

Such efforts have many similarities with literature-based quantitative research synthesis, but differ in some important ways. I begin with several examples of such work, and investigate the challenges, potential pitfalls, advantages, and issues involved in using meta-analysis for the synthesis of such large-group collaborative experimental work.

EXAMPLES

A number of examples of groups with similar experiments seeking to synthesize their results occur in the ecological literature. In some of them, meta-analytic techniques have been used, in others, more conventional primary statistical analysis was carried out. We discuss several of these to illustrate the kinds of problems that have been addressed. A good example is the National Fire and Fire Surrogate Study (FFS), in which ecological researchers at 12 forest sites carried out similar experiments with the goals of measuring ecological responses to common treatments used to reduce fire risk; they evaluated the effectiveness of these treatments, and determining modifying factors influencing those responses and effectiveness (McIver et al. 2008). A meta-analysis was published on the effects of fire and thinning treatments on carbon storage and sequestration (Boerner et al. 2008), in addition to papers with individual analyses on particular effects at specific sites (see *Forest Ecology and Management* Special Feature, May 2008).

A number of other multisite syntheses have also addressed practical and management issues in ecology. Stewart et al. (2008) discussed the effectiveness of control efforts for *Pteridium aquilinum* in 6 manipulative experiments conducted at 4 sites over 10 years. The results of this meta-analysis suggest that management plans based on single-site experiments are likely to fail; results of the control efforts differed among experiments and were related to both treatment differences and site characteristics. They also found small but highly significant effects of *P. aquilinum* control on resulting community species richness that might not have been as apparent in either single study results or in a narrative review.

Multisite studies concerned with plant community and ecosystem responses to climate change have been among the earliest to employ meta-analysis. One such example is an early synthesis of experiments conducted through the International Tundra Experiment (ITEX), where open top chambers are used to warm small areas of tundra, and a variety of responses have been measured over time. A meta-analysis of responses at 13 of the 28 ITEX sites (Arft et al. 1999) synthesized plant phenology, growth, and reproduction responses to warming. A subsequent meta-analysis of 11 ITEX sites (Walker et al. 2006) found additional responses at the community level, and predicted widespread losses in tundra biodiversity as a result of warming. Walker et al. (2006) point out that most previous studies were of short duration and were based on results at single locations, while narrative attempts at comparisons among studies faced various limitations and obstacles that hindered reaching general conclusions. These were largely overcome by using meta-analysis techniques to provide a global overview of experimental responses and predictions for future responses to climate change.

Rustad et al. (2001) carried out a meta-analysis of soil ecosystem and plant productivity responses to experimental soil warming at 32 sites in North America and Europe (with one site in South America). They found that across all sites, soil warming significantly increased mean soil respiration, organic horizon nitrogen (N) mineralization, and plant productivity, while decreasing soil moisture. However, there were also interesting patterns of variation in these parameters across sites. A surprising finding was that there were no significant differences between warming technologies (e.g., in-ground cables, infrared lamps) in the responses despite considerable controversy in the field over which technology was best. It may be that differences in responses to other factors such as site and vegetation were greater and/or more consistent than to different warming techniques, or due to limited power to detect such differences (Chapter 22).

Other multisite meta-analyses have been concerned with long-standing fundamental questions, similar to literature-based meta-analyses in ecology. Vehviläinen et al. (2007) used meta-analysis to synthesize the results of seven long-term experiments testing the effects of tree species diversity on herbivore abundance and damage in forests in England, Sweden, and Finland. They found that overall, tree diversity across these sites did not reduce insect abundance or damage to the trees, but rather, insect herbivory was highly variable and dependent upon the biology of the specific interactions between hosts and insects. These conclusions were far more apparent and robust as the conclusions of a meta-analysis than they would have been from examining individual studies or from a narrative summary comparing the results in different forests.

ISSUES

Meta-analysis of results contributed and put together by a research group that carried out the primary studies is in many ways similar to any other research synthesis; however, it also presents some unique issues, problems, challenges, and opportunities. How does such an effort differ from a literature-based research synthesis? What are the advantages and disadvantages of using meta-analytic methods over primary analytic methods on such data? In Chapter 1 we introduced the distinction between meta-analysis and systematic reviews. The synthesis of primary results of a research group is clearly not a systematic review, and may or may not use meta-analysis. Because such a synthesis only includes a narrow and predefined set of studies, it is subject to a number of limitations.

Potential biases

One major issue in interpretation of the results of such syntheses is that of potential bias in the data set. In a literature-based meta-analysis, there are many ways to limit or estimate the magnitude of various sorts of bias (e.g., Chapters 14, 15, and 17). On the one hand, there is no publication bias per se in combining results of a group of researchers because all of the results are at least theoretically available, and may consist of both published and unpublished data. On the other hand, the set of studies will be inherently circumscribed, and is almost always smaller than that in a literature-based synthesis in ecology and evolution.

Some systems and organisms will be overrepresented and others underrepresented ("research bias" sensu Gurevitch and Hedges 1999; see also Chapter 17). The data may be more accurately represented by a fixed-effects analysis, in which the results are not generalized beyond the current group of studies, rather than by a random or mixed model (Chapter 8). Although by definition publication bias is not going to be an issue, sometimes some of the sites are omitted from the synthesis. For example, in the ITEX synthesis (Arft et al. 1999) there were actually 28 sites in which the experiments had been carried out, but results from only 13 of these were available for inclusion in the meta-analysis. Various types of problems can interfere with data from a site being used in the synthesis. Most commonly, the researchers at that site are not able to supply the data by the group's deadline. However, if there are systematic reasons that data from some sites are not used, this is a potential source of bias, and the synthesists should carefully examine those reasons to determine if and how this could influence the conclusions.

Another difference between this type of meta-analysis and a literature-based synthesis is that the questions tend to be chosen earlier, at the time the individual experiments were conceived and set up, and thus are more predefined and less open to development at the synthesis stage. This is inherently a systematic approach in the sense of more strictly predefining the terms of the study. This also may make the synthesis easier, because there is likely to be general agreement on what the goals of the individual studies are, and on the questions and goals of their synthesis. Because the number of studies is likely to be quite limited in most cases, this will constrain developing more novel questions at the synthesis stage due, at the very least, to a limited data set.

Technical and statistical issues

One of the most compelling reasons to use meta-analysis rather than ANOVA, unweighted regression, or similar familiar statistical procedures, is that data sets compiled from the literature are likely to have vastly different sample sizes and thus vastly different sampling errors (Hedges and Olkin 1985). Consequently, using unweighted ANOVA or regression is likely to seriously violate the homogeneity of variances (HOV) assumption, resulting in unreliable tests (Gurevitch and Hedges 1999). In the multisite studies cited above and in other group efforts of this sort, sample sizes (i.e., numbers of plots or other individual experimental units in each study) may be much more similar among studies, because the studies may have been designed in coordination, with similar goals, motivations, and hypotheses. In that case, combining the results using unweighted tests is much less likely to result in a serious violation of HOV assumptions, and so use of meta-analytic methods may not be as crucial. Nevertheless, there are several advantages to adopting these methods.

One advantage of meta-analysis is that the structure of the analysis incorporates the separation of the two sources of variance involved in combining studies: the within-study and between-study variances. An examination and analysis of the relative magnitude of these two sources of variance can offer important insights into the results, and lead to subtle or not so subtle differences in interpretation of the larger picture afforded by the synthesis (Hedges and Olkin 1985, Gurevitch and Hedges 1999). Meta-analysis also provides the framework and context for a clearer overview of the results across studies; this is its goal and what it is designed to do. If the results of the different studies are on somewhat or largely different scales, using standardized effect sizes and meta-analysis will avoid an overemphasis of the results that are expressed on a larger scale and will present a more interpretable conclusion. For example, imagine that one is examining the effects of experimental changes in growing season length on productivity across different biomes. The responses might be much larger proportionally in an arctic than temperate system, but much larger in absolute magnitude in the temperate system. While various data transformations could be used to adjust for this, the interpretation of the results may be more straightforward using an effect size measure that standardizes the results, and a meta-analysis to examine the homogeneity or sources of variation (i.e., moderators or covariates) affecting those results across studies.

In addition, carrying out a meta-analysis may reduce the requirement for obtaining full raw data sets from each participant, asking rather only for the summary data (as effect sizes). Such data may be much easier to obtain from participants (see also the next section), allowing for a more complete data synthesis across a greater number of study sites. One potential disadvantage of using meta-analysis rather than ANOVA is that more complex designs may be easier to analyze using ANOVA, although techniques and software for more complex modeling of meta-analytic data are now becoming more available (e.g., meta-regression, Chapter 16). However, when different experimental designs are used in different studies, this may present challenges for research synthesis using any approach.

Practical problems and challenges

Some of the issues involved in carrying out a meta-analysis are practical ones. It is necessary to obtain all of the needed data and meta-data from all of the participants in a timely manner. It is critical that a clear set of protocols for data reporting be established well in advance, with the group agreeing and all members "buying in" to the format, required information, and the deadline. Data quality and reporting may be highly variable. As the participants see the synthesis effort in process and understand and contribute to its goals, they are likely to be reassured about the process and more enthusiastic about contributing.

Another practical problem is the protection of publication rights of authors of the individual studies. Authors may be reluctant to share data, particularly if not all of it is already published, until and unless they see that the synthesis will not impinge upon their rights to their own data. The mix of synthesis across studies and papers based on individual sites in the special feature on the FFS study in *Forest Ecology and Management* (May 2008) is an excellent example of how both the individual author's rights can be protected and a group synthesis effort can be successfully carried out. It may be more likely for people to be willing to contribute effect size measures in a prearranged format than to share their raw data. Meta-analyses often have excellent citation rates (e.g., Patsopoulos et al. 2005) and may increase the attention to and citations of the publications on individual studies. For example, the early meta-analysis on the ITEX results discussed above (Arft et al. 1999) and the Rustad et al. (2001) paper have both been cited hundreds of times. Finally, the ability to see one's own work, which is typically focused on a single site, in a larger context, and also seeing how the results fit in on a larger scale can be so interesting and satisfying that the experience may be rewarding enough intellectually to encourage future efforts to continue this sort of synthesis.

A related issue with such meta-analyses is whether contribution of data is sufficient for authorship; also, there is the problem of how to distinguish such contributions from those of the (almost always) smaller group or individual actually doing most of the work of the synthesis and writing the paper. In ordinary meta-analysis based on a literature search, authors of individual primary studies included in the analysis are not expected to be included as coauthors in the synthesis paper (although how to cite these papers can be a perplexing problem). But in a project where many participants contribute data, all are likely to expect to be coauthors. Rustad et al. (2001) handled this by listing individually the collaborators who participated in the data synthesis and paper writing, while the large group that contributed data was identified by the group name in the citation, and then the members of the group were listed in a footnote. Authorship issues have to be discussed early on in such collaborative projects to avoid conflicts. It is more problematic to assign credit appropriately after the paper is published; for example, if there are 14 authors, three of whom did the analysis and wrote the paper, how are those three to receive credit for their greater contributions in citations and on CVs (*Curriculum vitae*)?

Numerous other specific practical issues face those attempting to synthesize the work of a group of researchers working on similar problems on different sites. Almost always, the various studies are started at different times, last for different durations, and make different measurements. The solution to such problems requires both statistical and ecological expertise. If the pattern of responses over time is important, this could be taken into account by using a repeated measures approach (Chapter 16). If the actual year of the study is the more important variable—for example, when weather or climate plays a large role and is synchronized across studies—a different approach will be necessary. For example, if a certain year was very dry and prone to fires, it will be more important to combine studies by actual date, and to adjust in some other way for differences in study lengths. In many cases the studies will have different durations, and this is of course problematic. One option is to compare them based on how many years they have lasted (e.g., synthesizing results for all of the studies that have lasted at least three years up to the third year, then synthesizing results for all studies lasting up to five years, and so on). Another common approach is comparing only the studies' endpoints.

There are other technical statistical issues as well. For example, if some of the measurements scale arithmetically and some multiplicatively, even if they are measuring similar underlying phenomenon, great care must be taken in the choice of effect size and how the results are combined. In all cases, there is no substitute for thoughtful and well justified decisions in carrying out the quantitative synthesis, as we have emphasized throughout this handbook.

If there are different measurements made in the different studies, as is common in most meta-analyses in ecology, the scientific judgment of the meta-analyst is essential as to whether it is scientifically defensible to combine them. If the measurements are different ways of getting at the same underlying ecological phenomenon, meta-analysis techniques are an excellent way to put them on the same scale and combine them to ask general questions. Of course, if more than one measurement is made within studies, the issue of nonindependence becomes important (Chapter 16). Clearly, including several different measurements from individual studies on the same experimental units (e.g., soil respiration and rate of litter decomposition measured on the same plots) is likely to introduce serious problems of nonindependence (see Chapter 16 for suggestions on how to handle these issues). The problem is made worse when some sites have only one measurement and others have many, and all are thrown into the hopper; clearly the studies with many measurements would then be overweighted in the final synthesis results.

Another issue with multiple measurements occurs when each is fundamentally assessing a different phenomenon. In that case, it is probably meaningless scientifically to combine them into a grand mean, and interpretation of the results of such a synthesis is likely to be problematic at best, and nonsense at worst. On the other hand, there is room for disagreement about what makes sense to combine—when a synthesis is combining "apples and oranges" and when it is about fruit (Chapter 17).

As an example, one might wish to synthesize insect responses to plant characteristics where those responses in different studies are expressed in terms of various insect performance parameters (growth rate, consumption, survival, and weight). The meta-analyst might wish to include all of these studies in the same analysis to see if, for example, responses in terms of survival are different from those expressed as growth rate. Obviously the grand mean will not be a useful or meaningful parameter, but one might calculate means for each response parameter and compare them by looking at between-group heterogeneity. One problem with this approach is a technical one; particularly if you are using mixed models, the variance estimates will be determined by the variation between the effect sizes calculated from all of the measurements (growth, survival, etc.), and therefore may be messy and uninterpretable. One alternative approach is to analyze the studies on different measures in separate meta-analyses. Their means and confidence intervals can then be compared. Better yet is to use the more advanced modeling approaches discussed elsewhere in this book (e.g., Chapter 16).

A related and common problem in this type of research synthesis occurs when many more variables have been measured than there are independent replicates. Another problem occurs when there are many hypotheses of interest to test on the same modest data set. In the former case, multivariate meta-analysis (Chapter 16) may be a good alternative. The problem of testing many different hypotheses on the same data is not unique to these cases, but is likely to be particularly tempting because the different participants will want to emphasize and test different questions in the synthesis.

Analyses of multisite experiments in ecology often have relatively small numbers of studies compared to literature-based meta-analyses, and it may be difficult to decide on the level and identity of units to synthesize. For example, there may be different independent experiments at different sites, with each experiment involving different specific treatments and with different types of plots within treatments. In addition, the data from each site may be measured in different ways and sometimes on different entities (e.g., on different functional types, populations, or individual organisms). The data synthesists must decide at what level(s) to pool data before the actual meta-analysis is conducted, and then document and justify those decisions (Vehviläinen et al. 2007). Failing to clarify these issues or avoiding the use of meta-analysis methodology because of inherent challenges does not solve these problems.

If the studies have used different manipulations to ask some of the same questions (e.g., whether the responses to soil warming depend upon the method used for experimental warming; Rustad et al. 2001), one can use meta-regression or homogeneity tests to evaluate differences among studies that used different methods. One can then combine the results if warranted, or analyze them separately if not warranted (taking into account the limited power to detect differences). If the studies used very different methodology, measured different variables, or differed in other important ways, one needs to ask if there is a level of generality at which it makes sense to ask questions. If so, it may be reasonable and defensible to carry out data synthesis at that level. On the other hand, when studies are very similar, this can reduce the ability to generalize to other kinds of studies or other systems. Reduced generality may be perfectly satisfactory (one only wanted to ask about forests, and had no desire to evaluate responses in grasslands, deserts, or tundra) or it may be limiting. For example, if all studies used the same technology, but use of another technology results in systematically different answers, the conclusions of the individual studies as well as the synthesis can be misleading because they only apply when the first technology is used. Uniformity can also be limiting when all studies are of short duration and long-term data are lacking, and if there are large, nonlinear, and/or qualitative differences when studies are continued long term. However, it is often the case in multisite, collaborative experiments that the researchers at the different sites have some common understanding of methodology and approaches. One advantage, then, is that "mixing apples and oranges" and various other sources of extraneous heterogeneity are less likely to be problematic than in literature-based meta-analyses.

The greater the number of studies, the greater the possibility that one can generalize not only statistically but at a higher spatial scale ecologically (whether one asks questions about forests in the Adirondack and Catskill Mountains in New York State, or forests in eastern North America, or North American forests, or American, European, and Asian deciduous forests, for example). One can also ask more about modifying factors (duration of the experiment, latitude, diversity) when the number of studies combined is larger. While there are advantages to using homogeneous experiments (as in medicine; e.g., Chapter 25), a limited number of studies ("points") will limit the ability to interpolate geographically or conceptually between them if you wish to infer what is going on over the entire geographic or conceptual surface covered by the sites.

In many instances of this type of synthesis, whether at one point or yearly, the group will meet to discuss results, make plans, and sometimes to synthesize data and prepare results for reports or publication. It is important to introduce some of the goals and ideas the group leaders have for the meta-analysis before the meeting, and to also devise a data entry form well in advance; it can then be fine tuned with participants, and the leaders can then make sure it is filled out and returned by all participants well before the meeting. During the meeting, the complete data set can be distributed to all participants and they can be encouraged to conduct both exploratory data analysis, and to handle parts of the data synthesis individually or in groups. Group participants may be enthusiastic about the experience and the results they find, and may wish to take on part of the wrap-up of the synthesis after the meeting as well.

Expansion of the meta-analysis

While the primary goal of the meta-analysis might be to determine the current state of knowledge about the particular experiments in this group, data synthesists may become curious about placing their results in a wider context. Should new sites and new experiments be included? It will almost always greatly strengthen the robustness of the meta-analysis if the analysis is extended to include results from a literature search. Collaborative reviews in medicine are common, and occur when researchers conducting independent trials decide to pool either their raw data (Chapter 20) or their results, as in multisite ecology meta-analyses. In some cases, researchers choose to use central collection of data on all individual patient measurements. For example, the Early Breast Cancer Trialists' Collaborative Group (EBCTCG) overview combines data on trials of the treatment of breast cancer (EBCTCG 1992; Clarke and Ghersi 1997). This project is updated approximately every five years, and the results are also included in the Cochrane Collaboration (http://breastcancer.cochrane.org/oxford-overview-process). Building in a mechanism to systematically update the results of meta-analyses to include the results of additional experiments is a powerful extension of meta-analysis protocol in the medical litera-ture that would be potentially useful to emulate in ecology and conservation biology. Various efforts in the social sciences also exist. For example, the Collaborative for Academic, Social, and Emotional Learning (CASEL, http://casel.org) collects data and conducts meta-analyses of research on interventions that promote children's social and emotional development, using scientifically and statistically rigorous methodology.

Interpretation of results and limits to generality

The outcome of a meta-analysis of a group of coordinated experiments is certain to provide insights into, and an overview of, the outcome of those experiments that adds considerable value to the original work. The whole will almost certainly be greater than the sum of the parts, and the parts—the results at individual sites—can certainly be published separately in any case. Because such experiments will often be unique to that group of researchers, the information provided by the research synthesis is likely to be quite unique as well, providing data not avail-able elsewhere. It is very likely to be easier to obtain data from the different sites and to validate them than it would be to contact individual researchers and ask for their data, or to extract the data from the literature.

On the other hand, authors have to be cautious about generalizing from the results of their synthesis. The sites are likely to be a haphazard collection of places chosen for all sorts of prac-tical reasons, and are unlikely to provide complete coverage of or even a good representation of ecological systems or organisms. The "sampling" provided by the sites is also likely to be limited and statistically unbalanced (that is, there is probably more of one type of system than another). For example, the Rustad et al. (2001) research synthesis included many sites in North America and Europe, but only a single site in South America. So, the results can tell us some-thing about the responses to experimental treatments at this collection of sites, but perhaps might provide a misleading picture of nature in general. Of course, these issues are not unique to efforts to synthesize coordinated groups of researchers, but apply to a greater or lesser extent to literature-based meta-analyses as well.

OPPORTUNITIES

One of the most valuable contributions of meta-analysis is that it can also change primary re-search, as participants in the meta-analysis realize the value of data synthesis and seek to expand the known data. From formal, tightly linked groups of researchers to more loosely allied indi-viduals working on areas of common interest, the incorporation of a research synthesis perspec-tive and goals can open many new opportunities for future individual research and collaboration. This has happened with the Cochrane and to some extent with the Campbell research collabora-tions on a large scale in medicine and the social sciences, respectively, and there are attempts to introduce such efforts in conservation biology (Chapter 25). Because so much interest in ecol-ogy and evolutionary biology is now at landscape, regional, continental, and global scales, it is an almost inevitable step that interest in quantitative research synthesis will continue to grow exponentially in these fields, and that new methods and new approaches will be devised to ad-dress the fundamental and applied questions that are so pressing in these areas.

Presentation and Interpretation of Results

Quality Standards for Research Syntheses

Hannah R. Rothstein, Christopher J. Lortie,
Gavin B. Stewart, Julia Koricheva,
and Jessica Gurevitch

WHAT MAKES A QUANTITATIVE research synthesis good or flawed? How can authors improve the quality of their review at various stages in the process of planning and carrying out a research synthesis? What criteria can editors and reviewers use to assess whether a quantitative synthesis should be accepted for publication, revised, or rejected? How can readers of published syntheses determine how to evaluate the quality of what they are reading, and in doing so decide whether or not to trust its results and their interpretation? In this chapter, we present guidelines to address these questions. We outline these guidelines in the order of the stages involved in conducting a synthesis. It is important both for those conducting research syntheses and also for reviewers, journal editors, and readers of research syntheses to be aware of these aspects of carrying out a meta-analysis, along with the potential pitfalls and problems involved.

Our approach is based on the view, as expressed throughout this volume, that the systematic review and synthesis of a research literature is itself a scientific research activity. As such, it should follow the scientific method, and thus can be evaluated using relatively straightforward criteria.

Protocols for research syntheses have become fairly standardized in medicine and some of the social sciences, with guidelines published in both the *Cochrane Handbook for Systematic Reviews of Interventions* (Higgins and Green 2011), and elsewhere (e.g., Cooper 2009). When adapting these to research synthesis in ecology and evolutionary biology, the steps involved are as follows:

(1) Defining the review question
  (a) Defining the appropriate scope of study
  (b) Developing inclusion criteria
(2) Collecting the data
  (a) Developing search criteria
  (b) Searching for studies
(3) Evaluating the data
  (a) Evaluating the retrieved studies for inclusion in the synthesis, and extracting data
  (b) Checking for nonindependence, missing data, and other structural problems in the data
(4) Conducting the meta-analysis (when appropriate)
(5) Testing the robustness of results (sensitivity and publication bias analyses)

(6) Presenting the results

(7) Interpreting the findings and drawing conclusions

The remainder of this chapter will be based on a heavily adapted checklist of questions initially presented by Cooper (2009) for use in evaluating the quality of a research synthesis (Box 20.1). In addition to reviewing the questions that should be asked at each stage of the synthesis, we will describe ways in which poor choices at each stage can compromise the integrity of the review; we will also provide examples of good and poor practice. Table 20.1 lists common errors, and how they relate to the quality of a meta-analytic research synthesis.

BOX 20.1.
A checklist of questions for assessing the quality of a research synthesis (adapted from Cooper 2009).

DEFINING THE QUESTION

(1) Is it clear why the central problem is important?

(2) Are there clear conceptual definitions for the key variables, and are the variables available from the primary studies appropriate given the conceptual definitions above?

(3) Are there a sufficient number of studies available for addressing the primary research questions, but not so many that it will be impossible to synthesize their results with a reasonable effort in a reasonable amount of time?

(4) Is it clear which research designs are appropriate for answering the questions?

(5) Have decision rules been developed for inclusion/exclusion of studies?

COLLECTING THE DATA

(6) Are multiple search strategies and online databases used to find relevant studies?

(7) Are alternative terms for each key variable used in searches?

(8) Is the set of retrieved studies assessed for relevance by consistent application of the a priori decision rules for inclusion/exclusion?

(9) Was extraction of information from included studies done rigorously and reliably?

(10) If different types of studies are included, are they identified by appropriate coding information?

(11) Are the methods used to extract data from studies justifiable, clearly documented, and repeatable?

(12) Have appropriate methods been used for dealing with missing data (e.g., variances, sample sizes)?

SYNTHESIZING THE DATA IF A META-ANALYSIS IS PERFORMED

(13) Is an appropriate effect size metric used? Is this decision clearly explained and justified by the problem and/or data structure? If a nonstandard effect size metric is used, are its statistical properties and distributional assumptions understood, clearly explained, and justified?

(14) Are average effect sizes and confidence intervals reported? Is an appropriate model used for the analyses?

(Box 20.1. continued)

(15) Were the studies weighted by the inverse of the variance? If no weighting or another weighting scheme was used, was it explained and justified acceptably?

(16) Is the degree of heterogeneity of effect sizes examined?

(17) Are critical features of studies tested as potential moderators of study outcomes? If more than one moderator is tested, has that been taken into account appropriately in the analyses? Has the data structure, including possible confounding of moderators, and nonindependence of moderators or outcomes been taken into account?

TESTING THE ROBUSTNESS OF RESULTS

(18) Are sensitivity analyses conducted and used in interpreting the results?

(19) Is the possible impact of missing data or overrepresented data in the evidence base considered?

PRESENTING THE FINDINGS

(20) Are clear, high quality graphs and tables used to clarify the results?

(21) Were the procedures, results and conclusions of the research synthesis clearly and completely documented, and justified?

INTERPRETING THE RESULTS

(22) Are the generalizability and limitations of the findings discussed?

(23) Is a distinction made between study-generated and review-generated evidence?

(24) Are the findings and their interpretation considered in light of their biological, and/ or practical significance?

(25) Are major areas identified where more primary studies are needed?

DEFINING THE REVIEW QUESTION: DETERMINING THE APPROPRIATE SCOPE OF STUDY AND DEVELOPING INCLUSION CRITERIA

Research reviewers must clearly and unambiguously define the problem they are addressing, identify the scope of the review, clarify the questions being asked to address the problem, and define both the moderators and outcome variables involved. This is not all that different from what is done in a primary study, and should be quite familiar to researchers in this field. Questions can be posed at different scales and levels of specificity, and the choices made here will affect the scope of the review, the set of studies to be defined as relevant, and the interpretation of its results (Chapter 3). For example, a research reviewer who is interested in the problem of the effects of wind farms on bird species can pose research questions broadly, as in: "What are the effects of wind farms on bird species?" The researcher could also pose the question more narrowly, as in: "What are the effects of offshore wind farms on the local abundance of eider ducks?" More narrowly still, one could ask: "What are the effects of high power (>3 megawatts) offshore installations on mortality of eider ducks?" The question might be posed globally, regionally, or locally. A clear conceptualization of the question and variables of interest will help define the set of studies relevant to the review.

TABLE 20.1. Common errors and their relation to the assessment of meta-analysis quality.

Common error	Questions (Box 20.1)	Action/solution
Biases		
Study Selection bias	5–11	Broad search including multiple terms tested, sensitivity analyses and subgroups demarcated.
Publication bias	6, 8	Broad search not limited to published literature, and consistent application of a priori inclusion criteria. Explore impact.
Epistemological		
Lack of repeatability	2, 5	Clear definitions of constructs and methods.
Ignoring potential moderators of effect size; ignoring confounding among moderators	10–13, 16, 17	Define correct sampling universe and combine data appropriately. Examine heterogeneity and assess intercorrelation of potential moderators.
Causality undefined/spurious relations (results poorly interpreted and qualified)	1, 2, 4, 5, 10, 15–18, 23, 24	Examine any errors that can lead to spurious relations, including methodology and heterogeneity.
Multiple tests of hypothesis without controlling family-wide error rate	17	Control family-wide error rate or plan a priori comparisons.
Reporting		
Underreporting results.	13–15, 20, 22, 24	Completely reported meta-analytic results and justify interpretations. Report question definition, evidence collection, critical appraisal, and data synthesis.

The research reviewer should also provide a clear description of the relevant variables. In many cases, the primary studies may reflect a narrower (or in any case, different) version of the question than the one posed by the reviewer. Ecological meta-analyses are likely to be based on data structures with various types of characteristics and limitations. For example, there are likely to be many studies in some areas or kinds of locations (Europe, North America, heavily populated regions) and few in others. Similarly there are often many studies on some species, and scattered, smaller numbers of studies on other species (see also discussion below). In cases such as these, the research reviewer should either narrow the question, or clearly note the limitations of the included studies, and be very careful about extrapolations or generalizations for which the available data are not a representative sample. Otherwise, the review is in danger of being interpreted as being more widely applicable than is justified by the actual evidence.

In other cases, the reviewers may find that the research available is broader than they originally believed, and they may decide to expand the scope of the problem addressed. Keeping the question too narrow and excluding studies risks overlooking valuable evidence. Ecological meta-analyses tend to be broad and more general rather than very narrow and specific, although either approach can be result in a legitimate and important review (Chapter 25).

The type of problem addressed has implications for the research designs that can be used to test the hypothesized relationships, and thus for the types of study designs that are appropriate to include in the review. Specifically, one needs to consider (a) whether the research question concerns the description of a phenomenon (dead birds are found near wind farms), an association between phenomena (more dead birds are found around wind farms than in adjacent areas), or a causal explanation of a phenomenon (birds were observed flying into wind turbines); and (b) whether the research question focuses on the development of a single outcome over time (birds declined in abundance following the construction of a wind farm), or on comparing differences (bird abundance declined more in offshore wind farms than inland wind farms). When the question posed is a causal question, studies whose designs allow causal inferences to be made should be the first priority for synthesis (assuming they exist).

The criteria used to include or exclude a study from a systematic review will follow logically and directly from the research question. The more clearly the question is framed, the easier it will be to both decide what is relevant and to generate criteria for selecting studies to include. These criteria should be explicitly defined before the literature search begins, although it is sometimes necessary to modify them during the search if studies with unanticipated features are encountered. While other research reviewers or readers might have chosen different criteria for inclusion and exclusion, it should always be possible to replicate both the search, and the inclusion and exclusion of studies, based on the information provided in the review.

A good systematic review will provide a scientific, historical, and/or practical context for the review, note if there are conflicting theories or hypotheses, and identify potential covariates. Two syntheses may review the same set of studies, but could conceivably come to different conclusions if one review evaluated moderators that the other review did not consider.

COLLECTING THE DATA: DEVELOPING SEARCH CRITERIA AND SEARCHING FOR STUDIES

This stage of the systematic review process involves selection of the search terms and information sources that will be used to locate studies for the review. These must be completely and clearly stated in the paper being prepared. The objective of the search for relevant studies is to be as thorough as possible, given the resources available. This means attempting to maximize the coverage of the search, while minimizing biases, so that as many as possible of the studies potentially relevant to the question of interest, both published and unpublished, are retrieved. Identification and retrieval of studies exclusively by electronic searching of bibliographic and full-text databases is rarely, if ever, sufficient to locate all relevant research; in no case should a reviewer rely on a single scientific database to locate studies. In addition to electronic searching of databases, a thorough search for literature will generally include many or all of the following: handsearching of key journals, checking reference lists and citation searching, contacting researchers, examination of conference proceedings, using the World Wide Web, and (for some questions) checking research registers.

One of the most common mistakes made by researchers starting a literature search is underestimating the time and effort it takes to conduct a thorough search and to retrieve and screen potentially relevant studies. Whether or not a research synthesis includes a quantitative component (meta-analysis), its integrity and the validity of its conclusions are highly dependent upon the quality of information retrieval. If the literature located by the reviewer is biased (i.e., unrepresentative of all the studies conducted on the topic), the validity of the review is likely to be compromised (Rothstein et al. 2005, Hopewell et al. 2007; Chapter 14). No ecological or evolutionary meta-analysis can be free of research bias, the tendency to overrepresent some organisms and systems and omit or underrepresent others (Gurevitch and Hedges 1999). Thus, the conclusions of the meta-analysis are often going to be constrained as a result of these limitations in scope. If no studies on large mammals are included in a particular meta-analysis, it may be inappropriate to extend the conclusions to large mammals. If most studies are done in terrestrial systems or in temperate biomes, this will limit the applicability of the results to wetlands, marine systems, deserts, or tropical environments.

Clearly, a thorough and unbiased review requires sufficient human and financial resources to carry it out. Potential causes of study selection bias include abbreviated searches, inappropriate search criteria (e.g., only searching certain journals, as was universally done in ecological reviews prior to the availability of online scientific databases), ignoring disciplines outside one's own when the question at hand is multidisciplinary, and searching only English-language literature. The elements of an inclusive literature search and examples of inclusive searches are provided in Chapter 4. A high-quality review should present clearly articulated and defensible search strategies. Full, transparent, and precise documentation of each step undertaken during the search, including data sources searched and terms used, is needed to allow readers to see what is (and is not) covered by the review. This allows replication or updating of the review, and, not least, helps research reviewers keep track of what they have done.

EVALUATING THE PRIMARY STUDIES FOR INCLUSION IN THE SYNTHESIS, AND EXTRACTING DATA

The critical appraisal of each study to be included allows the assessment of the fit between the study and the questions being addressed in the research synthesis. Given the wide scope of the discipline, including the diversity of research questions, approaches, systems, and data reporting in ecology, it is unlikely that a well-conducted ecological meta-analysis will adopt a rigid hierarchy of study quality, This is unlike medicine, in which randomized, double-blind manipulative (controlled) experiments are (almost) always superior. In any case, reliance on a formal methodological quality scale is not advised (Jüni et al. 1999), as the various quality scales that have been developed correlate very poorly with each other; they were generally developed for applications in health care, focus nearly exclusively on internal validity considerations, and allow research reviewer biases to creep in (see Chapter 5 for a discussion of quality scales).

Systematic examination of methodological features of each included study and consideration of how these features relate to the purposes of the systematic review is a better approach because (1) it focuses attention on aspects of experiments (and other study types) likely to influence the study outcome (e.g., minimal or no replication, various methodological system-specific flaws or strengths); (2) it recognizes the sources of bias that are salient to ecology and evolutionary biology; (3) it enables later testing of the degree to which specific methodological features are related to study outcomes; and (4) it is more transparent and replicable.

Sources of possible bias might, for example, include the nature of the "control" units, including degree of baseline similarity of treatment and comparison (control) units, the reliability and validity of the outcome measure (e.g., how outcomes were measured), the sampling methodology, the duration of the study, and whether adjustments for covariates were made. The comparability and rigor with which the treatment was implemented will also be of concern in many ecological research syntheses. Decisions on exclusion of studies based on design or implementation flaws are known to vary substantially from researcher to researcher. If a meta-analysis does exclude certain studies a priori, it is important that the exclusion rules will have been developed before individual studies are read. If not, there is the potential that the rules will be influenced by the outcomes of the studies. Such rules must also be consistently applied across the body of retrieved studies.

A useful approach may be to analyze all of the data regardless of quality first, and then compare those results with an analysis of only higher-quality data; if the conclusions do not change greatly, it may support the robustness of the analysis using the entire data set. Sometimes studies vary widely in quality, but lower-quality studies contain uniquely valuable information, sometimes for very good and obvious reasons. For example, there may be an especially large number of well-replicated experiments on annual plants and a much smaller number of more poorly replicated studies on long-lived trees, but the data on the responses of trees might be invaluable. Studies on responses of predators are likely to be more common and far better replicated on odonates and spiders than on jaguars and tigers, but throwing out the large mammal data would be foolish.

Our view is that a relatively inclusive approach is often the most reasonable in ecology and evolutionary biology due to the breadth of the questions and the diversity of relevant organisms and systems. We caution, however, that inclusiveness is not a synonym for lumping all studies together uncritically, and we note again that it is sound practice to investigate whether methodological features as well as other moderators are related to study outcomes. Research reviewers should qualify their conclusions appropriately when the evidence base does not permit making strong causal inferences.

Even in cases with is a strong experimental evidence base, there is merit to adopting an inclusive approach. Consider the following hypothetical example, where the research question is: "What is the impact of increased CO_2 on plant growth?" Some well-replicated studies may have been conducted in growth chambers on a very small number of plants considered to be model systems; these studies possibly were of relatively short duration or included few levels of elevated CO_2. The use of a restricted set of manipulations (treatments), a narrow sample of subjects (species), and limited duration are not unusual in experimental ecology. While these true randomized, controlled experiments may have shown that elevated CO_2 can greatly increase plant growth, they may have limited generalizability and may be dangerous to extrapolate to responses expected from natural terrestrial plant communities. Inclusion of other types of studies in the meta-analysis might expand the information available, and thus add breadth and depth to the picture painted by the experimental studies. If the less rigorous evidence is consistent with that provided by the experiments, it may be reasonable to suggest that the experimental results generalize beyond the specific conditions, species, and outcomes studied. If the results from the different types of studies do not converge, research reviewers must be careful to note the limited generalizability of the experimental studies.

An additional aspect of study screening is the identification and elimination of duplicate data. Duplication may be unintentionally introduced by the reviewer who searches multiple data sources, because some studies appear in more than one data source. This type of duplication is generally easy to identify; the reviewer has to carefully screen the retrieved studies *across* sources to eliminate multiple occurrences of the same title. A careful reviewer will also screen for other types of duplication, such as multiple papers based on the same set of data, in order to ensure that any dependencies across studies are located and taken into account when the data is analyzed (Gurevitch and Hedges 1999, Wood 2008).

Once the set of included studies has been defined, rules and procedures for extracting and coding information from these studies must be developed and documented (if not done earlier).

These rules and procedures will allow the information needed for the review to be extracted and coded from each study in a consistent and unbiased manner. In addition to information about the study outcomes (expressed as effect sizes), important features of the study (e.g., study design; taxa; measured variables such as biomass, counts, etc.; field vs. lab; experimental vs. observational; location) will also need to be recorded. While some of the information to be extracted is relatively straightforward, the coding of other data often requires considerable judgment. Research reviewers should work from standardized coding instructions developed prior to the data extraction, and should specify what these were and how they recorded study information. Some means of assessing coder reliability and agreement are recommended, particularly when people doing the coding have varying degrees of experience. High-quality reviews will demonstrate that the decisions made by research reviewers were based on credible and replicable procedures, and document the coded items for each study. Detailed tables explaining how each item was coded are ideally included in online appendices. While they might be noted as available by contacting the authors, this can be unreliable and unsustainable over long time periods. Providing this information is particularly useful when the data extraction is complex, and in any case, complete transparency is the only way to ensure repeatability. This is one of the important ways in which systematic reviews and meta-analysis practices are an advance over traditional narrative reviews.

In medicine and the social sciences, it is now common for two or more coders/raters to extract information from studies. It is also often good practice to list both the included and the excluded studies, along with reasons for the exclusion (Chapter 4). Excluded studies are almost never listed in the ecological and evolutionary reviews. Much of this information could be included in online appendices to published papers.

EVALUATING NONINDEPENDENCE, MISSING INFORMATION, AND OTHER STRUCTURAL PROBLEMS IN THE DATA

Several issues concerning the structure of meta-analysis data are of particular concern in ecology and evolutionary biology due to the nature of the primary data, the kinds of questions asked in research synthesis, and the general broad approach taken in this field. These are discussed elsewhere in this volume, but are emphasized here because dealing with them appropriately is essential for the quality of the outcome of the meta-analysis. Two of the most frequently encountered of these issues are the various kinds of nonindependence among the study outcomes, and missing information in the primary publications.

Problems of nonindependence can arise in a number of ways that affect either the relationships among moderators (independent variables) or the effect sizes (the outcomes, or dependent variables). These include the confounding of moderating factors, studies that report multiple outcomes collected on the same subjects, studies that report effects at multiple time points (longitudinal analyses), and studies that have multiple treatments but a single control condition.

Often, moderator variables are confounded or correlated with one another, and it can be difficult or impossible to disentangle their effects. Confounding of moderators (explanatory factors) is clearly a common problem for factors such as habitat, taxon, trophic level, and functional group. It is important for the research synthesist to examine the data for such confounding (and possibly even test for it), and to be aware of potentially confounded moderators; otherwise, they risk serious misinterpretation of the results. If there is significant association between moderators, the most important thing to do is to consider this dependency in interpreting the results of the synthesis, and acknowledge that effects of two moderators cannot be fully separated. A possible solution to this problem, at least in a large meta-analysis, can sometimes be to find a group in which the moderators are not confounded; then it can be tested whether effects of a moderator within that group confirm the explanation for the effects of the moderators that are confounded in the larger group of studies. For example, Gurevitch et al. (1992) did this to assess the effect of caging (an artifact of experimental method) on competition; caging was confounded with organism and habitat in all but one group of organisms. Analysis of this one group of organisms (mollusks) confirmed the general impression (from the confounded data in the larger group of studies) that caging may artificially inflate the estimate of the effects of competition, which makes sense scientifically as well. Where this assessment is not possible, an important contribution of the research synthesis may be pointing out the problem and highlighting the need for studies that do not confound these factors. In any case, it is essential for synthesists to be aware of confounding, to report it, and to take it into account in interpreting the results of the meta-analysis.

A second and very common type of nonindependence is caused by the research synthesist. Because ecological meta-analyses are typically much more broadly cast than those in medicine (Chapter 25), and because ecological problems are typically difficult to reduce to simple unifactorial cause and effect (e.g., *what is the effect of aspirin on reducing the incidence of myocardial infarction?*), ecologists often wish to test the effects of many different moderators on the outcome in meta-analyses. They may approach this by assembling a data set with many different moderators, and test the effects of each of these factors in turn on the outcome. How do habitat type, the presence of predators, target organism trophic level, taxon and functional group, experiment duration, community diversity, and so on, affect the response to the effect of interest (e.g., response to climate change)? This approach is ubiquitous in ecological meta-analyses, and is likely to be staunchly defended by authors.

However, this approach can also result in several statistical problems. Multiple tests on the same data are subject to nonindependence of the tests and to potential bias. Probability levels are compromised, and confidence limits underestimated, when one tests the same data (the effect sizes) repeatedly, each time for the influence of a different moderator. Another problem with testing the same data many different times, each time with a different moderator, is the well-known one of "data fishing." If only the most "promising" (i.e., statistically significant) results are reported, the outcome is seriously biased (Chapter 14). These problems are not easily dismissed, because in ecology and evolutionary biology many moderators may potentially be affecting the outcome and may be important in the questions being addressed and the literature one is attempting to summarize.

When there is a very large data set and the number of moderators being tested is relatively small, the compromises to statistical rigor are minimal and the results are likely to not be much affected. Where this is not the case (many moderators and a smaller number of studies), there are a number of approaches to resolving the problem. The first is to only test one moderator. This is highly unsatisfactory because it will fail to address many of the most pressing ecological and evolutionary questions being asked in the literature, making the meta-analysis uninformative and wasting a lot of the information contained in the data. Another approach is to use Bonferroni or other corrections to modify significance levels. Limitations of this approach are that Bonferroni and similar corrections are typically overly conservative (i.e., they underestimate significance), and are an inelegant approach because they are not very informative; that is, they are not a good way of getting the most information out of the data. Another approach is to test moderating factors sequentially and hierarchically (Hedges and Olkin 1985, Gurevitch et al. 1992). In this approach, the most important factor or the one which best separates the largest number of studies is tested first. Then, tests of other factors are conducted within each of those categories; this can be repeated until one runs out of studies. For example, differences in outcome among taxonomic families might be tested first; then within each family, differences in effect sizes might be tested among species; and within species, differences might be tested for different methodology. There are a number of serious limitations with this approach, the most obvious one being that one runs out of studies very quickly. It is also rather clunky and inelegant. An alternative to testing moderators sequentially is to include them in the analysis simultaneously by using hierarchical models or factorial models (Chapter 8). A third approach is to model the outcome using maximum likelihood or Bayesian methods (Chapters 10 and 11). This has the advantage of being more informative and well-designed because the solution is tailored specifically to the problem; however, it is limited by the data available, and a model based on small numbers of studies fitting particular criteria is subject to very rough parameter estimates.

Another common issue in meta-analysis in ecology and evolutionary biology is the nonindependence of study outcomes (the dependent variables) due to the data structure of the literature being summarized. For example, one published paper may contribute 12 outcomes to a meta-analysis (e.g., if the experiment was conducted on 12 species), while another paper contributes only 1 outcome. It is not always the case that the effects from one study are nonindependent (that is, more similar to one another than to the measures in other studies), but clearly this is very possible. Research reviewers have adopted various approaches to handling dependent effect sizes. In some cases, each effect is treated as if it were independent of the others. Depending upon the degree of correlation among the effects, this might be fairly reasonable, or completely compromise the results by giving too much weight to the studies that contribute multiple effects, overestimating the precision of the analyses.

Another approach is to compute the mean or median of the effects from studies that report more than one effect and use this average value as the single outcome from each study. While this approach successfully addresses the problems of overweighting and false precision, it can create problems for conducting moderator analysis, and may make the results completely meaningless. For example, if one study tests the responses of 12 species to elevated CO_2 levels while other studies test individual species, combining the results of the 12 species to one measure of outcome will not be comparable to the results of tests on individual species. A third approach is to select only one of the effect sizes from each study, either randomly or based on a sensible a priori rule. In most cases eliminating all but one effect size per publication is far too conservative and sacrifices too much information. Borenstein et al. (2009) provide useful formulas for computing composites from correlated outcomes or time points, and for dealing with multiple comparison groups.

Other dependencies may exist in the data structure of the meta-analysis data set, including multiple studies from the same lab. Whether this is a problem or not will depend on the degree to which the effect sizes from such studies are more similar than any two other effects drawn randomly from the same population of studies. Sensitivity analyses can be used to explore the effects of issues related to data structure on the outcome of the meta-analysis. Uncertainty surrounding such decisions can be further explored via Bayesian meta-analyses (Chapter 11). Various ways of dealing with nonindependence between study outcomes are discussed in detail in Chapter 16.

Missing data can present a problem, whether in the form of entire studies that were not retrieved, in the form of missing information that prevents a research reviewer from calculating an effect size, or in the form of selective reporting of outcomes in primary studies. Such missing data can have a profound impact on the results of a systematic review. Research reviewers are unlikely to find every study on their topic of interest, no matter how thorough their search for literature has been, and there is evidence to suggest that in some cases the undiscovered studies may have different results than those that were included in the review. A particularly important type of sensitivity analysis is to see how robust the results are in relation to different assumptions made about (1) the extent of missing data, and (2) the results of the missing studies.

CONDUCTING THE META-ANALYSIS

Meta-analysis should be the synthesis method of choice whenever the goal of a systematic review is to summarize empirical research evidence in order to estimate the size of the relationship between variables and to identify moderators of that relationship. As emphasized throughout this book, vote counting is essentially never justified (although people try). Nevertheless, statistical combination of study results is not always appropriate (see Cooper 2007b for examples). In other cases, meta-analysis might be appropriate, but not possible because the studies in the evidence base do not present their findings in a usable manner (e.g., effect sizes cannot be calculated from the data). Further reasons for not doing a meta-analysis include deciding that all of the primary studies are of poor quality (this may be a matter of opinion), or that the research question is multivariate or uses complex research designs (meta-regression, Bayesian meta-analyses, network meta-analyses, and generalized evidence synthesis can deal with some, but not all of these situations). An additional reason or excuse for not conducting a meta-analysis is that that the studies are too heterogeneous. According to several experts (Ioannidis et al. 2008, Valentine et al. 2010; Chapter 1), this last reason is used more often than is actually warranted, and is frequently replaced with a more subjective and less transparent summarization of results across studies.

A special case exists where there are too few studies to do a meta-analysis. In fact, meta-analysis is mathematically possible with only two studies; however, statistical power will be low, there will be a lot of uncertainty in the estimate of the outcome, and the effects of moderators cannot be examined. When research reviewers opt not to use meta-analysis to combine and compare study results, justification for their decision should be provided. Even in this case, it may be possible to calculate effect sizes, standard errors, and confidence intervals for individual study results, and then present these in a table or forest plot (Valentine et al. 2010). In any case, the methods that are used to summarize the set of studies in the review should be clearly spelled out, and sufficient information must be provided so that the reader may appraise the techniques employed.

Effect sizes, weighting, and model choice

The choice and calculation of effect size metrics, and the reasons for weighting effect sizes in meta-analyses, are discussed in detail in Chapters 6 to 8. Effect sizes should be kept or converted to the metric that makes the most sense given the research question. The distribution of effect sizes in the data set being synthesized should be examined for possible outliers, as it is not unusual to find anomalously large (or small) values that can have a disproportionate and misleading effect on the analysis results. This is especially true in subgroup analyses where an extreme value may appear in a subgroup with only a few effects. Analyses with and without the outlier should be conducted, to see whether the outlier "makes a difference" in the overall results or conclusions; however, exclusion of outliers from the meta-analysis is generally not defensible. Choice of a meta-analytic model (fixed, random, or mixed) should be based on the researchers' beliefs about the nature of the underlying data. For most real world ecological syntheses, random- or mixed-effects models are most plausible. If, however, the number of studies is very small, then the estimate of the between-study variance will have poor precision.

While the random-effects might still be the appropriate model, the information needed to apply it properly is not available, and can be misleading (Borenstein et al. 2010b).

Heterogeneity

Quantifying and exploring heterogeneity is an important component of meta-analysis. Because the Q-test has low power, a meta-analysis based on a small number of studies with large within-study variance might yield a nonsignificant P-value even when the variation between studies is substantial. In order to quantify the dispersion of effects, a well-conducted meta-analysis will present a variety of heterogeneity statistics, as each captures different information. The I^2 statistic indicates the ratio of true to total variance (Higgins and Thompson 2002), while Tau (τ) represents the standard deviation of the distribution of true effects and Tau-squared (τ^2) represents its variance. The prediction interval describes the distribution of true effects around the mean effect. Of course, these statistics do not say anything about the substantive meaning of the heterogeneity of effect values; that is a matter of scientific rather than statistical interpretation.

ASSESSING BIAS AND THE ROBUSTNESS OF RESULTS

Exclusion criteria are an important potential source of biases. Publication bias is a special case of this general phenomenon and may occur when only published studies are targeted for inclusion (Chapter 14). One way to reduce bias is to use a variety of methods and sources to identify potentially relevant studies. The possible impact of publication bias should be assessed statistically as part of the meta-analysis (Chapter 14), preferably supplemented by visual inspection of contour-enhanced funnel plots (Peters et al. 2008).

Sensitivity analyses should be conducted in which alternative choices are systematically substituted to determine whether findings are robust in the face of these changes. Some examples of what can be examined using sensitivity analyses include the effects of altering inclusion criteria, excluding outlying values, comparing the results of different effect size metrics (e.g., ln R vs. Hedges' d), fixed- and random-effects meta-analyses, the consequences of "imputing" missing data, and bootstrapping or jackknifing (i.e., repeatedly re-running the meta-analysis excluding one study or one effect size at a time, or even excluding categories of studies in large meta-analyses).

PRESENTATION OF FINDINGS

Proponents of systematic reviews have made the claim that these reviews are superior to traditional narrative reviews because they are more objective, transparent, and replicable. It follows directly from the objectives and claims of systematic reviews that the methods used to conduct the review, the body of literature it summarizes, and its results must be reported clearly and with sufficient detail to allow readers to assess the validity of its conclusions.

It is also important to realize that the overall structure of systematic reviews and meta-analytic reports is more similar to that of the primary research paper than to that of narrative reviews (e.g., the latter practically never has a "Materials and Methods" section specifying how the data for the review were obtained and analyzed). Similar to primary research papers, meta-analysis publications are often subdivided into the introduction, materials and methods, results, and discussion sections. This organization may or may not fit the material in any particular review, but the content must include these elements (e.g., methods are critical to include in a meta-analysis).

Several sets of guidelines have been developed to structure the format and content of systematic reviews in various disciplines. They include the PRISMA (Preferred Reporting Items for Systematic Reviews and Meta-Analyses) statement (Moher et al. 2009), the MOOSE (Reporting Meta-analyses of Observational Studies in Epidemiology) statement (Stroup et al. 2000), the MARS (Meta-Analysis Reporting Standards) guidelines (APA 2008), and sections of the *Cochrane Handbook for Systematic Reviews of Interventions* (Higgins and Green 2011). Although these are for the most part relevant for ecological and evolutionary analyses, there are some aspects of paper format which are specific to meta-analyses in ecology and evolutionary biology. Below we briefly highlight these issues with respect to each part of the literature review.

Introduction: An introduction must clearly spell out the background and objectives of the review. It has to provide an overview of the research question (theoretical and methodological history). It also must discuss previous attempts at syntheses of the research topic, if any; these can be either qualitative or quantitative reviews. Just as a primary research paper has to justify the need for the study in the introduction, a quantitative review paper must justify the need for a new synthesis on the topic. The introduction should end with clearly stated aims and the scope of the review.

Methods: As mentioned above, unlike narrative reviews, systematic reviews and quantitative research syntheses always contain some type of methods section or sections. Similar to the methods section in a primary study, the methods section in the systematic review or meta-analysis should be sufficiently detailed as to permit replication. This is becoming important both when more than one meta-analysis is conducted on the same topic, and when meta-analyses are updated (Chapter 15).

The authors should provide details of the search methods and criteria (including the list of reference databases and keywords used, years or journals to which search was limited, etc.). Criteria for inclusion and exclusion can be followed by a brief description of the resulting database (e.g., "final database included 70 studies published during 1966–1999 that contained 124 estimates of the effect for 55 different plant species"), and how the authors dealt with potentially nonindependent response variables. Details of statistical analyses should include the effect size metric used and why it was chosen (including statistical assumptions), how data needed for calculation of effect sizes were extracted, the statistical model for the meta-analysis, methods used to assess publication bias, and so forth. The assignment of studies into groups for each analysis should be explained and the assessment of confounding moderators provided.

Results: This section may begin with some descriptive statistics (e.g., study organisms included in primary studies; range, average and median for sample sizes; duration of experiments, etc.). This information is important for a number of reasons, including generalizability of the results, identification of research gaps, and suggestion of directions for future studies. The overall effect size, its 95% CI, and indices of heterogeneity among studies are presented next, followed by the results of any analyses of covariates (explanatory variables) of effect size. It is strongly recommended that the main results be presented graphically for clarity and to facilitate comparisons by readers (Chapter 21). The results of additional analyses (e.g., sensitivity analyses, analyses of publication bias, or temporal trends in effects) should also be presented here.

We strongly encourage the authors to provide a table with effect sizes extracted from each study and all other study-specific variables either in the paper itself (if the number of studies included in the meta-analysis is relatively small) or in an electronic appendix (most ecological and evolutionary journals now allow online supplementary materials associated with the papers). This ensures long-term availability of the data included in the meta-analysis for sub-sequent updates, reanalyzes, and direct comparisons with data included in other reviews on the same topic. Direct availability of the data in the paper or as supplementary material on the journal website is much preferred to the "available from the author upon request" statement (journal life expectancy usually exceeds researcher availability expectancy). Journal editors and peer reviewers should be vigilant in insisting that the complete database used for meta-analysis is provided.

Discussion: This section should provide an overview of the major results and their interpretation in terms of (1) magnitude of the mean effects and their statistical and biological significance, (2) heterogeneity and generality of findings (e.g., whether effects vary between the studied groups, which factors may explain this variation, or how broadly the conclusions of the meta-analysis can be generalized), and (3) consistency of meta-analysis findings with the existing theory reviewed in the introduction. Limitations of the review, including those imposed by the quality of the primary research studies, are presented in this section. It is also very important to identify gaps in the existing empirical research and to suggest directions for future primary studies; often this is a major contribution of the meta-analysis. The section should conclude with a consideration of implications for theory, policy, or practice.

INTERPRETING THE RESULTS

High-quality reviews will consider the role of specific decisions in influencing the outcome of the review. When sensitivity analyses show that the findings are robust to different assumptions or decisions, increased confidence can be placed in the conclusions. When sensitivity analyses show that results are highly dependent upon specific choices, the research reviewer should note this as a limitation of the review.

An important and somewhat controversial issue is the extent to which evidence from a meta-analysis can demonstrate causal relationships. The implications of this issue are discussed in some depth in Chapter 24. One of the strengths of meta-analysis is that it allows the research reviewer to investigate a potential moderator even if it was not explicitly included in the analyses of the primary studies; this is possible if there is variation in this factor across studies. Comparing the average effect sizes in different subgroups can provide evidence as to whether these variables need to be examined more explicitly in future primary studies. Both substantive and methodological features of studies can be examined in this way.

Cooper (1998) identified two different types of evidence obtained from research syntheses; he called these study-generated and review-generated evidence about the research problem. Study-generated evidence comes from individual studies that have directly tested the hypothesis of interest. Review-generated evidence, on the other hand, comes from examining the effects on the outcome of factors that vary across studies, but which were not necessarily tested experimentally in all or even any of the individual studies. Cooper (1998) argued that observed relationships between study characteristics and effect sizes cannot reliably be used to support causal inferences, just as observational data in primary studies cannot be used to test hypotheses. Many meta-analysts agree with this assertion. Certainly, the synthesis of study-generated evidence from experimental primary research provides the strongest basis for causal conclusions in a meta-analysis. In a sense, review-generated evidence is often observational by nature, even when the primary studies in the review are all randomized experiments, because the experiments were not randomly assigned to the study characteristics of interest. (However, this is not the case when summarizing the overall effect across a body of experimental studies that all tested that effect.) When examining the influence of moderators that vary among studies, it is possible that they are confounded with the actual cause of the between-study differences of effect magnitude.

Philosophers of science disagree to some extent about the support for causal inference from observational data, including that from review-generated moderators in meta-analysis. Even the authors of this chapter disagree on it! This somewhat esoteric argument is very important for ecological and evolutionary meta-analyses, however. One argument against the more conservative view of what one can conclude from research syntheses (and from observational evidence in general) is that if evidence is highly consistent across many different sources, one can reasonably reach conclusions regarding likely causes. This can be valid, this argument maintains, even for moderators in research syntheses that were not tested in the individual experiments. The ability to assess the effect of such moderators is particularly important in meta-analyses in ecology and evolutionary biology. In these disciplines, the temporal and spatial scale at which a question is asked often precludes the use of an experimental approach in the individual studies, but evidence across studies may sometimes be brought to bear in such cases (Chapter 24). Responses across very large spatial scales, large numbers of species and higher taxa, and trends across long periods of time may be inaccessible within individual studies, but may be critically important to understanding large-scale patterns of responses (e.g., broad patterns of responses to global climate change). If such observational evidence was always invalid, much of geology and astronomy would be off limits to science, for example. Some meta-analysts in ecology and evolutionary biology see the evaluation of these kinds of large-scale patterns that cannot be tested experimentally in individual studies as one of the most important contributions meta-analysis can make in these fields (Chapters 24 and 26).

In any case, when attempting to draw causal inferences from review-generated evidence, one must tread carefully. Ideally, support for causal claims should not be based solely on the results of a single review. Similarly, review-generated findings confirm a priori predictions about moderators more strongly than post hoc searches for moderators. A very valuable outcome of research reviews can be to identify important new topics for primary research based on the results suggested by the meta-analysis.

Consideration of magnitude of effect is challenging in ecology and evolutionary biology, because the distribution of effect sizes resulting from different meta-analyses may be different from those in the social sciences (Pullin and Stewart 2006). Ecologists and evolutionary biologists should nevertheless attempt to interpret the biological significance of the results, including the magnitude of mean effect sizes and the true dispersion around the mean. A priori consideration of meaningful magnitudes is particularly valuable. Provision of absolute measures of difference may also help interpretation. For example, a systematic review of raptor displacement from roads (Martínez-Abraín 2008) reported a pooled effect size of $\ln R$ 0.22 (bootstrap 95% CI 0.036 to 0.423). This was interpreted as a modest effect sensu Lipsey and Wilson (2001), but provision of the mean absolute magnitude of nest displacement (663.46 ± 389.23 m) in addition to reporting the pooled effect allows ecologists to consider population level effects in relation to the density of a road network.

CONCLUSIONS

The aim of this handbook is to promote correct and thoughtful use of meta-analysis in ecology and evolutionary biology. It has been noted that standards of research synthesis in these disciplines so far fall short of those in medicine (Gates 2002, Roberts et al. 2006) and, as peer reviewers of many ecological and evolutionary meta-analyses, we do see a number of recurring mistakes in the manuscripts that we review. In the wise words of Ingram Olkin (in Mann 1990): "Doing a meta-analysis is easy. Doing one well is hard." We hope that by adhering to the recommendations described above, budding research synthesists will avoid common pitfalls. However, higher standards for meta-analyses in ecology and evolution will not be achieved without the cooperation of primary researchers, peer reviewers, and journal editors. All have a stake in ensuring the quality of meta-analyses published in these fields.

Graphical Presentation of Results

Christopher J. Lortie, Joseph Lau,
and Marc J. Lajeunesse

DATA ARE THE CURRENCY of science. Visualizations of data are thus one of the most compelling means to effectively communicate ideas in science (Cleveland 1985, Ellison 2001, Tufte 2001). Graphs present data in a visual form enabling the reader to read values, identify patterns, assess the outcome of a statistical technique, or analyze relationships within or between variables (Tukey 1972, Higgins and Green 2011). Not every graph has to serve all these functions; however, most sets of best practices for visualization in science also apply to meta-analyses. Unfortunately, there is a tendency to underreport data in ecology, erring on the side of presenting only average effects. Solutions to this problem include presenting average effects for groups of studies where relevant, including data in text form on graphs, and including raw data or additional analyses in online appendices.

While providing raw data has numerous advantages, including the opportunity for additional analysis by others, it is the responsibility of the authors of a paper to analyze and interpret their own data. Graphics should enable readers to visualize the trends and effects of modifying factors (e.g., how plant growth conditions might alter responses to elevated CO_2) as well as the central tendency with variances (i.e., overall means across studies and their confidence limits). Meta-analysis is particularly suited to effective visualization since its primary objective is to present standardized data synthesized across studies including the important modifying factors (Lipsey and Wilson 2001, Cooper et al. 2009).

Most of the best practices for presenting quantitative information pertain to meta-analysis. A general issue for effective data displays is that there is a dynamic tension between simplicity and communicating adequate information accurately (Tufte 1990). Clarity, structure, and choice of representation method should also highlight the patterns of differences detected statistically (Higgins and Green 2011), and a good rule of thumb is that a figure should include enough information to convey relationships without the need to describe the figure in the text (see Box 21.1 for best practices). General principles of graphical presentation, such as the degree of resolution of the data presented in a graph (coarse vs. very detailed) should be carefully considered in meta-analysis, as in primary data papers (Tufte 2001).

Effective visualizations of meta-analyses have been discussed extensively in the evidence-based medical literature (Light et al. 1994; Lau et al. 1998, 2006; Wang and Bushman 1999; Sterne et al. 2001; Barrowman and Myers 2003; Higgins and Green 2011) and to a lesser extent in ecology and evolutionary biology (Gurevitch and Hedges 1999, Rosenberg et al. 2000, Møller and Jennions 2001, Gates 2002, Lortie and Callaway 2006). The two most common meta-analysis plots are derived from the social sciences and include (1) modified error bar plots called forest plots (Lewis and Clarke 2001) used to summarize and compare weighted

---

BOX 21.1.
Best practices for effective meta-data visualization.

(1) Ensure that plots can stand alone within a publication.

(2) Include text in meta-analytic figures listing study weight, number of studies, variance, groups of data, or meaning of treatment effects.

(3) Use reference lines of "no effect" within plots.

(4) Visualize the studies used in the meta-analysis in some capacity and include information on the groups of studies. Histograms clearly communicate frequency or number of studies in different categories or number of studies reporting different magnitudes of effect size.

(5) Explore heterogeneity using visualizations and statistics. Graphics include funnel, trim and fill, and meta-regression plots.

(6) If covariates (explanatory moderating factors) are available and meaningful, publish graphics to communicate important factors that account for heterogeneity between studies. This provides the reader with a direct tool to assess generality and predictability of the hypothesis in question.

(7) Forest plots are common and very accessible to readers as they are simple and clearly convey central tendency and relative variation. At a minimum, use forest plots to summarize comparison of different groups of studies within a meta-analysis.

(8) Since meta-analyses are a synthetic endeavor promoting cross comparison of data between studies, always include tables in the electronic appendices with effect sizes for all individual studies.

Sources: Higgins and Green 2011, Tufte 2001

---

mean effects, and (2) meta-regression plots (scatterplots with significant fit lines) used to show the relationship between main effects and covariates (Lau et al. 1998, Thompson and Higgins 2002). In this chapter, we describe these two standard meta-analysis plots and provide sample graphics to illustrate usage. Details are also included for the use of simple histograms and funnel plots. A list of all meta-analysis plots is provided for the reader (Table 21.1); however, only those that are most commonly found in ecology and evolutionary biology are explained in detail in this chapter.

We note that standard meta-analysis software (Chapter 12) generally provides the capacity to plot the graphics discussed in this chapter. However, we also recommend that ecologists and evolutionary biologists consider using dedicated plotting applications for publication-quality graphs. Meta-analysis plots should emphasize relationships among studies, facilitate assessment of the relative contribution of each study (how it was weighted in the meta-analysis), and show the magnitude of the mean effect across studies and its confidence limits. To this end, meta-analysis plots often include the effect sizes and confidence limits of each study, summarize average effects for different categories or groups of studies, show central tendency for the overall model, and in some instances examine the influence of continuous modifier variables by using scattergraphs of effect sizes plotted against modifying variables (Higgins and Green 2011). Several of the data sets common throughout the book are used to illustrate the various graphics described herein.

TABLE 21.1. A list of common, useful meta-analysis plots in ecology. "Source" refers to whether the plot is used in its standard statistical form, derived from another plot type, or specifically developed for meta-analyses. "Figure number" refers to the notation in this chapter.

Plot	Source	Purpose	Figure
Histogram	Standard	To visualize frequency or distribution.	21.1
Funnel & Trim and fill	Modified scatterplots	To explore heterogeneity and bias.	21.2
Forest	Error bar plots	To summarize central tendency of effect size measures.	21.3
Meta-regression	Weighted regression	To explore heterogeneity and assess importance of ecological context.	21.4
Stem and leaf	Standard	To visualize central tendency and distribution of data.	
Normal quantile	Standard	To visualize central tendency and distribution of data.	
Box plot	Standard	To compare central tendency and distribution of data.	
Galbraith	Meta-analysis	To explore heterogeneity and outliers.	
L'Abbé	Meta-analysis	To compare event rates in treatment and control groups.	
Raindrop plot	Meta-analysis	To visualize central tendency, likelihood, and distribution of data.	
Response surface	Standard	To visualize effects of multiple factors.	
Bland-Altman plot	Standard	To compare two methods measuring the same parameter.	

A DESCRIPTION OF META-ANALYTIC VISUALIZATIONS

First, the studies should be graphed to show the factors and the groups relevant to the meta-analysis. Next, heterogeneity and publication bias should be explored. Plots that reveal patterns of heterogeneity are often useful to the reviewer even if they are not included in the final publication. Effect sizes for individual studies, groups of studies, or mean effects for from each study are graphed, with covariates as needed to elucidate the relative importance of modifying factors (i.e., temperature, latitude, etc. of each study) on study outcomes. Meta-analysis graphs can help identify groupings of studies and alternative modifying factors (Lau et al. 1998), in addition to presenting overall effects.

COMMON GRAPHS USEFUL IN ECOLOGICAL META-ANALYSIS
Histogram

Purpose: Show distribution of effect sizes from individual studies included in the meta-analysis.

Histograms effectively communicate central tendency, variation, and distribution (i.e., normality). Typically, the frequency of studies or effect sizes used in the meta-analysis is shown using histograms in their standard form (Fig. 21.1).

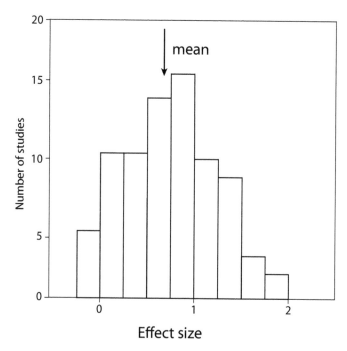

Figure 21.1. A comparison of the frequency of studies by effect sizes. This is an example of a conventional histogram to illustrate the distribution of effect sizes used in a sample data set of a meta-analysis (Torres-Vila and Jennions 2005). The arrow denotes the mean effect size.

Application

(1) Use histograms to visualize the distribution of studies within a meta-analysis.

(2) If possible, include groups, covariates, or important factors that might be important sources of heterogeneity or useful explanatory variables.

(3) Bar charts can be used to plot effect sizes but we recommend forest plots for this purpose.

Funnel Plots and Trim and Fill Plots

Purpose: Show the relationship between effect sizes and some measures of variances from individual studies.

Funnel plots are derived from scatterplots and have been used extensively in meta-analysis to examine publication bias (Egger, Davey Smith, and Phillips 1997; Light et al. 1994). In a funnel plot, some measure of study size or weight, such as sample size, standard error, or inverse variance is plotted on one axis while effect size is plotted on the other axis (Fig. 21.2). Visual estimates of asymmetry are used to assess publication bias wherein a gap (an absence of studies) shows overrepresentation of negative effect sizes for studies with small sample sizes (Fig. 21.2) (Light and Pillemer 1984). Trim and fill plots are formal statistic models using the asymmetry of a funnel to estimate bias (Chapter 14). The most extreme effect sizes are trimmed out and the gaps are replaced using iterative nonparametric estimates (Copas and Shi 2000, Sutton et al. 2000, Song et al. 2002). This process generates an estimate of relative publication bias and the plots are simply "trimmed" funnel plots; this means that they are identical to

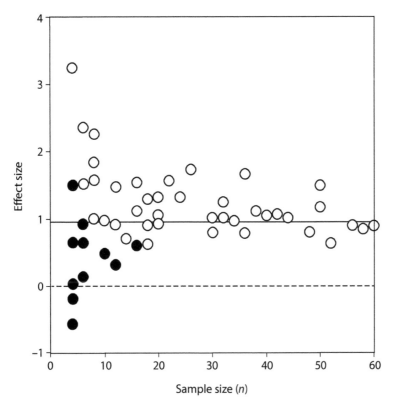

Figure 21.2. Typical funnel plot used to explore patterns of heterogeneity among studies within a meta-analysis. Each point represents the mean effect size for a single study. Note the overall funnel (or inverted funnel) shape to the plot. Open symbols indicate actual values, and filled symbols denote potential missing values that could suggest publication bias. The effect measures can be plotted on either axis.

funnel plots but with potential gaps filled and outliers removed. Graphics can also sometimes include simple scatterplots of the number of primary studies used in the meta-analysis plotted against the relative publication bias. The latter is calculated from the trimming or the probability that the true effect sizes are captured (Peters et al. 2007). However, both the visual and statistical assessments of funnel plots suffer from significant limitations including uneven treatment effects in small studies, subjectivity, and other small-study effects not related to bias, and thus may be very limited in their capacity to detect publication bias (Sterne et al. 2001, Lau et al. 2006). As such, we recommend that this approach be used cautiously. Further methods to explore publication bias are discussed in Chapter 14.

Application

(1) Use funnel plots only as a very preliminary means to explore the weighting of studies within a meta-analysis or gaps in the available literature.

(2) Use alternative and multiple methods to more thoroughly explore publication bias or heterogeneity between studies, such as sensitivity analyses or meta-regression models.

Forest plots

Purpose: Show effect sizes and confidence intervals from individual studies as well as group or overall mean effects and confidence intervals.

Forest plots are the meta-analytic equivalent of error bar or confidence interval plots, and are useful for showing mean effect size and variation among studies or groups of studies (Light et al. 1994; Egger, Davey Smith, and Phillips 1997; Lewis and Clarke 2001). Perhaps curiously to ecologists, meta-analysis forest plots are often drawn with the dependent (Y) axis horizontal, while the vertical axis is nonquantitative (individual studies). Typically, the symbol denotes effect sizes, and the horizontal lines show the confidence intervals (Fig. 21.3). The grand or pooled mean is often presented above or below individual points (Higgins and Green 2011). A dotted vertical reference line at zero is usually provided as a visual guide.

Text can be incorporated into forest plots to show number of studies, experimental levels or groups, author names, or other identificators for each study, percent weighting of study, or the effect size measure in numbers with confidence intervals adjacent to each point. Showing

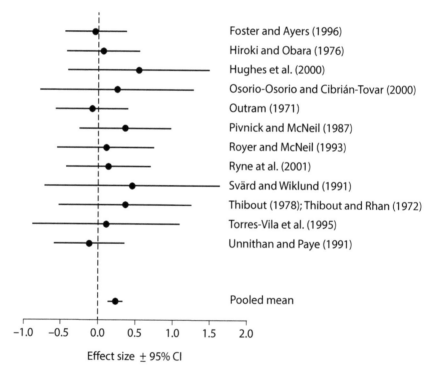

Figure 21.3. Forest plot showing effect sizes and 95% confidence intervals for individual studies included in a meta-analysis on mating history in Lepidoptera (Torres-Vila and Jennions 2005). As described in the text, at times it is desirable to present one set of studies from a larger meta-analysis; in this instance, data are from the monandrous studies only—otherwise there would have been too much to plot in a single, clear forest plot. The effect sizes are denoted by the circles, confidence intervals by the horizontal lines, and a dashed vertical line denotes no effect, or a mean of 0. The pooled mean is usually plotted slightly apart from individual studies and labeled as such.

information from each individual study (authors, year, effect size, sample size, etc.) is feasible only if the number of studies is relatively small; otherwise, a simplified version of forest plot with group means and overall means only is shown. The complete forest plot of individual studies can be provided in an electronic appendix (as is common in medical studies). On the horizontal axis, the effect size meaning can be stated with "favors treatment" or "favors control," or in ecology as a test result, such as "herbivore effect," "exclusion effect," and so forth. The results of cumulative meta-analysis (Chapter 15) can also be plotted in the form of a forest plot with year of publication listed vertically, and adjacent to each point. A variation of forest plots that commonly occurs in ecological meta-analyses could be best termed as "range" plots (Curtis and Wang 1998). Range plots resemble forest plots in almost every respect, but show the effect sizes from individual studies sorted by their sign and magnitude.

Application

(1) Forest plots are an excellent tool to communicate individual and mean effect sizes, weighting of studies, and variation or likelihood.

(2) Use grouping and distance on the vertical scale to facilitate comparison between groups or studies.

(3) Include a reference line of "no effect" and show the grand mean.

(4) Enhance forest plots with text on vertical axis further describing the data or groups, and label the effect size axis with its metric and meaning.

Meta-regression

Purpose: Show relationship between studies, overall results, and relations to covariates.

Meta-analyses may be carried out with continuous moderating factors as well as with categorical moderators. In both cases, weighted analysis is carried out that accounts for the precision of each data point (i.e., study sample size or variance). Meta-analysis data using weighted regressions on continuous moderators can be plotted to explore the effects of the moderators and to communicate results in published papers (i.e., relationships between an ecological covariate and the effect sizes or effect signs detected in a meta-analysis) (Fig. 21.4).

Typically, a scatterplot with a fitted line is used to visualize the meta-regression with the factor or covariate on the horizontal axis and effect size on the vertical axis. Confidence intervals for each point (or the fitted line), and study weightings denoted by relative symbol size can be shown. Considerations on what to plot include choice of covariates (Lau et al. 1998) or availability of data necessary to describe and model the heterogeneity between studies (i.e., productivity, nutrient levels, etc., may not be reported for all studies in the meta-analysis). Reference lines of "no effect" can be used as a visual guide. Between-study heterogeneity can also be derived from meta-regressions by comparing realized and predicted values, and calculating a metric of heterogeneity such as I^2 (Higgins and Thompson 2002, Higgins et al. 2003). In summary, it is usually clear whether there is overlap or divergence between studies or groups of studies when values are spread out on meta-regression plots of key factors.

Application

(1) Weighted regression analyses of ecological factors and effect sizes can be useful to explore heterogeneity and to visually describe ecological context.

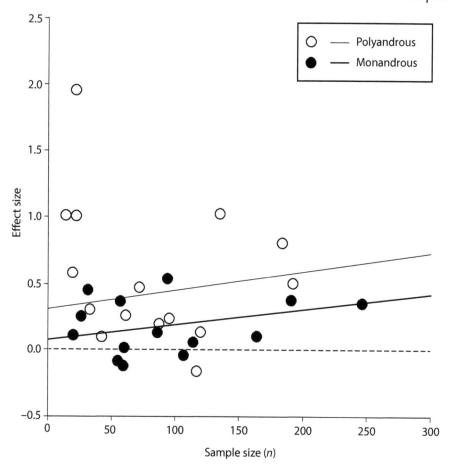

Figure 21.4. An example of a meta-regression showing the relationship between the effects size and a relevant factor (polyandrous or monandrous) for a meta-analysis on mating history in Lepidoptera (Torres-Vila and Jennions 2005). Each point represents a single study and the two different groups of studies are denoted by filled circles and the width of line.

(2) The scatterplot with fitted line can help elucidate differences between studies and assess the importance of groups of studies.

(3) Reference lines, confidence intervals, and weighting by symbol size can be included in these plots to increase the information content.

OTHER GRAPHS USED IN META-ANALYSIS

There are a number of additional meta-analysis graphics used to communicate either central tendency, variation, or a combination of both. Nonetheless, few of these are used in ecology and evolution, and are in general also becoming scarce in contemporary meta-analyses in medicine as forest plots, funnel plots, and regression plots become more widely adopted. Hence, a brief description here is provided in the unlikely event that an ecologist or evolutionary biologist encounters any of the following six alternative plot types.

The stem and leaf, normal quantile, and comparative box plots are useful to show central tendency, variation, and distribution of effect sizes (Lipsey and Wilson 2001). They all have a similar function to forest plots. These plots are used in their conventional form to present meta-analytic data, and to clearly provide visual effect size data for studies or groups of studies—provided the distributions do not encompass very large effect sizes or overly numerous entries into the meta-analysis (Wang and Bushman 1999, Lipsey and Wilson 2001). Their capacity to assess heterogeneity is however limited, particularly in the case of the stem and leaf plots; forest plots that include weighting are preferable.

Galbraith or radial plots are an alternative means to visually assess heterogeneity within a meta-analysis (Galbraith 1984, 1988); these have a similar function to funnel and forest plots. These plots are used to assess heterogeneity by showing the inverse of the standard error (horizontal axis) versus the study effect size divided by its standard error (vertical axis). This provides a means to assess asymmetry since values closer to origin have a higher standard error and are thus less precise. Galbraith plots have many positive attributes but are less intuitive to interpret and serve the same function as forest plots.

L'Abbé plots are used to compare event rates in treatment and control groups (L'Abbé et al. 1987)—there is no analog yet in ecology and evolutionary biology, but we can expect to see it in future. The event rate describes a set of dichotomous outcomes (common in medicine) for each study on a scatterplot showing control group rate on the horizontal axis and treatment group rate on the vertical axis. L'Abbé plots can serve as an alternative means to explore heterogeneity among studies within a meta-analysis that measures dichotomous outcomes (Song 1999), and may be of similar use in ecological and evolutionary meta-analyses but have not been used to date.

Raindrop plots are similar to forest plots in that effect size is plotted on the horizontal axis and symbols for each study or group are plotted on the vertical axis; however, these plots also include an estimate of log-likelihood (Barrowman and Myers 2003). Response surfaces can be used to visualize meta-analytic relationships, as between a treatment effect measure and more than one covariate or factor (Lau et al. 1998), but require adequate data on appropriate ecological covariates.

Finally, Bland-Altman plots show the average of two potentially correlated response measures, such as plant height and biomass, by plotting the difference and fitting a slope (Bland and Altman 1986). This type of plot could be used to compare two effect size measures in ecology, but in conventional statistics it is essentially the analog of plotting a correlation or residuals from a regression analysis.

CONSIDERATIONS AND GENERAL APPLICATION

There has been extensive discussion and exploration of best visualization practices for meta-analysis in the field of evidence-based medicine. Since ecological systems can be highly complex and variable, the context or potential covariates associated with each study may be important. Consequently, the meta-analysis graphics commonly used in other disciplines will certainly be useful in ecological and evolutionary syntheses, but multiple visualization approaches should be explored. This will allow ecologists to further refine interpretations and communicate meta-analytic findings effectively. While there is no need to reinvent the wheel, approaches to visualizing heterogeneity in meta-analyses are constantly evolving, and ecologists and evolutionary biologists should similarly strive to model this variation and report and visualize useful attributes in both primary and secondary studies.

Power Statistics for Meta-analysis:
Tests for Mean Effects
and Homogeneity

Marc J. Lajeunesse

A COMMON JUSTIFICATION FOR meta-analysis is the increased statistical power to detect effects over what is obtained from individual studies (Miller and Pollock 1994, Arnqvist and Wooster 1995a). A classic example in the medical sciences illustrates this advantage. Multiple independent studies were conducted to determine the effect of a blood-clot medication on the likelihood of surviving a heart attack; however, only 6 of 33 studies detected a statistically significant effect of this medication on patients. Pooling these studies with meta-analysis, however, Lau et al. (1992) found a significant and important overall effect: patients treated with this medication had a 20% reduction in the odds of dying. This is an impressive example of the value of meta-analysis given that 81% of the studies were unable to detect an effect. Individually, they lacked statistical power.

The statistical power of meta-analysis is also important for ecologists and evolutionary biologists because effect sizes are usually relatively small in these fields (Møller and Jennions 2002), and experimental sample sizes are often limited for logistic reasons (Arnqvist and Wooster 1995b). Consequently, many studies lack sufficient power to detect an experimental effect should it exist (Jennions and Møller 2003). Given that many studies lack power individually, how does a meta-analysis of these studies achieve greater statistical power? It is often assumed that the statistical power of meta-analysis is mainly a function of the number of studies included in the review; including more studies in meta-analysis results in a greater likelihood of detecting an effect should it exist (Hedges and Olkin 1985). However, recent advances in calculating the power of pooled effect sizes and homogeneity tests (Hedges and Pigott 2001, 2004), and subsequent surveys and simulations applying these calculations (e.g., Cohn and Becker 2003, Sutton et al. 2007), indicate a much more nuanced complexity to the statistical power afforded by meta-analysis.

Here I provide a brief overview of the factors that determine the statistical power of meta-analysis, and present statistics for calculating the power of pooled effect sizes to evaluate nonzero effects, and the power of within- and between-study homogeneity tests. With these statistics, I emphasize a "soft" retrospective philosophy where predetermined hypotheses about effect sizes and magnitudes of heterogeneity are used to estimate power. Finally, I survey ways to improve the statistical power of meta-analysis, and end with a discussion on the overall utility of power statistics for meta-analysis.

STATISTICAL POWER AND SAMPLING ERROR IN
EXPERIMENTS AND META-ANALYSIS

The power (ρ) of a statistical test is the probability of finding a significant result when it exists (Cohen 1988, 1992). It is explicitly defined as $\rho = 1 - \beta$; the complement to the probability of failing to detect this existing result (β or the type II error). Power is tied directly to the standard error of the data (SE), the degree to which the biological phenomenon measured through experimentation exists (known as the effect size or δ), and the significance criterion of the statistical test (the alpha level or α). Statistical power varies directly as a function of SE, δ, and α, and a way of illustrating how these parameters interact is:

$$\rho \propto \frac{\delta \cdot \alpha}{SE}. \tag{22.1}$$

Here, the statistical power of an experiment will be high when (1) the effect size is large, (2) the data are not variable and have a small standard error, and (3) the stringency of the statistical test is lenient (i.e., $\alpha > .05$). Given these interactions, if δ, SE, and α are known, then statistical power can be calculated (for a more extensive discussion, see Cohen 1988).

The parameter of interest here for meta-analysis is the standard error. The standard error estimates the variability in sampling error, and is determined by the sample size of a study (n) and the variance of the data (σ^2):

$$SE = \sqrt{\frac{\sigma^2}{n}}. \tag{22.2}$$

When n becomes large, the variability in sampling error becomes very small. A simpler interpretation of this relationship is that larger studies are more "precise" and have greater statistical power than smaller studies (Hedges and Olkin 1985). The simulation presented in Figure 22.1 illustrates the increased power of significance tests in larger studies. This is due to studies with a large n sampling a greater fraction of the population; with fewer data, random sampling can yield a fraction of the data that under- or overestimates the true population (predicted) effect. In effect, having a large sample size reduces the error associated with random sampling. This results in a more sensitive hypothesis test with greater statistical power.

This within-study sensitivity of statistical tests to sampling error is the primary justification for using meta-analysis over vote-count methods (Chapter 1). Vote counting relies heavily on average counts of significant and nonsignificant studies to provide summaries of research. This overlooks statistical issues within studies that help distinguish between true "biological" null results and erroneous null results due to low statistical power. For example, should a vote-count review pool the significance tests found in Figure 22.1, it would have counted 370 of 1000 simulated studies as null—even though a good proportion of these statistical tests were "null" because of small sample sizes. Vote counting will thus yield a biased synthesis because a greater number of studies will be treated as supporting the null hypothesis, resulting in a summary that underestimates the overall experimental effect across studies should it exist (Hedges and Olkin 1980). It is also important to note that increasing the number of studies pooled by vote counting (or simply having a large review sample size) does not improve statistical power, but in fact further increases the probability of making a review-level type II error (e.g., concluding a null outcome when one actually exists; see Hedges and Olkin 1985, Hunter and Schmidt 1990).

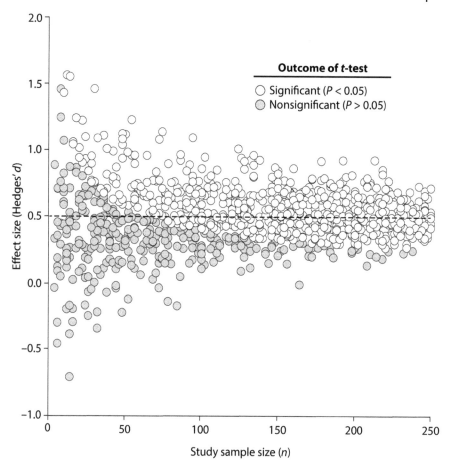

Figure 22.1. A simulation on how sampling error influences the precision of 1000 studies. Each data point is a simulated study attempting to detect an existing effect between a control and treatment group. The only parameter modified in this simulation is the sample size (n) of each study, which equals the sum of the number of random samples for the control (C) and treatment (T) groups, ($n = n_C + n_T$). Random samples (X) for each group were drawn from the following distributions: $X_C \sim N(0, 0.5)$ and $X_T \sim N(0.5, 0.5)$. For example, data for the control group were sampled from a normal distribution (N) with a mean of zero and variance of 0.5. Since the treatment group will yield on average a mean of 0.5, the "predicted" effect size for each study should also be 0.5 (*dashed line*). The "observed" effect size for each study was estimated using Hedges' d (Chapter 6). These observed effect sizes deviate from the predicted effect (0.5) because of sampling error. Finally, a t-test was used to evaluate whether the control and treatment groups significantly differed for each study. Here, the statistical power of the t-test increased with n.

Although this seems counterintuitive to sampling theory, this loss of power when pooling many studies is due to the likelihood of vote-count reviews including many more "null" studies that were incorrectly counted as null because of low within-study power.

Meta-analysis has greater statistical power than vote-count methods because it attempts to provide a synthesis free from the sampling error within studies. Meta-analysis achieves this in two ways. First, it quantifies each study using an effect size—an estimate of the magnitude and direction of a study outcome (e.g., Hedges' *d*). Having effect sizes (as opposed to vote counts) as the unit of review has the advantage that the significance tests of studies are not used to quantify results. This avoids the overrepresentation of "null" studies (incorrectly specified as "null" due to low statistical power) as in vote-count approaches. Second, meta-analysis takes advantage of the predictable differences in sampling error that occur among studies with dissimilar sample sizes; it does this by applying a weighting scheme that downweights studies with large sampling error. For example, Figure 22.1 illustrates a simulation of 1000 studies measuring the same "predicted" effect in a given population (here $\delta = 0.5$), but each study differs slightly from this "expected" effect because of sampling error. Sampling theory predicts that these slight differences in sampling error are normally distributed with a mean of zero (Hedges and Olkin 1985). The outcome of this distribution is the characteristic funnel shape of effect sizes—where smaller studies show greater variation than larger studies (Fig. 22.1). Here meta-analysis compensates for the low statistical power of individual studies because it downweights these studies with a large sampling error (using the inverse variance of each effect size as the weight; see Chapters 8 and 9). This weighting scheme is one way in which meta-analysis can achieve greater statistical "power," because it pools the actual experimental outcomes of numerous independent studies relative to their sampling error; it is therefore less sensitive to false negative or false positive outcomes due to low statistical power within each individual study.

However, similar to primary research (e.g., experiments, correlative studies), meta-analyses are also subject to sampling error. Here, each study (quantified as an effect size) is treated as a separate data point, and a meta-analysis with too few data points can have a biased synthesis of these data. This review-level error is referred to as second-order sampling error (Hunter and Schmidt 1990), and influences the ability of meta-analysis to detect effects when review-level sample sizes are low. Figure 22.2A shows how the number of studies included in a meta-analysis (K) relates to this review-level sensitivity to sampling error. A more formal way of illustrating this relationship and the statistical power of meta-analysis to detect an overall (pooled) effect ($\bar{\delta}$) that deviates from zero (δ_0) is:

$$\rho \propto \frac{(\bar{\delta} - \delta_0) \cdot \alpha}{SE(\bar{\delta})}, \qquad (22.3)$$

where α is stringency (alpha level) of the nonzero test, and *SE* is the standard error of the pooled effect size ($\bar{\delta}$). The meta-analysis *SE* is defined as:

$$SE(\bar{\delta}) = \sqrt{\sigma^2(\bar{\delta})} = \sqrt{\frac{1}{\sum_i^K [\sigma^2(\delta_i) + \tau^2]^{-1}}}, \qquad (22.4)$$

where $\sigma^2(\bar{\delta})$ is the sample variance of $\bar{\delta}$, and τ^2 is the between-study variance. What should be apparent from this review-level SE is that the number of studies included in a meta-analysis (K) does not directly influence the statistical power of meta-analysis. For comparison, see the contribution of the sample size *n* of an experiment in Equation 22.2. Instead, K affects *SE* indirectly because more studies included in the meta-analysis results in a greater sum of the weights used to penalize each study when pooling their effect sizes (e.g., the sum of all the inverse within-study variances). Thus, with every additional study included in a meta-analysis, another inverse variance (or weight) is included in the overall variance estimation. Figure 22.2B shows how the inclusion of additional studies in a review significantly decreases $\sigma^2(\bar{\delta})$ and consequently the 95% CI (confidence interval) of the pooled effect size also decreases. Thus, the addition of numerous studies to a meta-analysis results in a more sensitive review-level hypothesis test (greater statistical power to detect an effect) because the 95% CI used to evaluate nonzero effects (e.g., $\bar{\delta} - \delta_0$) becomes narrow.

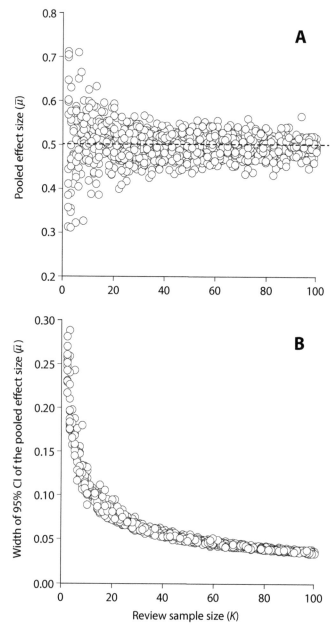

Figure 22.2. A simulation on how within-study sampling error influences the between-study sampling error of 1000 meta-analyses. Each data point is a simulated meta-analysis pooling K number of effect sizes ($\bar{\mu}$) as generated in Figure 22.1 under a fixed-effects model (Chapter 9), where **A** depicts the second-order sampling error of these meta-analyses, and **B** indicates the width of estimated 95% confidence intervals of each pooled effect size. The only parameter modified in this simulation was the sample size (K) of each meta-analysis. What is apparent from this simulation is that meta-analyses based on few studies are more likely to have pooled effect sizes that deviate from the expected effect (here 0.5) and have difficulty detecting this nonzero effect (e.g., broad 95% CI).

Another important parameter affecting the SE of meta-analysis is the between-study variance, τ^2. The between-study variance estimates the amount of heterogeneity observed among effect sizes, and is applied in meta-analysis when a random-effects model is assumed (see Chapters 8 and 9). In a fixed-effects model, $\tau^2 = 0$ because it is assumed that sampling error is the only source of variation among studies (Chapter 8). The random-effects model is important for meta-analysis in ecology and evolutionary biology, because it assumes that there are multiple population effects contributing to variation among effect sizes. For example, visualize three overlaid funnel distributions (like the one found in Figure 22.1), but with each centered about different population effects (e.g., 0.2, 0.5, and 0.8). It has been argued that the random-effects model is the most biologically relevant model for ecological meta-analysis (Gurevitch and Hedges 1999).

However, assuming a random-effects model (where $\tau^2 \neq 0$) will yield a meta-analysis with less statistical power relative to the fixed-effects model (Cohn and Becker 2003). This decrease in power is due to each individual study carrying less weight in the overall estimation of $SE(\hat{\delta})$ (see Equation 22.4). This change in the weighting scheme will increase the width of the 95% CI because the weight of each study (e.g., the inverse variance) becomes smaller with the addition of τ^2. For example, the weight of a fixed-effects model will always be larger than a random-effects model when $\tau^2 \neq 0$:

$$\frac{1}{\sigma^2(\delta_K)} > \frac{1}{\sigma^2(\delta_K) + \tau^2}, \tag{22.5}$$

and taking the inverse of the sum of these random-effects weights across studies will yield a larger SE than the sum based on a fixed-effects model (see Equation 22.4). However, the magnitude of this decrease in power depends on the size of the between-study variance. Thus, for a random-effects model, the potential statistical power gained by including numerous studies in a meta-analysis might be offset by a large amount of heterogeneity (τ^2) among these pooled effect sizes (Cohn and Becker 2003, Sutton et al. 2007).

This difference in statistical power between fixed- and random-effects models should not, however, be used as justification for assuming a priori only a fixed-effects analysis, or used as sole justification for considering one effect-size model over another. The efficiency of the weighting scheme used by a fixed-effects model is dependent on whether the data represent a homogenous collection of effect sizes (e.g., whether they share a common effect, as depicted in Fig. 22.1), and any violation of this assumption will result in a loss of statistical power (Chapter 8). This is one reason why Gurevitch and Hedges (1999) have argued that the random-effects model may be more appropriate for ecological meta-analysis because it is unclear how ecological data meet this assumption of homogeneity—the conservatism of the random-effects model is meant to balance this uncertainty (Chapter 8).

Explicitly testing the assumption of homogeneity for a fixed-effects model using a within-group Q-test (Chapters 8 and 9) has been a justifiable way to evaluate its appropriateness for meta-analysis (Hedges and Olkin 1985). This homogeneity test evaluates whether effect sizes vary beyond the predicted sampling error, as seen in Figure 22.1, and whether moderator variables or covariates should be explored to explain this variation. Not detecting significant homogeneity, however, is not evidence that it does not exist entirely—homogeneity tests themselves will have low statistical power to detect heterogeneity when review-level sample sizes are small (Hunter and Schmidt 1990). A loss of power is also an issue for the between-group Q-test when few studies are further parsed into subgroups to test hypotheses (Hedges and Olkin 1985; Hunter and Schmidt 1990). Alternatively, although uncommon in ecological meta-analyses, if a random-effects model must be assumed a priori, then a statistical test can be performed to evaluate whether $\tau^2 \neq 0$ (Hedges and Olkin 1985).

The issue as to how the between-study variance influences the statistical power of meta-analysis can be further exacerbated by how the between-study variance is itself estimated. The most common approach in ecological meta-analyses to date, and the one advocated by Hedges and Olkin (1985) and adopted by MetaWin (Rosenberg et al. 2000), is the DerSimonian-Laird estimator of τ^2 (see Chapters 8 and 9). However, now there is evidence that other estimators, such as the restricted maximum-likelihood (REML) based alternative, may provide more efficient and unbiased estimates of τ^2 under elaborate statistical modeling (e.g., meta-regression models; Viechtbauer 2006). Again, given the strong influence of τ^2 for determining the power of meta-analysis, an accurate and reliable estimate of τ^2 is important to minimize the potential for type I and type II errors.

In summary, the statistical power of meta-analysis increases when there are large nonzero population effects, when the statistical significance levels of the hypothesis test are lenient (i.e., $\alpha > 0.05$), and when there are many studies included in the review. However, statistical power decreases if there is large sampling error within each study and/or if the between-study variance among these studies is large.

INTERPRETATION AND APPLICATION OF POWER ANALYSIS

Before beginning with this section, I urge the reader to consider the vast literature on the use and utility of power analyses in primary studies (e.g., Hoenig and Heisey 2001, Jennions and Møller 2003, Nakagawa and Cuthill 2007). The main points of these papers are also germane for meta-analysis—given that null review-level results occur and need explanation (Peterman 1995). This is because there are pitfalls in power analysis that the reader should be aware of before attempting power analyses with their data. For example, below I outline one of the most grievous pitfalls in power analyses—this is the "hard" power analysis where the observed effect size of a statistical test is used to calculate its power. In the remainder of the section, I outline the "retrospective" power statistics for the "classic" statistical tests of meta-analysis; these are tests for nonzero effects and the within- and between-study homogeneity tests. These statistics are called "retrospective" because they are calculated after an experiment was designed and analyzed (or in our case after an analysis with meta-analysis), and should be distinguished from the "prospective" power analyses commonly used in the design of experiments (e.g., estimating adequate sample sizes).

Retrospective power analysis

In response to Arnqvist and Wooster's (1995a) influential review of meta-analysis in ecology and evolution, Peterman (1995) raised the following issue in regard to the power of inference of meta-analysis: How should researchers interpret a meta-analysis that fails to reject a null hypothesis? This question was timely given the already considerable debate in primary research on how to distinguish between a true biological null (zero effect) and a null outcome due to low statistical power (see Peterman 1990a,1990b). The problem relates to the application and interpretation of tests for statistical power; there is a null (nonsignificant) statistical outcome, and a power test is applied to evaluate whether this outcome is false negative (type II error) or a true "biological" null. One way to achieve this is to use the observed (calculated) effect size of the study as the basis for estimating statistical power (Gerodette 1987, Petterman 1990a, Taylor and Gerodette 1993). This approach is referred to as a "hard" retrospective power analysis. However, it is now known that this type of power analysis is uninformative because it will always indicate that a null study has low power. This is because the study's P-value and power are both statistically dependent on the observed effect size (Colegrave and Ruxton 2003; also see Equation 22.1). A "hard" power analysis simply restates the statistical significance of the test (Thomas and Juanes 1996, Foster 2001). In other words, a nonsignificant statistical test will always have low power when the observed effect size is used to calculate power.

To avoid this problem, a "soft" retrospective approach is preferred where a set of hypothesized population effect sizes (e.g., small, medium, and large effects) is used to evaluate the power of the study (Cohen 1988). This small, medium, large "t-shirt" approach can also be used to evaluate the power of meta-analysis by using expected variances given the number of studies synthesized by meta-analysis (see below; and Hedges and Pigott 2001). However, hypothesizing small, medium, and large effects is also not without criticism because without prior knowledge of the potential magnitude of actual population effects (which are largely unknown for most ecological systems), the above "t-shirt" hypotheses might not provide a meaningful frame of reference.

Another way to generate more meaningful population effect size (or variances) estimates for power analysis is to use the pooled effects from previously published meta-analyses (Hedges and Pigott 2004). For example, there have already been five meta-analyses synthesizing studies testing for local adaptation in parasites (Van Zandt and Mopper 1998, Lajeunesse and Forbes 2002, Lively et al. 2004, Greischar and Koskella 2007, Hoeksema and Forde 2008). Future meta-analyses (and primary research) in this field should use the pooled effect sizes from these previous meta-analyses as a range of potential hypotheses to evaluate statistical power. Alternatively, a more continuous range of population effects could be explored analytically or with simulations to evaluate statistical power—however, for the purpose of the chapter, I focus entirely on the "t-shirt" approach because of its simplicity and ease of calculation.

When is there sufficient power?

The convention for many fields is that there is sufficient power when $\rho \geq 0.8$. This is based on Cohen's (1977) initial recommendation that a type II error rate of 0.2 is acceptable for experiments in the social sciences (remember, $\rho = 1 - 0.2$). For meta-analysis, Hunter and Schmidt (1990) argue for a type II error rate of 0.25 ($\rho \geq 0.75$). This more conservative error rate is justified due to the inability of a meta-analyst to experimentally manipulate the number of studies available for synthesis (or the representation of individual studies within moderator groups to test hypotheses), thus recognizing a limited means to improve statistical power. This of course assumes that the meta-analyst conducted a rigorous search of the literature for all the relevant studies (Chapters 3 to 5). Meta-analysts in applied ecology might want to consider even more conservative type II error rates given how the high cost of not rejecting a false null hypothesis can adversely influence policy decisions and management practices (Peterman 1990a, Di Stefano 2003; Chapter 26).

STATISTICAL METHODS FOR EVALUATING POWER

This section outlines the statistical methods for calculating the power of meta-analysis, and illustrates their application using Torres-Vila and Jennions' (2005) meta-analysis on female fecundity of moths and butterflies when mated with virgin or nonvirgin males. For additional examples and further details on the derivation of these statistics, see Hedges and Pigott (2001, 2004). It is also important to note that these power statistics are applicable to any effect size metric (e.g., Hedges' d, response ratios, correlation coefficients, etc.), and that these equations apply to both fixed- and random-effects models for pooling effect sizes. Finally, many of these tests of statistical power for meta-analysis are available in a SAS macro (Cafri and Kromrey 2008); however, for the more simple statistical tests, I report how to calculate these using Microsoft Excel equations.

Power statistics for mean effect sizes

A frequent test in meta-analysis is to determine whether a mean (pooled) effect size ($\bar{\mu}$) differs from a null effect (μ_0). Typically, for most effect size metrics this null effect is zero ($\mu_0 = 0$). Hedges and Olkin (1985) describe a direct way of testing whether $\bar{\mu} = \mu_0$ using the following two-tailed statistic:

$$Z = \frac{\bar{\mu} - \mu_0}{\sqrt{\sigma^2(\bar{\mu})}}, \text{ where if } |Z| > c_{\alpha/2} \text{ then } \bar{\mu} \neq \mu_0. \tag{22.6}$$

Here, $\sigma^2(\bar{\mu})$ is the variance of $\bar{\mu}$, and $c_{\alpha/2}$ is the critical value where Z rejects the hypothesis that $\bar{\mu} = \mu_0$. This critical value is defined as the $100(1 - \alpha)$ percentile of the inverse standard normal distribution, and equals $c_{0.05/2} = 1.96$ when the conventional significance level (α) of 0.05 is assumed. For other significance levels, $c_{\alpha/2}$ can be calculated in Microsoft Excel using the following equation: = NORMSINV $(1 - \alpha/2)$. Finally, the P-value of this Z test is calculated in Excel using: $= 2 * (1 - \text{NORMSDIST} (Z))$. The one-tailed version of this Z test assumes a priori that the difference between the mean and null effect size has a known and predicted direction. For example, $\bar{\mu}$ will always be greater or equal to μ_0. The critical value for this one-tailed test of $Z > c_\alpha$ becomes $c_{0.05} = 1.645$.

Assuming a significance level of 0.05, the statistical power (ρ) for the above Z test is:

$$\rho^Z_{\text{two-tailed}} = 2 - \Phi(1.96 - \tilde{Z}) - \Phi(1.96 + \tilde{Z}). \tag{22.7}$$

Here, $\Phi(x)$ is the standard normal cumulative function, and is calculated in Excel using: = NORMSDIST (x). Whereas the statistical power for the one-tailed Z-test, with a priori knowledge on the direction of $\bar{\mu}$ relative to μ_0, is

$$\rho^Z_{\text{one-tailed}} = 1 - \Phi(1.645 - \tilde{Z}). \tag{22.8}$$

Note that Equations 22.7 and 22.8 do not use the Z value calculated directly from Equation 22.6. This would result in a "hard" retrospective power analysis (see the previous section for why this should be avoided). Instead, "soft" estimates of Z (referred to here as $\tilde{Z}$), that are independent from the observed pooled effect size $\bar{\mu}$, are hypothesized.

Estimating $\tilde{Z}$ to perform a "soft" retrospective power-analysis requires both the observed sampling variance of the pooled effect $\sigma^2(\bar{\mu})$ and a hypothesis about the true population effect ($\tilde{\mu}$). Estimating the sampling variance is straightforward, and can be calculated directly using meta-analysis (Chapters 8 and 9). Hypothesizing the population effect $\tilde{\mu}$ is less straightforward but can be derived from published information or by surveying a range of magnitudes of effect when calculating power (see previous section; Muncer et al. 2002, 2003). When published information on $\tilde{\mu}$ is missing, a reviewer can hypothesize small, medium, and large values of $\tilde{\mu}$ to evaluate the sensitivity in power of meta-analysis under these different magnitudes of effect. Cohen (1988) describes an effect size (measured as a standardized mean difference) of 0.2 as small, 0.5 as medium, and 0.8 as large. Although these values seem arbitrary, effect sizes like Hedges' d are measured in units of standard deviations, such that an effect of 0.2 actually indicates a difference of a fifth of a standard deviation. To integrate $\tilde{\mu}$ into power analyses, $\tilde{\mu}$ replaces $\bar{\mu}$ from Equation 22.6; this yields the appropriate $\tilde{Z}$ for Equations 22.7 and 22.8. Alternatively, a meta-analyst can use the range of observed effect sizes as good hypotheses on

TABLE 22.1. The statistical power of the pooled effect sizes ($\bar{\mu}$) from Torres-Vila and Jennions' (2005) meta-analysis on the fecundity of female moths and butterflies when mated with virgin or nonvirgin males. This analysis also includes mating system (e.g., polyandrous or monandrous) as a moderator variable to explain variation in research outcomes among the moths and butterflies used to test this hypothesis. Indicated in bold are the tests with low power (e.g., $\rho^{Z}_{\text{two-tailed}} < 0.75$).

Grouping	K	Effect size			Statistical power ($\rho^{Z}_{\text{two-tailed}}$)		
					Small	Medium	Large
		$\bar{\mu}$	95% CI	$\sigma^2(\bar{\mu})$	$\bar{\mu} = 0.2$	$\bar{\mu} = 0.5$	$\bar{\mu} = 0.8$
Fixed effects							
all studies	25	0.36	0.27–0.45	0.0020	0.994	1.0	1.0
polyandrous	12	0.48	0.35–0.61	0.0038	0.900	1.0	1.0
monandrous	13	0.25	0.13–0.37	0.0043	0.862	1.0	1.0
Random effects							
all studies	25	0.33	0.21–0.46	0.0045	0.847	1.0	1.0
polyandrous	12	0.46	0.28–0.64	0.0085	**0.583**	0.997	1.0
monandrous	13	0.22	0.05–0.39	0.0077	**0.625**	0.999	1.0

small and large effects—where for example, the smallest and largest observed effect size from the meta-analysis are treated as hypotheses of $\bar{\mu}$ (Hedges and Pigott 2001).

Table 22.1 shows the results of hypothesizing Cohen's small, medium, and large values of $\bar{\mu}$ when estimating the statistical power of Torres-Vila and Jennions' (2005) meta-analysis on the fecundity of female moths and butterflies when mated with virgin or nonvirgin males. This study had reasonable power to detect nonzero effects using a fixed-effects model, but under a random-effects model the tests had lower power to detect small effects. This result is expected given that a random-effects model will always have lower power relative to a fixed-effects model (Cohn and Becker 2003), and also given that the 95% CI nearly overlapped the null effect. A significant nonzero effect with low power should be interpreted cautiously. It offers only partial evidence that there is a true effect. Here, more data are needed to fully justify that the observed effect has a biological basis and that it is not due to sampling error.

Confidence intervals for assessing power: Advantages and limitations

Confidence intervals (CI's) are a measure of statistical power—they estimate study uncertainty and provide information on the underlying population parameters, such as variance (Hayes and Steidl 1997, Hoenig and Heisey 2001, Colegrave and Ruxton 2002). However, more typically CI's are used to test whether mean effect sizes ($\bar{\mu}$) differ from zero (the null prediction or μ_0). If the 95% CI does not include the value specified by the null hypothesis (usually zero), then μ_0 can be rejected (Steidl and Thomas 2001). The breadth of the CI around the mean effect size ($\bar{\mu}$) indicates the statistical power of the test, because estimates of CI are related to estimates of power (Nakagawa and Cuthill 2007). For example, Figure 22.2B shows the rapid reduction in breadth of the 95% CI as the review sample size increases. This is also why confidence intervals, much like power analysis, can be used *prospectively* in primary research to determine ideal sample sizes for experimentation (see Goodman and Berlin 1994). Unfortunately, an equivalent statistic to confidence intervals is not available for homogeneity tests, so the meta-analyst must apply the "soft" approach to assess the power of these tests.

Power statistics for homogeneity tests

Homogeneity tests, such as Q statistics, are used to determine whether a group of studies share a common effect size. These analyses are important because they help assess the appropriateness of the fixed-effects model, and whether studies should be parsed among different moderator groups (Chapter 8). Below, I present the power statistics for the homogeneity test of fixed-effects (ρ^Q) and random-effects (ρ^{Q^*}) models. I only briefly describe the theory behind these homogeneity tests. For further details of these models, see Chapter 8. Homogeneity tests can also be applied in an ANOVA style meta-analysis where studies are parsed among moderator groups to evaluate within- and between-group variation (Hedges and Olkin 1985). Here, between-study homogeneity (Q_B) is calculated to test whether the pooled effect sizes among these groupings differ. These moderator groupings are important to test hypotheses and serve to explain variation among effect sizes beyond what is expected due to sampling error.

Fixed-effects model: Within-study homogeneity

Homogeneity statistic Q_W tests whether the effect sizes (δ) across a group of K studies share a common effect (e.g., $\delta_1 = \delta_2 = \ldots = \delta_K$) and is calculated as follows for the fixed-effects model:

$$Q_W = \sum_{i=1}^{K} \frac{(\delta_i - \bar{\mu})^2}{v_i}, \text{ where if } Q_W > \chi^2_{K-1} \text{ then } \delta_1 \neq \delta_2 \neq \ldots \neq \delta_K. \tag{22.9}$$

Here, v_i is the observed variance of each ith effect size. A significant Q_W indicates that there is variation (heterogeneity) among effect sizes that is not explained solely by sampling error. The power of this homogeneity test is

$$\rho^{Q_W} = 1 - F(c_\alpha | K - 1; \tilde{Q}_W) \tag{22.10}$$

Power is calculated using the noncentral chi-square (χ^2) cumulative distribution function, described here as $F(x | b; \lambda)$. The parameters of this function are as follows: x is c_α, the $100(1 - \alpha)$ percent point of the central χ^2 distribution with b degrees of freedom (here $b = K - 1$); and the noncentrality parameter (λ) is estimated with $\tilde{Q}_W$. Below, I describe how to hypothesize different values of $\tilde{Q}_W$ to evaluate power. Microsoft Excel can be used to estimate c_α: CHIINV (α, $K - 1$). Unfortunately, Excel does not have a noncentral χ^2 cumulative distribution function, but statistical software like SAS and SPSS provide such functions; these are PROBCHI (c_α, $K - 1$, $\tilde{Q}_W$) and NCDF.CHISQ (c_α, $K - 1$, $\tilde{Q}_W$), respectively. This function can also be found on Casio Computer Company online calculator (see http://keisan.casio.com/).

Estimating $\tilde{Q}_W$ requires making hypotheses about plausible levels of heterogeneity. Again we can apply a "rule of thumb" developed in the social sciences for possible values for these hypotheses. Hedges and Pigott (2001) describe a convention of three magnitudes of heterogeneity:

$$\tilde{Q}_W = \begin{cases} \text{small} = .33(K-1) \\ \text{medium} = .66(K-1) \\ \text{large} = K-1. \end{cases} \tag{22.11}$$

These values are based on a review of meta-analyses which found that the ratio of between- and within-study variance is typically 0.33, but rarely exceeds 1 (Schmidt 1992). The Q_W test can be seen as a crude evaluation of this ratio; where the top half of Equation 22.9 is the between-studies variance ($\delta_i - \bar{\mu})^2$ and the bottom v is within-study variance. An application of these hypotheses of $\tilde{Q}_W$ to evaluate the statistical power of homogeneity tests is found in Table 22.2. In general, Torres-Vila and Jennions' (2005) $\tilde{Q}_W$ tests had low power to detect heterogeneity, and in particular the low power of the Q_W test among monandrous taxa indicates that its nonsignificance is susceptible to a type II error. Of course, low power to detect significant heterogeneity does not necessarily indicate that such heterogeneity exists, only that the test remains inconclusive given the available data.

Fixed-effects model: Between-study homogeneity

The between-study homogeneity statistic (Q_B) evaluates whether a collection of pooled effect sizes parsed among m moderator groups differs (e.g., $\bar{\mu}_1 = \bar{\mu}_2 = \ldots = \bar{\mu}_m$). Note that Q_B is an omnibus test and needs only one of the grouped pooled effects to differ in order to be significant. When $m > 2$, post hoc contrasts are needed to evaluate which moderator groups differ (see Hedges and Pigott 2004). The between-study homogeneity test for the fixed-effects model is calculated as

$$Q_B = \sum_{j=1}^{m} \frac{(\bar{\mu}_j - \bar{\mu})^2}{\sigma^2(\bar{\mu}_j)}, \text{ where if } Q_B > \chi^2_{m-1} \text{ then } \bar{\mu}_1 = \bar{\mu}_2 = \ldots = \bar{\mu}_m, \quad (22.12)$$

and $\bar{\mu}$ is the pooled effect size across all studies. The statistical power of Q_B is evaluated with:

$$\rho^{Q_B} = 1 - F(c_\alpha \mid m - 1; \tilde{Q}_B), \quad (22.13)$$

where $F(x \mid v; \lambda)$ is the same noncentral χ^2 cumulative distribution function used in Equation 22.9.

Hypothesizing $\tilde{Q}_B$ for power analysis when there are more than two groups ($m > 2$) is difficult and requires assumptions about the expected magnitude of differences between each group (Hedges and Pigott 2001). For simplicity, I only consider the case where $\tilde{Q}_B$ evaluates the difference between two groups. Here $\tilde{Q}_B$ can be estimated as follows:

$$\tilde{Q}_B = \frac{(\tilde{\mu}_B)^2}{\sigma^2(\bar{\mu}_1) + \sigma^2(\bar{\mu}_2)}. \quad (22.14)$$

In this equation, $\tilde{Q}_B$ is based on the observed variances of the pooled effect sizes of both groups, and a hypothesis about the possible magnitude of difference between each group ($\tilde{\mu}_B$). Again, small, medium, and large values for $\tilde{\mu}_B$ can be hypothesized following Cohen's (1988) rule of thumb. Using these hypotheses for $\tilde{\mu}_B$, Table 22.2 describes the power tests for the

TABLE 22.2. The statistical power of the within- and between-study homogeneity tests of the pooled effect sizes presented in Table 22.1. Indicated in bold are the tests with low power (e.g., $\rho^Q < 0.75$).

Grouping	Homogeneity tests			Statistical power (ρ^Q)		
	Q	df	P	Small	Medium	Large
Within-group *						
All studies	46.5	24	0.004	**0.284**	**0.608**	**0.833**
polyandrous	30.7	11	0.001	**0.188**	**0.383**	**0.573**
monandrous	9.2	12	0.688	**0.196**	**0.404**	**0.601**
Between-group **						
fixed effects	6.6	1	0.010	**0.602**	0.999	0.999
random effects	3.4	1	0.064	**0.350**	0.976	0.999

* Within-group hypotheses on small $\tilde{Q}_W = 0.33(K - 1)$, medium $\tilde{Q}_W = 0.66(K - 1)$, and large $\tilde{Q}_W = K - 1$ amounts of heterogeneity.

** Between-group hypotheses on small $\tilde{\mu}_B = 0.2$, medium $\tilde{\mu}_B = 0.5$, and large $\tilde{\mu}_B = 0.8$ differences between polyandrous and monandrous group effects.

between-group fixed-effects homogeneity statistics. These tests show that Torres-Vila and Jennions' (2005) meta-analysis had low power to detect a difference between polyandrous and monandrous groups, should the "true" difference between these groups be small.

Random-effects model: Between-study homogeneity

Testing for within-study homogeneity is uninformative in the random-effects model because it does not assume that studies share a common population effect size (see Hedges and Olkin 1985). However, the between-study homogeneity test under the random-effects model (Q_B^*) is still useful for evaluating differences among moderator groups. The homogeneity test under the random-effects model uses an estimate of the between-study variance (τ^2) to adjust the within-study variances of each effect size (v). Under this model, the variance of each study will equal $v_i^* = v_i + \tau^2$, and the new random-effects variance v^* replaces v in Equation 22.9. This between-study variance is estimated as follows: when Q_W from the fixed-effects model is less than $K-1$, then τ^2 will equal zero. Otherwise, the DerSimonian-Laird estimator is

$$\tau^2 = \frac{Q_W - (K-1)}{\sum v_i^{-1} - (\sum v_i^{-2})(\sum v_i^{-1})^{-1}}. \tag{22.15}$$

The statistical power of Q_B^* is evaluated with the following:

$$\rho^{Q_B^*} = 1 - F(c_\alpha | m-1; \tilde{Q}_B^*) \tag{22.16}$$

where $F(x | v; \lambda)$ is the same noncentral χ^2 cumulative distribution function used in Equation 22.9. As in the fixed-effects model, different hypotheses about the magnitude of heterogeneity are useful to evaluate the statistical power of Q_B^*. These hypotheses of $\tilde{Q}_B^*$ are found in Equation 22.11. Table 22.2 shows that the between-study homogeneity test assuming a random-effects model had even lower power to detect small differences between monandrous and polyandrous than the fixed-effects model did. This difference in power is expected given that the random-effects model adds τ^2 to the variances of each effect size. This additional variance component results in a less sensitive, but more conservative, hypothesis test (see above section; also see Cohn and Becker 2003, Hedges and Pigott 2004).

Limitations of power statistics for meta-analysis

An important limitation of these power analyses is that their behaviour is unknown when the studies included in meta-analysis violate the normality assumption of statistical models. For example, this occurs when estimates of power that are based on noncentral distributions of test statistics (e.g., the noncentral χ^2 distributions used in Equation 22.7) are tied directly to the statistical inference procedures used for data sampled from normal populations. Because of these assumed distributions, power analyses are more sensitive to violations of statistical normality than significance tests, which rely only on central distributions of test statistics (Hedges and Pigott 2001).

PRACTICAL WAYS TO IMPROVE POWER

Given the several factors contributing to the statistical power of meta-analysis, Table 22.3 outlines a few approaches that can improve the statistical power of meta-analysis and homogeneity tests. These suggestions rely mostly on improving the representation of studies among moderator groups, and on improving the quality of data used to estimate effect sizes and variances.

TABLE 22.3. A few practical ways to improve the statistical power of meta-analysis.

Do the following	Advantage	Disadvantage
Collate more studies.	Decreases variances and 95% CI of pooled effect sizes. Also allows for greater representation among moderator variables to test hypotheses, and diminishes potential for publication bias.	Time and resource intensive (see Chapters 4 and 5 for more details). May not improve the estimation performance of the random-effects model for pooling effect sizes.
Decide on an appropriate experimental outcome to quantify.	Using experimental measurements that are closely aligned with the biological effect of interest will increase the likelihood of detecting effects, and to detecting the difference between means or the strength of correlations between traits.	With a more explicit outcome, fewer studies will fit this definition, and thus fewer effect size data will be available for meta-analysis.
Use an appropriate effect size metric to quantify results.	Effect size metrics differ in the amount of information required to quantify an effect size. The more precise the effect size metric, the better the estimate of the population effect will become.	Using an effect size metric that is too stringent will decrease the number of studies that will fit the requirements for metric. Using effect size metrics that are too liberal will decrease the precision of the effect size to evaluate the underlying effect.
Convert effect sizes into a metric that satisfies model assumptions (e.g., transform correlations to Fisher's z).	Improves variance estimates and narrows the 95% CI.	No disadvantage! Although, Fisher's z are more difficult to interpret and less intuitive than raw correlations.
Use an estimate of the effect size variance that requires more information from individual studies.	More information is included for each study and thus increases the efficiency of the weights when pooling results.	Incomplete information and variable reporting of the required statistics can make this difficult (see Chapter 6). Thus variances are estimated using resampling methods or simplified surrogates of variance.
Avoid or account for the nonindependence of effect sizes (see Chapters 15 through 17).	Improves the estimate of pooled effect sizes and their variances, and avoids "pseudoreplication" that can inflate the conclusions drawn from the review.	Accounting for nonindependence requires knowledge on how effect sizes are correlated. This information is rarely available, but can be extrapolated from other sources.
Exclude outliers.	Decreases the deviation of the pooled effect size from the null expectation, and improves the evaluation of study heterogeneity.	Limits the scope of the review, and decreases the amount of potential studies parsed among moderator groups.

(*continued*)

TABLE 22.3. *Continued*

Do the following	Advantage	Disadvantage
Impute missing data (see Chapter 13).	More studies are included in the overall analysis, despite not having complete information. Including studies is always better than excluding studies!	Requires making assumptions of the statistical distribution of missing data. These assumptions may not have a biological basis.
Use moderator groupings or covariates to explain variation among research outcomes.	Removes "noise" among effect sizes, decreasing the pooled variance. Can help distinguish between experimental and biological effects.	Including additional parameters in models decreases the performance of statistical tests (e.g., the df's of Q-tests). Moderator groups will have smaller sample sizes and thus are more prone to sampling error, and have greater variances.

For example, when effect size metrics that require multiple pieces of information to quantify research outcomes (e.g., Hedges' d) are used over less restrictive metrics (e.g., $\ln R$), studies lacking the necessary information are often excluded from the meta-analyses. Smaller sample sizes will decrease the statistical power of meta-analysis. However, Lajeunesse and Forbes (2003) found that meta-analyses based on few, but high precision, effect sizes (relative to the same number of studies estimated with coarse effect sizes) had improved error rates because more within-study information was used to control for bias. A balance must be met between restricting analyses to studies with precise information and expanding the scope of the review by including studies with coarse effect size data. Increasing the precision of effect size data is one way to improve the statistical power of meta-analyses. The closer the effect size estimate is aligned with the biological phenomenon of interest, the stronger the hypothesis test becomes.

CONCLUSION AND PROSPECTUS

The scope of power analysis for meta-analysis spans a greater variety of statistical issues than those typically covered in primary research. Power statistics in primary research are more commonly used to evaluate null outcomes, but here with meta-analysis they can also serve as a tool to assess the detection of heterogeneity among effect sizes. I believe this latter application will be the real strength of power analysis for meta-analysis—given that confidence intervals are already heavily used and preferred over power statistics for evaluating null effects (Nakagawa and Cuthill 2007). Assessing heterogeneity is central to meta-analysis, because it provides the basis for hypothesis testing and exploring whether moderator variables (or covariates) explain variation in research outcomes (either experimental or ecological). Providing confidence to these tests is essential given the potential risk of overlooking or underemphasizing biologically meaningful explanations for variation in research because of low statistical power.

Finally, this chapter focuses entirely on the *retrospective* use of power analysis for meta-analysis. Another application uses power statistics *prospectively* for planning research and designing experiments where replication or funding is limited (Møller and Jennions 2001, 2002). The *prospective* approach applied this way will have limited use for conducting meta-analyses, given that meta-analysts cannot experimentally manipulate the number of published/unpublished studies available or "design" the ideal meta-analysis. Here, a meta-analysis is a retrospective endeavor, limited to the diversity and abundance of published literature—these data cannot be manipulated to increase or maximize the power of meta-analyses.

However, one potential application of *prospective* power analysis is to determine when to stop adding studies to a meta-analysis (see Sutton et al. 2007); that is, to detect when there is minimal gain by including more research outcomes. This application could be useful when time and resources to process data for meta-analysis are limited, and exploration of moderator effects is a minor component of the review. For example, Figure 22.2A shows that reviews with more than 40 studies do not show a significant improvement for estimating the population effect of 0.5. Another application of the *prospective* approach would be to calculate the minimum number of studies needed to detect an existing effect (this is not the same as calculating the fail-safe number; Chapter 14). This can be applied to preliminary or exploratory meta-analyses with thousands of potential studies where effect sizes need to be extracted. This application requires a working hypothesis of the between-study variance (τ^2). Here, the between-study variance can be estimated from a random subsample of the studies available for review. Hedges and Olkin (1985) provide a test to evaluate whether τ^2 is nonzero. If there is little heterogeneity among effect sizes (between-study variance), then a smaller review sample size could be justified.

An important goal for performing power analyses in meta-analysis should be to inform future research. Meta-analyses may potentially identify new and interesting effects, or may provide inconclusive evidence for important hypothesis tests of ecological or evolutionary theory. However, if these review-level outcomes are identified as having low statistical power, then this serves as an important stepping point for new experiments to test the validity of these findings. Alternatively, if large effects with strong statistical power are identified, then these effects should be used for the prognostic calculation of statistical power when designing new experiments.

ACKNOWLEDGEMENTS

I thank Terri Pigott, Alex Sutton, and Michael Jennions for comments. Work on this chapter was supported by the University of South Florida and a National Science Foundation grant to the National Evolutionary Synthesis Center (EF-0423641).

Role of Meta-analysis in Interpreting the Scientific Literature

Michael D. Jennions, Christopher J. Lortie, and Julia Koricheva

THE PROBLEM OF INFORMAL "EXPERT" ASSESSMENT OF RESEARCH FINDINGS

SCIENTISTS OFTEN DEAL WITH vast amounts of data, and the ability to summarize this information effectively is a major asset. Therefore, researchers make a considerable effort to acquire the necessary statistical skills to rigorously analyze each empirical data set that they collect. The same thoroughness should occur when writing up work for publication, which ideally requires synthesis of the scientific literature for each question that is answered (i.e. statistical test conducted) to place the results in context. This synthesis is a real challenge. One thing that makes it challenging for ecologists and evolutionary biologists to stay up to date with research findings is that, in so doing, they usually try to place their own results in a much broader context. This means that they often do not confine their frame of reference to studies of the same species, taxon, or ecosystem. There is, not unexpectedly, evidence that those working on a taxon studied by relatively few researchers are more likely to cite studies of other more popular study systems. For example, herpetologists more often cite studies of other classes of vertebrate than those studying mammals or birds cite reptilian studies (Bonnet et al. 2002; see also Taborsky 2009). In ecology and evolutionary biology, one reason for the inclusion of citations following specific research results is to illustrate the extent to which the author's results agree or disagree with the findings of others. Arguably, the most common way researchers assess the "level of agreement" is to consult reviews, including meta-analyses, to reach a general conclusion (e.g., most studies report a positive finding), which they can then claim their own study supports or contradicts.

The validity of any assessment of general findings depends on the rigor of the review used to inform this judgment. If the review is a meta-analysis, one can be fairly certain that the confidence intervals for the mean trend provide a reasonable quantitative summary of the available studies. If it is a narrative review, then the potential for a subjective bias on the part of the reviewer, including reliance on "expert opinion," which is surprisingly often flawed or erroneous (Surowiecki 2004), is more worrisome. The general conclusions drawn have a strong bearing on how research findings are presented (i.e., whether we describe results as refuting or supporting a general trend), and therefore have a major impact on the future direction of a researcher's own work and that of their colleagues (e.g., few researchers will ask a question if publications repeatedly state that we already know the answer).

In practice, reviews usually cover broad rather than specific questions. For example, it is easy to locate a review of the evolution of female mate choice, but finding one on the exact relationship between male advertisement call pitch and female mating preferences in frogs is more difficult. Consequently, attempts to summarize the findings of previous studies that tackle a specific question are rarely based on consultation of a quantitative, or even narrative, review. This can result in a highly idiosyncratic data synthesis process because each author must conduct their own separate review for every finding they wish to comment upon. This can lead to problems. Extrapolating from our own behavior and that of colleagues, many researchers tend to compile lists of papers that tackle a specific question. They then categorize these as reporting a significant positive or negative relationship, or failing to do either. Some researchers are disciplined and maintain spreadsheets of categorized publications, others are content with simply relying on their memory. If these lists were simply drawn upon to cite studies that have looked at the same question there would be no real concern. The problem, however, is that there is a temptation to tally up studies in an informal "vote count" (Chapter 1) and draw conclusions about general patterns. Aside from the problem that vote counting is a poor method to calculate trends, there is the underlying concern that the studies being tallied are a biased sample of those that have been conducted.

For many research questions (especially those subsidiary to the main focus of a study), it is probably fair to state that a good number of researchers simply consult a few recent papers and make a judgment as to the average outcome based on how many studies are cited as supporting or refuting a hypothesis. This raises a question: Are the cited studies a random sample of all those conducted? The answer in almost all cases is "no." Another commonly used shortcut to identify general trends is to simply accept at face value a published statement by a colleague that empirical support for a predicted outcome is rare or common. This is based on the assumption that he or she is an expert who has been more systematic in their review of the literature. This kind of copying can readily lead to a positive feedback loop (because authors rely on citations in the published literature that they themselves contribute to), and the emergence of fads and fashions with no link between the popular consensus and reality if the experts that initiate the feedback are wrong (Bikhchandani et al. 1992). Expert opinion is notoriously unreliable because many experts simply rely on their own biased and qualitative assessment of the literature (e.g., Antman et al. 1992).

A pragmatic approach to synthesizing the literature based on brief consultation of a subset of the published studies is understandable given time constraints. Even so, the potential for this to lead to misinformation should be readily apparent if citation and publication practices (which also inform "expert opinion") are associated with research findings (Chapter 14). First, judgment calls as to where the weight of evidence lies are often based on such factors as the relative ease with which one can recall studies that report positive and negative results. Unfortunately, humans have a propensity to recall certain events more readily than others; they are also prone to a range of other cognitive biases (Piatelli-Palmarini 1994). This can generate substantial memory bias as to the rate of occurrence of different types of events, such as pleasurable and painful experiences (Gilbert 2006). Although we are unaware of any formal investigation, it is worthwhile considering whether there is a greater likelihood of recalling a study that reports a highly significant relationship than one that fails to do so. Second, the findings of a study appear to influence the ease with which it is located in the literature, so careful attention must be paid to the potential for sampling bias. For example, the papers that appear to be more readily remembered and cited are those published in English in high-profile journals, and written by influential researchers, large research teams, or those working in the same country (Leimu and Koricheva 2005b, Wong and Kokko 2005). This is a potential problem because, due to publication bias, some of these factors are associated with effect sizes and/or their variances (i.e., our statistical confidence in the estimate; Chapter 14). This could create a large discrepancy between the "conventional wisdom" of what has been shown by previous studies based on an informal summary of findings from more readily located or remembered work, and the outcome of a quantitative meta-analysis based on a well-defined sampling protocol (Chapters 3 to 5).

Some might argue that errors made during an informal assessment of a field are only a short-term problem because the truth will eventually prevail when a formal quantitative analysis is conducted. We think this attitude is counterproductive. In ecology and evolutionary biology, a decision on whether or not to test a hypothesis is largely dictated by the decisions of individual researchers, rather than panels or committees that determine policy and direct research. Unlike the sponsorship of research in some areas in the health sciences, natural science funding bodies do not require a formal meta-analysis as part of a grant application. They take it on good faith if the applicants state that they will work on a poorly studied or unresolved issue. At best, they seek confirmation from peer review and rely on "expert opinion" that, as already noted, is often flawed (Antman et al. 1992). This can lead to an enormous waste of resources if studies are designed to ask questions that have already been satisfactorily answered (although perhaps not in the same study system, but in a more general sense). In the medical sciences, for example, the use of cumulative meta-analysis has shown that costly large-scale trials have sometimes been conducted long after the efficacy of a treatment could be established through meta-analysis (Chapter 15). Conversely, systematic reviews that identify gaps in usable data (e.g., Stewart, Pullin, and Tyler 2007) or the potential for strong publication bias (Palmer 2000), reveal cases where phenomena that are widely described as well-established turn out to be unproven or disputable and therefore worthy of closer study. Anyone who has conducted a meta-analysis is familiar with the regularity with which well-known papers purporting to demonstrate a phenomenon either fail to do so, or do so without providing reliable information about the biologically relevant magnitude of the effect.

In the health sciences there are obvious ethical concerns about delays in the implementation of effective treatments and the associated problem of unnecessary research and badly presented research. For example, it was shown that studies continued to be conducted and patients assigned to control groups long after the efficacy of a treatment was demonstrated statistically (e.g., Fergusson et al. 2005 found that 52 more studies than necessary were conducted in one area of medicine; Chapter 15). This concern led the editors of the prestigious medical journal *The Lancet* to change the submission requirements for studies describing the results of new clinical trials. The authors are now asked to include a *"clear summary of previous research findings, and to explain how their trial's findings affect this summary."* The journal also asks for the following: *"The relation between existing and new evidence should be illustrated by direct reference to an existing systematic review and meta-analysis. When a systematic review or meta-analysis does not exist, authors are encouraged to do their own. If this is not possible, authors should describe in a structured way the qualitative association between their research and previous findings"* (Young and Horton 2005, 107). Although it seems unlikely to happen soon, a similar policy in ecological and evolutionary journals would be welcome.

Ideally, continually updated meta-analyses would be available for every research question posed for which sufficient information exists. Clearly this is not going to happen, but are there practical measures that can be implemented now? To start, we need to acknowledge that our potentially subjective and informal approach to synthesizing past work has created a scientific culture with an undue emphasis on P-values, simply because they allow studies to be designated as valuable, or ignored as inconclusive. An unfortunate byproduct of this practice is an unwarranted reliance on the information that can be extracted from an individual study, especially one that reports a highly significant result. Specifically, we tend to treat studies with low P-values as being correct (or at least irrefutable), and discard studies with less clear-cut or nonsignificant findings when considering past work. This is a poor practice because a high percentage of positive (i.e., $P < 0.05$) results might actually be false; for a fascinating review, see Ioannidis 2005c, and see Ioannidis 2008 for a review of the related issue of inflated estimates. Ioannidis (2005c) derives simple equations to predict the poststudy probability that a statistically significant result is true (i.e., that the actual effect differs from the null value). A key equation is that a positive result is more likely true than false when $(1 - \beta)R > \alpha$. Here $1 - \beta$ = power (i.e., $1 -$ type II error rate), R = ratio of true effects to no effects in the field of study, and α is the type I error rate (usually 0.05). Consequently, the proportion of false positive results is higher when sample sizes are small and true effects are weak (as these decrease statistical power). Ioannidis (2005c) similarly presents equations to show how both a publication bias toward positive results (e.g., due to multiple testing and selective reporting) and increased numbers of researchers testing the same question further increase the proportion of false positive results. These insights should be of particular concern in ecology and evolutionary biology because in these fields (1) studies often have low sample sizes (Jennions and Møller 2003); (2) true effect sizes are often small (Møller and Jennions 2002); (3) numerous relationships are usually tested because studies are often exploratory so R can be small (e.g., whether extinction risk is related to body size, population size, sexual dimorphism, diet, clutch size, and so on); (4) statistical approaches are not formalized even when data sets have identical structures, which encourages "statistical fishing"; (5) several outcome variables are usually examined (e.g., effects of elevated CO_2 on growth of different plant parts, osmoregulation, or rates of photosynthesis) all of which can, post hoc, be described as important; and (6) in some "hot topic areas," such as climate change, numerous research teams are each testing the same hypotheses (e.g.. whether the onset of breeding moved forward in species X in the last 40 years).

What can be done to help researchers better summarize developments in ecology and evolution? Readers of Chapter 1 will hopefully agree that a modern meta-analysis offers a research summary that is superior to vote counting based on critical P-values or to (worse still) reliance on statements about the strength of relationships taken from narrative reviews. Of course, when a published meta-analysis is unavailable a researcher must rely on their own synthesis of the field. Papers report P-values rather than effect sizes, so the path of least resistance is to filter primary studies through the sieve of threshold P-values. The solution is two-pronged, and will ensure a more mature approach to the assessment of statistics and create conditions that should make quantitative reviews easier to conduct. First, ecological and evolutionary journals must encourage the reporting of a wider spectrum of outcomes (e.g., 95% confidence intervals for effect sizes, investigation of sources of heterogeneity among studies), rather than P-values (if the 95% CI does not overlap the null value, the reader immediately knows that $P < 0.05$ anyway). Second, researchers must learn to evaluate studies in terms of this wider spectrum of outcomes, rather than P-values. Figure 23.1 illustrates the interacting impact of a range of measures on a meta-analysis, including not only the effect estimate, but also sample sizes and the corresponding confidence intervals.

The approach we take in the remainder of this chapter is motivated by our own real world experiences as ecologists and evolutionary biologists who have conducted meta-analyses. Collating data for a meta-analysis invariably leads to a shift from a worldview where the focus is on seeking the truth based on single "perfect" studies ("textbook examples," Chapter 1) to one that regards the literature as a population of studies, each with one or more effect sizes. These effect sizes are estimates of the "true" effect size so that their pooled magnitude, variance, and heterogeneity become the real focus of attention. We therefore begin with a brief review

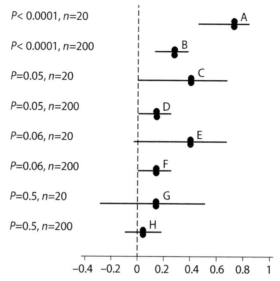

Figure 23.1. The relationship between *P*-value, sample size, estimated mean effect size (in this case *r*, Pearson's correlation coefficient), and confidence intervals showing the effect of a change in sample size with the same level of significance (modified from Nakagawa and Cuthill 2007).

of why effect sizes and their variances (usually expressed as confidence intervals) are more informative than *P*-values. We then discuss how meta-analysis promotes "effective thinking" (Nakagawa and Cuthill 2007) that can change approaches to several commonplace problems. Specifically, we address the issues of (1) exemplar studies versus average trends, (2) resolving "conflict" between specific studies, (3) presenting results, (4) deciding on the level at which to replicate studies, (5) understanding the constraints imposed by low statistical power, and (6) asking broad-scale questions that cannot be resolved in a single study.

In this chapter, we focus on estimating effect sizes as a key outcome of meta-analysis, but acknowledge that other outcomes might be of more interest in other situations. These could include, for example, comparisons between effect sizes, hypothesis testing, evaluation of moderators of effect sizes, and identifying other sources of heterogeneity between studies. However, we would argue that the points we make in the context of estimating effect sizes apply more generally to this wider meta-analysis spectrum. We should note that, for brevity, we often do not distinguish between parameter estimation and hypothesis testing. The standard null hypothesis in most areas of ecology and evolutionary biology is that a measured parameter (the "effect size" in a meta-analysis) has a mean value of zero or a nonzero theoretically predicted value (e.g., for allometric scaling, see Chapter 24). In some cases, however, parameters are estimated without being used to test a formal hypothesis (e.g., the annual rate of decline in coral cover).

EFFECT SIZES VERSUS *P*-VALUES

Ecologists and evolutionary biologists have continually been encouraged to switch from a frequentist approach of null hypothesis significance testing (i.e., whether *P*-values cross a threshold value like 0.05) toward other statistical approaches in order to summarize their research findings (e.g., Fernandez-Duque 1997, Johnson 1999, Stoehr 1999, Jennions and Møller 2003, Nakagawa 2004). These calls have largely gone unheeded. For example, adoption of

Bayesian approaches has been erratic and has only occurred in some subdisciplines (Garamszegi et al. 2009). This is probably because established biologists lack the time to master unfamiliar statistical theory, and user-friendly software is limited (although more recent introductory textbooks for ecologists might facilitate a shift; e.g., McCarthy 2007). In contrast, the suggestion that ecologists and evolutionary biologists summarize data by presenting effect sizes and their confidence intervals is a relatively undemanding request. Suitable software exists and effect sizes are readily conceptualized as "sample size corrected" versions of familiar test statistics such as F or t. In other words, these test statistics are formulated by combining effect and sample sizes (Rosenthal 1994; Chapters 6 and 7).

Why then are data in almost every ecological and evolutionary paper still summarized using P-values? Researchers are perhaps unaware of the benefits that reporting effect sizes offer. This is understandable when one considers that the benefits often accrue to the scientific community (e.g., the ability to conduct a meta-analysis) rather than to the individual author, who might even pay a cost. For example, if $P < 0.01$ and the average referee currently takes this as a sign of a "clean result," why reduce the chance of publication by reminding referees that the estimated effect size is small or the confidence interval wide? More practically, researchers often do not know how to calculate effect sizes and their variances. This might be easy for many standard test statistics, but it is trickier for others (Chapters 6 and 7; Nakagawa and Cuthill 2007). Reviewers are not known for their leniency toward those who say: "I did the right thing and used approach X, except when it was really tricky and would have taken me ages to work out how to do it." Editors would, however, be doing the fields of ecology and evolutionary biology a service if they at least required authors to provide effect sizes for a prescribed set of simple statistical tests such as unpaired t-tests or F-values from one-way ANOVAs. Of course, the growing use of meta-analysis (Chapter 1) might, by itself, stimulate a change in how statistical tests are reported. Experience suggests, however, that reporting effect sizes in primary empirical studies will not become widespread until journals make it a prerequisite. Otherwise, it merely adds another chore to the publication process with no obvious reward to the researcher.

So why report effect sizes? Presenting effect sizes and their confidence intervals allows for better interpretation of data than examination of P-values. The standard null hypothesis is that no relationships between variables or differences among groups exist. P-values simply indicate the likelihood that a relationship or difference as or more extreme than that observed will occur if the null hypothesis is true. In short, a P-value only tells us whether the 95% confidence interval for an effect size includes the null expectation. This creates a dichotomy that can be handy, but can mislead the unwary reader because it ignores both the width and central location of the confidence interval. These two extra pieces of information can lead to a radical reassessment of a result initially interpreted solely on the basis of P-values. For example, many researchers might agree with the statement that a relationship in their field of study is important if they are only told that $P < 0.01$. They would, however, probably revise their opinion if informed that the 95% CI is $r = 0.03$ to 0.16 (e.g., when $n \approx 1000$) because the fairly narrow confidence interval means that the true correlation is probably weak. Similarly, if $P < 0.01$, even if the estimated effect is large, say $r = 0.49$, researchers would probably not describe the relationship as strong when told that the 95% CI was $r = 0.12$ to 0.74 (e.g., when $n \approx 25$) because the estimate is so imprecise. Finally, even when a result is nonsignificant, any biological interpretation should be influenced by the width of the confidence interval. If large (e.g., $r = -0.30$ to 0.45), we recognize that the data are inconclusive; if small we can conclude that an effect is either weak or absent. In sum, P-values are only biologically meaningful when sample sizes are taken into account. (Even then, information on the direction of the relationship is essential. Though a seemingly obvious point, this is information that often goes unreported when presenting nonsignificant results (e.g., Cassey et al. 2004)). Graphical presentation of effect sizes is an efficient way to reveal the limits of using P-values (see Fig. 23.1 taken from Nakagawa and Cuthill 2007).

Are we exaggerating the problem? To drive home the danger of focusing on P-values it is worth answering a simple question: How often have you responded to a colleague's query about your latest study by saying something like, "great news, we got a significant result," without any mention of sample size and thus the effect size or its confidence interval? Most of us, if honest, can only reply, "I do that all the time." This illustrates the point that, in practice, we fail to distinguish between studies A to D, and often lump together studies E to G in Figure 23.1 when we present our research findings to others. The take-home message gleaned from casual conversations about other people's studies (whether it was a positive or negative result) is very similar to the information we retain when we finish reading a paper and rely on P-values to summarize what was reported. For those interested in an extended but readable account of the benefit of using effect sizes rather than P-values for ecologists we recommend Nakagawa and Cuthill (2007).

WHAT IS SO GREAT ABOUT YOUR STUDY?

It is useful to think of the practice of science as involving two competing tasks. First, to replicate precisely any study that produces an exciting result to validate the finding. Second, to make broad generalizations that can predict or account for events across a wider range of circumstances (Chalmers 1999). In some respects, these conflicting demands mirror the extent to which different researchers emphasize individual P-values or the distribution of effect sizes. This is because study replication is often associated with confirming that a prior study with a positive result was valid (for a more detailed discussion of what constitutes successful replication, see Kelly 2006), while generalization is usually associated with estimating the mean and variance (and sometimes sources of heterogeneity) in effect sizes across a range of studies (i.e., meta-analysis). The tension between these demands can be acute for ecologists and evolutionary biologists because (1) biological systems that are nominally the same actually vary spatially and temporally, so it is unclear what constitutes satisfactory replication of a study; and (2) the variety of available study systems is immense so that controversies arise as to the appropriate level at which to seek generalities. For example, Palmer (2000) has described all studies that are replicated using different species or systems as "quasi replication." Some workers go so far as to argue that ecology is "a highly idiographic science best served by amassing a catalogue of case studies" (Simberloff 2006b, 921), while others prefer to seek generalities. An example of the debate this engenders is seen in the contrasting viewpoint of Gurevitch (2006) and Simberloff (2006a, 2006b) on how best to study the interactions between native and invasive species in testing for the existence of "invasional meltdowns."

Meta-analysis is a quantitative framework to answer questions about the mean strength and sources of variation in relationships across studies. However, because it has been underutilized by ecologists and evolutionary biologists, greater emphasis has been placed on the value of individual studies. Unfortunately, and perhaps as a result, a bizarre practice has crept into many areas of ecology and evolutionary biology; this is to identify a key study (often based on little more than a small P-value and/or a large response) and then extrapolate from its findings to a wider set of circumstances. In its crudest form this results in the deification of classic studies in textbooks, which are treated as exemplars of a wider reality ("textbook examples," Chapter 1). This approach might be defensible in disciplines where the phenomenon under study is relatively invariant, but it is almost certainly inappropriate in ecology and evolutionary biology given the obvious biological differences among populations (with even more variation among species or higher taxa), the amount of background "environmental noise" in most studies, and large measurement error in many subdisciplines (e.g., evolutionary fitness is notoriously difficult to measure). The weakness of this approach to the literature is driven home if one calculates the probability that a single study will report the true mean effect size, even when we are only interested in knowing this for a single population under identical conditions. If one thinks in terms of a population of effect sizes measured with error, it is clearly improbable that this single study—especially given a publication bias favoring stronger findings—will report the true mean effect size. (For example, assuming a normal distribution, for any given study there is a 32% chance that its effect size will be more than one standard deviation from the true mean, and a 5% chance it is more than 1.96 standard deviation from the true mean.)

The medical literature provides numerous cautionary tales of the dangers of an overreliance on single studies, no matter how comprehensive. One critique of meta-analysis (which often combines estimates from smaller and larger studies) is that the final conclusion (e.g., of whether a medical intervention is effective) sometimes differs from that reached in large-scale, controlled randomized trials ("megatrials"); the latter have historically been seen as the preferred "gold standard" (comparisons are summarized in Ioannidis et al. 1998, Lau et al. 1998). It has, however, been pointed out that the results of megatrials can differ from each other as much as they do from the pooled estimates derived from a meta-analysis (Furukawa et al. 2000; for case studies of this phenomenon in conservation biology see Chapter 26). Large-scale clinical trials draw on a homogenous pool of research subjects (a single species), use the same rigorous methodologies (double-blind trials), and have very large sample sizes (1000 to 10,000 subjects). It is therefore apparent that even when uniformity is maximized, there are still unknown sources of heterogeneity that affect the outcome of a treatment. This problem of study heterogeneity is likely to be far greater in ecology and evolutionary biology.

Given the obvious biological variation among ecosystems and species, and research budgets of ecologists and evolutionists that usually preclude sample sizes in the ten thousands, it is foolhardy to conclude too much from the outcome of any single study no matter how comprehensive it is. The long-term studies of the life histories and demography of red deer on the Isle of Rhum and Soay sheep in Scotland (Clutton-Brock and Coulson 2002), or the demographics of rainforest trees on the 50 ha plot in Panama (Condit et al. 1995) are model examples of the very best ecological data sets we have from single studies. Even these studies are, however, unlikely to estimate precisely the average strength of relationships if we want to build up a picture for a broader range of species or forest types. Worse still, due to temporal variation, even these studies have reported different effects depending on the time interval over which data were analyzed. For example, the effect of maternal body condition on offspring sex ratio was eventually shown to vary with population density in red deer (Kruuk et al. 1999), and demographic patterns vary among rainforest sites only short distances apart within Panama (Condit et al. 2005).

ARE WE REALLY SO DIFFERENT?

Preoccupation with exemplary studies can generate a mythical quest for the "ideal" study. Many researchers will flatly state that they conducted their study because previous tests of a hypothesis yielded contradictory outcomes. An unflattering but plausible interpretation of their statement is that they implicitly believe that the earlier studies were flawed, and a new study is required in which all confounding variables are controlled to obtain the true answer. If one is interested in making generalizations, this view is patently absurd. There is no one set of conditions that create the perfect study, unless one wants to generate a hypothesis that is confined to one species, in one place, at a single point in time. This is a pointless task for ecologists and evolutionary biologists.

The myth of the "perfect study" creates a mindset where the main aim of a researcher might be to refute the findings of an earlier study that has gained prominence (i.e., it is thought to be "counter-intuitive"). Instead of trying to accumulate a body of evidence in the form of a population of effect sizes, researchers directly pit their results against those of other studies. This is most obvious whenever acrimonious disputes arise between research teams whose results lie on opposite sides of the $P = 0.05$ divide. Implicit in such a dispute is that one of the studies must be erroneous. These disputes are not confined to studies that have been closely replicated, and can arise even when studies are on different populations, species, or ecosystems (quasi replication sensu Palmer 2000).

Closer inspection of effect sizes often reveals that perceived conflicts among studies are illusionary. There is no reason for ecological studies with identical protocols to produce identical estimates of effect sizes (for a review, see Kelly 2006). First, average effect sizes and sample sizes in ecology and evolutionary biology are sufficiently low that sampling error can generate considerable variation in estimated effects (Møller and Jennions 2002). Summarizing using threshold P-values creates problems because studies that fail to refute the null hypothesis are treated as contradicting those that do (see Kalcounis-Rüppell et al. 2005). The absurdity of this is shown in Figure 23.2. Here the significant correlations in studies B and C are treated as being in agreement with each other, but in conflict with the findings from study A, even though studies A and B yielded very similar estimates of the effect size (Stoehr 1999). Low statistical power makes it likely that attempts to replicate a study that reported a significant effect will produce a nonsignificant result (Jennions and Møller 2003). Study D illustrates this argument graphically. If this study is replicated at the same scale, the confidence intervals will be similarly wide. Even if the estimated mean effect in Study D is the true mean, almost half the new estimates will be nonsignificant because they will fall to the left of study D with 95% confidence intervals that then overlap zero. It is instructive to view a graph plotting the function that relates the likelihood that a null hypothesis is rejected in an exactly replicated study, to the P-value obtained in an initial study; see Greenwald et al. (1996), or Figure 1 in Kelly (2006).

Before researchers embark on accounts of why their results differ from those of another study, they should first test whether sampling error alone provides a parsimonious explanation. In our experience, broad 95% confidence intervals in ecology and evolutionary biology mean that nonsignificant and significant results rarely differ more than expected by chance (i.e., effect size estimates overlap). But if two studies do differ, what should we conclude? If they are truly identical studies then this can *only* be due to chance. If they are not, then any of the innumerable

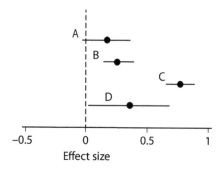

Figure 23.2. Effect sizes (means and 95% confidence intervals for Pearson's correlation coefficient) for four hypothetical studies. The null hypothesis (effect size = 0) is indicated by a vertical line.

factors that differ between them could account for the difference. We lack the data to decide which is the case because each study yields one effect size for a given hypothesis. This should be self-evident and seems trivial, but inspection of almost any recent journal will show that researchers still attempt to explain why two particular studies differ. Indeed, this is an almost compulsory component of the Discussion section of papers. Editors and reviewers will invariably ask authors to speculate as to why their findings differ from those of previous studies.

It is almost impossible to perfectly replicate biological studies, particularly those conducted in the field. This is because populations have different histories of selection and dispersal (Garant et al. 2005); genetic and environmental effects during development mean that individuals vary in their responses to identical stimuli (e.g., David et al. 2000); and responses often vary with time of year, physiological state, weather conditions, and so on (e.g., Qvarnström et al. 2000). Ecologists and evolutionists work with organisms, not atoms (clones or inbred lines in the laboratory are the closest analog we have). From this standpoint alone, researchers will achieve more for their field if they view their own work as contributing toward the ability to generalize, rather than seeing it as an attempt to validate or refute another result. Given modest statistical power, there is also limited ability even to reject a null hypothesis internal to their own study with high confidence, again suggesting that studies are best viewed as contributing to the wider picture.

HOW TO PRESENT RESULTS

Researchers who embrace meta-analysis can find it difficult to write Discussion sections of papers. Traditionally these are a forum where one must inflate the conclusions that can be drawn from a study in order to ensure its publication. When studies are viewed as contributing single data points to larger data sets, there is no incentive to overinterpret individual results. This can be daunting, but researchers should remind themselves of what they already know: their results are probabilistic. The outcomes of individual studies vary due to sampling error, and as a result of the genuine variation in the strength of causal factors under different conditions. No single study, no matter how large, is guaranteed to produce a universally correct answer (Ioannidis et al. 1998). Pragmatically, we suggest that authors use the discussion section to highlight the accuracy of their estimates (i.e., sample sizes and the attention given to reducing measurement error); the quality of their experimental design (i.e., how well they have controlled for confounding variables); and the extent to which the tested relationships have been studied by others.

One suggestion is for authors to include small-scale meta-analyses that estimate the mean effect size for studies that have asked the same question(s) that is/are the focus of their own study; see Young and Horton (2005) for a formal requirement to do so in some journals. A comprehensive meta-analysis is a major task (Chapters 3 to 5). So, to make the task manageable authors can narrow their coverage to include only studies of the same species, taxonomic group, or ecotype. Even if the resultant meta-analysis is imperfect and does not conform to the strict protocols of a systematic review (Chapters 3 to 5; Roberts et al. 2006), it will still be of greater value than the current practice of citing a few papers that did or did not report a significant result in the same direction. Tables can be used, but if results are presented graphically (as in Fig. 23.2), the extent to which there are discrepancies between studies, and how power issues affect the likelihood of reporting a significant result (Colegrave and Ruxton 2003) is more easily grasped. In practice, there is precedence for this approach as one occasionally encounters primary research papers where the authors have tabulated "vote counts" of other studies that have asked the same question (e.g., Reynolds and Jones 1999). It is a small step to shift from reporting P-values to effect size estimates. Given the need for researchers to have incentives for this extra effort, it is worth noting that publications that include this type of value-added quantitative information are more likely to be cited. They are a ready source of information for those compiling larger data sets to produce a more rigorous meta-analysis, and are usually formally acknowledged by then being cited.

META-ANALYSIS AND DECISIONS ABOUT STUDY REPLICATION

Once meta-analysis is embraced, "researchers see their piece of research as a modest contribution to the much larger picture in a research field" (Nakagawa and Cuthill 2007). From this perspective the goal is to ensure that multiple studies are available for subsequent analysis. This raises a question: How should studies be replicated? This answer is important because it influences the activities of individual researchers and of funding bodies. For example, is it better to fund a researcher who has reported an intriguing finding in an earlier pilot study and wishes to replicate the study with a larger sample size, or should funding be directed toward others who can ask the same question in other study systems (quasi replication)? One view is that some subdisciplines, such as behavioral ecology, have been damaged by a failure to precisely replicate studies that make exciting but controversial claims (for reviews, see Palmer 2000, Kelly 2006). If high-profile studies are false, they have undue influence in the long term (making them harder to dispel) but they can be dismissed quickly if they are rigorously scrutinized and swiftly replicated. Here we will make the counterargument that quasi replication is actually a more profitable approach for ecologists and evolutionists. There is, however, a caveat. This is only true *if* it is combined with a shift toward using the results of meta-analyses to guide the interpretation of published studies.

The appropriate form of study replication depends on how general we want to make our conclusions. We use a hypothetical case study to make our point. Consider the skepticism sometimes felt when a study produces an unexpected but impressive result. For example, it might be shown that feral cats with more symmetric whiskers sire more sons than daughters ($P = 0.02$, $n = 50$, $r = 0.33$; 95% CI: 0.04 to 0.62). This could lead to a plethora of "copy cat" studies (forgive the pun). The question of appropriate replication depends entirely on whether we are concerned that this specific study has miscalculated the biology of feral cats, or whether we want to test whether the finding is indicative of a wider phenomenon. We might be equally skeptical in both cases because a single study is the source of a whole new hypothesis (i.e., that paternal whisker symmetry predicts offspring sex ratios).

In disciplines like physics, researchers tend to accept that everyday studies produce "correct" results. The exceptions usually arise in areas at the forefront of theory where the requisite instrumentation or software is often at the limits of our technology. (For example, there is currently much debate about a recent result, using the CERN particle accelerator, indicating that muon neutrinos travel faster than the speed of light. Most physicists appear to assume that the speed was incorrectly measured because, if this is not the case, the finding would require major reorganization of established and well-corroborated theory). Greater confidence in the reliability of other researchers' findings is partly attributable to lower stochasticity and a better understanding of the effects of confounding variables (e.g., temperature, density) so that estimates of effect sizes tend to be more precise and replicable. In fields like physics greater attention can be given to testing whether a theory or phenomenon can be generalized (e.g., determining what classes of materials display superconductivity). Disinclination to repeat an original study might also be attributable to researchers having less affinity with knowing about a specific chemical compound rather than, say, learning about the general properties of a class of materials. In contrast, many biologists really want to know the truth about cats (or dogs or orchids). They define themselves by the organisms they study, rather than the theories they are testing. In such cases, our hypothetical cat study will trouble these types of researchers if they believe that the pattern is spurious. They will remeasure the whisker symmetry-sex ratio relationship in other cat populations but, as we have already noted, it is impossible to replicate ecological studies with perfect precision. Unless researchers can demonstrate fraud, a flawed statistical analysis, or mismeasurement, it will not be possible to "disprove" the original study. There are too many uncontrolled factors. Eventually, however, cat researchers might accumulate sufficient studies to conduct a meta-analysis. If the weighted mean correlation across cat studies is close to zero we can, without making any direct judgments about the validity of the original study, conclude that it was unrepresentative.

We have discussed this example at length to make the following point: Does anyone really care that much about cats? Maybe not, at least when they are acting as scientists rather than pet owners. Close replication of a study is probably motivated more by the knowledge that, given current practices, it could become influential and be presented as "correct" simply because $P < 0.05$ without being independently verified. It might even become a textbook exemplar. One response is therefore to subject this single study to intense scrutiny. If it proves to have grossly miscalculated the average effect then there is scientific progress. It is rather limited progress though. In our case, we only end up knowing a lot more about cats. And what if the original cat study was a good estimate of the mean effect? A more economical approach is to test for a general rule. We can, for example, productively ask whether whisker (or other aspects of body) symmetry predicts sex ratios in felids by conducting studies on lions, cheetahs, pumas, and so on. If a meta-analysis indicates a weighted mean effect close to zero, we have learned that symmetry tends not to predict offspring sex ratios in felids. With hindsight, we can either infer that cats are unrepresentative of felids (which could then be tested) or that the original study reported a false positive. If there is a significant mean effect, however, our confidence that symmetry predicts sex ratios has been expanded to cover the average feline. Of course, in so doing we accept that there is less robust evidence available to confirm whether we have a good estimate of the predictive value of whisker symmetry in any given species. Generality trades off with specificity. The extent to which ecologists and evolutionary biologists can accept this trade-off is a major source of conflict between those who embrace and reject the use of meta-analysis.

The use of meta-analysis should ameliorate the view that quasi replication is uninformative about the validity of an earlier study. The only caveat is that we must ensure that quasi replication does not increase selective reporting compared to that for true replication (Palmer 2000, 470; e.g., truly replicated studies will be reported if they contradict the original study, while quasi-replicated studies that fail to detect an effect might go unpublished). There is currently a disinclination to reject the positive findings of an earlier study as inaccurate unless precise replication shows that they are anomalous (Kelly 2006). However, meta-analysis allows one to draw inferences about a specific study if one is prepared to generalize across disparate studies. For example, if a focal study reports a significant relationship, but meta-analysis of a wider range of studies shows that the mean effect is close to the null hypothesis, a meta-analyst would probably conclude (in the absence of additional information) that the results of the focal study were due to a type I error. In our case study, whisker symmetry in cats *might* have predictive powers that are not apparent in other felids, but if one takes a broader perspective the external evidence does not support this claim. Of course, if a specific study is an outlier with respect to the general distribution of effects, then it might be worthy of further investigation because it could have a genuine biological basis. Exactly the same population level approach can be taken to detect scientific misconduct if some authors consistently report larger effect sizes than their coworkers, again with the caveat that some researchers might work on systems where measurement error is smaller, or tighter experimental designs are possible, or they might be more skilled at statistically controlling for confounding variables.

The points we have raised about study replication might seem trivial, but they are not. Embracing a meta-analytic perspective requires a profound cultural shift in how research findings are presented. If individual studies are given undue prominence then precise replication to challenge controversial findings will remain the favored response to controversial studies (Kelly 2006). Hopefully, however, greater use of meta-analysis will shift this mindset so that researchers are more interested in "the average study" rather than those at the extremes of the effect size distribution. Whether this will actually happen is still unclear. First, ecologists and evolutionary biologists are trained to observe and emphasize discontinuities in nature (e.g., to assign individuals to a species, or habitats to ecotypes). It requires a conceptual leap to switch to a worldview where, to pursue our case study one last time, one is comfortable speaking of symmetry predicting offspring sex ratio in felids, but can refrain from making a follow-up statement that this is true in feral cats but not in, say, lions if the single lion study had a value of $r = 0.10$ (95% CI: -0.19 to 0.39, $P = 0.50$, $n = 50$). The average felid is intangible, while lions and cats are real. Second, it would be amiss not to acknowledge that evolutionary biology (and perhaps ecology) is an unusual science because contingency and rare events matter. For example, although many patterns and rules have been detected that allow us to predict the direction of adaptive evolution (e.g., a shift in life history strategies toward earlier sexual maturation in response to predation; Roff 2002), sometimes a fascinating adaptation only evolves once. For example, one species of burrowing owl (*Athene cunicularia*) collects and places dung in front of its burrow (Levey et al. 2007). Experiments show that this behavior significantly increases the rate at which the owl's preferred prey of dung beetles are consumed. There is no way to generalize this result, because no other species use dung in this fashion, and the only replication possible is to validate the positive effect of dung placement on owl foraging success. In some respects, evolutionary biology is akin to economics where general laws can be formulated but rare events (which are common enough as a class) often lead to unique outcomes making predictions difficult (for a popular account of the "dismal sciences," see Taleb 2007).

THE ADVANTAGES OF "EFFECTIVE THINKING"

Meta-analysis and the use of effect sizes can improve ecological and evolutionary studies by allowing researchers to focus on questions that they could not previously answer either in practice or in principle. Nakagawa and Cuthill (2007) have coined the apt phrase "effective thinking" for the resultant mindset. Here are some advantages of "effective thinking":

(1) Power and wasteful explanations: It is a truism that sample sizes in ecology and evolutionary biology are small. For example, empirical studies that involve tracking the life histories of individuals to measure their reproductive success, growth rates, survival rates, or that attempt to estimate share of paternity using microsatellites, suffer severe logistic and funding constraints. Small sample sizes result in low statistical power and frequent failure to reject false null hypotheses. Although some ecological journals encourage the presentation of power analyses, this is still uncommon. Doing so voluntarily could punish researchers because, without a baseline reference for average power in a field (e.g., Jennions and Møller 2003), reviewers are more likely to reject papers when power appears low, say, $< 30\%$. Authors are therefore under pressure to discuss negative results as though they are conclusive. This is wasteful and generates spurious arguments. Presenting effect sizes and their confidence intervals (even though they convey

similar information) is a gentler way to remind readers about the extent to which they can draw inferences from specific tests (Colegrave and Ruxton 2003). In the long run, it also makes it easier to assess the repeatability of studies, by comparing the location and precision of effect size estimates (e.g., Fig. 23.2).

(2) Detecting trends: Given low statistical power, vote counting of significant studies is a very weak method to detect general biological trends. The use of effect sizes makes it far easier to detect patterns. In Figure 23.1, for example, if the criterion for significance was $P < 0.01$, only two of eight studies rejected the null hypothesis. Inspection of the graph shows, however, that all eight studies reported a positive relationship. This leads to a very different interpretation than that reached if one were to extract the eight P-values from the text of the paper (e.g., 0.001, ns, ns, ns, 0.001, ns, ns, ns). If the mean effect differs from the null value, deciding whether there is a causal relationship depends on the design of the original studies (see Chapter 24 for further discussion of how to interpret mean effect sizes).

(3) Future study design: Information about the average effect size can provide post hoc insight into why many published studies did not obtain significant results (e.g., due to low power to detect an average sized effect). It also ensures that future studies testing for the focal relationship in a specific context are designed with adequate statistical power. In addition, it creates the necessary benchmark against which comparisons can be made (see no. 5, below). However, there is a caveat; if there is a tendency for earlier studies to report inflated estimates of effect sizes (Chapter 15), then using these studies to design future work will lead to an overestimation of statistical power that, in turn, will increase the proportion of significant findings that are false positives (Ioannidis 2005c).

(4) Identifying sources of variation: Different studies, even those as precisely replicated as a biological system allows, rarely produce identical results. Compiling a data set of effect sizes allows us to ask why. We can first test whether the heterogeneity in effect size estimates is greater than expected by chance due to sampling error. If it is, then we have genuine conflict among studies and a new world to explore that was hidden when we only focused on significance testing in the original studies. The next step is to undertake exploratory studies to identify potential correlates of effect sizes. How these are interpreted depends on whether studies were randomly assigned with respect to the variables of interest. If they were (and this is often a judgment call), then we can tentatively posit a causal relationship between these factors and the relationship (effect size) under study. For example, if the correlation between body size and fecundity is stronger in deep water than shallow water marine species, we might causally attribute this to a depth effect. However, we cannot discount the possibility that the available species were drawn nonrandomly from the two habitats (e.g., it was harder to obtain data from larger bodied animals in deeper water). Also, because we have not experimentally manipulated depth, we cannot exclude the possibility that a correlate of depth (e.g., light levels or temperature) is responsible for the variation in effect sizes. Nonetheless, through judicial data exploration much progress can be made. For example, comparisons of effect sizes obtained from studies of colder and warmer waters at the same depth can corroborate or diminish an argument that temperature rather than depth affects the size-fecundity relationship. The use of such "natural experiments" is an unavoidable component of ecological and evolutionary research because some questions are simply not amenable to formal experimental manipulations.

The search for sources of variation in effect sizes is more likely to be important in ecology and evolution that in other areas of sciences. This is because there is greater variation in the range of study systems for which we want to draw general conclusions, the methods used to collect data and test hypotheses are more variable, and our ability to control confounding variables (especially in field studies) is limited (Chapter 25). The use of meta-analysis models that include predictor factors or continuous covariates to explore variation in research findings is arguably one of the key advances that a shift to effect size thinking can deliver for ecology and evolutionary biology, and the synthesis of these two disciplines.

(5) Ranking the importance of factors: Within a single study researchers often test how well a range of factors (or experimental manipulations) predict changes in a response variable. If results are only reported in terms of P-values, it is not easy to rank their relative importance. In contrast, presenting effect sizes and measures of their variability offers a simple way for readers to compare the influence of different factors or treatments (see Fig. 24.9 in Chapter 24). The identical approach can be used to compile data from separate studies to identify which factors are strong or weak predictors, or to identify those factors where large confidence intervals for effect size estimates suggest that we need more data before we draw any conclusions. One could argue that when sample sizes are the same, P-values can be used to rank factors. This is true, but in ecology and evolutionary biology sample sizes are almost never identical, and may be consistently smaller for some variables because they are more costly or difficult to measure. For example, a sexual difference in body size is easier to measure than one in immune system effectiveness (one might also question whether it is reasonable to combine such different responses in a meta-analysis). A study of correlates of bib size in male sparrows by Nakagawa et al. (2007) is a nice case study illustrating how pooling effect sizes across studies can inform the direction of future research (see Fig. 24.8 in Chapter 24).

(6) Should I ask the same question? The use of cumulative meta-analysis allows us to test whether estimates of the mean effect size have stabilized (Chapter 15). If so, this implies that future studies of a similar nature are unlikely to meaningfully alter our conclusions. This encourages researchers to ask new questions, or to direct their attention to exploring finer-scale variation in the strength of an effect under different circumstances (see no. 4, above).

(7) Effect sizes as new variables: Effect sizes are themselves data points that can be used as either predictor or response variables in statistical analyses. We have already described their use as response variables whenever attempts are made to predict sources of heterogeneity in effect size estimates. The comparative method has been enormously effective in studies looking at the evolution of adaptive traits (Felsenstein 1985) and, to a lesser extent, in asking higher-level questions in ecology, such as those about community composition (e.g., Losos 1996, Cardillo et al. 2008). Comparative tests have led to major advances in our understanding of how traits coevolve and what drives the evolution of specific life histories and body shapes. Now that phylogenetic comparative analyses are becoming available for effect sizes (Chapter 17; Adams 2008, Felsenstein 2008, Lajeunesse 2009, Hadfield and Nakagawa 2010) we should see increased interest in studying patterns of coevolution between effect sizes and fixed traits or even between pairs of effect sizes. Researchers have asked why morphological traits like relative testes size are so much bigger for some species than others. (It is due to intense sperm competition in species where females mate with multiple males.) Equivalent questions can now be posed in the same way for more "dynamic" properties captured by effect sizes that once seemed less quantifiable, such as how boldness or shyness relate to fitness (Smith and Blumstein 2008), or the extent to which body size increases as temperature decreases (Adams and Church 2008). It is also worth remembering that, although still rarely done,

effect sizes can be used as predictor variables (Chapter 24). We can also ask questions about how effect sizes coevolve. For example, in species where mating with nonvirgin males has a more detrimental effect on female fecundity (Torres-Vila and Jennions 2005), are females more discriminating about mating with virgins? Effect sizes calculated using the proportion of females that choose virgins over nonvirgins in two-choice trials would allow this idea to be tested; that is, are the two effect sizes correlated?

(8) Improving meta-analyses: Reporting effect sizes in primary studies would greatly facilitate the extraction of effect sizes for meta-analyses. It would reduce the risk of transcription and calculation errors when compiling a data set for a meta-analysis, and would result in greater replicability for the meta-analyses based on these data.

(9) Management and Policy: In applied areas of ecology and evolutionary biology, those unfamiliar with the details of scientific methodology are often required to develop management strategies and formulate policies based on scientific findings. Reporting effect sizes is likely to reduce the likelihood that the potential magnitude of a given practice will be incorrectly estimated due to over-reliance on P-values.

CONCLUSIONS

In our view, publication practices in ecology and evolutionary biology overemphasize the value of individual studies. The resultant focus on P-values has led some researchers to believe their task is to confirm or refute isolated null hypotheses. However, on closer inspection this is almost never their real long-term goal. Even those who only seem interested in understanding their small corner of the natural world tend to have greater aspirations. No working biologist ever presents results in isolation. Invariably other studies, often on different species, taxa, or ecosystems, are cited. Why? Either there is an expectation that there is a general rule, so that studies detecting the same pattern or experiments identifying the same causal factor are cited; or the researcher thinks that his/her study differs from previous ones in a way that will influence causation, so that failure to obtain the same result is worth highlighting. Given this practice, even those ecologists and evolutionary biologists who are primarily interested in working out the details of their own study system should be happy to accept some responsibility for presenting data in a form that makes it easier to conduct meta-analyses.

We believe that the intellectual goal of most ecologists and evolutionary biologists is to uncover general rules in nature, and to identify the exceptions that push research in new directions. This goal is only achievable when we work on a scale that is larger than our own research projects. A science that seeks only to test an isolated hypothesis is merely a program to catalogue nature in a piecemeal fashion. Some empiricists have long accepted the reality that individual studies are small pieces of a big picture. In evolutionary biology, the advances in understanding that have come from the use of the phylogenetic comparative method perfectly illustrate this process. Biologists have learned to accept that grueling fieldwork is often boiled down to a single data point for a species in a phylogenetic regression. We should be equally comfortable with the fact that the real value of the statistical tests that we calculate is often not to confirm the occurrence of a phenomenon in our own study system (although this might be of great interest to ourselves and a few others), but rather to generate an effect size that can be pooled to explore trends at higher levels of analysis.

Finally, we should acknowledge that many of the points we have made in this chapter address issues that are beyond the immediate control of many meta-analysis practitioners. They lie in the domain of editors, funding agencies, and so on. Even so, today's young biologist is tomorrow's chief editor or funding agency executive. This chapter is ultimately a work of advocacy that can hopefully be invoked by, for example, those querying editorial decisions or challenging the "conventional wisdom" of reviewers whose opinions are not always substantiated by valid quantitative analysis of the literature. It is worthwhile questioning current publication practices, because change does not occur without dissent and debate.

Using Meta-analysis to Test Ecological and Evolutionary Theory

Michael D. Jennions, Christopher J. Lortie,
and Julia Koricheva

THE USE OF META-ANALYSIS by ecologists and evolutionary biologists to tackle both large and small controversies was established from the beginning of the method's application. In two of the earliest ecological meta-analyses, Järvinen (1991) quantified how female age affects laying date and clutch size in two bird species (a very focused question), while Gurevitch et al. (1992) asked the "big picture" question of what evidence there was that competition shapes communities. Subsequent promotion of meta-analysis (Arnqvist and Wooster 1995a, Gurevitch et al. 2001) and the development of software aimed at biologists (Rosenberg et al. 2000) led to rapid growth in the use of meta-analysis in both applied ecology (Chapter 26) and theory-driven areas of ecology and evolutionary biology (Fig. 24.1). The use of meta-analysis to test theory is now widely accepted. For example, by 2009 at least 15 ecological and evolutionary meta-analyses have appeared in *Science* and *Nature*. Some of these simply quantify patterns, but others involve direct tests of hypotheses that are components of complex theories, including those related to sex ratios (West and Sheldon 2002), the evolution of cooperation (Griffin and West 2003), species diversity (Worm et al. 2002, updated by Hillebrand et al. 2007), food webs (Brett and Goldman 1997), and ecosystem function (Cardinale et al. 2006).

Ecological and evolutionary questions are particularly amenable to meta-analysis because constraints on the ability of researchers to answer a question definitively within a single study invariably generate controversies (Arnqvist and Wooster 1995a). First, the relationships being studied are often weak (Møller and Jennions 2002). Second, studies have low statistical power because of natural and/or logistic constraints on sample sizes. Third, measurement error is often large (e.g., it is hard to measure evolutionary fitness or calculate tree biomass). Fourth, multiple causal factors usually *interact* to influence outcomes but only a few factors can be manipulated in a single study, and key factors often differ in their magnitude among studies. Fifth, there are genuine biological differences across species and among populations. Sixth, adaptive evolution and changes in environmental factors can be sufficiently rapid to generate temporal variation. For example, humans as predators and modifiers of the environment have selected for rapid changes in animal phenotypes due to phenotypic plasticity (Hendry et al. 2008) and differential survival (Darimont et al. 2009). The net result of these six phenomena is that the reported relationships between traits of interest can differ across studies when the underlying causal relationships are similar, or even identical.

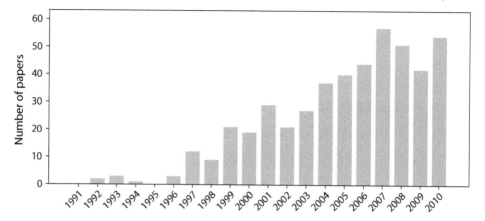

Figure 24.1. To illustrate the increased interest in meta-analysis among ecologists and evolutionary biologists, we performed a search in Web of Science for publications that contain *meta-analy** or *metaanaly** in their title, abstract, or as a keyword. To ensure the papers located were relevant, we confined our search to ten representative ecological and evolutionary journals (*American Naturalist, Ecology Letters, Ecology, Evolution, Global Change Biology, Journal of Evolutionary Biology, Marine Ecology Progress Series, New Phytologist, Oikos,* and *Oecologia*).

When results vary among studies with respect to the direction and/or the statistical significance of a relationship, and estimates of relationships are imprecise (with wide confidence intervals) this leads to a few basic types of questions and controversies among researchers:

(1) Does the biological relationship really differ from the null expectation?

(2) Do the data support one hypothesis (or even an entire theory) more strongly than another, given competing predictions about the strength/direction of a relationship?

(3) When results differ among studies, is this due to chance, or can some heterogeneity be explained to partly resolve the seemingly contradictory findings? This is arguably the most exciting aspect of meta-analysis because identifying moderator variables can generate new hypotheses, and sometimes even new theories.

(4) When several factors determine the outcome of an experiment, which of these is the more influential? Equivalently, in observational studies where several factors are correlated with a variable of interest, do some factors consistently explain more variation than others?

Here we describe nine case studies that illustrate how meta-analysis has contributed to theoretical developments in basic research in ecology and evolution. We do not provide a formal systematic review but use real examples to illustrate how meta-analysis has been able to address the problems listed above. The main research topics we cover are maintenance of biodiversity (Case 1); sexual selection (mate choice/fighting behavior) (Cases 2, 8, 9); sex ratio theory (Case 3); allometric scaling (Case 4); the invasiveness of exotic plants (Case 5); seed size and plant abundance (Case 6); and the role of competition and predation in structuring communities (Case 7). Comparable examples from conservation biology and environmental management are provided in Chapter 26. Our hope is that these case studies will resonate with the reader and provide "templates" for ways to conduct comparable tests on analogous controversies in their own fields of research.

We conclude the chapter with some brief comments on constraints that, at least to date, ecologists and evolutionary biologists have described as limiting their ability to conduct appropriate meta-analyses. These include the need to control for phylogenetic effects, an inability to construct models with several predictor variables, nonrandom sampling of nature, publication bias, and selective reporting of statistics in the literature. Most fundamental, however, is the simple lack of data when questions are asked at a level where each study/species contributes a single data point (e.g., when testing which factors account for heterogeneity in effect sizes among studies).

TESTING RULES AND THEORIES IN ECOLOGY AND EVOLUTION

Before describing our case studies, we must briefly discuss how two key terms ("rules" and "scientific hypotheses") are used by many ecologists and evolutionary biologists. We provide our own pragmatic insights into how theories are tested. Philosophers of science might not like what they read, but we have strived to provide an honest account of current working practices.

Biological rules

Meta-analysis can test whether *biological rules* have empirical support. By "rules" we mean patterns that occur repeatedly in nature. In other scientific disciplines consistent natural patterns are treated as correlative laws (see O'Hara 2005 for a mini-review of laws in ecology). A biological rule is essentially an empirical generalization (Mayr 1963). For example, Bergmann's rule states that organisms tend to be larger in cooler climates (Bergmann 1847). There are no formal criteria for how widely a rule should apply, but it has been suggested that it should minimally apply in more than 50% of species (Blackburn et al. 1999). This raises the practical concern that there is usually insufficient statistical power to confirm whether a relationship exists in any single study. Meta-analysis offers an alternate route to test the validity of rules. Given problems with vote counting (Chapter 1) a modern meta-analytic approach should be used. This means that the criterion for whether a rule holds is that the mean effect is nonzero or exceeds some agreed upon criterion (e.g., $r > 0.3$ or $d > 0.5$). Once such criteria are established, meta-analysis provides a standardized method to test rules. These criteria might vary among disciplines, however, because the regularity of a pattern is ultimately less important than accounting for variation in focal traits. For example, if three rules used in conjunction can explain most of the observed variation, then the fact that each of these rules "fails" when viewed in isolation in more than 50% of the cases would seem to be unimportant.

A *putative rule* is a conjecture about an unverified natural pattern. There are many widely reported patterns that are not yet labeled "rules." For example, putative rules may be based on direct observation of seemingly consistent natural patterns that have not yet been formally quantified (e.g., the "owner advantage effect" whereby owners usually win territorial fights against intruders; Kokko, Lopez-Sepulcre, et al. 2006; Case 9). Putative rules may also be based on deductions stemming from theoretical considerations (e.g., when either natural or sexual selection is stronger, we expect to see more refined adaptations). For example, in cooperative breeders we expect the precision of a parent's ability to skew the offspring sex toward the helping sex to be stronger when helping confers greater kin-selected benefits to parents (Griffin and West 2003; Griffin et al. 2005; Case 3). Finally, putative rules may be based on the combination of other rules (e.g., relative size is a strong predictor of fighting success in most taxa so, given an "owner advantage effect," we predict the rule that owners of territories are larger than nonowners). Rules are simply tests of empirical correlations, so meta-analysis of purely *observational* data can be used to test their validity.

Scientific hypotheses and theories

Meta-analysis can quantify the extent to which competing *hypotheses* explain a biological pattern (often a rule-like phenomenon). The transition from a rule to a scientific hypothesis requires conjecture about a causal agent (e.g., that the success of alien plant species as invaders depends on the extent to which local herbivores have experience with functionally similar species; Case 5). The hypothesis therefore generates predictions that can be empirically tested (e.g., that species X is more likely than species Y to invade habitat W). In general, hypotheses are formulated by applying concepts from a body of theories. Here we are using *theory* in the scientific sense to refer to a set of ideas about causality (e.g., that body size affects metabolic rate); in conjunction with baseline assumptions about the study system (e.g., that all mammals are endotherms), they are used to predict (hence "explain" because they provide a causal account) a wide range of phenomena in the real world (e.g., clines in mammalian body size with altitude, latitude, or global warming). Most theories contain far too many components to be falsified with a single empirical test. Instead, theories are evaluated on the basis of how often they succeed in generating testable hypotheses that predict/explain real world phenomena. A good theory is repeatedly successful in generating hypotheses that can be tested and are then empirically supported. A theory is considered especially valuable when (1) it can successfully predict phenomena for which it was not originally formulated (e.g., the application of sex ratio theory to haplodiploid, eusocial insects, such as ants, led to the prediction, which was verified a few years later, of "split sex ratios" whereby colonies specialize in producing either female or male reproductives); (2) it yields an even better prediction when new information comes to light (e.g., the fit between predicted and observed offspring sex ratios in fig wasps was improved when it was later discovered that cryptic species exist, with the result that the number of mothers of a given species laying eggs in a single fig had been overestimated) (see West 2009 for tests of sex ratio theory); and (3) it yields predictions that are precise rather than general (e.g., the sex ratio will be 75% female, not just female biased).

Within a given discipline, theories tend to be hierarchically arranged so that specific types of natural events are often explained with the assumption that a higher level theory is correct. This means that the higher level theory is very rarely directly tested. For example, the theory of natural selection has been so well tested that it is now accepted by all mainstream biologists. Consequently the idea that animals will exhibit traits that increase their fitness is the starting point for lower level theories (e.g., sex ratio theory assumes that selection favors mothers who produce offspring sex ratios that will increase the number of grandchildren produced). If a test of a hypothesis yields a negative result, the lower level theory is more likely to be questioned than is the higher level theory.

In biology, multiple factors usually contribute to shaping a given relationship; thus one problem, especially in areas of biology where theory is less well developed, is that even when hypotheses are based on different theories, they sometimes fail to generate mutually exclusive predictions (this seems to be more common in ecology than evolutionary biology). Likewise, most tests of hypotheses require a suite of background assumptions about the study system so that when a prediction derived from a hypothesis fails, it can be difficult to determine whether the underlying theoretical framework is flawed, or if one or more assumption is erroneous. Instead, differences in the predictive power of hypotheses derived from competing theoretical propositions are often used as a measure of relative success, although this will vary depending on the extent to which confounding/interacting variables are taken into account (e.g., Case 7).

Ideally, meta-analysis of *experimental* data is necessary to test hypotheses. What constitutes causation is a topic fraught with philosophical pitfalls, but most biologists take the view that causation is demonstrated through experimentation. If a factor is manipulated and this leads to a meaningful change in a response variable, then it is said to have caused the change. It should be noted, however, that many biological phenomena are resistant to direct experimentation (e.g., the timeframe of the expected response is too long, or the scale of the experiment prohibitively large). Consequently, there is a tradition in ecology and evolutionary biology of using phylogenetic comparative methods (Chapter 17) and "natural experiments" to provide independent tests of hypotheses (and the theories that underpin them) (e.g., Sax et al. 2007). These *correlational tests* are always imperfect because other factors might covary with the "manipulated" factor. For example, the causal role of temperature (rather than, say, summer day length) on the latitudinal cline in body size that led to Bergmann's rule could be tested by looking for altitudinal clines at the same latitude, because temperature declines with elevation. This controls for day length but, of course, other factors change with altitude that might influence body size and confound interpretation. Similarly, phylogenetic regressions reveal correlations between traits and selective forces/environments, but they can never directly prove that the putative selection force has driven the evolution of a trait (Harvey and Pagel 1991). Even so, if a sufficient number of natural experiments are consistent with theoretical predictions, then a theory is often considered to have been corroborated by the weight of a variety of independent, albeit individually flawed, tests.

   The ways in which theories develop are often pivotal to how they are tested. Theories sometimes make predictions about natural patterns that have not yet been investigated. In such cases a theory is usually first tested by examining whether the predicted outcome exists. To return to an earlier example, split sex ratio theory made the prediction that in social insects, differences among colonies in the asymmetry of relatedness between workers and male versus female reproductives will result in some colonies specializing in producing male and others female reproductives (Boomsma and Grafen 1990). Meta-analysis of natural correlations confirms that this theoretical prediction holds when relatedness asymmetry varies due to either differences in the number of queens per colony or in the number of matings by a queen in single queen colonies (Meunier et al. 2008). In other cases, theories account for a pattern that is already well documented (e.g., a known rule), and there are usually several alternate theoretical explanations for the pattern. Tests depend on direct experimental manipulation of a causal factor (which is usually equivalent to testing an assumption of the theory), or on the use of correlational tests from innovative "natural experiments."

Testing theoretical hypotheses using meta-analysis

Meta-analysis probes the wider applicability of hypotheses originally tested at a smaller scale. For example, instead of asking a question about Lake Victoria or Malawi, one asks whether species diversity generally declines with depth in lakes. Meta-analysis can similarly detect factors that explain heterogeneity in effect sizes, which often generates new hypotheses if these patterns were not apparent in the original studies. For example, is the rate at which diversity declines with depth correlated with lake size? This is a question that could not, even in principle, be asked in the original studies (assuming one lake per study) because it is at a larger scale of analysis. Of course, larger-scale questions have different sampling units so the result of a single meta-analysis should be viewed with caution. For example, even if the rate of diversity decline with depth increases with lake size in East Africa, it will require several meta-analyses based on different sets of lakes to confirm that this is a universal rule. This is similar to the use of repeated estimates of allometric scaling coefficients from different taxa (Case 4). Using meta-analysis to tackle larger-scale questions is underappreciated by ecologists and evolutionary biologists (a few existing examples are described in Chapter 26). The more familiar cross-study approach to hypothesis testing is the phylogenetic comparative method (Chapter 17). This can, however, only detect coevolution of traits or repeated evolution of traits in specific contexts. Causality (evolution in response to a stated force of selection) is not directly tested, although it can be inferred based on plausibility (Harvey and Pagel 1991). In contrast, we can test for causality in some meta-analyses.

Causality can be inferred in a meta-analysis if the original studies involved random assignment of individuals to experimental groups, and the meta-analyst is asking whether there is a treatment effect. For example, Roberts et al. (2004) could draw conclusions about the effect of testosterone on immune function because testosterone levels were experimentally manipulated in primary studies (Case 2). This is an example of what Cooper (1982) has called "study-generated evidence," where the original studies directly test for a causal relationship. He juxtaposed this with "review-generated evidence," where the meta-analyst tests whether study-level characteristics explain variation in effect sizes (e.g., methodology used, taxon tested, geographic or temporal trends). Here it is not possible to infer causality because the moderator variable might be correlated with factors that are the true causal agents. For example, Sheldon and West (2004) found that the effect of maternal condition on the production of sons ($r_{condition}$) depended on whether condition was measured before or after conception (Case 3). It is possible that the methodology used was a cause of the difference in effect size, but it is also possible that it covaries with other factors (e.g., preconception assessment might be more likely in captive than wild populations).

For ecological studies it will often be possible to subsequently subject review-generated hypotheses to experimental tests by manipulating putative causal factors at a higher level, and then conducting another meta-analysis. For example, the methodology used to measure condition can be randomly assigned to species and the effect size for $r_{condition}$ then compared between studies using the two methods. The distinction between "study-generated" and "review-generated" evidence is therefore likely to be respected by ecologists (but see the discussion of this issue in Chapter 20). As a practical issue, it is often the case that primary studies are mixed with respect to the evidence used to test a hypothesis. Consider studies of the effect of a soil nutrient on plant toxicity. In some studies, nutrient levels are experimentally manipulated, whereas other studies simply investigate the correlation between nutrient levels and toxicity. The two types of studies must be analyzed separately if one wants to use a meta-analysis to infer a strict sense causality.

However, many evolutionary (and some ecological) questions cannot, even in principle, be tested using experiments because they may involve (1) inferences about past selection (i.e., the causal agent of change), (2) large spatial scale patterns, or (3) many different species, and so on. When evolutionary biologists conducting a meta-analysis detect study level correlates of effect sizes, they are therefore more likely to describe them as causally responsible for variation in effect sizes, especially if there is an already well-supported theory predicting that these moderating factors affect selection. For example, consider the interpretations of the moderating influence of estimates of natural selection's strength on levels of kin discrimination and offspring sex skew provided by Griffin and West (2003) and Griffin et al. (2005) (Case 3). This kind of causal inference is technically incorrect because it is not based on experiments involving randomly assigned subjects. Even so, reliance on the weight of evidence from multiple independent correlational tests in natural experiments and phylogenetic comparative analyses is a standard feature of many evolutionary explanations for trait evolution.

DOES THE PREDICTED RELATIONSHIP EXIST?

Many false theories are highly appealing because they agree with conventional wisdom (see Levitt and Dubnar 2005 for some entertaining examples). Whether the underlying data support them is often unclear. In the absence of a formal quantitative analysis, it is easy to promote studies that corroborate hypotheses based on a favoured theory and downplay others. Recent meta-analyses have shown, however, that there is little evidence for some widely accepted hypotheses in ecology and evolutionary biology. We first offer two illuminating examples of meta-analyses that respectively show that there is very weak evidence that a predicted causal relationship exists or that a fundamental assumption of a hypothesis holds. We then describe a third case where a series of meta-analyses show that predictions based on a specific theory are actually supported, even though many researchers have vigorously disputed this due to conflicting results from individual studies.

Case 1: The Janzen-Connell hypothesis

The Janzen-Connell hypothesis (JCH) is a hugely influential and widely cited explanation for the maintenance of high species diversity in forests (Janzen 1970, Connell 1971). It still forms the conceptual basis for numerous primary studies, especially in tropical plant biology. The main argument is based on the empirical claim that species of (adult) trees are more widely dispersed than expected by chance; the causal statement is that this is because seed-to-adult survival is lower near parents. One proximate causal mechanism for distance-dependent recruitment is that parents are a reservoir of species-specific harmful pathogens (or possibly herbivores) that increase mortality rates for nearby seeds/seedlings. A key prediction is therefore that seed survival to adulthood increases with distance from the parent. To test this prediction, Hyatt et al. (2003) compiled data from 152 studies on 75 species that *experimentally* planted seeds or seedlings either at different distances from adults or in habitats where conspecific adults were present or absent, and then measured survival. The mean log odds ratio was close to zero (95% CI: -0.13 to 0.27) (a mean log odds ratio > 0 indicates survival is greater farther from, rather than closer to, the parent). There was no significant heterogeneity among studies, and effect sizes did not differ between studies of different habitat types and did not depend on the criteria used to categorize the distance of seedlings from parents.

Hyatt et al. (2003) provide no support for one important prediction of the JCH, although at least three other key predictions still need to be tested. Given the large number of primary studies that have tested the JCH, it will be interesting to see if researchers continue to conduct similar studies, or if Hyatt et al.'s study will change the nature of future primary research on the JCH and how it is described. The practice of placing a high value on studies that report strong, clear results (Chapter 23) means that there could be a trend to continue to prominently cite studies that have reported large effects consistent with the distance-dependent prediction of the JCH alongside the findings of Hyatt et al. (2003), even when the latter included the former in their database (e.g., Svenning and Wright 2005). Hyatt et al. (2003) were careful to note that some species could genuinely have recruitment patterns where seedling survival is lower close to adults. Even so, the bigger picture is that, based on the statistically significant positive effects of distance on seed survival, the number of species identified as showing lower seedling recruitment when closer to the parent is no more than what is expected by chance. The onus is therefore on researchers who reported significant results for a given species to confirm their validity through study replication (Kelly 2006).

Case 2: The immunocompetence handicap hypothesis

Sexual selection is one of the processes most intensively researched by evolutionary and behavioral ecologists (Jennions and Kokko 2010). A central issue is whether extravagant male traits are "honest" signals of quality, so that females benefit from choosing elaborately ornamented mates (Kokko, Jennions, et al. 2006). Unless the efficiency with which a signal is produced varies among males, then all males should invest equally in signals (Getty 2006). This eliminates the very variation among males required to make female choosiness worthwhile. Folstad and Karter (1992) introduced the immunocompetence handicap hypothesis (ICHH) to partly explain how signal honesty is maintained, and it has stimulated hundreds of studies. They noted that testosterone appears to mediate the expression of sexual traits in male vertebrates, but that there was also evidence that testosterone is an immunosuppressant. They therefore argued that only "high quality" males in good condition can produce sufficient testosterone to generate large sexual signals and maintain an effective immune system. If true, sexual signals are honest indicators of quality, because testosterone mediates a trade-off between immunity and attractiveness.

It is inadvisable to use the direction of phenotypic correlations between specific life history traits to detect trade-offs if males vary in their ability to acquire resources that are then differentially allocated to a suite of traits (Reznick et al. 2000, Hunt et al. 2004). For example, there is an obvious financial trade-off between money spent on houses and cars. Even so, rich people can afford nice houses and flashy cars, so there is a positive correlation between house and car value in societies where people vary widely in income levels. A better way to detect trade-offs is through experimental manipulations. Researchers have altered testosterone levels by castrating males or by surgical insertion of silastic implants; the effect on immune function or parasite loads was then monitored to test the key ICCH assumption that testosterone is immunosuppressive. Roberts et al. (2004) calculated 36 effect sizes from 21 *experimental* studies of 24 species. Using species as the unit of analysis, the mean effect did not differ significantly from zero (Hedges' $d = 0.32$; 95% CI: -0.04 to 0.68). This surprising finding undermined a basic tenant of a hypothesis ubiquitously invoked to explain the evolution of male sexual ornaments. It has prompted consideration of other mechanisms that could resuscitate the ICCH. Interestingly, this generated a second meta-analysis that reviewed a neglected proximate mechanism that acts in the reverse direction: namely, that immune system activation might suppress testosterone production (and hence sexual trait expression). This trade-off could also maintain sexual signal honesty. Boonekamp et al. (2008) conducted a meta-analysis of studies that experimentally elevated immune system activity by injecting nonpathogenic antigens and then measured testosterone levels. They found a strong positive effect ($r = 0.52$; 95% CI: 0.41 to 0.61), but they could only locate 13 studies of 6 species. Their meta-analysis reveals both an intriguing pattern and a shortage of critical data. If additional studies confirm that immune system activation depresses testosterone production, and this effect is stronger than any effect of testosterone on immune system function, then the two meta-analyses of Roberts et al. (2004) and Boonekamp et al. (2008) will have overthrown a key assumption of the ICCH that has been widely accepted for over 15 years.

Case 3: Sex ratio theory and the Trivers-Willard hypothesis

Sex ratio theory predicts the extent to which selection favors parents that can bias the sex of their offspring, and also predicts by how much they should bias allocation towards one sex (Frank 1990, West 2009). It has been spectacularly successful in explaining sex ratio skew in insects with haplodiploid sex determination where mothers control offspring sex by deciding whether to fertilize eggs (unfertilized eggs develop into males) as they are laid (Hardy 2002). There is debate, however, about the extent to which vertebrates with chromosomal sex determination can adaptively bias offspring sex (Cockburn et al. 2002). There is also skepticism about most of the reported statistically significant results. So many post hoc tests are conducted trying to link offspring sex ratios to other variables that significant findings might simply reflect type I errors (Palmer 2000, Ewen et al. 2004).

Three key predictions about parental optimization of offspring sex ratios have been the subject of meta-analyses that primarily used *observational* data from vertebrates. The first is that females breeding with attractive males should produce more sons because, if attractiveness is heritable, the offspring have above average reproductive success (Burley 1981). The second prediction is that in cooperative breeders, parents should bias the sex ratio to the helping sex when the number of helpers in the group is suboptimal (Pen and Weissing 2000). (This is a case of local resource enhancement.) If the optimum number of helpers per group is high, there should also be a population level sex bias toward the helping sex This was meta-analyzed by Silk and Brown 2008, but it should be noted that there are theoretical problems with the use of population sex ratios to test individual optimization (see West and Sheldon 2002). The third prediction is that females in better condition should bias offspring production toward the sex that benefits most. This hypothesis was introduced by Trivers and Willard (1973) (TWH) in a landmark paper that stimulated most of the subsequent hypotheses about other factors that could lead to parents optimizing offspring sex (e.g., the first and second predictions, above). For polygynous mammals (in which males can have several mates), the TWH predicts that females in good condition will produce more sons. This is based on the observation that there is usually greater variance in male than female mating success, and the twin assumptions that the former is largely determined by male-male competition for mates and that females in good condition produce more competitive sons. This prediction has some unstated assumptions about the effect of maternal condition on other offspring traits (Leimar 1996). For example, high female social rank in primates might improve condition, but rank is inherited down matrilines so production of high status offspring could favor mothers in high condition producing more daughters (Silk et al. 2005). Despite the pitfalls of testing background assumptions, the TWH remains hugely influential.

West and Sheldon (2002) conducted the first meta-analysis to demonstrate a positive correlation between male attractiveness and production of sons ($r = 0.19$, 95% CI: 0.04 to 0.34, $n = 8$ species), and between the number of helpers in a group and production of the helping sex ($r = 0.40$, 95% CI: 0.19 to 0.57, $n = 4$ species). Griffin et al. (2005) revisited the issue of sex ratio skew in cooperative breeders. Using a larger data set of 11 species, they failed to find a significant relationship between production of the helping sex and the number of helpers in a group ($r = 0.17$, 95% CI: -0.01 to 0.35) (but see below for a twist). West et al. (2005), however, reported a significant positive mean effect using an even larger data set.

Sheldon and West (2004) then conducted a meta-analysis to test the TWH. They confined their analysis to ungulates whose breeding biology is most consistent with the underlying assumptions (e.g., polygynous mating systems and small litter sizes; see Frank 1990). They reported a significant mean positive effect of female condition on the production of sons ($r = 0.09$, 95% CI: 0.04 to 0.14, $n = 7$ studies). This effect is small, explaining < 1% of variation in sex ratios, but it was far higher for some subsets of the data (see below). Sheldon and West (2004) were careful to note that their findings are consistent with the TWH, but do not constitute a definitive test because other explanations for a link between maternal condition and offspring sex ratios are possible (i.e., the data analyzed are observational). It should be noted, however, that a few primary studies have experimentally manipulated female condition and have shown that this affects the offspring sex ratio (for a review, see Hardy 2002).

Taken together, these meta-analyses have increased confidence that sex ratio theory can explain variation in offspring sex ratios in vertebrates. Significant findings do not appear to be solely due to type I errors (compare to claims by Palmer 2000). The challenge is still there, however, to explain why the variation explained is so small (e.g., 1 to 16%).

WHAT DO MEAN EFFECT SIZES OR PARAMETER VALUES REVEAL?

Stochastic events mean that ecological and evolutionary data sets are often sufficiently "noisy" that simply using meta-analysis to test whether there is a nonzero effect is a major goal. As a research field matures, however, theories should reach the stage where precise predictions are made about the form of a relationship or, at the very least, competing hypotheses or explanations can be distinguished based on a difference in the predicted direction of an effect. Here we outline two case studies. The first meta-analysis involves estimation of a parameter, rather than an effect size (Chapter 7) to test a biological rule (and thereby determine which of two theories is more consistent with the data). The second meta-analysis involves a comparison of effect sizes between two "treatments" that allows for a direct test of competing theoretical hypotheses.

Case 4: Allometric scaling coefficients

In recent years there has been much controversy about how body size scales to metabolic rate (and other traits) (White et al. 2007). Given $MR = aMb$ (where MR = metabolic rate; M = body mass; a = constant, which might be taxon specific; b = allometric scaling coefficient), what is the value of b? The long-held "Euclidian geometric" hypothesis is that b is 2/3 (White and Seymour 2003), based on a decrease in the surface to volume ratio as body size increases; this is because surface area increases as a square, and volume (i.e., mass) as a cube, of linear size measures. Recently, however, high-profile theories have emphasized the mechanics of energy and material transport in branching delivery systems with fractal dimensions (e.g., arteries and plant veins). This "fractal geometric" theory predicts that $b = 3/4$ (Enquist et al. 1998, West et al. 2002). Determining the value of b is a substantive issue because strong claims have been made about the extent to which a metabolic theory of ecology based on a 3/4 scaling rule can explain the organization of nature at levels ranging from individuals to ecosystems (Brown et al. 2004).

At least five recent meta-analyses have estimated b for basal metabolic rate (other measures of MR are also considered in some analyses). These studies have used different data sets, taken slightly different analytic approaches and reached contrasting conclusions. Dodds et al. (2001) and Glazier (2005), for example, analyzed data from birds and mammals and concluded that there was little evidence favoring 0.75 over 0.67. In contrast, Savage et al. (2004) concluded that $b = 0.75$ in mammals, and Farrell-Gray and Gotteli (2005) argued strongly that data from birds and mammals respectively are 7 and 105 times (based on likelihood ratio tests) more consistent with 0.75 than 0.67, but for reptiles 0.75 is only twice as likely as 2/3. So, despite the potential for meta-analysis to yield a definitive answer, the situation remains unclear. There are some posited explanations for these differing conclusions. For example, the difference between earlier studies of mammals reporting $b = 0.67$ (White and Seymour 2003) and the 0.75 of Savage et al. (2004) is that the latter included ruminants in their analysis. Basal MR is difficult to measure in ruminants (McNab 1997), and their exclusion yields an estimate of 0.67 (White et al. 2007). This example is a reminder that the conclusions of meta-analyses, as with primary studies, can vary depending on the reliability of the data used. It always pays to inspect the data set used in a meta-analysis to see whether the inclusion criteria, however objective, lead to the use of inappropriate data on biological grounds.

To date, White et al. (2007) have conducted the most comprehensive meta-analysis on the scaling of metabolism, using 127 published data sets covering birds, mammals, and several ectothermic taxa. They also tried to control for potential sources of variation by taking into account the type of metabolic rate measured (e.g., exercise, daily, or field) and the statistical method by which b was calculated. As noted by others, they reported a marked difference between ectotherms and endotherms. For regressions, the mean b was significantly greater than 0.75 for ectotherms and 0.67 to 0.75 for endotherms. After correcting for phylogeny the results generally agreed with $b = 0.67$ for endotherms and $b = 0.75$ for ectotherms.

White et al. (2007) make the general point that given differences in b among taxa it is misleading to argue for a universal scaling based on 1/4's because of fractal dimensions. Although models have been developed that incorporate variation in scaling exponents (e.g., Kozlowski et al. 2003), and some fractal models predict deviation from 0.75 scaling (e.g., West et al. 1997) these models do not make strong a priori predictions. The onus is therefore on theoreticians to develop new mechanistic models that account for the discrepancies and differences revealed by meta-analysis. For example, although Farrell-Gray and Gotteli (2005) found greater support for 0.75 than 0.67 for endotherms, the 95% CI for b excluded 0.75. They therefore developed a statistical model showing how measurement error creates a downward bias in slope estimates that might account for this discrepancy.

White et al. (2007) conclude that determining b is an issue of major relevance because many cross-species studies correct for body mass using "standard" formula based on fixed values of b that are clearly inappropriate for some taxa. They note: "Savage et al. (2004) suggested that a century of science was distorted by trying to fit observations to an unsatisfactory surface law ($b = 2/3$). Given the apparent widespread acceptance and application of $b = 3/4$ it seems history is in danger of repeating." White et al. (2007) provide an interesting test case of the extent to which a large-scale meta-analysis is viewed as a comprehensive summary of the available data, or as "just another study" akin to a primary study that can be cited as supporting or refuting a theoretical prediction. If it is the former, as one would hope, there should be major reassessment of the uncritical use of $b = 3/4$ (see also White et al. 2009).

Case 5: Herbivory and plant invasions

Invasive exotic species have long been studied (e.g., Elton 1958), not only because of the needs of environmental managers and conservationists (Chapter 26), but also because they provide "natural experiments" that offer insights into basic evolutionary and ecological processes, such as rates of adaptation or how community structure affects ecosystem function (Sax et al. 2007). A long-standing and largely unresolved question is why some species are more invasive than others. Darwin (1859) predicted that alien plants are more likely to become established in an area if it *does not* contain native plant species to which they are closely related (e.g., from the same genus), because the invader would then encounter fewer predators (herbivores) capable of eating it, and would be subject to less interspecific competition. An alternate hypothesis is that alien plants are more likely to invade an area that *does* contain closely related plants because they will be preadapted to herbivory imposed by generalist herbivores (e.g., the alien plants' secondary metabolites will be better able to deter native herbivores), even if there is greater interspecific competition with related species.

Ricciardi and Ward (2006) distinguished between these hypotheses using data from 18 *experimental* studies comparing alien plant survival (or, in a few studies, another measure of performance) between herbivore exclusion enclosures and control plots. They found that the positive effect of exclusion of native herbivores on survival is six times greater for alien plants in an area where they are novel genera, than in those where they are not novel (Fig. 24.2). This meta-analysis reveals one of those rare instances where Darwin appears to have made a mistake. Ricciardi and Ward (2006) conclude that the likelihood of a plant species invading is at least partly attributable to the extent to which the herbivore community has previously had experience with functionally similar species.

Ricciardi and Ward's analysis was prompted by an earlier meta-analysis by Parker et al. (2006) who showed that the previous evolutionary experience of plants with generalist herbivores might be a good predictor of the likelihood they invaded a community. Parker et al.

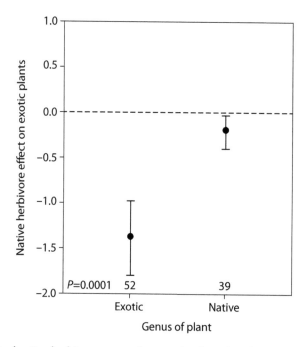

Figure 24.2. Effect of native herbivore removal on exotic plant abundance or survival. Values below the dashed line indicate a decrease in relative abundance in the presence of native herbivores. The effect of herbivore presence is more detrimental to plants belonging to exotic genera (not in the same genus as a native species in the invaded region) than to native genera (in same genus as a native species). Effect size—ln(H+/H−)—is presented as mean ± 95% CI (bootstrapped and bias corrected). Numbers below the confidence intervals indicate the number of experiments contributing to the mean. The *P*-value is for the test that the effect size for exotic and native genera is the same (redrawn from Ricciardi and Ward 2006).

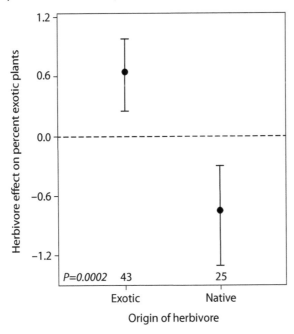

Figure 24.3. Effect of native and exotic herbivores on the relative abundance of exotic plants. Values below the dashed line indicate a decrease and those above indicate an increase in exotic plants' relative abundance in the presence of herbivores. Effect size—$\ln(H+/H-)$—is presented as mean $\pm$ 95% CI (bootstrapped and bias corrected). Numbers below the confidence intervals indicate the number of experiments contributing to the mean. The P-value is for the test that the effect size for exotic and native herbivores is the same (redrawn from Parker et al. 2006).

(2006) reached this conclusion after testing for a difference in the effect of native and alien herbivore exclusion experiments on the relative abundance of exotic to native plants (i.e., a measure of the extent to which invasive species predominate). Using data from 35 studies they showed that the *direction* of the effect differed significantly between the two types of herbivores (Fig. 24.3). The relative abundance of alien plants was suppressed by native herbivores and enhanced by alien herbivores. The former suggests that native herbivores preferentially consume alien plants because they are less resistant to native herbivory. However, the fact that alien herbivores actually enhance alien plant success only makes sense if the alien plants and herbivores used in these studies have a shared evolutionary history. Indeed, this was the case, with 88% of alien plant species sharing the same native ranges on their continent of origin as the alien herbivores.

A major implication of this meta-analysis is that it can partly account for the perplexing fact that European plants have more often invaded other parts of the world than plants from these areas have invaded Europe. This is because European herbivores have been widely introduced worldwide (e.g., pigs, rabbits, horses), while the reverse direction of introductions is rarer. In addition, Europeans often decimated native herbivores on other continents (e.g., bison, kangaroos). Based on the findings of Parker et al.'s (2006) meta-analysis, this should simultaneously increase the detrimental effect of alien herbivores on native plants, and reduce the cost to alien plants from exposure to native herbivores (at least mammalian herbivores) outside Europe.

Given the general failure of researchers to predict which species will invade or, in a natural context, be a successful colonist, the findings of these two meta-analyses are impressive. At the heart of both is a planned comparison of effect size between two treatments/classes to distinguish between competing hypotheses. In the absence of a meta-analysis it is unlikely that these patterns would have been so readily apparent, and we therefore predict that both these meta-analyses will be highly influential in the years to come.

EXPLORING HETEROGENEITY IN EFFECT SIZES

One of the most valuable uses of meta-analysis is to modify existing theories or even generate new theories. First, this occurs when a putative biological rule is confirmed (e.g., mean effect is nonzero or exceeds a threshold value), which then spurs the search for a causal explanation for the natural pattern it describes. Second, and more interestingly, meta-analysis can reveal greater heterogeneity in effect sizes than can be attributed to sampling error. Identifying previously neglected sources of variation can lead to new hypotheses, refine existing approaches to data collection (Case 6), and provide correlational tests of existing theories (Case 3). Third, meta-analysis can test whether there is a consistent interaction between factors that affect response variables, so that the interaction itself becomes the object of theoretical explanations (see Case 7). Here we focus on the last two phenomena, and illustrate them by revisiting Case 3, and by introducing two more case studies.

Case 3 revisited: Variation in the extent of sex ratio skew

If a meta-analysis shows that the mean effect size does not differ from the null value, this does not preclude looking for moderating factors. Tests for heterogeneity are still worthwhile because a null value is compatible with significant differences among groups that respond in opposite directions to the same experimental treatment or natural factors. An illustrative case study is provided by Ewen et al. (2004), who conducted a meta-analysis of studies that investigated correlations between offspring sex ratios and a range of parental and temporal factors that have been theoretically linked to biased sex ratios in passerine birds (e.g., female age, male quality, clutch size, and laying date). The mean effect size was close to zero ($r = 0.01$, 95% CI: -0.01 to 0.02, $n = 139$), and Ewen et al. concluded: "Facultative control of offspring sex is not a characteristic biological phenomena in breeding birds." Later, the authors discovered a spreadsheet mistake that led them to conclude erroneously that there was no heterogeneity in effect sizes (but the mean effect was unaffected) (Cassey et al. 2006). Subsequent inspection of subgroups in the data showed a small but significant positive effect of male quality on sex ratios ($r = 0.08$, 95% CI: 0.03 to 0.14, $n = 17$) in the same direction reported by West and Sheldon (2002), as well as significant effects for some temporal traits (e.g., egg laying sequence and date in breeding season).

A meta-analysis of correlates of sex ratios by Griffin et al. (2005) provides an even more interesting example of the value of looking for moderator variables. In cooperative breeders, parents should bias the offspring sex ratio toward the helping sex when the number of helpers in the group is suboptimal, and vice versa. The observed relationship between helper number and the sex ratio skew (r_{sex}) should therefore be positive. Griffin et al. (2005) initially noted that the mean value of r_{sex} did not differ from zero, but closer inspection showed that a continuous moderator variable could explain much of the variation in r_{sex}. This moderator was a second effect size, namely, the influence that an increase in the number of helpers had on a parent's reproductive success (r_{help}). This is an exciting finding because it fits with the general premise that adaptations are more likely to be fine-tuned when variation in a trait has a strong effect on fitness. The precision of the adaptations (i.e., proximate mechanisms that allow parents to skew the sex ratio) that generate positive values of r_{sex} are positively associated with the fitness benefits that accrue from so doing (i.e., r_{help}). The relationship between r_{help} and r_{sex} therefore provides a correlational test that corroborates the theory of natural selection by showing that an adaptation (extent to which parents skew offspring sex) is more precise when selection is stronger. As a cautionary aside it is worth noting that there was no significant heterogeneity in effect sizes. It was only the creative insight of applying a theoretical expectation to control for a potential confounding factor, which could create problems in assessing the value of r_{sex} (i.e., inappropriate inclusion of studies where "helpers" do not actually help), that prompted this finding. In a similar analysis, Griffin and West (2003) also corroborated kin selection theory by showing that the extent to which helpers preferentially assist close relatives is also positively correlated with r_{help}. We are sure that comparable use of effect sizes as moderator variables in meta-analyses on many other topics will be similarly illuminating.

Meta-analysis of sex ratios also illustrates how identifying a source of heterogeneity can refine how theories are tested. Sheldon and West (2004) found a significantly greater mean effect size when the effect of female condition on the production of sons was measured before versus after conception ($r = 0.17$ versus 0.05, $n = 15, 17$). This makes sense post hoc if gestation and/or feeding sons rather than daughters has a more negative effect on maternal condition so that the link between preconception condition and the production of sons is weakened when postconception condition is used. Future studies should therefore measure condition before the breeding season starts. Similarly, there was a far stronger link between measures of condition based on behavioral dominance than morphological indices ($r = 0.25$ versus 0.06, $n = 11, 26$). Again there is a plausible post hoc explanation for this finding (e.g., dominance is a better predictor of longer term access to resources), but it is important to remember that such conjectures must be tested. Nonetheless, it seems sensible for future studies to collect behavioral and morphological data to test the TWH. Then, planned paired comparisons of effect sizes for the two measures of condition can be conducted to provide an independent test of the repeatability of the trends reported by Sheldon and West (2004).

Case 6: Seed size and plant abundance

Numerous studies have asked whether seed size can account for variation in plant abundance (e.g., Guo 2003). If true, it could explain differences in plant rarity at a range of spatial scales. The relationship is not straightforward, however, because the size and lifespan of adult plants affects the trade-off between seed size and number (Moles et al. 2004). Size and lifespan both affect the density and amount of cover provided by adults. Murray et al. (2005) conducted a meta-analysis to test the relationship between seed size and abundance from 18 communities. The mean effect was $r = 0.001$ (95% CI: -0.16 to 0.16). Again, however, exploration of sources of variation based on prior considerations showed that this null outcome obscured two strong effects. When abundance was based on density measures there was a significant negative correlation between seed size and abundance, but when it was based on percentage cover it was significantly positive (Fig. 24.4). Thus, seed size is an important predictor of aspects of plant community structure, but only when conceptual considerations are taken into account (i.e., whether by species abundance we are referring to the number of plants or the area that they cover). This result has theoretical implications because these findings are more consistent with some theoretical models than others (see Murray et al. 2005). It is, however, slightly disconcerting to note that, as with Griffin et al. (2005), there was no significant heterogeneity in

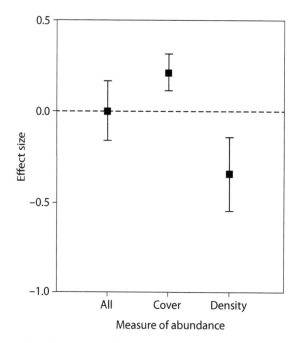

Figure 24.4. Effect sizes for the relationship between seed size and abundance within all plant communities, and when these are divided into those where abundance was measured as either density or cover. The horizontal line at zero indicates the absence of an effect. Effect size (Fisher's z-transformed r) is presented as mean ± 95% CI (bootstrapped and bias corrected) (redrawn from Murray et al. 2005).

the pooled sample of effect sizes. Heterogeneity tests based on the Q statistic have low statistical power when the number of studies is small, so a failure to detect heterogeneity does not mean that the data are homogeneous (see Chapters 9 and 22 in this book and Chapter 16 in Borenstein et al. 2009). This is a reminder that thoughtless application of meta-analysis protocols can mislead. There is always scope for scientific creativity and taking advantage of insights from other sources. In short, if there is theoretical reason to test for the influence of a moderator variable it is worth doing so, even if a heterogeneity test suggests otherwise.

Case 7: Competition and predation

Heterogeneity in effect sizes can arise due to interactions between causal factors (e.g., moderators; Chapters 8 and 20) whose magnitude differs among studies. Although rarely used by ecologists, factorial meta-analytic techniques can directly test for significant interactions. Full statistical models that replace multiple comparisons of effect sizes have a number of advantages (Chapter 8). In this example, each factor was subdivided into different classes based on levels of another factor (e.g., test for an effect size difference for the effect of the presence/absence of factor A in samples where factor B is present and when B is absent; repeated for factor B, with subdividing based on factor A). Gurevitch et al. (2000) introduced factorial meta-analysis to measure the effect of interactions between predation and competition on community structure. An earlier meta-analysis of field experiments showed that interspecific competition has a strong effect on growth and survival (Gurevitch et al. 1992) and a qualitative review suggested that the effect of predation is also strong (Sih et al. 1985). However, a review of matched comparisons of the relative effects of predation and competition was lacking (i.e., predator and competitor removal experiments in the same community). This comparison is of theoretical interest because ecologists differ in the emphasis placed on predator-driven "top down" and competition-driven "bottom up" processes in structuring natural communities. Quantifying predation-competition interactions is also of interest because nonequilibrium theories of community structure suggest that predation is a "disturbance" that reduces the deterministic outcome of competition to promote competitor coexistence.

Gurevitch et al. (2000) meta-analyzed 39 experiments from 20 studies that manipulated levels of *both* predation and competition and measured their effects on prey growth, survival, density, and mass. The overall effect of predation and competition differed depending on the response variable. Competitor removal has a modest positive effect on growth and mass, while predator exclusion had an effect of comparable magnitude in the opposite direction. In contrast, predator exclusion had a much larger effect than competitor removal on survival (although both in the same direction), while there was no difference in the effect of the two factors on density (Fig. 24.5). The relative importance of bottom up and top down processes in communities is therefore trait specific. However, most community structure theories emphasize population density so it is fair to conclude that predation and competition have comparable effects.

The theoretical prediction that predation reduces the effects of competition was strongly supported because the effect of competitor removal was always stronger when predators were absent (Fig. 24.6). Formal tests showed significant effect sizes for the competition-predation interaction for growth and density but not for mass or survival. This finding yields the practical

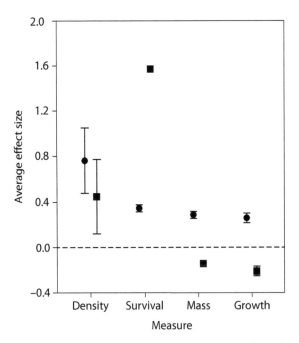

Figure 24.5. Effect sizes for competitor presence/absence (*circles*), and predator presence/absence (*squares*) on organisms' growth, mass, survival, and density. Values above the dashed line indicate an increase in the absence of competitors or predators. Effect size (Hedges' *d*) is presented as mean ± 95% CI (redrawn from Gurevitch et al. 2000).

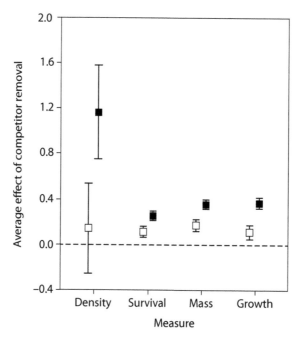

Figure 24.6. Effect size of competition in the presence (*open squares*) and absence (*closed squares*) of predators on organisms' growth, mass, survival, and density. Values above the dashed line indicate an increase when competitors are removed. Effect size (Hedges' *d*) is presented as mean ± 95% CI (redrawn from Gurevitch et al. 2000).

conclusion that accounting for natural levels of predation in communities should explain variation in effect sizes among different competitor removal experiments.

Gurevitch et al. (2000) provides an illustrative example of how heterogeneity testing to reveal differences in outcomes among groups can profitably direct future research. To take just one example, the interaction term for survival was significantly positive for anurans ($d = 0.69$, 95% CI: 0.39 to 0.99) and significantly negative for plants ($d = -2.57$, 95% CI: -3.13 to -2.02); competitor removal in the presence of predators had a positive effect on plant survival and a negative effect on anuran survival (Fig. 24.7). Gurevitch et al. (2000) proposed proximate explanations for this, and several other taxon-specific effects that now need to be tested experimentally (e.g., that in anurans competitor presence diverts predation pressure). Indeed, following on from the pioneering work of Gurevitch et al. (2000), more sophisticated analyses have taken into account the extent to which predation-competition interactions are affected by such factors as productivity (Worm et al. 2002), ecosystem type, and dominance structure of plant assemblages (for a meta-analysis see Hillebrand et al. 2007).

RANKING THE IMPORTANCE OF DIFFERENT FACTORS

The magnitude of almost everything ecologists and evolutionary biologists measure is influenced by multiple factors. For example, many traits evolve under multiple selective forces so that no single factor can be described as driving the evolution of a trait. Given this, it is worthwhile determining which factors are better predictors of experimental outcomes or are more strongly correlated with a focal variable. Assessment can be based on consideration of both the

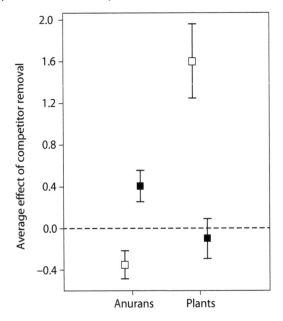

Figure 24.7. Effect sizes of competitor removal on survival in the presence (*open squares*) and absence (*closed squares*) of predators. Results are shown separately for plants and anurans. Values above the dashed line indicate an increase in survival when competitors are removed. Effect size (Hedges' *d*) is presented as mean ± 95% CI (redrawn from Gurevitch et al. 2000).

mean and variability in effect sizes. Here we offer two case studies to illustrate this approach in both between- and within-study situations.

Case 8: Badge size in house sparrows

Many bird species have sexually dimorphic chest markings. In house sparrows, for example, males have black bibs that are often referred to as "badges of status" because of evidence that they predict social dominance. There is, however, also evidence that badges are under sexual selection through female choice. If so, there is an expectation that badge size will signal information about traits of value to females. The potential benefits females gain by being choosy include greater male parental care (incubation assistance or feeding nestlings), or benefits that accrue to their offspring if fitness-enhancing male traits are heritable. The strength of the correlation between badge size and mating success (or reproductive success) provides indirect evidence about the relative importance of sexual selection (male-male competition and/or female choice) in the evolution of badges. Similarly, the strength of the correlation with specific male traits provides an index of the relative information content of badges (i.e., whether badge size reliably signals specific male qualities).

Nakagawa et al. (2007) conducted a meta-analysis of house sparrow studies examining phenotypic correlations between bib size and six other male traits ($n = 9 - 20$ studies/trait) (for an erratum, see Nakagawa et al. 2011). The exact information conveyed by house sparrow badges is disputed; studies have shown that several male life history traits are correlated with badge size, but there is disagreement among studies. The traits examined were status (male's social rank during fights), parental ability, age, body condition, cuckoldry (extent to which males lost paternity in their own nest), and reproductive success (net offspring production).

The correlation of each trait with badge size is shown in Figure 24.8. There are two key findings. First, badge size is most strongly correlated with fighting ability (status), moderately correlated with male age, and only weakly correlated with body condition. It is not significantly associated with any other traits. Second, there is no correlation between badge size and male reproductive success. There are three main conclusions to draw from this. First, badge size is a good signal of fighting ability (and is probably only spuriously correlated with age because fighting success improves with age). This suggests that badges have evolved under selection imposed by male-male competition (see also Andersson 2006). It is also consistent with a widely made claim that black plumage traits requiring melanin more often evolved through selection imposed by male-male competition than female choice (but see the meta-analyses by Griffith et al. 2006 and Santos et al. 2011 who reach a different conclusion). Second, there is no evidence that badge size signals prospects for male parental care to females. Third, even though badge size signals fighting ability, and badges might have evolved in response to this, there is no evidence that badge size is *currently* under sexual selection because it does not correlate with reproductive success. This implies that there must be strong counterselection against badge size or a correlated trait that negates the obvious benefits of greater fighting ability (for a meta-analysis, see Meunier et al. 2011). This immediately suggests that it will be fruitful for researchers to identify these "hidden" costs. Similarly, the meta-analysis suggests that female choice for larger badge size is weak. Nakagawa et al. (2007) point out that this is a genuinely unexpected finding because house sparrows are used as a model species for the

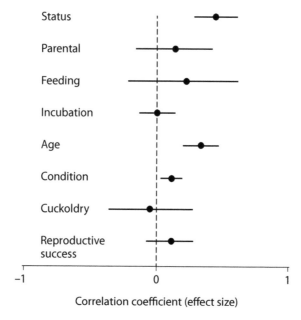

Figure 24.8. Effect size (*r*) presented as mean ± 95% CI for correlates of badge size in house sparrows. Status = male's social rank during fights; parental = investment in offspring (subdivided into feeding young and incubating eggs); condition = measure of body index (size-corrected mass); cuckoldry = extent to which males lost paternity because their own female mated with rivals; reproductive success = net offspring production (redrawn from Nakagawa et al. 2007).

study of sexual selection. This meta-analysis could therefore lead to a shift toward other species when researchers are interested in studying female choice for a male trait.

Case 9: Fighting in fiddler crabs

It is sometimes useful to compare the effect of different treatments within a single study. For example, Fayed et al. (2008) conducted four experiments on fiddler crabs where wandering males fight residents for a burrow. Their study attempted to account for factors contributing to the well-established rule-like pattern in many taxa that owners tend to win territorial fights against intruders (the "owner advantage effect"; for a review, see Kokko, Lopez-Sepulcre, et al. 2006). In each experiment Fayed et al. experimentally manipulated a single factor which had been hypothesized to cause owners to win more often. For practical reasons, the sample sizes varied among experiments, so confidence interval width differed. Only one null hypothesis was significantly refuted (Fig. 24.9): if the mechanical advantage an owner has when he uses his burrow for leverage is removed, this significantly reduces the proportion of fights that owners win ($P = 0.003$). Inspection of mean effect size estimates and their confidence intervals show, however, that (1) an owner's tenure on a territory might be equally important because the 95% CI widely overlaps that for a mechanical advantage; (2) owner knowledge of relative food availability on a territory probably has a smaller effect, but with a larger sample size it might be a significant contributing factor; and (3) differences in the relative fighting ability of intruders and owners are unlikely to be a major factor in generating an ownership advantage because the mean estimate is close to zero with a narrow 95% CI. This modest example illustrates the shift in perspective provided when attention is paid to effect sizes and their variance, rather than P-values alone (Chapter 23).

COMMON PROBLEMS

Several concerns have repeatedly been raised when ecologists and evolutionary biologists discuss the results of meta-analyses. A first concern is that effect sizes from different species are

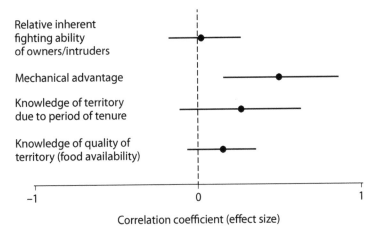

Figure 24.9. Comparison of effect size estimates from four experiments looking at the same outcome: the change in the proportion of fiddler crab fights won by owners of a territory (data from Fayed et al. 2008).

being treated as statistically independent data points. This reflects longstanding recognition of the need to correct for phylogenetic constraints, which has been inculcated by those using the comparative method (Harvey and Pagel 1991). Fortunately, methods to control for shared evolutionary history in meta-analyses are now available (e.g., Adams 2008, Lajeunesse 2009, Hadfield and Nakagawa 2010; Chapter 17). That said, these methods could be difficult to implement if a suitable phylogeny is unavailable, or if the taxa being considered are so diverse that it would be necessary to invest considerable effort in combining phylogenies to create a "supertree" that contained all the taxa under consideration.

A second concern is that only univariate tests of moderating factors are available in MetaWin, which is the software most often used by ecologists and evolutionary biologists. (More recently, however, there has been a rapid uptake of R and there are already several R packages that can be used to conduct meta-analysis, see Chapter 12.) In this chapter, we have pointed out that tests for interactions between factors have already been around for some time (Case 7), and that general modeling approaches that can incorporate multiple moderator variables are now available (Chapters 8 to 11, 16, 17, and 20). That said, there is currently no easy way to decide which model provides the most parsimonious explanation. The development of a model selection protocol for meta-analysis analogous to the use of the Akiake information criterion would be a useful advance (Helmut Hillebrand, pers. comm.).

A third concern is that a research bias toward disproportionately studying certain species or systems affects the conclusions of meta-analyses (Gurevitch and Hedges 1999; Chapter 14). This is a valid concern, so caveats should be placed on the extent to which results can be generalized. In our experience, because meta-analysts take a quantitative approach, they are more likely to raise concerns about biased use of certain taxa, or types of study systems, than are those producing a purely narrative review.

A fourth concern is that publication bias and selective reporting generate misleading conclusions. Again, however, formal tests for publication bias within a meta-analysis are a better way forward than simply despairing that the scientific literature is uninterpretable (Chapter 14).

Finally, the most self-evident problem with published meta-analyses is that heterogeneity in effect sizes is usually large. Often there are too few studies available for many key groups to test with reasonable statistical power (Chapter 22) whether a given factor is a moderating source of variation. When questions are asked at the study or species level it soon becomes clear, despite the growth in the scientific literature, how little data we actually have at our disposal to tackle specific topics. This will be readily apparent if one inspects almost any of the case studies we have described (e.g., see Roberts et al. 2004). The only way forward is to conduct more primary studies. One advantage of publishing a meta-analysis is that it can indicate which types of studies are most likely to advance a field.

CONCLUSIONS

Although met with initial scepticism, meta-analysis has become increasingly influential in testing theory. If a meta-analysis is done well and there are sufficient data to reach robust conclusions, then its findings set the stage for future empirical work. If, however, statistical power is low (Chapter 22), the data set is a biased sample of the real world (or the available studies), and/or the conclusions are weak (i.e., sensitive to the inclusion of new studies) or show temporal trends (Chapter 15), a meta-analysis can stymie legitimate research by creating an illusion of consensus where none should exist. Primary researchers must maintain a healthy skepticism when assessing published meta-analyses (e.g., consider the different conclusions reached in Case 4). Each meta-analysis can be viewed as a single study sampling a population of effect sizes. As with primary studies, chance events, type II errors, outlier data points, points with excess leverage, and unknown sources of bias can lead to false conclusions. There should be a healthy feedback between conclusions reached by meta-analysts and subsequent research by empiricists. Fortunately, most meta-analyses in ecology and evolutionary biology are conducted by active empirical researchers so there is some reassurance that this feedback will occur.

ACKNOWLEDGMENTS

We thank Megan Higgie for redrawing the figures.

Contributions of Meta-analysis
in Ecology and Evolution

History and Progress of Meta-analysis

Joseph Lau, Hannah R. Rothstein,
and Gavin B. Stewart

META-ANALYSIS WAS FIRST INTRODUCED in medicine and the social sciences, and was used extensively in these fields decades earlier than in ecology and evolutionary biology. In this chapter we review the development of meta-analysis in medicine and the social sciences in order to illustrate its background and compare its application in these fields to those in ecology and evolution. For the purpose of this chapter, by "medicine," we mean all aspects of health care and the biomedical sciences, including the diagnosis and treatment of individual patients, public health policy, health care financing and decision making, and basic and clinical biomedical research. Social science, as defined in this chapter, covers a variety of disciplines, including social, clinical, and organizational psychology, education, social welfare, criminology, business management, and economics. It encompasses both basic and applied research, and may focus on theory formulation and testing, or on informing policy or practice.

HISTORY OF META-ANALYSIS IN MEDICINE AND SOCIAL SCIENCES
Early efforts and criticisms

The statistician Karl Pearson is generally acknowledged as having published the very first meta-analysis; he summarized the effects of enteric fever (typhoid) vaccines given to British soldiers over the course of several military campaigns (Pearson 1904). After a long lull during which there was little development, the field expanded significantly starting in the 1970s. The term "meta-analysis" was coined only in 1976 to describe "the statistical analysis of a large collection of analysis results from individual studies for purposes of integrating the findings" (Glass 1976, 3). Several seminal meta-analyses were published in the late 1970s in both medicine and the social sciences. Thomas C. Chalmers published one of the earliest modern medical meta-analyses in the United States in his 1977 paper, "Evidence favoring the use of anticoagulants in the hospital phase of acute myocardial infarction" (Chalmers et al. 1977). This study combined six randomized controlled trials, five of which did not demonstrate statistical significance, and found a significant reduction of overall mortality in patients treated with anticoagulants. In the United Kingdom, Sir Iain Chalmers (unrelated to Thomas Chalmers) began work in 1978 on the Oxford Database of Perinatal Trials, which sought to identify and synthesize clinical trials on the effects of perinatal interventions. This database, with accompanying systematic reviews and meta-analyses of the included trials, can be seen as the forerunner of today's Cochrane Collaboration (discussed later in this chapter).

Among the earliest meta-analyses in the social sciences were those on the effectiveness of psychotherapy (Smith and Glass 1977), the effects of expectations on behavior (Rosenthal and Rubin 1978), the relationship between class size and academic achievement (Glass and Smith 1979), and the cross-situational validity of employment tests (Schmidt and Hunter 1977).

Meta-analysis was almost immediately met with biting criticism after its introduction, and was highly controversial for a time in both medicine and the social sciences. Hans Eysenck famously called meta-analysis "an exercise in mega-silliness" and stated that meta-analysis of psychotherapy interventions by Smith and Glass (1977) represented an "abandonment of scholarship" (Eysenck 1978). Similarly, in medicine, the meta-analysis by Chalmers et al. (1977) was criticized by Goldman and Feinstein (1979) in a commentary entitled, "The problems of pooling, drowning, and floating." The criticisms centered on the appropriateness of combining results obtained from different independent studies that were conducted in different settings, used somewhat different treatment protocols, were compared with different controls, and assessed somewhat different outcomes. More fundamentally, the controversy centered on changing ideas about reaching generalizations from data, and the nature of scientific evidence.

The issues of generalizability and heterogeneity in meta-analysis have since been a central topic of discussion and debate among researchers in disciplines in which meta-analysis has been applied. Numerous articles of empirical research and methodological innovations have been published over the past three decades on how best to handle heterogeneous data in meta-analyses and interpret their results, as well as how to best reach general conclusions based on available evidence. While heterogeneity remains an important topic for meta-analyses in medicine and the social sciences, this issue is now much better understood and is no longer considered controversial in either areas, as the methods have become both more sophisticated and widely accepted (Glasziou and Sanders 2002).

In addition to heterogeneity, dealing with the varying quality of individual studies has been a topic of heated discussion from early in the use of meta-analysis. Some research synthesists (e.g., Slavin 1986) have argued that only "the best" studies should be included; at the opposite end of the spectrum, others (e.g., Hunter and Schmidt 2004) have argued that all studies (except perhaps those with clearly fatal flaws) should be included, and that quality should be considered only as a moderator of effect. More recently, there has been a growing consensus in social science (e.g., Petticrew and Roberts 2006) that specific features that might lead to particular biases should be critically examined, rather than using overall quality as a criterion. Overall quality is considered by many to be highly subjective, and it has been found that most quality scales do not correlate well with one another (cf. Jüni et al. 1999). Whether this should be done a priori, or as a covariate analysis, is still the subject of some disagreement.

Critical appraisal of the evidence is considered essential for meta-analyses in medicine. Individual studies are assessed according to predefined criteria and quality grades are often applied. Poorer quality studies may be rejected from meta-analyses or their impact may be assessed in sensitivity analyses (see Chapter 7).

Progress since the 1970s

Despite early objections, meta-analysis was quickly taken up by medical and social scientists resulting in a quick increase in the number of publications since the 1970s (Figs. 25.1A and 25.1B). Many of the reasons for the rapid expansion of the number of meta-analyses are common to almost all scientific disciplines (Table 25.1). In many fields, available data increased exponentially as the number of published studies (often with small sample sizes) grew, often

A.

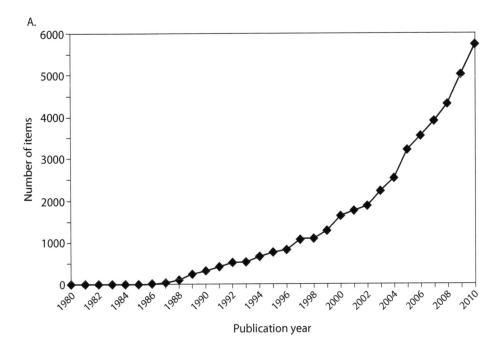

B.

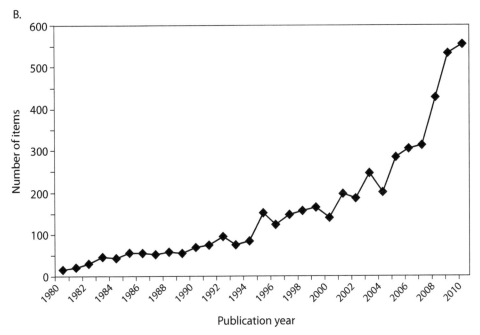

Figure 25.1. A. Number of articles per year indexed as "meta-analysis" in "publication type" in Medline. **B.** Number of studies indexed in the Social Science Citation Index with "meta-analysis" in the title; three variants of the spelling (meta-analysis, metaanalysis, meta analysis) were used in the search.

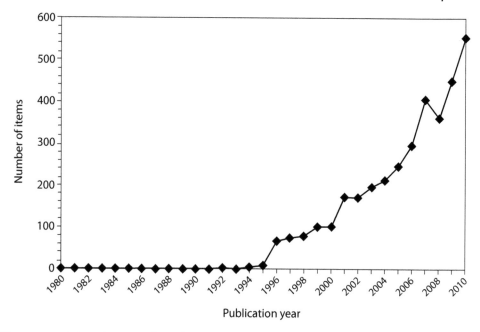

Figure 25.1. C. Number of ecological and evolutionary studies indexed in the Web of Knowledge with "meta-analysis" or "meta analysis" in title, abstract, or keyword. (Note that articles labeled as meta-analyses by keywords are often not meta-analyses *sensu stricta*. Approximately half the studies labeled as meta-analyses involve combination of effect sizes.)

without yielding consistent results. Narrative reviews typically ended with a call for more research, no matter how many studies had already been conducted. Researchers began to realize that despite their limitations, meta-analyses are more objective than are narrative reviews, and also more informative. Meta-analyses provide a systematic quantitative synthesis of an entire body of evidence, and produce a numerical estimate of the overall effect and its variance, as well as quantifying factors influencing the magnitude of effects across studies. Additionally, it was demonstrated that meta-analyses can address research questions beyond those posed by the individual studies in the analysis, such as identifying larger-scale patterns and finding research gaps in areas where insufficient evidence is available for reaching general conclusions.

Thousands of meta-analyses now appear each year in the medical literature (Fig.25.1A). As of January 2012, the National Library of Medicine in the United States, which indexes over 5500 biomedical journals published worldwide in the Medline database, had indexed over 48,000 items related to meta-analysis. Over 5700 items were indexed in 2010 alone. Because of the central role of treatments in patient management, early meta-analyses in medicine dealt mainly with randomized, controlled trials of interventions. While meta-analyses of therapeutic trials still predominate, there are increasing numbers of meta-analyses of observational studies of risk factor associations and studies evaluating diagnostic test accuracy; more recently, meta-analyses of genetic tests and associations have increased, as well as those concerned with health economics.

Similarly, meta-analysis has become ubiquitous in psychology, education, and other social sciences. For example, by the end of 2010 the Social Sciences Citation Index database contained nearly 5000 articles with meta-analysis in the title (Fig. 25.1B). As in medicine,

TABLE 25.1. Impetus for the need to synthesize (systematic review/meta-analysis) evidence that are common in ecology, medicine, and social sciences/education.

Need to make sense of data available for synthesis (coping with information explosion).
Need to quantify effects across all known information on those effects.
Need for critical appraisal (variable quality and conclusions of primary research studies).
Need to understand sources of variation in outcome among studies.
Need for interpretation: Lack of training by end users (e.g., managers, policy makers) on information to access and on interpreting primary literature or synthesizing research findings.
Need for objectivity (experts can be biased and narrative review articles unreliable).
Need to inform policy based on data (evidence-based medicine, evidence-based conservation, etc.).
Desire to extend understanding over a wider array of research conditions than possible in individual studies (i.e., at a larger geographic scale).
Need to identify knowledge gaps and inform future research agenda.

meta-analyses are generally replacing narrative reviews in the social sciences. Meta-analytic work has influenced policy and practice in the areas of education, psychotherapy, business management, and criminal justice in countries throughout the world (Cooper 2007a, Sherman et al. 2002, SIOP 2003, Weisz 2004). A large proportion of meta-analyses in social sciences are conducted to evaluate the effectiveness of a treatment or other intervention. However, many questions of interest to social scientists and their audiences cannot be addressed by conducting experiments, because they would be unethical, infeasible, or both, and meta-analyses of observational studies are common as well.

Evidence-based medicine and the Cochrane and Campbell Collaborations

Much of the increase in use of meta-analysis in medicine has been concomitant with the adoption of the evidence-based medicine (EBM) paradigm, beginning in the early 1990s (Evidence-based Working Group 1992). A popular definition of EBM is offered by David Sackett et al.: *"Evidence based medicine is the conscientious, explicit, and judicious use of current best evidence in making decisions about the care of individual patients. The practice of evidence based medicine means integrating individual clinical expertise with the best available external clinical evidence from systematic research."* (Sackett et al. 1996). Systematic reviews and meta-analyses of the scientific medical literature are the fundamental tools of EBM, and a synergistic relationship has developed among them, with the development of the EBM paradigm fueling the need for better methods of synthesis and interpretation of health care evidence; improved meta-analysis methods in turn are accelerating the acceptance and growth of EBM. Many decisions on patient care, health care policy, and financing in developed countries are increasingly based on these analyses.

Development and applications of meta-analysis in medicine and the social sciences is supported by several organizations and working groups. The Cochrane Collaboration (www .cochrane.org) is an international group, established in the 1990s; thousands of volunteer investigators worldwide currently receive training and support in conducting systematic reviews and meta-analyses. Volunteers maintain registries of clinical trials to facilitate the work of those conducting specific reviews. They also hand search journals to identify incorrectly tagged articles. Methods groups have been formed within the collaboration to advance the methodologies to conduct systematic reviews and meta-analyses. These are published online in the Cochrane Library and are disseminated worldwide. As an indication of the active growth in this work, a search conducted in February 2009 found 3737 completed systematic reviews and 1939 protocols included in the Cochrane Library; a new search conducted in December 2011 found 4892 systematic reviews and 2075 protocols. This work represents about 20% of the total number of medical systematic reviews (based on the proportion of Cochrane reviews in the Database of Abstracts of Reviews at the Centre for Reviews and Dissemination). In addition, over 564,000 clinical trials are included in the Cochrane Central Register of Controlled Trials.

The Campbell Collaboration (www.campbellcollaboration.org) was started by an international group of social scientists in 1999 who were inspired by the success of the Cochrane Collaboration; their goal was to produce, maintain, and disseminate systematic reviews of research evidence on the effectiveness of social interventions. The Campbell Collaboration currently has three active subject groups. These include social welfare, education, and criminology, as well as an umbrella methods group, with subgroups dedicated to research design, statistics, information retrieval, equity issues, economics, and training. The collaboration maintains a register of works in progress and completed reviews, as well as a database of randomized controlled trials in social welfare, education, and criminology, and a register of initiated trials.

Several work groups in medicine have developed guidelines to standardize and improve the reporting of meta-analyses of both randomized controlled trials and observational studies. These include QUOROM—Quality of Reports of Meta-analyses (Moher et al. 1999), now renamed PRISMA for Preferred Reporting Items for Systematic Reviews and Meta-Analyses; and MOOSE—Meta-analysis Of Observational Studies in Epidemiology (Stroup et al. 2000). Recommendations from the above working groups include checklists of elements on reporting information in the abstract, introduction, methods, results, and discussion, with the aim of improving the usefulness of meta-analysis. Recently, a website devoted entirely to enhancing the quality and transparency of health research was started; this can be viewed at www.equator-network.org. Similar steps to improve the quality of reporting practices are being taken in the social sciences and education; most notably, these include the introduction of meta-analysis reporting standards (MARS) in the sixth edition of the American Psychological Association's publication manual (American Psychological Association 2010).

General characteristics of meta-analyses in medicine and the social sciences

The vast majority of meta-analyses in medicine involve studies that directly compare two or more groups of patients (i.e., studies to evaluate the effectiveness of interventions or associations of factors with health outcomes), or two or more tests in the same group of patients (i.e., studies to evaluate diagnostic tests). Intervention studies may be carried out using an experimental (i.e., randomized controlled trial) or observational (i.e., cohort or case-control) design. Thus, both experimental and observational data may be used in meta-analyses in medicine, although usually those types of studies are analyzed separately, according to their study design. Dichotomous outcomes (e.g., dead or alive, cured or failed) are the most common type of data combined in meta-analysis in medicine. Most medical meta-analyses report an overall effect (e.g., relative risk) across studies. There are also many meta-analyses of continuous measures; for example, these include serum cholesterol level (mmol), blood pressure (mm of mercury), quality of life measurements (dimensionless unit). Metrics of effect size, such as standardized mean differences or correlations (Chapter 6), are seldom used in medicine. Meta-analyses in medicine also often explore heterogeneity in effect sizes to ascertain potential covariates explaining variation in effects.

In the social sciences, the central goal of a meta-analysis is usually to make sense of the overall body of data on a particular question. In some cases this will mean focusing on the summary effect (overall mean effect) across studies, while in others it will mean assessing the variation in effects across studies. Continuous outcome measures are most frequently used in the social sciences, since the underlying variables they measure are conceptually continuous (e.g., reading ability, on the job productivity, marital satisfaction); however, dichotomous outcomes are also used (e.g., rearrest within six month of completion of treatment, drug use, compliance). Meta-analyses have been conducted based on a wide variety of research designs, including randomized experiments, quasi experiments, and observational studies. In some cases, meta-analyses are done based on a single research design, but in other cases, results of studies using different designs are combined (with design features treated as covariates). Meta-analyses in the social sciences are expected to first report a summary effect (standardized mean difference, correlation coefficient) and a measure of the heterogeneity of the effect sizes in the analysis, and then to investigate potential explanations for variation in effect across studies when appropriate and feasible.

In general, meta-analyses in medicine try to achieve as much homogeneity as possible in the data being combined so that the answer will be meaningful for the management of patients with specific characteristics. To inform medical decision making, one needs to know the specific characteristics of the population, the setting, and the comparative efficacy and harms of available treatments; the results of the treatments can then be interpreted and applied in a meaningful way. This is possible in medicine because the number of primary studies is often very large; orders of magnitude more primary studies are published each year in medicine than in ecology and evolutionary biology or the social sciences. However, despite having more than 21 million items indexed in the Medline database, a frequent conclusion of systematic review in medicine is that there is an insufficient body of specific and high quality studies from which to draw definitive conclusions. Medical research also is generally conducted in more tightly controlled conditions in comparison with many primary studies in social sciences and ecology (see below).

Meta-analysts in the social sciences generally accept the idea that heterogeneity (variation) in effect sizes across studies is inevitable, even when the question addressed by each of the studies is essentially identical. This is because it is quite likely that the studies will differ in various ways (e.g., different populations, interventions, experimental design, and/or implementation). Therefore, in addition to computing a summary effect in the social sciences, a critical part of the meta-analysis is an assessment of the degree and potential causes of variation in effects across studies. Meta-analysts in the social sciences recognize that in their research domains, important variables may be complex and multifaceted, and that even clearly focused research questions involving the same variables can be posed either broadly or narrowly. In contrast to medicine (but similar to ecology and evolutionary biology), heterogeneity in the set of studies makes it necessary to examine the effect of plausible moderators (covariates) of effect size that exist in the real world. A narrow specification of participants, intervention (or other independent variable), comparator, and outcome would maximize the opportunity to study the causal efficacy of the intervention under well-controlled and highly standardized conditions; however, the results of such a meta-analysis would have limited applicability to the range of situations to which the analysts would like to generalize in the real world. This problem is exactly analogous to issues in ecological meta-analysis.

The PICO (population, intervention, comparator, and outcome) method is commonly used in medicine to formulate answerable research question and to guide data collection, analyses, and interpretation of results (Counsell 1997; Chapter 3). An example of a well-focused question for a meta-analysis of intervention studies might be: "What is the five-year overall mortality (*outcome*) in adults with hypercholesterolemia (*population*) taking statins (a cholesterol lowering drug) (*intervention*) compared with those treated with a low-fat diet (*comparator*)?" In this question, all elements of PICO are present. Further clarifications of the PICO elements (i.e., type and dose of statin) as well as specification of the study design (i.e., randomized controlled trials or cohort studies) would be needed to identify studies eligible for the meta-analysis.

As in medicine, social science meta-analyses increasingly use the PICO method to focus the research question and guide data collection. A well-focused question for a meta-analysis of intervention studies might be: "What is the improvement in reading ability (*outcome*) for sixth- to eighth-grade students who are reading below grade level (*population*), and who receive at least three months of weekly one-on-one after-school tutoring in reading (*intervention*), compared with those who participate in an after-school remedial reading class for the same amount of time (*comparator*)?"

Meta-analyses of randomized controlled trials of treatments may initially identify hundreds or thousands of studies, but after careful selection criteria are applied, they will ultimately include from as few as two to many dozens of studies. The average meta-analysis of health care interventions synthesizes about 10 studies (Engels et al. 2000). Meta-analyses involving more than a few dozen studies are uncommon in medicine. There are several reasons why the number of studies addressing a specific medical question, and hence their availability for meta-analysis, tends to be relatively small. Partly this is a result of asking a very specific question and reducing heterogeneity among study designs as much as possible. In addition, the approval of a human study evaluating treatments is made on the premise of equipoise, which is the principle that the conventional treatment and the new or experimental treatment are believed to be equally efficacious and have similar adverse effects. Otherwise, it would be unethical to conduct such studies. Once enough evidence on efficacy and safety has accumulated, it is considered unethical to conduct additional studies. Furthermore, clinical trials on humans are often complex, require a long follow up, and are expensive to conduct. Also, certain populations tend to be difficult to recruit for trials (e.g., children). Because of these limitations, many questions in medicine remain unanswered or inadequately addressed by clinical trials.

Observational studies are free from most of the above constraints, but since they cannot control for confounding factors, much more care is needed in selecting them for the meta-analysis and interpreting the results. Estrogen replacement therapy in women (Stampfer and Colditz 1991) and antioxidant supplements to reduce overall mortality (Bjelakovic et al. 2007) offer well-known examples where individual observational studies and their meta-analyses reported significant beneficial results that were not later confirmed by large randomized trials. (However, meta-analysis has been able to decisively resolve effects by combining results of small randomized trials before conclusive results were available otherwise; see Lau et al. 1992). Results from meta-analyses of observational studies in medicine are generally most useful for hypothesis generation. Of course, there are many cases in which randomized control trials are impossible, and conclusive evidence is established from observational studies (e.g., direct and indirect effects of smoking). In the absence of evidence from randomized controlled trials and when clinical practice or policy decisions are needed, the strength of recommendations based on meta-analyses of observational studies is not as strong as it would be when based on randomized controlled trials (Guyatt et al. 2008).

As in medicine, a comprehensive literature search for social science meta-analyses typically yields hundreds to thousands of citations that are evaluated for potential relevance. The final number of studies included in the meta-analysis may range from five to several hundreds. It is hard to estimate the number of studies in a "typical" meta-analysis in the social sciences, because the average varies across specific disciplines.

Most of the meta-analyses in the social sciences use inverse-variance weighting and analyze the observed effects from studies based on the methods introduced by Hedges and Olkin (1985). Some, however, particularly in the areas of industrial psychology and organizational science, use an approach called psychometric meta-analysis (Hunter and Schmidt 1990, 2004). This approach focuses attention on the attenuating effects of measurement error, restriction of range, artificial dichotomization of continuous variables, and other "artifactual" sources of error on observed effects; it analyzes effects that have been corrected for these psychometric errors (Borenstein et al. 2009).

META-ANALYSIS IN ECOLOGY AND EVOLUTION

Meta-analysis was introduced to ecology in the early 1990s (Table 25.2). Since then there has been a dramatic increase in the number of papers published annually that are relevant to meta-analyses in ecology and evolutionary biology (Fig. 25.1C).

Although meta-analysis is becoming more widely adopted across the fields of ecology and evolutionary biology, it is difficult to measure its impact at this point. Thirteen ecological and evolutionary meta-analyses have been of sufficient importance and general interest to be published in *Nature* or *Science* (as of July 2008). These include analyses of climate change impacts (Root et al. 2003, Parmesan and Yohe 2003), biodiversity loss (Myers and Worm 2003, Gardner et al. 2003), evolutionary mechanisms (West and Sheldon 2002, Griffin and West 2003), and ecosystem functioning (Worm et al. 2002). Meta-analyses have also been influential in the policy arena as demonstrated by a meta-analysis of vital rates for spotted owls (Boyce et al. 2005) and burning management in upland habitats (Stewart et al. 2005). The former was used as evidence in a public enquiry that resulted in federal protection for spotted owls in the United States, while the latter was cited multiple times in revision of the upland burning code by the United Kingdom government (Glaves and Haycock 2005). A casual search (December 2011) found over 20 papers cited over 200 times that were concerned with meta-analysis across a wide range of topics in ecology and evolutionary biology.

Meta-analysis techniques have been employed for a variety of goals in ecology and evolutionary biology. The utility of the approaches is illustrated by examples throughout this book but synthesis across multiple studies has allowed us to draw some important conclusions. For example, one of the most intractable problems in ecology is scaling up from small-scale studies to large-scale processes and phenomena. Meta-analysis can be very useful in this context by often identifying large-scale processes and patterns, even when these are obscured by local factors. Identification of a globally coherent fingerprint of climate change impacts across natural systems (Parmesan and Yohe 2003; Root et al. 2003) and identification of regional coral declines across the Caribbean (Gardner et al. 2003) represent two such examples. Meta-analysis has also been useful in resolving fundamental controversies in ecology, such as consumer versus resource control of species diversity and ecosystem functioning (Worm et al. 2002; see also

TABLE 25.2. Summary of meta-analyses in various disciplines.

Field	Estimated no. of published meta-analyses	First published meta-analysis	Applications of meta-analyses	Average number of primary studies included in meta-analyses	Focus of the meta-analyses	Typical focus of primary studies
Ecology	$\sim 10^3$	1991	• Estimating strength of treatment effects in experiments, or relationships between variables in observational studies • Quantifying and accounting for heterogeneity	5–20, often with complex nesting (multiple effect sizes from individual studies)	Sources of heterogeneity (diverse species, systems)	Individuals, populations, species, communities
Social Sciences	$\sim 10^4$	1977	• Benefits and harms of interventions; • Estimating strength of relationships • Psychometrics • Theory development	Varies by discipline	Estimate overall effect, take advantage of heterogeneity to model variations in effect size	Individuals, groups, organizations
Medicine	$> 10^4$	1904; modern applications, 1970s	• Benefits and harms of interventions • Accuracy of diagnostic tests • Association studies (risk factors, genetics) • Prognosis • Health economics	~10 randomized controlled trials for treatment ~20 for diagnostic test accuracy	Typically as homogeneous as possible to estimate common effect for specific populations or settings of interest	Individual patients

examples discussed in Chapter 26). In the applied ecology and conservation biology literature, meta-analysis has been employed to assess the effectiveness of habitat management (Newton et al. 2009) and ecological restoration programs (Stewart et al. 2008, 2009).

Many of the most noteworthy contributions of meta-analysis in ecology and evolutionary biology tend to concern either identification of consistent large-scale trends or exploration of causes of heterogeneity. Heterogeneity in research synthesis in ecology and evolution tends to be large (e.g., many species rather than one, wide variation in environments in field studies) and is itself often the focus of interest; understanding the magnitude and sources of heterogeneity is often the primary goal of ecological and evolutionary meta-analyses. The general approach to meta-analysis in ecology and evolutionary biology is therefore similar to that in the social sciences, and largely contrasts with the general approach in clinical medicine. In this last field, narrow domains of interest are specified using PICO components, and the presence of heterogeneity across randomized, controlled trials may sometimes be considered a reason for not conducting a meta-analysis (Table 25.1). In contrast, meta-analysis in other areas of medicine increasingly depends on the robust analysis of observational or quasi-experimental data, and is more akin to both the social science and ecological and evolutionary domains.

As the quality and number of primary studies in ecology and evolutionary biology continues to grow, opportunities to look for general patterns (and departures from them) using meta-analysis will become more numerous. The increased sophistication of the statistical techniques should also make detection of patterns easier and more robust. While pioneering meta-analysts in ecology have demonstrated the potential of the techniques, it is likely that many future developments and refinements will be needed to fully realize the potential of meta-analysis in both ecology and evolutionary biology.

General approach of meta-analysis in ecology and evolution

Meta-analysis in ecology and evolutionary biology typically involves combining the outcomes of data from multiple independent studies but can also involve combining primary data from different studies (Chapter 18), using "individual patient" meta-analysis techniques (Stewart and Tierney 2002). Results can be extracted from the literature or contributed by members of a group of researchers working on similar problems (Chapter 19). Most ecological and evolutionary meta-analyses also involve at least some exploration of how an effect varies across some hypothesized source of heterogeneity, such as different species or geographic areas. Unlike meta-analysis in medicine, the increase in the power to detect small effects that can be derived by pooling results is not usually the primary focus of meta-analyses in these fields, with some notable exceptions (e.g., Tonhasca and Byrne 1994). However, given the low statistical power and relatively small magnitude of the effects in ecology and evolutionary biology (Møller and Jennions 2002, Jennions and Møller 2003), the benefits of detecting small but real and potentially biologically significant effects by pooling results should not be ignored.

Despite the focus on exploring heterogeneity, the validity of meta-analyses in ecology and evolutionary biology has often been challenged on the basis of inappropriate pooling of heterogeneous studies (e.g., Markow and Clarke 1997), just as it was earlier in medicine and the social sciences. Clearly, pooled effects derived across heterogeneous studies, sites, or species require careful interpretation, but the analysis of such variation can be rigorously defended. The existence of controversy regarding the interpretation of the results of a body of studies is a frequently cited rationale for undertaking meta-analysis. This is because analyzing the sources of variation and relating the high apparent heterogeneity in the outcome to explanatory covariates may resolve controversy. The general approach of quantifying and explaining heterogeneity using random- or mixed-effects models has served ecology and evolutionary biology well, and forms the basis of many of the analytical approaches discussed in this book (Chapter 8).

Ecological and evolutionary meta-analyses use a number of different effect size metrics (Chapters 6 and 7). Continuous measures of outcome (e.g., biomass, density, fitness) are more common in ecology and evolutionary biology than binary measures; therefore, correlation coefficients, standardized mean differences and response ratios (Chapter 6) are most frequently used as metrics of effect size in ecology and evolutionary biology, whereas effect size metrics based on dichotomous outcome data (risk ratios, odds ratios, relative risk or relative benefit) are used relatively rarely (but see Hyatt et al. 2003, Maestre et al. 2005, and Newton et al. 2009). Correlations are more often used in evolutionary than ecological studies, and in observational rather than experimental studies, whereas standardized mean differences are more common in experimental studies (e.g., those employing factorial designs with treatment and control groups). The log response ratio (Hedges et al. 1999) was developed specifically for ecological meta-analyses and is very commonly used to summarize the outcomes of both experimental and observational studies.

The definition of PICO components, broad systematic searches for relevant studies, and appraisal of study quality covariates adopted in medical and social science systematic reviews have not yet been widely adopted in ecological meta-analyses. There is a growing recognition of the need for broad searches (Chapter 4), appraisal of study quality (Chapter 5), and sensitivity analyses to test statistical assumptions (Chapter 7). Critics of ecological meta-analyses often point to problems of publication bias (e.g., Kotiaho and Tomkins 2002), although it has only rarely been demonstrated (Chapter 14). This should encourage exploration of possible publication (or other) bias, and broad, systematic searches. Ecological meta-analysts rarely incorporate gray literature in meta-analyses, and so conclusions on comparisons between published and unpublished studies cannot be reached. A serious, major limitation for ecological meta-analysis remains poor reporting in primary literature (Chapters 1 and 13). Meta-analysts may also be guilty of poor reporting of methods and results. Currently, standards for conducting and reporting results of meta-analyses in ecology are extremely variable in the published literature, and papers calling themselves "meta-analyses" that do not meet any of the standards or criteria set out in this volume continue to be published (Chapter 27). Improved reporting standards and the use of robust methods and repeatable rules to derive and combine effect sizes are essential for progress (Chapter 20).

IMPACT OF META-ANALYSIS IN MEDICINE, THE SOCIAL SCIENCES, AND ECOLOGY AND EVOLUTION

Systematic reviews and meta-analyses in medicine are routinely relied upon to inform patient management decision making and formulate health policies. In the social sciences they are used to inform treatment decisions by psychologists, social workers, and members of the criminal justice system, as well as to inform policy decisions in education and other fields. In all disciplines in which they have been applied, systematic reviews and meta-analyses have also been used to educate, elucidate scientific thinking, generate new hypotheses, identify knowledge gaps, plan future studies, and improve future research methods.

The number of meta-analyses continues to grow and shows no signs of leveling off. Once discussed primarily in a few academic centers and subject to criticism and even ridicule, systematic reviews and meta-analyses are now routinely relied upon to provide information for informed decision making by government agencies and others worldwide. Meta-analyses in medicine have been found to garner twice as many citations as narrative review articles on the same topic (Mickenautsch 2010); in the social sciences meta-analyses are also much more heavily cited than are primary studies in the same area. The term "evidence-based" originated in medicine about 20 years ago and is now established not only as part of the medical lexicon; the concept (EBM) has now been adopted by many other disciplines, including conservation biology (Pullin and Stewart 2006). Systematic reviews and meta-analyses, as the foundation of EBM and evidence-based social and educational policy, and as a pillar of cumulative scientific knowledge in many different areas, have had a major impact in both the biomedical and social sciences. From all appearances, their impact will only increase in the future. In ecology and evolutionary biology meta-analyses have identified broad-scale patterns, resolved (and created) controversy, helped to generate new hypotheses, identified both environmental impacts and effective means of mitigating them, and suggested new research agendas for primary studies.

THE FUTURE OF META-ANALYSIS IN ECOLOGY AND EVOLUTIONARY BIOLOGY

Science cannot progress by merely accumulating facts. New information must be understood in the context of the existing knowledge. Data based on one species or site may or may not be sufficient to generalize conclusions to other species and sites without considering the responses more broadly. Understanding large-scale patterns and responses over many species and ecological systems is urgently needed for both fundamental and applied problems in ecology and evolutionary biology. Ecologists, conservation biologists, and evolutionary scientists therefore have compelling and fundamental reasons for increasing their use of meta-analysis, and for continuing to develop meta-analysis methodology for their data. Studies cited above suggest some of the utility of meta-analysis for examining large-scale patterns and resolving controversy. There is also a pressing need to develop and improve the standards and methodology for meta-analysis in these fields. To some extent, we can use appropriate methodology from the social sciences and medicine to improve standards for data searching (Chapter 4), for use of inclusion criteria and critical appraisal of quality (Chapter 5), to explore heterogeneity (Chapters 8 and 9), to control for nonindependence (Chapters 16 and 17), for the imputation of missing data (Chapter 13), and to explore bias (Chapter 14). Ecologists and evolutionary biologists should be encouraged to continue to increase the rigor of their analyses and make more realistic, precise, and unbiased estimates or predictions concerning ecological and evolutionary patterns and processes. The natural world is wonderfully varied. Meta-analysis can contribute to our appreciation and understanding of the extent and importance of this variation.

As long as ecologists and evolutionary biologists continue to accrue more information, tools will be required to synthesize it. We hope that this book will provide an impetus not only for increased use of meta-analysis, but also for an increase in the robustness of the research synthesis methodologies employed in ecology, evolution, and conservation biology.

Contributions of Meta-analysis to Conservation and Management

Isabelle M. Côté and Gavin B. Stewart

META-ANALYSIS WAS FIRST BROUGHT to the attention of conservation biologists by Fernandez-Duque and Vallegia (1994). Syntheses of conservation-related literature had largely been, until then, narrative reviews. Fernandez-Duque and Vallegia (1994) outlined several advantages of meta-analytical techniques in terms of guiding conservation decisions. In particular, they highlighted the fact that committing a type II error (e.g., assuming that a particular human action has no effect when it in fact does) can have more serious consequences for conservation than making a type I error (e.g., assuming that an action has an effect when it really does not). When the action is potentially positive, a type II error leads to the oversight of possibly useful methods in the conservation biologist's tool kit. However, when the action is potentially negative, as are many anthropogenic habitat alterations, a type II error leads to the erroneous conclusion that the activity can continue because it appears to be benign. The tighter control of type II errors offered in meta-analyses (Arnqvist and Wooster 1995a) is therefore a significant advantage in conservation and management (see Mapstone 1995 and Field et al. 2004 for similar arguments).

In the first decade of the application of meta-analysis to conservation questions, applied ecologists largely overlooked the guidelines developed in the medical and social sciences, particularly in relation to literature searching and reporting. For example, after comparing ecological reviews (including conservation and environmental management reviews, but not limited to meta-analyses) and medical systematic reviews using 27 detailed criteria well established in medicine, Roberts et al. (2006) concluded that ecological reviews were more likely to be prone to bias and to lack details in the methods used to search for studies. The authors also concluded that ecological reviews were less likely to have assessed the relevance of studies and quality of the original experiments. Overall, ecological reviews were of lower quality and showed greater variation in reporting style and review methods. Similarly, Gates (2002) reviewed 29 ecological meta-analyses (including a few on applied or conservation topics) and found shortcomings in the majority when it came to evaluating data quality, providing information on search methods, and testing for evidence of publication bias and for nonindependence.

There has now been a second "call to arms," urging conservation biologists to embrace the methodologies pioneered in the social and medical sciences and adopt a robust evidence-based approach to the management of biodiversity. This encouragement was voiced almost simultaneously by Pullin and Knight (2003), Fazey et al. (2004), Pullin et al. (2004), and Sutherland et al. (2004). The Centre for Evidence-Based Conservation, established in the United Kingdom in 2003, has led the way in developing protocols and carrying out systematic reviews in applied ecology.

In this chapter, we look back through the relatively short history of meta-analyses and systematic reviews of conservation interest and highlight four areas to which these methods have contributed significantly. Note that our chapter is not meant to be a systematic review, but is very much biased by our own work and interests.

CHALLENGING CONVENTIONAL WISDOM

There is often little science underpinning practical conservation. For example, only 2% of the knowledge sources used by conservation practitioners in Broadland, one of the most important and highly managed wetland systems in the United Kingdom, were from scientific primary literature (Sutherland et al. 2004). By contrast, more than 77% were anecdotal (e.g., common sense, personal experience, speaking to other managers). There is certainly much to be gained from incorporating informal knowledge and understanding into conservation assessment and decisions (e.g., Dulvy and Polunin 2004, Fazey et al. 2006); however, objective tests of management interventions have occasionally shown that the conventionally held "best practices" were in fact less effective, and sometimes plainly more damaging, than less commonly used alternatives (e.g., Cowie et al. 1992, Ditlhogo et al. 1992, Ausden et al. 2001).

Meta-analysis has been used to challenge some conventionally held notions about management effectiveness. Côté and Sutherland (1997), for example, in one of the earliest conservation-related meta-analyses, tested the effectiveness of controlling predators for the protection and enhancement of bird populations. There is a long history in most parts of the world of killing small carnivorous mammals and corvids (i.e., crows and relatives), which can prey on game birds or bird species of conservation concern. Although predation is but one of many sources of bird mortality, it is one that is widely perceived by game and conservation wardens as being both important and controllable. But does controlling predators really result in more birds? By meta-analyzing the results of 20 predator removal programs, Côté and Sutherland (1997) found that eradicating predators did indeed result in higher hatching success of eggs and higher post-breeding (autumn) bird densities, but there was *no* detectable effect on breeding bird populations. Predator control therefore is successful at meeting the goal of game management, which is to increase harvestable postbreeding populations, but whether or not it meets the goal of conservation management, which is to increase breeding bird population size, is still debatable (see Smith et al. 2010 for a more recent meta-analysis on the same topic).

Conventional wisdom is not limited to assuming the effectiveness of entrenched interventions; it also includes beliefs held about the functions and ensuing ecological importance of different habitats or ecosystems, which can indirectly affect conservation decisions. One example of the usefulness of meta-analysis in this context concerns the importance of mangroves as nursery habitat for juvenile fish and invertebrates. There is near universal acceptance of this function of mangrove habitats, which was until recently untested. Two predictions arise from the hypothesis that mangroves provide important nursery habitat. First, the density of juvenile fish and invertebrates (collectively referred to as nekton) should be higher in this habitat than in adjoining nonmangrove habitats. Second, the survival of nekton should also be higher in mangroves than elsewhere. Sheridan and Hays (2003) tested these predictions and found that mangroves held significantly *lower* densities ($d = -4.6$, 95% CI: -9.9 to -0.9) of nekton than alternative nonmangrove habitats (i.e., coral reefs, seagrass, and nonvegetated habitat) combined, although there was no detectable difference in density in pairwise comparisons between mangroves and each individual habitat. By contrast, survival of nekton was significantly higher in mangroves than in adjoining habitat ($d = 0.8$, 95% CI: 0.2 to 1.4). The authors caution that their results are preliminary because of the limited availability of some of the data, particularly for survival (only 8 between-habitat comparisons compared to 114 for density). Of course, mangroves may be very important in some areas, for some species, and when examined at specific spatial scales (e.g., Mumby et al. 2004, Manson et al. 2005). Nonetheless, this initial meta-analytical assessment of the overall importance of mangroves as nursery habitat appears to challenge conventional wisdom.

MEASURING EFFECTIVENESS OF MANAGEMENT INTERVENTIONS

One of the key potential uses of systematic reviews, and meta-analyses in particular, in conservation is to allow a quantitative assessment of the effectiveness of management interventions. Here, we highlight two examples of such an application: (1) marine protected areas, and (2) managing invasive species.

Marine protected areas

Marine protected areas (MPAs, also called marine reserves) are areas of the sea that receive, at least in theory, some protection from exploitation. More than 4000 MPAs have been declared around the world (UNEP-WCMC Protected Areas Database), most in the past two decades; as a result, they are one of the most popular forms of marine protection. There is a plethora of studies reporting densities and sizes of fish in and out of individual reserves, creating an opportunity to evaluate the overall effectiveness of MPAs in fulfilling one of their common objectives, which is the enhancement of fish abundance and biomass.

Mosqueira et al. (2000) were the first to use meta-analysis to synthesize the MPA literature in a quantitative manner. Their search appeared systematic; they consulted three databases, undertook manual searches of published literature as well as gray literature and unpublished reports, and contacted authors to obtain unpublished data. The selection criteria were clear but restrictive in terms of study design (only in/out comparisons included), level of protection afforded to the MPAs (only no-take areas were considered), and how comprehensive the fish surveys were (all species had to have been surveyed). As a result, 31 MPAs were retained in the analysis, although 12 of these were amenable only to a vote-counting analysis due to missing data (e.g., sample sizes or variances). The 19 other MPAs yielded abundance estimates for nearly 350 marine fish species. With analyses carried out at the species level, Mosqueira et al. (2000) concluded that the overall abundance of fishes inside reserves was, on average, 3.7 times higher than outside reserve boundaries, and that this enhancement was mainly the result of an increase in abundance of species that are the target of fishing outside reserves. Protection had no effect on nontarget species, although large-bodied species responded more to protection, irrespective of their fishery status. A meta-analysis of the same data using reserves as the unit of analysis suggested that the beneficial effect of MPAs on abundance is much smaller (28% increase within boundaries) and restricted to target species (Côté et al. 2001). Overall, there was a nonsignificant increase in fish abundance within reserves, and only two of the individual MPAs showed a significant effect. However, reserves enhanced fish species diversity, by 11% on average, within their boundaries (Côté et al. 2001).

Two subsequent quantitative reviews have essentially upheld the conclusions of these two initial studies. Micheli et al. (2004) meta-analyzed abundance estimates for 376 fish species from 31 no-take MPAs and associated control areas. Most trophic groups showed significant positive effects of protection. Piscivores in particular were, on average, 80% more abundant in reserves. However, 19% of individual species—mostly small-bodied, site-attached fishes—were negatively affected by protection, perhaps as a result of increased predation or competition from species that are more abundant in reserves.

Halpern (2003) considered 81 MPAs and control areas that contributed at least one of four measures of outcome (density, biomass, size, and diversity). These reserve-level analyses differed from the previous ones because species within each reserve were aggregated, resulting in an "ecological bias" (Berlin et al. 2002), but preventing quasireplication. Nevertheless, the diversity of fish and invertebrate communities and the mean size of organisms within reserves were 20–30% higher than in unprotected areas. The density of organisms was roughly double in reserves, while biomass was nearly triple.

Taken together, these quantitative reviews suggest that MPAs are management interventions that may work spectacularly well; however, they also highlight the fact that there is great variability and uncertainty regarding MPA performance. Seeking to explain this variability, in terms of reserve characteristics and socioeconomic features of design and implementation, is now an important and active area of research.

Managing invasive species

Ecologists have used meta-analysis to explore the resistance of ecosystems to invasion (Levine et al. 2004, Liu and Stiling 2006), the impact of invasive species (Vila et al. 2004, McCarthy et al. 2006), and the management of invasive species (Tyler et al. 2006; Stewart, Pullin, and Tyler 2007). Clearly, in an applied context, the last is the primary focus of concern. It is therefore unsurprising that the effectiveness of control interventions for rhododendron (*Rhododendron ponticum*; Tyler et al. 2006) and bracken fern (*Pteridium aquilinium*; Stewart, Pullin, and Tyler 2007), two highly invasive plant species, have been investigated using quantitative systematic reviews incorporating meta-analysis.

Rhododendron ponticum is an invasive species in many countries, posing a threat to native flora and fauna because it is capable of altering entire seminatural communities through its vigorous spread. The meta-analysis of Tyler et al. (2006) found that postcut application of the herbicide Glyphosate, or applying herbicides containing metsulfuron-methyl or imazapyr, can effectively control *Rhododendron ponticum*. However, their review was hampered by lack of robust evidence and small sample sizes, which was highly surprising given the magnitude of the efforts underway in numerous European nature reserves and national parks to control rhododendron. The systematic review of bracken control methods undertaken by Stewart, Pullin, and Tyler (2007) demonstrated that asulam application effectively controlled bracken, and quantified the increased effectiveness of repeated applications of the herbicide.

In both cases, the necessity for an objective systematic review and meta-analysis was considered dubious by some because both topics were thought to be already well covered. However, both analyses shed light on previously unknown relationships, and clearly identified discrepancies between empirical knowledge and practice based on experience (see below).

REPLACING LARGE-SCALE MONITORING BY SMALL-SCALE SURVEYS?

One of the strengths of meta-analysis in conservation (as well as in other fields) is the potential to use multiple small-scale studies to provide insights into patterns and processes occurring at much larger scales. Can meta-analyses eventually replace large-scale studies altogether? Medical scientists have been grappling with this question for more than a decade by comparing the results of large randomized, controlled clinical trials—the "gold standard" in evaluating the efficacy of clinical interventions—with those of meta-analyses of several smaller studies (Cappelleri et al. 1996, LeLorier et al. 1997; Chapter 25). In conservation, there have so far been few opportunities to compare regional patterns detected from large-scale monitoring with those deduced from meta-analyses of smaller-scale surveys. The two exceptions we are aware of both concern coral reefs.

Hughes et al. (2002) combined coral recruitment data from 21 single-reef studies undertaken on the Great Barrier Reef (GBR) over a 16-year period and compared the results to those obtained from a 2-year-long, large-scale, standardized survey of 33 reefs distributed across the GBR. The two approaches show very similar large-scale patterns. Recruitment by spawning corals peaked in the central GBR and declined steadily with increasing latitude by up to more than 20-fold, whereas recruitment by brooding corals declined 3- to 5-fold from north to south.

In the other case study, Gardner et al. (2003) used a meta-analysis of short, single-reef-derived data on coral cover to piece together the trajectory of ecological change on reefs over three decades in the entire Caribbean basin. During this period, coral cover declined, on average, from 50% to approximately 10%, with some subregional variation in the rate of coral decline. The data stemmed from studies that had used a variety of survey methods, generating concern about potential biases. However, Côté et al. (2005) specifically tested for and failed to find methodological differences in rates of coral decline. Moreover, they compared the overall rate of coral decline derived from the meta-analysis of small-scale studies to that derived from a Caribbean-wide reef monitoring program in which annual standardized surveys on selected reefs are carried out. Over a matching time period (1993–2000), both methods generated remarkably similar mean rates of coral decline (meta-analysis: 10% per annum, 95% CI: 8% to 12%; monitoring program: 10 % per annum, 95% CI: 2% to 18%), although the error associated with the monitoring program was larger owing to the smaller number of reefs included. Côté et al. (2005) also investigated the sensitivity of meta-analytically generated rates of coral decline to sample size by subsampling the original data set. Rates of change in coral cover that deviated acceptably from the whole data set (i.e., <1% in absolute terms, or <10% in relative terms) could be obtained with minimum sample sizes of about 125 reef sites (or approximately 40% of the whole data set).

Côté et al. (2005) concluded that meta-analysis offers a powerful tool to quantitatively estimate current rates of ecological change and to reconstruct recent patterns of change on coral reefs, and perhaps in other habitats too. By using existing data collected for a variety of purposes, meta-analysis removes the expense of setting up new coordinated monitoring programs and the need to wait for these programs to generate initial evidence of ecosystem stress; nevertheless, such monitoring programs—just like large, randomized controlled trials in medicine—are still perceived as the best sources of information about environmental change (Ford 2000).

IDENTIFYING GAPS

Systematic reviews and meta-analyses in applied ecology can identify gaps in knowledge. This is especially true when the analyses are based on questions formulated by the needs of conservation practitioners or policy makers, rather than when the analyses are developed by scientists who may be interested in synthesising data for a multitude of reasons. Two kinds of knowledge gaps are routinely encountered, and there is often high value in objectively identifying these deficiencies.

The first type of knowledge gap is where there is simply insufficient evidence for a meta-analysis to be undertaken at all. This is relatively common. For example, of the 15 systematic reviews published (as of July 2007) by the Centre for Evidence-Based Conservation, 6 (40%) reported insufficient information for critical quantitative appraisal. These reviews ranged widely in topics, from the effect of hedgerow corridors for plant population viability to the effectiveness of trapping to eradicate introduced mink. In some of these cases, a number of relevant studies were available but were unsuitable for inferential meta-analysis; reviewers then resorted to creating a tabular summary of the existing data and multivariate analyses of the few existing data points (e.g., Stewart et al. 2005). While such work is of limited utility in terms of guiding practice, it is clearly very important for the research community to have such knowledge gaps identified, as well as the necessary assembly of information that will one day lead to a more conclusive review. Ideally, this should promote increased funding that will enable researchers to address the problems robustly.

Other knowledge gaps that can be identified when carrying out a meta-analytical review are related to unexplained heterogeneity in the results. For example, the bracken review discussed above could not relate variation in effectiveness to herbicide concentration, geographical area, or habitat, all potential correlates of the effectiveness of herbicides in controlling the troublesome fern, because samples sizes were small relative to the number of potential effect modifiers (Stewart, Pullin, and Tyler 2007). This, again, is a fairly common problem—more than half (8/15) of the CEBC reviews could not examine sources of heterogeneity in effect sizes in part because of small sample sizes, but also because potential correlates of effect sizes were often not reported. Highlighting that unexplained variation in effect sizes exists among studies can be useful in terms of informing future research directions and for guiding future experimental designs.

CONCLUSION

Meta-analysis, and systematic reviews in particular, have a very short history of use in conservation and applied ecology. Despite this, we have shown through the examples presented above, that these techniques have already contributed to answering a number of important applied questions. However, although these answers often have clear implications for conservation policy and legislation, they currently remain the focus of primarily academic interest. Given a little more time, we anticipate that evidence bases that are collated using systematic reviews, particularly in association with meta-analysis, will increasingly influence decision making in the environmental sector, just as they have in medical fields.

Conclusions: Past, Present, and Future of Meta-analysis in Ecology and Evolution

Jessica Gurevitch and Julia Koricheva

WHERE WE CAME FROM AND WHERE WE STAND NOW

META-ANALYSIS HAS BECOME ESTABLISHED and widely applied in ecology, evolutionary biology, and conservation ecology (Gurevitch et al. 2001, Stewart 2010) since being first introduced in the early 1990s, as we have seen throughout this volume. During the past two decades since its introduction, methodology has been developed and refined, standards have become established, and the utility of these techniques has become more widely known among scientists in these fields. As in other disciplines, there was initially considerable controversy and resistance to the implementation of meta-analysis in ecology and evolution, followed by more general recognition of its value. We are now in the process of moving research synthesis to a new level in these fields, with new methods being developed and implemented and higher standards in their application, and we hope that this handbook will contribute to that process.

The introduction of meta-analysis has changed the way data are understood and interpreted in other disciplines, and has begun to have an impact on the way ecologists and evolutionary biologists view scientific information as well. Traditionally, students learned about ecology and evolution from case studies that tell us how things work and what is important, with subsequent experiments confirming, refuting, or more likely adding nuance to the original understanding. Meta-analysis offers us a different view: studies can be seen as members of a population of related studies, and like any population, can be characterized statistically to reveal generalities not obvious from examination of each one separately.

One of the most compelling factors in the broad acceptance, use, and development of meta-analysis in ecological and evolutionary research is the proliferation of data and the expanding ability to access them. Simply put, at one time there was little information to work with, and the large body of data that has accumulated over the past several decades has given research synthesists a lot to work with. Furthermore, there is a compelling need to make sense of this accumulated knowledge to answer both applied and fundamental questions. What does a body of literature on a particular issue tell us? What do we know about a topic, and what is still not understood? What has been inadequately studied? Meta-analysis offers powerful tools to assess the current state of knowledge. A meta-analysis can also tell investigators when no additional information is needed to resolve a question. When that is the case, to continue to do experiments that test essentially the same thing is at best wasteful and at worst may hinder effective action. In medicine, this may mean lost lives; in ecology and conservation biology it might mean lost opportunities to save natural systems or species. The ability to minimize type II errors—reducing failure to detect a real effect, particularly when sample sizes are small, as is ubiquitous in these disciplines—is one of the real strengths of meta-analysis, and one that has very high value in this field.

While once we talked about the importance of increased access to computers and computing power in facilitating meta-analysis in our discipline, this is no longer an issue. Any student and certainly any research ecologist in North America, Europe, Australia, or in many other regions, has access to sufficient computing power and speed to do any meta-analysis on their desktop or laptop computer. Access to scientific databases remains an issue, and access to the scientific literature even more so. It is impossible to predict if open access journals and articles will change the access to scientific publications in the future. However, free online access for some of the best scientific journals has been made available in developing countries through the HINARI (WHO), AGORA (FAO), and the OARE (UNEP) initiatives in recent years, and we hope that this access will only increase.

While we are clearly strong advocates of the application of meta-analysis in our fields, it is important to recognize that meta-analysis is not always appropriate; it is useful for some applications and not in others, as with any other statistical or conceptual tool. It is inappropriate when it makes no sense to combine studies. (In the soon to be famous words of our collaborator Gavin Stewart, while we may wish to cautiously combine apples and oranges, we should not be combining apples and ferrets.) Other cases in which meta-analysis is not a useful tool include those where data are insufficient. This can occur when there are too few studies, when studies cover too narrow a focus to answer the questions posed, and when the data quality or reporting is so poor that it becomes impossible to reliably incorporate the information (although one wonders in this last case why the primary studies were published). There are situations in which one wishes to obtain a quick overview of what has been done on a topic and is not able to invest the time and resources in a meta-analysis. When studies use fundamentally different approaches or designs, meta-analysis may (or may not) be inappropriate. We emphasize that it is of the greatest importance to think about the scientific issues first, and then to make sure your approach is statistically robust. It takes a deep understanding of the research field in question to have a scientifically meaningful meta-analysis.

CURRENT CHALLENGES FOR META-ANALYSIS IN ECOLOGY AND EVOLUTION

Various factors have hindered or limited the reach and impact of meta-analysis in these fields. As we have seen in this book, one of the most troubling is poor reporting standards for data in the ecological and evolutionary literature. Poor reporting standards remain ubiquitous even after decades of pressure to improve them. Even if they were corrected tomorrow, the literature published up to now would still suffer from this poor reporting, creating problems for research synthesis. Most common and most problematic is publishing P-values alone, which not only limits the usefulness of a study for meta-analysis, but also makes it almost impossible to interpret the outcome even of just that study. Another common problem is providing effect estimates (e.g., calculated sample means for the experimental and control groups being compared) but not sample sizes or any measure of variation around the estimate. Experimental designs may be described briefly or poorly, making it difficult to know how to interpret or use the data in a meta-analysis. All of these issues limit the value of the primary literature, and compromise efforts to synthesize those data. While it should be unnecessary by now to call for the improvement of reporting standards for publication in this discipline, we do so once more;

recent evidence suggests that reporting may in fact be improving in the ecological literature (e.g., Corrêa et al. 2012).

Vote counts (Chapter 1) remain persistent in the ecological literature, continue to be published, and continue to be justified as a legitimate statistical method (Gurevitch and Mengersen 2010), despite their well-known limitations (Chapter 1; Gurevitch and Hedges 1999). One of the most common justifications for conducting vote counts is poor data reporting—that is, since there are not enough data available in published papers to carry out a weighted meta-analysis, it might appear that the next best thing is to do a vote count. While dealing with missing data is a complex and difficult issue (Chapter 13), using poor methods for research synthesis will likely only compound the problems.

Strong objections to the use of meta-analysis and systematic research synthesis have been voiced since their introduction in ecology and evolution, and continue in some quarters (e.g., Hurlbert 1994, Simberloff 2006b, Whittaker 2010). The categorical opposition to using meta-analysis in ecology and evolutionary biology has decreased markedly over the last 15 years, although limited understanding of its objectives and methods among ecologists is not uncommon. Of course, meta-analyses vary greatly in quality, rigor, and robustness, and it is certainly reasonable to criticize particular studies or approaches on methodological or other scientific and statistical grounds. However, continued accumulation of new research without the evaluation and synthesis of the existing studies on a topic will impede, rather than accelerate, new understanding and knowledge; this is a profoundly inefficient use of resources (e.g., Chapter 23). Research synthesis in ecology, evolution, and conservation still remains to be fully acknowledged (or funded) as a serious, legitimate, and major avenue of research and an invaluable tool for making progress, as it is in medicine and the social sciences.

Fortunately, awareness and understanding of research synthesis methodology, including meta-analysis, have greatly increased in ecology and evolution in recent years. More people are mastering the issues and methods necessary to conduct good research syntheses, and the number of valuable quantitative syntheses in these disciplines has grown substantially. We hope that this volume will help in that regard.

As in other disciplines, meta-analyses in ecology and evolution may focus on mean effects and be inattentive to other ways of presenting the results. Moreover, they may fail to focus on the important issue of accounting for and explaining heterogeneity between studies, as discussed throughout this book. This issue is not unique to research synthesis, but is exaggerated in meta-analysis. The analysis of heterogeneity may also be much more important in meta-analyses in ecology and evolution than in other disciplines (Chapter 25). Typically, experiments in our discipline have many different sources of heterogeneity and are less controlled than in medicine or the physical sciences. In this regard, ecology and evolution share certain commonalities with the social sciences (Chapter 25). Other challenges and problems in the application of meta-analysis to the ecological and evolutionary literature are similar to issues in primary analyses. For example, confounded or otherwise nonindependent moderators present very similar problems in research synthesis as in primary analyses (Chapters 16 and 20). As in primary analyses, incomplete reporting of methods and results in meta-analyses unfortunately also occurs. The same criteria apply to reporting standards for research synthesis as for primary studies: methods have to be reported in sufficient detail (e.g., exact search terms and inclusion criteria) to allow replication of the study. Meta-analyses which do not report the methods used to collect the studies and analyze the data should simply not be published.

Conducting a meta-analysis can be an enormously rewarding experience, but it is also a slow and often painful process, and it should not be done in haste or without a great deal of reflection. We hope that this volume offers a contribution to the thoughtful use of these techniques, and to improving these standards for authors, reviewers, and editors, and for the publication standards of journals (Chapter 20).

FACTORS FACILITATING PROGRESS OF META-ANALYSIS IN ECOLOGY AND RELATED DISCIPLINES

Many factors have contributed to progress in meta-analysis in these research fields. Emphasis on careful framing of the question (Chapter 3), more systematic and repeatable approaches to literature searches (Chapter 4), systematic methods for data appraisal (Chapter 5), and methods for evaluating various potential biases, estimating their magnitude, and correcting for them (Chapter 14), lead to improved quality, transparency, and repeatability in ecological and evolutionary meta-analysis and research synthesis. Other more recently available methods include those for recovering missing and partial data (Chapter 13), methods for phylogenetically independent meta-analysis (Chapter 17), and the application of meta-analysis for synthesizing ecological parameters using less familiar metrics across studies, where their statistical properties are well understood (Chapter 7). Some of the most important advances are the development and application of maximum likelihood (Chapter 10) and Bayesian methods (Chapter 11) for ecological meta-analysis. These may allow a great deal more flexibility in modeling than the more familiar methods of moments and least squares (Chapter 9).

A different kind of progress in ecological and evolutionary meta-analysis comes from interacting with and collaborating with researchers in other disciplines. The editors and authors contributing to this book come from a wide range of disciplines and geographic locations, and have brought different skills, knowledge bases, and questions, (not to mention personalities) to the collaboration. The interaction and variety of contributions have enriched all of us and led to many close friendships as well. Collaborative efforts can be enormously fruitful, fun, and rewarding, but can also be challenging (and sometimes disappointing). We would be remiss if we did not mention "the curse of collaboration"; even with the most congenial and motivated collaborators, just as one finally gets the time to focus fully on a joint project, they receive an automated e-mail reply from the other saying something like: "I have just left on an extended field study requiring a camel trek across central Mongolia, and I'll read your e-mail in 10 weeks when I return in late August." Of course, in late August the first collaborator is just about to start teaching Introduction to Basic Advanced Ecology to a class of 650 students for the first time.

One of the most rewarding and valuable, if occasionally difficult, sorts of collaboration can be between scientist and statistician. Ecologists and other scientists often mistakenly believe that the most valuable thing they can contribute to such a collaboration is data. This is not the case. Statisticians have no shortage of data. What scientists *can* contribute are interesting questions—problems with both statistical and scientific nuances that make them challenging and valuable from the perspective of both fields. Statisticians can offer new or better methodology, but can also provide better ways to frame research questions, among other contributions. Graduate students beginning a research synthesis may have difficulty because no one in their department is very familiar with meta-analysis, and not infrequently they may contact one of us asking for help analyzing their data. (Unfortunately, however much we would like to help, we rarely have enough time to be able to do so; see the FAQ.) A recommendation we frequently make is to encourage them to do an internet search on "Your Own University" (or, "Your city") and "meta-analysis"—often they will discover people working in their own institutions or cities with shared interests and expertise in the statistical problems they are trying to solve. Those people may be in departments that would ordinarily never intersect—but they can be valuable sources of highly productive collaborations, or just provide ordinary help over hurdles. Other collaborations work very well with electronic communication, although if possible it is fruitful to get together in person to make the most progress. Obviously there are many other things to say about collaborations in general, but we cannot do so here; we merely suggest this as a way to make real progress in this interdisciplinary enterprise.

A very specialized kind of collaboration exists in the form of working groups at national synthesis centers in the United States. This book began with an idea from JK, who contacted JG by e-mail (we had not met at that point); we decided to propose a working group to the National Center of Ecological Analysis and Synthesis (supported by the US National Science Foundation). This was funded and led to four meetings over a period of three years, and resulted in many friendships, the book you have in your hands, and various collaborative spin-offs. Funding for such workshops and working groups in these centers has provided a huge boost to the development of synthetic research areas, and has been an enormously successful investment.

A very different factor is also likely to contribute in important ways to the development of research synthesis in these fields and others in the future, but is not without great pitfalls. This is the increasing online availability of raw and summary data from published and unpublished sources, with meta-data explaining what the data mean and how they were collected. If the raw data were available in the future providing the supporting background for published papers, this would neatly solve some of the problems of poor reporting—researchers could themselves calculate standard deviations, do back transformations, and so on. This would allow more sophisticated methodology for modeling research syntheses. The pitfalls are many, including the obvious problem that the quality and usability of the data depend on the quality of the meta-data provided. Moreover, this would require substantive knowledge of characteristics of the study design and the way the experiment was conducted, and of the nature of the data. In any case, online data availability will not resolve the issues related to all of the work done before such online appendixes are made available. Ecology and evolution are subjects with an unusual emphasis on "legacy" data—the literature has a very long half-life, and data (as well as ideas) published years or decades ago may be of high current value. Nevertheless, the advent of accessible online data offers considerable promise for meta-analysis in our disciplines.

POTENTIAL AND FUTURE OF META-ANALYSIS IN ECOLOGY

This and other recent books introducing meta-analysis and more broadly, research synthesis, to a new generation and wider audience indicate that the utility of these methods continues to be increasingly widely recognized across a range of disciplines (e.g., Borenstein et al. 2009, Cooper et al. 2009). Meta-analysis in ecology, evolution, and conservation biology has always benefited from methods and approaches in other disciplines. Originally, meta-analysis in our disciplines borrowed chiefly from methods in educational psychology and closely related social sciences. However, even early on, methods have been developed specifically for the nature of the data and the questions in ecological and evolutionary research. This continues to be the case as statisticians and theoreticians become drawn to problems in research synthesis in these fields, and as practitioners continue to explore new questions that demand the development and refinement of methodology. The benefits of adopting and adapting techniques from other disciplines are obvious; there have been far more people working on meta-analysis in medicine and other disciplines for a longer period of time.

While we will continue to look to other disciplines for techniques and methodological advances, we will also perhaps make valuable contributions to research synthesis in those other fields. For example, methods being developed for phylogenetically independent meta-analysis (Chapter 17) may prove useful for other types of nonindependence in different disciplines. Some of the challenges that we face in doing meta-analysis in our discipline also exist (in perhaps more subtle forms) in other fields, and recognition and resolution of these problems may be helpful for research synthesists in those areas.

Meta-analysis continues to grow in application for a wide range of problems in ecology and evolution. It has been used to inform the fundamental understanding of physiological, population, community, and ecosystem ecology, as well as evolutionary biology, and to resolve long-standing controversies (Chapter 24). The introduction of evidence-based conservation practices, analogous to evidence-based medicine, promises to transform applied ecology, establishing a much more rigorous basis for management decisions and priorities (Chapter 26; Stewart 2010). Meta-analysis and systematic reviews may also serve as an "early warning system" when synthesizing data across many individual studies to detect immanent catastrophic changes, ranging from global collapse of the top tier of the ocean's food webs as a consequence of overfishing (Worm and Myers 2004, Myers and Worm 2005, Myers et al. 2007) to quantifying responses to climate change across taxa and at a range of spatial scales. Use of these statistical tools will enable us to determine patterns of response at large scales and across many different species much more quickly and accurately than impressions based on individual studies.

While there are obvious benefits in writing a textbook about an area of growing popularity, this very popularity is also a substantial drawback. The literature on meta-analysis is growing exponentially, with more advanced methods, more diverse applications, and more sophisticated ways of interpreting and presenting results. By the time this handbook is published, parts of it will be out of date. We hope that our handbook will continue to assist readers in understanding the principles of meta-analysis in ecology and evolution and facilitate understanding of related literature for some time to come. We encourage you to contribute to advances in the field, and we look forward to seeing your own work influencing future research and practice.

GLOSSARY

Aggregation bias—Situations in which some subgroups or individuals display very different responses as compared to the rest of the group, resulting in a misleading average when response is aggregated over the entire group. *See also ecological bias.*

Apples and oranges problem—A common criticism of meta-analysis that refers to the idiom "comparing apples and oranges" and questions the validity of combining results from different studies (e.g., studies that differ methodologically or have been conducted on different species). The main counterargument used by meta-analysts is "that it is a good thing to mix apples and oranges, particularly if one wants to generalize about fruit, and that studies that are exactly the same in all respects are actually limited in generalizability" (Rosenthal and DiMatteo 2001).

Assumptions of statistical models—Basic properties that we believe to be true regarding the nature of the data to which the model is applied (for example, independence of observations). Like all statistical procedures, meta-analysis procedures have assumptions that, if violated (that is, if they are not met), can call the conclusions into question.

Bayesian analysis—A school of statistical thought first proposed by Reverend Thomas Bayes in 1763. In a Bayesian approach to statistical inference, evidence or observations are used to update or to newly infer the probability that a hypothesis may be true. Thus, the Bayesian school interprets the concept of probability as "a measure of a state of knowledge" rather than a frequency, as used in a *frequentist approach.* Classical (frequentist) statistics fits data to statistical distributions, while Bayesian statistics fits statistical distributions to data.

Between-group heterogeneity (also called among-group heterogeneity)—Variation among groups or categories of studies. The categories of studies are defined by *moderators* (also called *covariates or explanatory variables*), which describe the nature of the different groups.

Between-study variation (or heterogeneity)—Another major component of variation in meta-analysis data (in addition to study-specific *sampling error*), which is due to between-study differences in effect size estimates resulting from study-specific characteristics such as species used, treatment intensity, or experimental procedure. In a meta-analysis with a hierarchical structure, between-study variation includes both between-group and within-group variation (heterogeneity).

Bias—A systematic error or deviation from the truth in results or inferences that distorts or limits the conclusions. In meta-analysis, two types of bias that are commonly identified are *publication bias* and *research bias.*

Bootstrapping—A nonparametric, iterative procedure used to generate confidence intervals around a given statistic (e.g., mean effect size) by randomly choosing (with replacement) n studies from a sample size of n, and generating a distribution of possible values for a statistic. The lowest and highest 2.5% values are chosen as 95% bootstrap confidence limits.

Combining probability levels—An old type of quantitative research synthesis in which probabilities from significance tests (P-values) in individual studies are statistically combined to obtain an overall significance test for the null hypothesis. This is a valid approach but it provides limited information about the responses.

Conceptual definition—A description of a concept which can be used to distinguish it from related, but not relevant concepts. *See also operational definition.*

Confidence interval—The range of values that is expected to contain the true value of the population parameter with a given probability; commonly used in frequentist statistics as a method to estimate the uncertainty of a given statistic.

Correlation coefficient—A measure of the linear association between the two variables. A correlation coefficient can range from -1 (perfect negative association), to +1 (perfect positive correlation), with 0 indicating no linear relationship between the variables. One of the most commonly used parametric correlation coefficients is Pearson's product-moment correlation coefficient, which is often used (when transformed using *Fisher's z-transformation*) as a metric of effect size in meta-analysis.

Covariates—*See moderator variables.*

Credible interval—In Bayesian analysis, an interval that contains an unknown quantity with a specific probability calculated from the posterior distribution.

Cumulative meta-analysis—A special form of meta-analysis in which effect sizes from individual studies are entered into the analysis sequentially, one study at the time, and are based on some predetermined order (most commonly chronological); the mean effect size and confidence intervals are recalculated at each step. Cumulative meta-analysis could be used for updating the results of previous meta-analyses, for finding out at what point in time the mean effect size became significant, or for detection of patterns of temporal changes in effect size.

Dissemination bias—Dependence of accessibility of research findings on the direction or strength of these findings.

Ecological bias, or ecological fallacy (better term is *aggregation bias*)—A term used in the medical (and sometimes social sciences) literature to refer to situations in which the responses of a subgroup are very different from that of the rest of the group (of individuals, studies, etc.), so that the average response aggregated over the entire group is misleading. For example, most of the effect sizes may be zero, while the responses of a subgroup are very strong, leading to an overall estimate of a slight response for the entire population. This is a very specific type of heterogeneity. Ecologists do not use this term.

Effect size—A dimensionless measure of study outcome or a measure that puts all responses across studies on the same scale. It indicates the magnitude of the effect or response in each study in a meta-analysis, providing a "common currency" for comparison of the results across studies. Different metrics of the effect size can be used depending on the research question and type of data to be analyzed and may include standardized mean difference, correlation coefficient, response ratio, odds ratio, or others.

Eligibility criteria—*See inclusion criteria.*

Empirical Bayes analysis—A quasi Bayesian analysis in which some of the unknown quantities are replaced by sample estimates from the data rather than being averaged over their prior distribution,

Evidence-based practices—Practices that are based on the current best evidence in making decisions, which in turn is based on a combination of individual expertise with the best available external evidence from research syntheses. Systematic reviews and meta-analyses are the fundamental tools of evidence-based practices. In medicine, evidence-based practices are widely used to make clinical recommendations for individual patients. In ecology, the emerging field of evidence-based conservation involves making decisions on appropriate management and policy actions based on systematic review of available evidence rather than on local or expert opinion alone.

External validity—External validity refers to the degree to which the results and conclusions of a study can be generalized to the larger population.

Exclusion criteria—Characteristics of the primary studies that make them ineligible for inclusion in a systematic review or a meta-analysis. These characteristics must be decided by the meta-analyst *before* the study selection has begun.

Fail-safe number (N_{fs})—The number of additional studies with an effect size of zero that would need to exist to negate the significance of an observed mean effect size or to reduce it to a specified minimal value. If N_{fs} is very large, it is often taken to indicate that the results of meta-analysis are robust to possible publication bias.

File drawer problem—A catchphrase referring to publication bias against nonsignificant results and describing the situation when statistically significant results get published, whereas nonsignificant results remain in the authors' file drawers (Rosenthal 1979).

Fisher's z-transformation—A transformation that converts Pearson's correlation coefficient to a normally distributed variable. In meta-analysis, *z*-transformed correlation coefficients are used in calculations, rather than untransformed Pearson's correlations.

Fixed-effects meta-analysis model—A statistical model used to combine effect sizes across studies, which assumes that all studies share the same true effect size and the observed variation in effect sizes between studies is due solely to sampling error.

Flat file database—A database that exists in a single file in the form of rows and columns, with no relationships or links between records and fields except in the table structure. A simple Excel spreadsheet is a flat file. *See relational database.*

Forest plot—A specific type of plot common in meta-analysis, in which the effect sizes and confidence intervals from individual studies are indicated by a point (for the mean) and line (for the confidence interval) for each study; these are typically aligned horizontally and placed above one another in a vertical column, together with the combined overall mean effect and its confidence interval. Individual studies on forest plots can be arranged in chronological order (*see cumulative meta-analysis*) or sorted by the sign and magnitude of their effect sizes (from lowest to highest).

Frequentist analysis (classical statistics)—The frequentist approach to *statistical inference* is based on the concept of a sampling distribution; it assesses the precision of an estimate by the frequency with which it would occur over a large number of repeated samples from the same population. Frequentist approaches to inference include methods such as *maximum likelihood, method of moments*, and the *least squares method*. In classical or frequentist statistics, the data are fit to statistical distributions such as the normal distribution. It contrasts with the approach taken in *Bayesian analysis*.

Funnel plot—A scatterplot of effect sizes against a measure of their variance (e.g., sample size or standard error) used as a visual tool for detection of *publication bias*. In the absence of publication bias, it is assumed that the plot should have the shape of a symmetrical funnel as the precision of the estimation of the effect size increases with the sample size. Publication bias may cause the systematic absence of certain values in the plot, resulting in plot's asymmetry.

Gray literature—Research that is publicly available but difficult to locate or retrieve (and thus is less likely to be included in a meta-analysis); this includes conference papers, technical and governmental reports, working papers, dissertations, theses, and so forth. It is often believed that results found in the gray literature are less likely to be subject to publication bias than data in published papers.

Hedges'd (also known as Hedges'g)—The standardized mean difference between two groups in a study (e.g., control and treatment)—that is, the difference between group means divided by their pooled standard deviation and corrected for small-sample bias. Hedges' *d* is one of a family of metrics of effect size based on standardized mean differences that differ slightly from one another. It is commonly used as a measure of effect size in ecological meta-analyses.

Heterogeneity—The variation in the effect size estimates among a set of studies. In ecological and evolutionary meta-analysis, total heterogeneity is usually subdivided into between-group and within-group heterogeneity.

Hierarchical model—A form of *meta-regression* model with nested subgroups, in which studies are combined into subgroups that may themselves be combined into higher level groups, and so on.

Homogeneity test (also known as *heterogeneity test*)—A test (usually *parametric* and based on statistics such as I^2 and Q) that compares the observed variation in effect sizes with the variation expected if only sampling error is causing the differences. If the value obtained is smaller than the critical value for the chosen level of significance, then we cannot reject the hypothesis that variance in effect sizes was produced by sampling error alone. If the value is higher than the critical, then the meta-analyst would usually proceed to examine which study characteristics (*moderators*) are associated with larger or smaller average effect sizes. Inferences based on homogeneity tests are subject to the power of the tests, which is typically low.

I^2 *homogeneity statistic (sometimes referred to as I^2 heterogeneity statistic)*—Used to quantify the degree of heterogeneity between effect estimates across studies. Variation across study estimates is assumed to arise from two sources: within-study sampling variation and between-study variation. The I^2 statistic calculates the percentage of the total variability across effect estimates that is due to between-study variation. The statistic varies between 0 and 1, with 0 indicating no between-study variation. (*See also Q homogeneity statistic*).

Inclusion criteria—Conditions that must be met by a primary study in order for it to be included in the research synthesis. (*See also exclusion criteria*).

Independence—See *statistical independence*.

Individual patient data (IPD) meta-analysis—A type of meta-analysis in medicine that uses raw data from primary studies, which enables analysis at the individual level rather than at the study level as in conventional meta-analysis. Analogous data structures also exist in ecology and evolution, but have been used less frequently than in medicine.

Inference—See *statistical inference*.

Internal validity—Refers to the degree to which the results of a study are free from confounding variables that may have an impact on the study's results and conclusions.

Kappa analysis—Test used to measure the degree of agreement between two reviewers when selecting studies or data for inclusion in a meta-analysis.

Least squares method—A method of estimating population parameters by choosing parameter estimates that minimize the squared residuals between the observed estimates and those predicted from the model. This method is used in both primary statistical analyses and meta-analysis.

Likelihood—A function of the parameters in Bayesian analysis of a statistical model that describes the probability of the observed data.

Mantel correlation—A statistic that measures the degree of similarity between two matrices using randomization test.

Markov chain Monte Carlo (MCMC)—A Monte Carlo simulation method for calculating joint posterior probabilities using the conditional distributions of each random variable, given all the others.

Maximum likelihood method—A method of estimating population parameters by choosing the parameters that maximize the probability of the data.

Meta-analysis—A set of statistical methods for combining magnitudes of the effects (*effect sizes*) across different data sets addressing the same research question. Meta-analysis is one of the types of quantitative research synthesis.

Meta-regression—An extension of the basic meta-analysis model in which *moderators* (categorical or continuous, independent, explanatory factors also known as covariates) are used, as part of the model, to explain between-study variation in effect sizes. An older and now-eclipsed use of the term "meta-regression" is a model with only continuous moderators.

Method of moments—A frequentist statistical method of estimating population parameters, such as the mean, variance, median, and so forth; it is carried out by matching the theoretical moments of interest with the corresponding estimates based on the data.

Missing data—Data representing either study outcomes or study characteristics that are unavailable to the meta-analyst either because they do not exist, or because they are not reported.

Mixed-effects meta-analysis model—A form of meta-regression model that encompasses both fixed and random effects; for categorical moderators, the effects of the groupings of experiments on effect parameters are considered fixed, while the variation within groupings is considered random.

Moderator variable—A factor or variable that affects the magnitude or sign of the effect size and is used in meta-regression models to explain between-study variation in effect sizes. Moderators may be categorical (e.g., type of treatment, trophic level, sex) or continuous (e.g. duration of the experiment, intensity of treatment, latitude or altitude of the study site).

Narrative review—A traditional, nonsystematic method of research synthesis based on expert opinion. This method reviews available studies on the topic without a statistical analysis of the study outcomes.

Negative findings—Results of a study that are either in the opposite direction to that expected, or not statistically significant.

Nonindependence—Nonindependence of effect size estimates is a common violation of an important statistical assumption in meta-analysis that may arise due to multiple estimates of effect size per study, several studies conducted by the same researcher/research lab, phylogenetic dependencies, and so forth. All types of nonindependence can lead to underestimates of the standard error of the mean effect size, and therefore to liberal evaluations of the statistical significance of effects.

Odds ratio—The ratio of the odds of an event in one group to the odds of an event in another group. Odds ratio is commonly used as a metric of effect size for data with dichotomous measures of outcome (e.g., dead/alive), particularly in medicine.

Operational definition—A definition of a concept that relates to the ways it can be assessed or measured. Different operational definitions of the same concept are possible. For example, the concept of plant resistance to herbivores can be operationally defined either in terms of manifestation of specific defensive traits (e.g., concentrations of secondary chemicals, density of trichomes, etc.) or as an inverse of herbivore densities and damage. *See also conceptual definition.*

Parametric statistical test—Statistical test based on parameters of a probability distribution that is assumed to describe the data.

PICO method—A method used to formulate a focused research question for systematic review and meta-analysis, particularly in medicine. PICO stands for population, intervention, comparator, and outcome, and each question has to define these four components.

Positive findings—Results of a study that are in the expected direction, or that are statistically significant.

Posterior distribution—In *Bayesian analysis*, an updated *prior probability distribution* that incorporates observed data.

Power—*See statistical power.*

Power analysis—A statistical analysis that estimates the likelihood of obtaining a statistically significant result given the magnitude of the effect in question, the sample size, and the desired type I error rate.

Precision—Measure of the certainty with which an unknown quantity is estimated; usually the inverse of the variance. When the precision of an estimate is small, the variance of that estimate is large, and vice versa. Large precision (small variance) is desirable when estimating effects and model parameters.

Primary research—Original research reports based on observational or experimental studies, which provide data for meta-analysis.

Prior probability distribution—In Bayesian analysis, a probability distribution representing knowledge, belief, or assumptions about properties of the data before any data have been observed.

Protocol—A plan or set of steps to be followed in a systematic review that describes the rationale for the review, the objectives, and the methods that will be used to locate, select, and critically appraise studies, and to collect and analyze data from the included studies.

Publication bias—Influence of magnitude, direction, and/or statistical significance of research findings on the publication fate of a study (e.g., whether, when, and where a study is published. If publication bias is present, the effect sizes included in a meta-analysis generate different conclusions than those obtained if effect sizes for all the appropriate statistical tests that have been correctly conducted were included in the analysis.

Q homogeneity statistic (sometimes referred to as *Q* heterogeneity statistic)—Measures the degree of variation of effect estimates across studies. The *Q* statistic is based on the sum of the squared differences between the effect estimate from each study and the overall effect estimate, weighted by the precision of the estimate. *See also I^2 homogeneity statistic.*

Quantitative research synthesis—A type of research synthesis that involves some form of statistical analysis of study outcomes. Quantitative research synthesis includes such methods as vote counting, combining probabilities, and meta-analysis (combining effect sizes).

Random-effects meta-analysis model—A statistical model used to combine effect sizes across studies that accounts for both random sampling variation in estimating the true effect for each study and variation in the true effect between the studies.

Randomization test—A nonparametric iterative procedure used to generate a statistical distribution of test statistics in order to determine their significance (based on data). For example, in a randomization test for between-group heterogeneity (Q_B), studies are randomly reassigned to groups in each iteration (keeping the number of studies in each group unchanged), a frequency distribution of possible values for a test statistic is generated, and the proportion of randomly generated statistics more extreme than the actual test statistic is taken to be the significance level for the data set. Randomization tests require no assumptions regarding statistical distribution of the data, although independence of observations is still assumed. Contrasts with *parametric statistical tests*.

Rate difference—The difference between two rates; it can be used as an effect size metric (e.g., to compare rates of responses in the control and treatment groups).

Rate ratio—The ratio of two rates. In meta-analysis, it is often the rate of response in the treatment group relative to that of the control group.

Relational database—Unlike a *flat file database*, a relational database consists of several data tables linked through the common fields. For example, a relational database may consist of a table containing bibliographic information about all individual studies included in a meta-analysis, a table describing details of the experimental design from each study, a table with effect size estimates, and a table with detailed information on the study species used. The above tables could be linked through the reference number field, which enables the combination of data from several tables for querying and analysis.

Resampling tests—Nonparametric techniques that allow the evaluation of a given test statistic's significance. They are often used when the data do not conform to the assumptions of the parametric tests, and include *bootstrapping* and *randomization tests*.

Research bias (sensu Gurevitch and Hedges 1999)—The tendency of scientists to study organisms or conditions in which one has high expectation of detecting statistically significant effects. Research bias may lead to overestimation of the magnitude of the overall mean effect size and the effect sizes for some groups of studies in meta-analysis.

Research synthesis—A review of primary research on a given topic with the purpose of integrating the findings and, for example, creating generalizations or resolving conflicting results or conclusions.

Response ratio—The ratio of one mean of outcome to another (e.g., the mean of some continuous variable measured in the experimental group to that in the control group). It is widely used in ecology and is a popular metric of effect size in ecological meta-analyses (Hedges et al. 1999). Meta-analysis is usually performed on the natural logarithm of the response ratio to linearize the metric and to normalize its sampling distribution.

Sampling error—The deviation between the effect size statistic and the population parameter that it estimates; it occurs because the statistic is estimated based on a sample rather than the entire population.

Scoping search—Initial, cursory review of literature performed to assess the feasibility and value of a proposed review, and prior to conducting a systematic literature search.

Selective reporting—Incomplete reporting of results, such as only reporting efficacious treatments or statistically significant estimates, or only publishing papers that report statistically significant positive outcomes.

Search strategy—The methods used by a reviewer to locate relevant studies for a meta-analysis or a systematic review. The search strategy could include key word searching of bibliographic databases, hand searching relevant journals, checking reference lists of relevant studies, contacting scientists involved in research on a particular topic, and so on.

Sensitivity—Attribute of a search strategy that relates to its ability to identify information of relevance. A high-sensitivity search captures virtually all relevant information, but in the process also captures much information that is not relevant. *See also specificity.*

Sensitivity analysis—An analysis used to determine how sensitive the conclusions of the meta-analysis are to changes in decisions or assumptions about the data and the methods that were used. For example, a sensitivity analysis may involve exclusion of studies that are perceived as potential outliers to see whether the estimate of the mean effect size changes significantly.

Shrinkage—Term in *Bayesian analysis* describing the amount by which an observed individual random quantity is moved toward its expected value or average by a model.

Significance level (*P*-value)—Indicates the probability of obtaining a value of test statistic that is the same or more extreme than the value observed in the sample, if the null hypothesis is true.

Significance test—A statistical test that estimates the probability of obtaining a value of test statistic that is the same or more extreme than the value observed in the sample, if the null hypothesis is true.

Specificity—Attribute of a search strategy that relates to its ability to obtain only relevant information. A high-specificity search would contain a very high proportion of "hits" that are relevant, but in the process may miss relevant information because of restrictive search terms. *See also sensitivity.*

Standardized mean difference—The difference between two estimated means divided by an estimate of the standard deviation. Standardized mean difference is one of the metrics of effect size commonly used in ecology and evolution (*see Hedges'* d).

Statistical independence—Statistical independence of effect sizes is an important assumption of meta-analysis. The assumption is that the collection of effect sizes pooled in a review does not share a correlated structure, and that each effect size represents an independent piece of information. *See also nonindependence.*

Statistical inference—The use of statistics to derive estimates, test hypotheses, and draw conclusions based on data. Two schools of statistical inference are the *frequentist approach* and *Bayesian analysis.*

Statistical power—The power of a statistical test is the probability of finding a significant effect when a true effect exists.

Study-generated evidence (sensu Cooper 1998)—The evidence uncovered through examination of sources of variation in effect sizes that have been directly tested in primary studies included in the analysis. This type of evidence allows the meta-analyst to make statements concerning causality of the effects (compare with *synthesis-generated evidence*).

Synthesis-generated evidence (sensu Cooper 1998)—The evidence uncovered through examination of sources of variation in effect sizes that have not been directly tested in primary studies included in the analysis (e.g., geographic location of the primary study or taxa of the study species). This type of evidence can be used to generate new hypotheses, but does not allow making statements concerning causality of the effects (compare with *study-generated evidence*).

Systematic review—The type of research synthesis on a precisely defined topic using systematic and explicit methods to identify, select, critically appraise, and analyze relevant research. Systematic review may or may not include meta-analysis of the data.

Trim and fill method—A nonparametric method used to test and adjust for *funnel plot* asymmetry; it is based on the estimation of the number of missing studies that might exist in a meta-analysis and the effect that these studies might have had on its outcome (Duval and Tweedie 2000a, 2000b). Trim and fill method is one of the indirect tests for *publication bias* and is a form of *sensitivity analysis.*

Vote counting—A primitive form of quantitative research synthesis in which all available studies are classified into several groups based on the statistical significance and/or direction of the research findings, and the numbers of studies in each group are counted and compared. The group with the most "votes" is considered to provide the best guess about the direction of the effect. Unlike meta-analysis, vote counting does not provide information on the magnitude of the effect and does not take into account variation among studies in sample size.

Weighting—A statistical procedure that allows observations with desirable characteristics to contribute more to the estimation of a parameter. In meta-analysis, individual studies are usually given different weights based on the precision of the effect size estimate. These weights are typically calculated as the inverse of the sampling variance, resulting in studies with small variance receiving larger weights and contributing more to the mean effect size estimate than studies with large variance.

FREQUENTLY ASKED QUESTIONS

What is the minimum number of studies required for a meta-analysis?

A meta-analysis can be conducted by combining as few as two studies (or one study if it contains multiple species or sites). Usually more than two studies will be combined, especially if your primary interest is not in the mean effect but in exploring sources of variation in outcomes, and if you are planning to examine the effects of moderators. The average number of studies in ecological meta-analyses has not been documented, but is in the range of 20 to 30 effect sizes, although it may contain many more.

When should I stop collecting data for a meta-analysis?

There are obviously diminishing returns for search efforts. Beyond a certain stage, each additional unit of time invested in searching returns fewer references that are relevant to the review, and there comes a point where the rewards of further searching may not be worth the effort required to identify the additional references. Also, the longer the search is extended, the more likely the paper published will become outdated before it is submitted! There is no empirical basis for guidance regarding the endpoint of a search or the amount of time one should spend locating studies for systematic reviews in ecology, although after the event, fail-safe numbers or tests of funnel plot asymmetry (Chapter 14) may illustrate potential shortcomings of the search strategy.

The Centre for Evidence-Based Conservation (www.cebc.bangor.ac.uk) recommends full viewing of the first 50 hits of any search and then checking for any relevant hits in the next 50, but for many search topics this arbitrary number may not be a useful guide. There are no similar rules of thumb for searching the gray literature, although it is sensible to stop when most of the references you turn up in a particular data source are duplicates of those already found. Ultimately, the decision as to how much time to invest in the search process depends on the question a review addresses and the resources available. It is, however, important that the process used is transparent, so that readers can judge its validity and to ensure repeatability.

How do I know I haven't missed anything important?

You can never be certain, but if your search was systematic and followed a comprehensive search strategy, it is less likely that you will have missed a large quantity of relevant data. Checking the bibliographies of recent relevant papers and reviews is an integral part of the search. Reviewers of submitted meta-analysis papers may helpfully point out papers you have missed, although this is not a recommended strategy for authors. If your database searching has not retrieved the majority of references identified in existing reviews and by experts, you will want to ascertain why, and may wish to modify the searches if they lack sufficient sensitivity.

How do I know that my data selection is repeatable?

The issue of interobserver reliability arises when more than one person works on study selection because the task is too large for one person to tackle. The only way to check whether two people are consistent in their decisions about study inclusion is to have both work on the same subsample of the list and compare their final lists (e.g., using the kappa analysis described in Chapter 4). This is also the best way to demonstrate that study selection is repeatable. More broadly, clarity in reporting of search and selection criteria, as discussed above, will allow others to repeat the research synthesis after it is published.

What do I do if my initial search returns an unmanageable number of hits?

If your initial search returns a huge number of hits (i.e., 20,000 or more), it most likely suggests that the specificity of your search was too low. You should repeat your search with more specific keywords. In a few cases, however, the large number of returns will be the result of a relatively specific search, reflecting truly abundant relevant literature. If your resources or time are too limited to tackle all papers, it will be necessary to narrow the scope of your review. Some search engines allow exclusion of papers on irrelevant topics with the same search term. For example, Web of Science will allow you to restrict the hits to particular subject areas. If you are searching the literature on the ecology of invasive species, for example, any search that includes a version of the term "invasive" will turn up a huge number of medical studies on cancer that you will want to drop from your search.

Is it enough to use only the Web of Science for search of the relevant papers?

This depends on the topic. A search strategy should generally include all sources of potentially relevant papers. These may not all be included in Web of Science, so other sources may need to be accessed. In all cases, the search strategy should be clearly stated as part of the write-up of the meta-analysis. In the medical field it is required that several scientific databases will be searched. In ecology and evolutionary biology the Web of Science is currently the most commonly used scientific literature database for research syntheses, but access to it is generally limited to those with connections to research universities. Agricola is also useful in some fields, particularly for research with a focus on applied ecology and concening plants and insects. Google Scholar has free access, and is increasingly being used by research synthesists in these fields; it has the benefit of accessing a greater number of gray literature sources and older papers, and the disadvantage of missing many recent scientific publications. It has a broader reach but is less systematic than the Web of Science in locating relevant papers. In general, we recommend searches using at least two different databases because they often retrieve somewhat different sources. Chapter 4 has an extended discussion of search methodologies.

Am I free to use any unpublished data I may find?

A great advantage of published information is that the data generally may be used with appropriate citation or other acknowledgement, particularly in a research synthesis where specific published data are not quoted directly. This is not the case for unpublished data or even data buried in the gray literature (e.g., government reports). Researchers will often agree to release their unpublished data, but may limit their use. For example, they may require that the data from their specific study sites be aggregated in overall means, that the raw data not be made available in any appendixes, and/or ask to become authors on any publications arising from the database. It helps to be able to reassure people that your use of their data will not prejudice their ability to publish them in the future. This is the case particularly if your systematic review or meta-analysis addresses a different question than the one which motivated collection of the primary data.

What if I have limited access to resources?

Systematic reviews are resource intensive. The first approach if you have limited resources should be the development of a collaborative network to undertake the review. Open access journal sources are increasingly available on the World Wide Web. Google Scholar, Scirus, Pubmed, and some publishers' websites can be freely searched, even if full text access is restricted. These databases may include sources in the gray literature that are not available through the Web of Science, for example. Limited access to resources is therefore not necessarily a major concern for the searching phase of review but may become critical during the information retrieval stages. Public libraries and the libraries of professional organizations may provide an alternative route to information in the United States and United Kingdom, but this is less likely to be an avenue for access in developing countries.

How do I decide whether to choose a flat data file or a relational database for my meta-analysis?

This depends on the size and complexity of the data set. While a flat data file is conceptually easy, it may become cumbersome if there are a large number of studies or a large number of variables to be coded per study at different levels (e.g., life history traits of individual species, experimental design characteristics of studies, bibliographic information) for each study.

Can I use more than one metric of effect size in a meta-analysis?

No. Within a single meta-analysis you must use the same measure of effect for all studies. However, it is often possible to convert one measure of effect size into other effect size metrics (Chapter 13), so this may not be a serious limitation in many cases. You can also conduct separate meta-analyses on variables that require different metrics of effect size (e.g., odds ratio for survival data and standardized mean differences for biomass data).

Can I convert one metric of effect size into another?

Yes, although not always. See Chapter 13.

Can I use a response ratio if the mean for either of my groups is zero, or if one is negative and one is positive?

No. The response ratio can only be used when neither of the means is equal to zero and they both have the same sign. It should also not be used when the standardized mean of the control group $\sqrt{n_C}\,\overline{X}_C/SD_C < 3.0$, where n_C, $\overline{X}_C$ and SD_C are sample size, mean, and standard deviation of the control group, respectively (Hedges et al. 1999).

Can I obtain a correlation coefficient from χ^2 with more than one degree of freedom, or F-statistics with more than one degree of freedom in the numerator?

Yes, but it is not a simple procedure. See Chapter 6 for an explanation and more information.

How do I choose the right metric of effect size when more than one could be used?

Choose the metric whose interpretation makes the most sense with respect to the data you have and the scientific questions you are trying to answer. The literature in your field of interest is sometimes a useful guide here.

Can I directly compare the magnitudes of different effect size metrics?

Generally, no. Each effect size measure has its own sampling distribution and the meaningful scale of each metric may be quite different. There is no meaningful comparison between a ratio and a difference, for example.

Can I use other, nonstandard metrics of effect size?

Yes, but it is not recommended unless you understand their sampling distribution very well. See Chapters 6 and 7 for more information.

What do I do when my estimate of the random-effects variance is negative?

The numerator of the moments estimate of the variance component is always a difference between two values; these are the error heterogeneity from the fixed-effects version of the model and its associated degrees of freedom (Chapter 9). It is possible for this value to be negative, leading to a negative estimate of the random-effects variance. Clearly, a negative variance estimate does not make sense. Because the estimate is imprecise, a negative variance estimate in this situation is usually interpreted as meaning that the random-effects variance is inconsequential (equal to zero) and the random-effects model collapses to a fixed-effects model meta-analysis. If one wishes to force a random-effects model meta-analysis (believing the problem is in the estimate rather than the model), a maximum likelihood or Bayesian approach should be used since the manner in which the random-effects model variance is estimated and used is different. See Chapters 10 and 11 for more information on these approaches.

Is it ever appropriate to use a fixed-effects model in ecological meta-analysis?

Yes, although the choice of a fixed- versus random/mixed-effects model requires careful consideration of a number of issues (Chapters 8 and 20). Generally, random- or mixed-effects models are more appropriate in ecology and evolution.

How important is knowledge of the scientific area one is reviewing, if one is familiar with meta-analysis protocols?

Scientific expertise and knowledge of the area one is reviewing is essential in order to correctly code research variables, develop the review protocol, and interpret the results of the analysis. Working with excellent collaborators can make a big difference in overcoming one's own limitations in one area or another of the review.

What is a quality meta-analysis?

This is discussed in detail in Chapter 20 and includes being thorough, careful, and transparent, among other characteristics. A valuable meta-analysis should be also be original and important, just as is true for a primary study.

Can I do a meta-analysis on nonexperimental research? What effect size can I use if I am not comparing experimental and control groups?

Yes, meta-analysis of observational data is possible. Various effect size metrics can be used, including correlation coefficients (Chapter 6) and other metrics (Chapter 7). Issues regarding the interpretation of the results, particularly causal inferences, require careful consideration (Chapters 20 and 24).

Can I test for the influence of a number of different study characteristics of interest (moderators) on the study outcomes (effect sizes)?

See the discussion in Chapters 16 and 20. It is generally not legitimate to test the same data set many times, each time testing a different factor for its potential effect on the study outcomes, unless (in some cases) appropriate measures are taken to avoid erroneous conclusions.

What are the most important things I should check initially as an editor or peer reviewer of an ecological or evolutionary meta-analysis to assess its quality and completeness?

You should check the questions in Box 20.1.

Should funnel plots or figures exploring publication bias be presented in the manuscript in preparation for publication?

Neither funnel plots nor trim and fill plots are commonly presented in the final publication; however, they could be provided in supplementary electronic material to the paper.

Is there a best figure for each type of effect size measure?

Yes and no. Most effect size measures can be presented using forest plots. However, if odds, rates, or ratios are used, effects should be plotted on a log scale. If covariates are continuous, line or scatterplots can be used, and if outcomes are dichotomous, L'Abbé plots can help to visualize frequency of occurrence or event rate.

Can conventional visualization tools and graphs be used for meta-analyses?

Yes, provided study weighting is addressed and incorporated.

To what extent should data be shown graphically versus in tabular format?

Similar to conventional statistics, use plots to show trends and facilitate comparison between groups. However, many forest plots incorporate tabular elements by listing text adjacent to the points on the vertical axis.

Which ecological and evolutionary journals publish meta-analyses?

Unlike long narrative reviews, which are often invited contributions, meta-analyses can be published not only in journals which specialize in review papers (e.g., *Biological Reviews*, *Quarterly Review of Biology*, *Annual Review of Ecology, Evolution and Systematics,* etc.), but also in most ecological and evolutionary journals that publish primary research papers (e.g., *Ecology, Evolution, American Naturalist, Ecology Letters*), as well as in general science journals like *Nature, Science, PNAS,* and so forth. When choosing a journal for submission of a meta-analysis, it is often helpful to identify journals in which primary studies included in your analysis are most commonly published, and then check whether these journals accept meta-analyses. Potential journals can also be identified through online searches of similar meta-analyses, inspection of reference lists in key journal articles, and citation searches for comparable papers.

Should I provide the complete database with effect sizes from individual studies and all the study-related variables when I publish my meta-analysis?

The amount of information provided to a journal or a reader depends on a number of considerations, including the requirements of the potential user, the project protocol, and/or funding agency guidelines, and so on. Some journals are now requiring that all papers, including meta-analyses, supply the corresponding data. Some funding agencies also require this. In any case, it is good practice to make the meta-dataset available to readers (e.g., in a form of electronic appendix to the paper). This will greatly facilitate subsequent meta-analytic updates on the topic.

Can we contact the editors and authors of this book for a free advice on how to conduct our own meta-analyses?

While we are happy to promote the use of meta-analysis in ecology and evolution (in fact, that is why we wrote this book), we are unable to answer to all the individual requests for help and advice. We suggest that in the first instance you try to locate and contact research synthesists from other departments at your institutions. You would be amazed to find that most of the medical and social sciences departments next door to you have one or several specialists in meta-analysis, and contacting them may lead to fruitful collaboration. If this search is not successful, contact one of the meta-analysts who have published reviews in your research area. They might be able to help you or redirect to someone who can help. In addition, the editors and authors of this book frequently give courses on meta-analysis for postgraduate students and postdoctoral researchers. We are happy to consider invitations to conduct these courses at your institution, especially if it is located in a country that is blessed with good climate, interesting flora and fauna, and exciting local cuisine. The list of preferred locations is available upon request from the editors.

REFERENCES

Adams, D. C. 2008. Phylogenetic meta-analysis. *Evolution* 62:567–572.

Adams, D. C. and J. O. Church. 2008. Amphibians do not follow Bergmann's rule. *Evolution* 62: 413–420.

Adams, D. C., J. Gurevitch, and M. S. Rosenberg. 1997. Resampling tests for meta-analysis of ecological data. *Ecology* 78:1277–1283.

Ades, A. E. 2003. A chain of evidence with mixed comparisons: Models for multi-parameter synthesis and consistency of evidence. *Statistics in Medicine* 22:2995–3016.

Adkison, M. D. and Z. Su. 2001. A comparison of salmon escapement estimates using a hierarchical Bayesian approach versus separate maximum likelihood estimation of each year's return. *Canadian Journal of Fisheries and Aquatic Sciences* 58:1663–1671.

Adler, P. B., E. W. Seabloom, E. T. Borer, H. Hillebrand, Y. Hautier, A. Hector, W. S. Harpole, et al. 2011. Productivity is a poor predictor of plant species richness. *Science* 233:1750–1753.

Agrawal, A. A. 2007. Macroevolution of plant defense strategies. *Trends in Ecology & Evolution* 22:103–109.

Akçay, E. and J. Roughgarden. 2007. Extra-pair paternity in birds: Review of the genetic benefits. *Evolutionary Ecology Research* 9:855–868.

Alatalo, R. V., J. Mappes, and M. Elgar. 1997. Heritabilities and paradigm shifts. *Nature* 385:402–403.

American Psychological Association. 2010. *The Publication Manual of the American Psychological Association*, 6th edition. American Psychological Association, Washington, DC.

Anderson, D. R., K. P. Burnham, and W. L. Thompson. 2000. Null hypothesis testing: problems, prevalence, and an alternative. *Journal of Wildlife Management* 64:912–923.

Andersson, T. R. 2006. *Biology of the Ubiquitous House Sparrow*. Oxford University Press, Oxford.

Antman, E. M. and J. A. Berlin. 1992. Declining incidence of ventricular fibrillation in myocardial infarction: Implications for the prophylactic use of lidocaine. *Circulation* 86:764–773.

Antman, E. M., J. Lau, B. Kupelnik, F. Mosteller, and T. C. Chalmers. 1992. A comparison of results of meta-analyses of randomized control trials and recommendations of clinical experts: Treatments for myocardial infarction. *Journal of the American Medical Association* 268:240–248.

APA (Publications and Communications Board Working Group on Journal Article Reporting Standards). 2008. Reporting standards for research in psychology: Why do we need them? What might they be? *American Psychologist* 63:839–851.

Arends, L. R. 2006. Multivariate Meta-analysis: Modelling the Heterogeneity. PhD diss., Erasmus University, Rotterdam, Netherlands. Available at http://repub.eur.nl/res/pub/7845/Proefschrift%20Lidia%20Arends.pdf.

Arends, L. R., Z. Vokó, and T. Stijnen. 2003. Combining multiple outcome measures in a meta-analysis: An application. *Statistics in Medicine* 22:1335–1353.

Arft, A. M., M. D. Walker, J. Gurevitch, J. M. Alatalo, M. S. Bret-Harte, M. Dale, M. Diemer, et al. 1999. Responses of tundra plants to experimental warming: Meta-analysis of the international tundra experiment. *Ecological Monographs* 69:491–511.

Arnqvist, G. and D. Wooster. 1995a. Meta-analysis: Synthesizing research findings in ecology and evolution. *Trends in Ecology & Evolution* 10:236–240.

Arnqvist, G. and D. Wooster. 1995b. Reply from G. Arnqvist and D. Wooster. *Trends in Ecology & Evolution* 10:460–461.

Attwood, S. J., M. Maron, A.P.N. House, and C. Zammit. 2008. Do arthropod assemblages display globally consistent responses to intensified agricultural land use and management? *Global Ecology and Biogeography* 12:585–599.

Auger, C. P. 1998. *Information Sources in Grey Literature (Guides to Information Sources)*, 4th edition. Bowker Saur, London.

Ausden, M., W. J. Sutherland, and R. James. 2001. The effects of flooding wet grassland on soil macroinvertebrate prey of breeding wading birds. *Journal of Applied Ecology* 38:320–338.

Aytug, Z., W. Zhou, H. R. Rothstein, and M. Kern. 2009. The conduct and reporting of meta-analyses in I/O psychology and OB: Standard or deviation? Poster presented at the 24th Annual Conference of the Society for Industrial and Organizational Psychology, New Orleans, LA.

Bagnardi, V., A. Zambon, P. Quatto, and G. Corrao. 2004. Flexible meta-regression functions for modeling aggregate dose-response data, with an application to alcohol and mortality. *American Journal of Epidemiology* 159:1077–1086.

Baker, R. and D. Jackson. 2006. Using journal impact factors to correct for the publication bias of medical studies. *Biometrics* 62:785–792.

Bank, E. M., P.A.L. Bonis, H. Moskowitz, C. H. Schmid, J.P.A. Ioannidis, C. Wang, and J. Lau. 1992. Correlation of quality measures with estimates of treatment effect in meta-analyses of randomized controlled trials. *Journal of the American Medical Association* 287:2973–2982.

Barrowman, N. J. and R. A. Myers. 2003. Raindrop plots: A new way to display collections of likelihoods and distributions. *American Statistician* 57:1–6.

Barrowman, N. J., R. A. Myers, R. Hilborn, D. G. Kehler, and C. A. Field. 2003. The variability among populations of coho salmon in the maximum reproductive rate and depensation. *Ecological Applications* 13:784–793.

Barto, E. K. and M. C. Rillig. 2010. Does herbivory really suppress mycorrhiza? A meta-analysis. *Journal of Ecology* 98:745–753.

Barzi, F. and M. Woodward. 2004. Imputation of missing values in practice: Results from imputations of serum cholesterol in 28 cohort studies. *American Journal of Epidemiology* 160:34–45.

Bauchau, V. 1997. Is there a "file drawer problem" in biological research? *Oikos* 79:407–409.

Bax, L., L. M. Yu, N. Ikeda, H. Tsuruta, and K.G.M. Moons. 2006. Development and validation of MIX: Comprehensive free software for meta-analysis of causal research data. *BMC Medical Research Methodology* 6:50.

Beal, D. J., D. M. Corey, and W. P. Dunlap. 2002. On the bias of Huffcutt and Arthur's (1995) procedure for identifying outliers in the meta-analysis of correlations. *Journal of Applied Psychology* 87:583–589.

Becker, B. J. 1992. Using results from replicated studies to estimate linear models. *Journal of Educational Statistics* 17:341–362.

Becker, B. J. 1994. Combining significance levels. In *The Handbook of Research Synthesis*, edited by H. Cooper and L. V. Hedges, 215–230. Russell Sage Foundation, New York.

Becker, B. J. 1995. Corrections to "Using results from replicated studies to estimate linear model." *Journal of Educational Statistics* 20:100–102.

Becker, B. J. 2005. Failsafe N or file-drawer number. In *Publication bias in meta-analysis*, edited by H. R. Rothstein, A. J. Sutton, and M. Borenstein, 111–125. John Wiley, Chichester, UK.

Becker, B. J. and M. J. Wu. 2007. The synthesis of regression slopes in meta-analysis. *Statistical Science* 22:414–429.

Begg, C. B. 1994. Publication bias. In *The Handbook of Research Synthesis*, edited by H. Cooper and L. V. Hedges, 399–409. Russell Sage Foundation, New York.

Begg, C. B. and M. Mazumdar. 1994. Operating characteristics of a rank correlation test for publication bias. *Biometrics* 50:1088–1101.

Beirinckx, K., H. van Gossum, M. J. Lajeunesse, and M. R. Forbes. 2006. Sex biases in dispersal and philopatry: Insights from a meta-analysis based on capture–mark–recapture studies of damselflies. *Oikos* 113:539–547.

Bender, D. J., T. A. Contreras, and L. Fahrig. 1998. Habitat loss and population decline: A meta-analysis of the patch size effect. *Ecology* 79:517–533.

Bender, R., C. Bunce, M. Clarke, S. Gates, S. Lange, N. L. Pace, and K. Thorlund. 2008. Attention should be given to multiplicity issues in systematic reviews. *Journal of Clinical Epidemiology* 61:857–865.

Benjamini, Y. and Y. Hochberg. 1995. Controlling the false discovery rate: A practical and powerful approach to multiple testing. *Journal of the Royal Statistical Society Series B* 57:289–300.

Berenbaum, M. R. 1995. The chemistry of defense: Theory and practice. *Proceedings of the National Academy of Sciences of the United States of America* 92:2–8.

Bergelson, J. and C. B. Purrington. 1996. Surveying patterns in the cost of resistance in plants. *American Naturalist* 148:536–558.

Berger, J. O. and D. A. Berry. 1988. Statistical analysis and the illusion of objectivity. *American Scientist* 76:159–165.

Bergmann, C. 1847. Über die verhältnisse der warmeökonomie der thiere zuihrer grosse. *Göttinger Studien* 1:595–708.

Berkey, C. S., J. J. Anderson, and D. C. Hoaglin. 1996. Multiple-outcome meta-analysis of clinical trials. *Statistics in Medicine* 15:537–557.

Berkey, C. S., D. C. Hoaglin, A. Antczak-Bouckoms, F. Mosteller, and G. A. Colditz. 1998. Meta-analysis of multiple outcomes by regression with random effects. *Statistics in Medicine* 17:2537–2550.

Berkey, C. S., F. Mosteller, J. Lau, and E. M. Antman. 1996. Uncertainty of the time of first significance in random effects cumulative meta-analysis. *Controlled Clinical Trials* 17:357–371.

Berlin, J. A., N. M. Laird, H. S. Sacks, and T. C. Chalmers. 1989. A comparison of statistical methods for combining event rates from clinical trials. *Statistics in Medicine* 8:141–151.

Berlin, J. A., M. P. Longnecker, and S. Greenland. 1993. Meta-analysis of epidemiologic dose-response data. *Epidemiology* 4:218–228.

Berlin, J. A., on behalf of University of Pennsylvania Meta-analysis Blinding Study Group. 1997. Does blinding of readers affect the results of meta-analyses? *The Lancet* 350: 185–186.

Berlin, J. A., J. Santanna, C. H. Schmid, L. A. Szczech, and H. I. Feldman. 2002. Individual patient-versus group-level data meta-regressions for the investigation of treatment effect modifiers: Ecological bias rears its ugly head. *Statistics in Medicine* 21:371–387.

Besag, J., P. Green, D. Higdon, and K. Mengersen. 1995. Bayesian computation and stochastic systems. *Statistical Science* 10:3–41.

Bikhchandani, S., D. Hirshleifer, and I. Welch. 1992. A theory of fads, fashion, custom, and cultural-change as informational cascades. *Journal of Political Economy* 100:992–1026.

Bjelakovic, G., D. Nikolova, L. L. Gluud, R. G. Simonetti, and C. Gluud. 2007. Mortality in randomized trials of antioxidant supplements for primary and secondary prevention: Systematic review and meta-analysis. *Journal of the American Medical Association* 297:842–857.

Björklund, M. 1997. Are "comparative methods" always necessary? *Oikos* 80:607–612.

Blackburn, T. M., K. J. Gaston, and N. Loder. 1999. Geographic gradients in body size: A clarification of Bergmann's rule. *Diversity and Distribution* 5:165–174.

Bland, J. M. and D. G. Altman. 1986. Statistical methods for assessing agreement between two methods of clinical measurement. *The Lancet* 8476:307–310.

Blenckner, T., R. Adrian, D. M. Livingstone, E. Jennings, G. A. Weyhenmeyer, D. G. George, T. Jankowski, et al. 2007. Large-scale climatic signatures in lakes across Europe: A meta-analysis. *Global Change Biology* 13:1314–1326.

Blettner, M., W. Sauerbrei, B. Schlehofer, T. Scheuchenpflug, and C. Friedenreich. 1999. Traditional reviews, meta-analyses and pooled analyses in epidemiology. *International Journal of Epidemiology* 28:1–9.

Blomberg, S. P., T. Garland, and A. R. Ives. 2003. Testing for phylogenetic signal in comparative data: A modeling approach for adaptive evolution. *Evolution* 57:717–745.

Boerner, R.E.J., J. Huang, and S. C. Hart. 2008. Effects of fire and fire surrogate treatments on estimated carbon storage and sequestration rate. *Forest Ecology and Management* 255:3081–3097.

Böhning, D., R. Kuhnert, and S. Rattanasiri. 2008. *Meta-analysis of Binary Data Using Profile Likelihood*. Chapman and Hall/CRC, Boca Raton, FL.

Böhning, D., U. Malzahn, E. Dietze, and P. Schlattmann. 2002. Some general points in estimating heterogeneity variance with the DerSimonian-Laird estimator. *Biostatistics* 3:445–457.

Bolker, B. M., M. E. Brooks, C. J. Clark, S. W. Geange, J. R. Poulsen, M.H.H. Stevens, and J.S.S. White. 2009. Generalized linear mixed models: A practical guide for ecology and evolution. *Trends in Ecology & Evolution* 24:127–135.

Bonduriansky, R. 2007. Sexual selection and allometry: A critical reappraisal of the evidence and ideas. *Evolution* 61:838–849.

Bonnet, X., R. Shine, and O. Lourdais. 2002. Taxonomic chauvinism. *Trends in Ecology & Evolution* 17:1–3.

Boomsma, J. J. and A. Grafen. 1990. Intraspecific variation in ant sex ratios and the Trivers-Hare hypothesis. *Evolution* 44:1026–1034.

Boonekamp, J. J., A.H.F. Ross, and S. Verhulst. 2008. Immune activation suppresses plasma testosterone level: A meta-analysis. *Biology Letters* 4:741–744.

Booth, A., M. Clarke, D. Ghersi, D. Moher, M. Petticrew, and L. Stewart. 2011. An international registry of systematic-review protocols. *The Lancet* 377:108–109.

Borenstein, M., L. V. Hedges, J.P.T. Higgins, and H. R. Rothstein. 2009. *Introduction to Meta-analysis*. John Wiley and Sons, New York.

Borenstein, M., L. V. Hedges, J.P.T. Higgins, and H. R. Rothstein. 2010a. *Computing Effect Sizes for Meta-analysis*. John Wiley and Sons, Chichester, UK.

Borenstein, M., L. V. Hedges, J.P.T. Higgins, and H. R. Rothstein. 2010b. A basic introduction to fixed-effect and random-effects models for meta-analysis. *Research Synthesis Methods* 1:97–111.

Borowicz, V. A. 2001. Do arbiscular mycorrhizal fungi alter plant-pathogen relations? *Ecology* 82:3057–3068.

Boyce, M. S., L. L. Irwin, and R. Barker. 2005. Demographic meta-analysis: Synthesizing vital rates for spotted owls. *Journal of Applied Ecology* 42:38–49.

Bracken, M. B. 1992. Statistical methods for analysis of effects of treatment in overviews of randomized trials. In *Effective Care of the Newborn Infant*, edited by J. C. Sinclair and M. B. Bracken, 13–20. Oxford University Press, Oxford.

Brett, M. T. and C. R. Goldman. 1997. Consumer versus resource control in freshwater pelagic food webs. *Science* 275:384–386.

Briggs, A., K. Claxton, and M. Sculper. 2006. *Decision Modeling for Health Economic Evaluation*. Oxford University Press, Oxford.

Britten, H. B. 1996. Meta-analyses of the association between multilocus heterozygosity and fitness. *Evolution* 50:2158–2164.

Brok, J., K. Thorlund, C. Gluud, and J. Wetterslev. 2008. Trial sequential analysis reveals insufficient information size and potentially false positive results in many meta-analyses. *Journal of Clinical Epidemiology* 61:763–769.

Brook, B. W., J. J. O'Grady, A. P. Chapman, M. A. Burgman, H. R. Akcakaya, and R. Frankham. 2000. Predictive accuracy of population viability analysis in conservation biology. *Nature* 404:385–387.

Brooks, S. P. and A. Gelman. 1998. General methods for monitoring convergence of iterative simulations. *Journal of Computational and Graphical Statistics* 7:434–455.

Brower, J. E., J. H. Zar, and C. N. von Ende. 1998. *Field and Laboratory Methods for General Ecology*. McGraw-Hill, Boston.

Brown, H. K. and R. A. Kempton. 1994. The application of REML in clinical trials. *Statistics in Medicine* 13:1601–1617.

Brown, E. N. and C. H. Schmid. 1994. Application of the Kalman Filter to computational problems in statistics. In *Methods in Enzymology*, edited by M. L. Johnson and L. Brand, 171–181. Numerical Computer Methods, Part B, Volume 240. Academic Press, New York.

Brown, J. H., J. F. Gillooly, A. P. Allen, V. M. Savage, and G. B. West. 2004. Toward a metabolic theory of ecology. *Ecology* 85:1771–1789.

Buck, S. F. 1960. A method of estimation of missing values in multivariate data suitable for use with an electronic computer. *Journal of the Royal Statistical Society Series B* 22:302–303.

Burley, N. 1981. Sex-ratio manipulation and selection for attractiveness. *Science* 211:721–722.

Bushman, B. J., H. M. Cooper, and K. M. Lemke. 1991. Meta-analysis of factor analyses: An illustration using the Buss-Durkee hostility inventory. *Personality and Social Psychology Bulletin* 17:344–349.

Bushman, B. J. and M. C. Wang. 1996. A procedure for combining sample standardized mean differences and vote counts to estimate the population standardized mean difference in fixed-effects models. *Psychological Methods* 1:66–80.

Bushman, B. J. and M. C. Wang. 2009. Vote-counting procedures in meta-analysis. In *The Handbook of Research Synthesis and Meta-analysis*, edited by H. Cooper, L. V. Hedges, and J. C. Valentine, 207–220, 2nd edition. Russell Sage Foundation, New York.

Butler, M. A. and A. A. King. 2004. Phylogenetic comparative analysis: A modeling approach for adaptive evolution. *American Naturalist* 164:683–695.

Cafri, G. and J. D. Kromrey. 2008. A SAS® macro for statistical power calculations in meta-analysis. *SESUG Proceedings*: paper 159–2008, http://analytics.ncsu.edu/?page_id=424.

Cappelleri, J. C., J.P.A. Ioannidis, S. D. de Ferranti, C. H. Schmid, M. Aubert, T. C. Chalmers, and J. Lau. 1996. Large trials versus meta-analyses of smaller trials: How do their results compare? *Journal of the American Medical Association* 276:1332–1338.

Card, N. A. 2012. *Applied Meta-Analysis for Social Science Research*. The Guildford Press, New York.

Cardillo, M., A. Purvis, and J. L. Gittleman. 2008. Global patterns in the phylogenetic structure of island mammal assemblages. *Proceedings of the Royal Society of London Series B* 275:1549–1556.

Cardinale, B. J., D. S. Srivastava, J. E. Duffy, J. P. Wright, A. L. Downing, M. Sankaran, and C. Jousseau. 2006. Effects of biodiversity on the functioning of trophic groups and ecosystems. *Nature* 443:989–992.

Carlin, B. P. and T. A. Louis. 2008. *Bayesian Methods for Data Analysis*, 3rd edition. Chapman and Hall, New York.

Cassey, P., T. M. Blackburn, R. P. Duncan, and J. L. Lockwood. 2005. Lessons from the establishment of exotic species: A meta-analytical case study using birds. *Journal of Animal Ecology* 74:250–258.

Cassey, P., J. G. Ewen, T. M. Blackburn, and A. P. Møller. 2004. A survey of publication bias within evolutionary ecology. *Proceedings of the Royal Society of London Series B* 271:S415–S454.

Cassey, P., J. G. Ewen, and A. P. Møller. 2006. Revised evidence for facultative sex ratio adjustment in birds: A correction. *Proceedings of the Royal Society of London Series B* 273:3129–3130.

CEBC (Centre for Evidence-Based Conservation). 2010. *Guidelines for Systematic Review in Conservation and Environmental Management*. Version 4.0. Centre for Evidence-Based Conservation,

Bangor University, Bangor, UK. http://www.environmentalevidence.org/Documents/Guidelines print.pdf.

Centre for Reviews and Dissemination. 2009. *Systematic Reviews: CRD's Guidance for Undertaking Reviews in Healthcare.* CRD, University of York. http://www.york.ac.uk/inst/crd/pdf/ Systematic_Reviews.pdf .

Chalmers, A. F. 1999. *What is This Thing Called Science?*, 3rd edition. University of Queensland Press, Brisbane.

Chalmers, T. C., R. J. Mattra, H. Smith, and A. M. Kunzler. 1977. Evidence favouring the use of anticoagulants in the hospital phase of acute myocardial infarction. *New England Journal of Medicine* 297:1091–1096.

Chaloner, K., T. Church, T. A. Louis, and J. P. Matts. 1993. Graphical elicitation of a prior distribution for a clinical-trial. *Statistician* 42:341–353.

Chan, A.-W., A. Hróbjartsson, M. T. Haahr, P. C. Gøtzsche, and D. G. Altman. 2004. Empirical evidence for selective reporting of outcomes in randomized trials: Comparison of protocols to published articles. *Journal of American Medical Association* 291:2457–2465.

Chan, A.-W., K. Krleza-Jeric, I. Schmid, and D. G. Altman. 2004. Outcome reporting bias in randomised trials funded by the Canadian Institutes of Health Research. *Canadian Medical Association Journal* 171:735–740.

Cheung, S. F. and D.K.-S. Chan. 2004. Dependent effect sizes in meta-analysis: Incorporating the degree of interdependence. *Journal of Applied Psychology* 89:780–791.

Cheverud, J. M., M. M. Dow, and W. Leutenegger. 1985. The quantitative assessment of phylogenetic constraints in comparative analyses: Sexual dimorphism in body-weight among primates. *Evolution* 39:1335–1351.

Chinn, S. 2000. A simple method for converting an odds ratio to effect size for use in meta-analysis. *Statistics in Medicine* 19:3127–3131.

Choy, S. L., R. O'Leary, and K. Mengersen. 2009. Elicitation by design in ecology: Using expert opinion to inform priors for Bayesian statistical models. *Ecology* 90:265–277.

Clarke, M. and D. Ghersi. 1997. Meta-analysis, collaborative overview, systematic review: What does it all mean? *Australian Prescriber* 20:93–96.

Cleveland, W. S. 1985. *The Elements of Graphing Data.* Wadsworth Advanced Books and Software, Monterey, CA.

Clutton-Brock, T. H. and T. Coulson. 2002. Comparative ungulate dynamics: The devil is in the details. *Philosophical Transactions of the Royal Society of London Series B* 357:1285–1298.

Cockburn, A., S. Legge, and M. C. Double. 2002. Sex ratios in birds and mammals: Can the hypotheses be disentangled? In *Sex Ratio: Concepts and Research Methods*, edited by I.C.V. Hardy, 266–286. Cambridge University Press, Cambridge.

Cohen, J. 1960. A coefficient of agreement for nominal scales. *Educational and Psychological Measurement* 20:37–46.

Cohen, J. 1969. *Statistical Power Analysis for the Behavioral Sciences.* Academic Press, New York.

Cohen, J. 1977. *Statistical Power Analysis for the Behavioral Sciences*, revised edition. Academic Press, New York.

Cohen, J. 1988. *Statistical Power Analysis for the Behavioral Sciences*, 2nd edition. Erlbaum, Hillsdale, NJ.

Cohen, J. 1992. A power primer. *Psychological Bulletin* 112:155–159.

Cohn, L. D. and B. J. Becker. 2003. How meta-analysis increases statistical power. *Psychological Methods* 8:243–253.

Colautti, R. I., I. A. Grigorovich, and H. J. MacIsaac. 2006. Propagule pressure: A null model for biological invasions. *Biological Invasions* 8:1023–1037.

Colegrave, N. and G. D. Ruxton. 2003. Confidence intervals are a more useful complement to non-significant tests than are power calculations. *Behavioral Ecology* 14:446–450.

Coleman, R. A., A. J. Underwood, L. Benedetti-Cecchi, P. Åberg, F. Arenas, J. Arrontes, J. Castro, et al. 2006. A continental scale evaluation of the role of limpet grazing on rocky shores. *Oecologia* 147:556–564.

Coltman, D. W. and J. Slate. 2003. Microsatellite measures of inbreeding: A meta-analysis. *Evolution* 57:971–983.

Condit, R., S. Aguilar, A. Hernandez, R. Perez, S. Lao, G. Angehr, S. P. Hubbell, and R. B. Foster. 2004. Tropical forest dynamics across a rainfall gradient and the impact of an El Nino dry season. *Journal of Tropical Ecology* 20:51–72.

Condit, R., S. P. Hubbell, and R. B. Foster. 1995. Mortality rates of 205 Neotropical tree and shrub species and the impact of a severe drought. *Ecological Monographs* 65:419–439.

Congdon, P. 2003. *Applied Bayesian Modeling*. John Wiley and Sons, New York.

Congdon, P. 2005. *Bayesian Models for Categorical Data*. John Wiley and Sons, New York.

Congdon, P. 2007. *Bayesian Statistical Modelling*, 2nd edition. John Wiley and Sons, New York.

Congdon, P. 2010. *Applied Bayesian Hierarchical Methods*. CRC Press, New York.

Connell, J. H. 1971. On the role of natural enemies in preventing competitive exclusion in some marine animals and in rain forest trees. In *Dynamics of Populations*, edited by P. J. Den Boer and G. Gradwell, 298–312. Pudoc, Wageningen, Netherlands.

Cooper, H. 1982. Scientific guidelines for conducting integrative research reviews. *Review of Educational Research* 52:291–302.

Cooper, H. 1998. *Synthesizing Research: A Guide for Literature Review*, 3rd edition. SAGE Publications, Thousand Oaks, CA.

Cooper, H. 2007a. *The Battle over Homework: Common Ground for Administrators, Teachers, and Parents*, 3rd edition. Corwin Press, Thousand Oaks, CA.

Cooper, H. 2007b. Evaluating and interpreting research syntheses in adult learning and literacy. Occasional Paper of the National Center for the Study of Adult Learning and Literacy, http://www.ncsall.net/fileadmin/resources/research/op_research_syntheses.pdf .

Cooper, H. 2009. *Research Synthesis and Meta-Analysis: A Step-by-Step Approach*, 4th edition. SAGE Publications, Thousand Oaks, CA.

Cooper, H., K. DeNeve, and K. Charlton. 1997. Finding the missing science: The fate of studies submitted for review by a human subjects committee. *Psychological Methods* 2:447–452.

Cooper, H., L. V. Hedges, and J. C. Valentine, editors. 2009. *The Handbook of Research Synthesis and Meta-Analysis*, 2nd edition. Russell Sage Foundation, New York.

Cooper, N. J., A. J. Sutton, A. E. Ades, and N. Welton. 2008. Addressing between-study heterogeneity and inconsistency in mixed treatment comparisons: Application to stroke prevention treatments for atrial fibrillation patients. Paper presented at 3rd Annual Meeting of the Society for Research Synthesis Methodology, Corfu, Greece.

Copas, J. and J. Q. Shi. 2000. Meta-analysis, funnel plots and sensitivity analysis. *Biostatistics* 1:247–262.

Copas, J. B. and J. Q. Shi. 2001. A sensitivity analysis for publication bias in systematic reviews. *Statistical Methods in Medical Research* 10:251–265.

Corrêa, A., J. Gurevitch, M. A. Martins-Loução, and C. Cruz. 2012. C allocation to the fungus is not a cost to the plant in ectomycorrhizae. *Oikos* 121:449–463.

Côté, I. M., J. A. Gill, T. A. Gardner, and A. R. Watkinson. 2005. Measuring coral reef decline through meta-analyses. *Philosophical Transactions of the Royal Society Series B* 360:385–395.

Côté, I. M., I. Mosqueira, and J. D. Reynolds. 2001. Effects of marine reserve characteristics on the protection of fish populations: a meta-analysis. *Journal of Fish Biology* 59 (Supplement A):178–189.

Côté, I. M. and W. J. Sutherland. 1997. The effectiveness of removing predators to protect bird populations. *Conservation Biology* 11:395–405.

Counsell, C. 1997. Formulating questions and locating primary studies for inclusion in systematic reviews. *Annals of Internal Medicine* 127:380–387.

Cowie, N. R., W. J. Sutherland, M.K.M. Ditlhogo, and R. James. 1992. The effect of conservation management of reed beds. II. The flora and litter disappearance. *Journal of Applied Ecology* 29:277–284.

Crawley, M. J. 2002. *Statistical Computing: An Introduction to Data Analysis Using S-Plus.* John Wiley, New York.

Crumley, E. T., N. Wiebe, K. Cramer, T. P. Klassen, and L. Hartling. 2005. Which resources should be used to identify RCT/CCTs for systematic reviews: A systematic review. *BMC Medical Research Methodology* 5:24.

Csada, R. D., P. C. James, and R. H. M. Espie. 1996. The "file drawer problem" of non-significant results: Does it apply to biological research? *Oikos* 76:591–593.

Curtis, P. S. 1996. A meta-analysis of leaf gas exchange and nitrogen in trees grown under elevated carbon dioxide. *Plant, Cell and Environment* 19:127–137.

Curtis, P. S. and X. Wang. 1998. A meta-analysis of elevated CO_2 effects on woody plant mass, form, and physiology. *Oecologia* 113:299–313.

Daehler, C. C. 2003. Performance comparisons of co-occurring native and alien invasive plants: Implications for conservation and restoration *Annual Review of Ecology and Systematics* 34:183–211.

Darimont, C. T., S. M. Carlson, M. T. Kinnison, P. C. Paquet, T. E. Reimchen, and C. C. Wilmers. 2009. Human predators outpace other agents of trait change in the wild. *Proceedings of the National Academy of Sciences of the United States of America* 106:952–954

Darwin, C. R. 1859. *The Origin of Species.* John Murray, London.

David, P., T. Bjorksten, K. Fowler, and A. Pomianowski. 2000. Condition-dependent signaling of genetic variation in stalk-eyed flies. *Nature* 406:106–108.

Davies, Z. G., C. Tyler, G. B. Stewart, and A. S. Pullin. 2008. Are current management recommendations for saproxylic invertebrates effective? A systematic review. *Biodiversity and Conservation* 17:209–234.

de Craen, A.J.M., H. van Vliet, and F. M. Helmerhorst. 2005. An analysis of systematic reviews indicated low incorporation of results from clinical trial quality assessment. *Journal of Clinical Epidemiology* 58:311–313.

Dear, K. 1994. Iterative generalized least-squares for meta-analysis of survival data at multiple times. *Biometrics* 50:989–1002.

Dear, K.B.G. and C. B. Begg. 1992. An approach for assessing publication bias in systematic reviews. *Statistical Methods in Medical Research* 10:251–265.

Deeks, J. J. 2002. Issues in the selection of summary statistics of meta-analyses of clinical trials with binary outcomes. *Statistics in Medicine* 21:1575–1600.

Demidenko, E. 2004. *Mixed Models—Theory and Applications.* John Wiley and Sons, New York.

DerSimonian, R. and N. M. Laird. 1986. Meta-analysis in clinical trials. *Controlled Clinical Trials* 7:177–188.

Detsky, A. S., C. D. Naylor, K. O'Rourke, A. J. McGreer, and K. A. L'Abbé. 1992. Incorporating variations in the quality of individual randomized trials into meta-analysis. *Journal of Clinical Epidemiology* 45:255–265.

DeZee, K. J., W. Shimeall, K. Douglas, and J. L. Jackson. 2005. High-dosage vitamin E supplementation and all-cause mortality. *Annals of Internal Medicine* 143:153–154.

Di Stefano, J. 2003. How much power is enough? Against the development of an arbitrary convention for statistical power calculations. *Functional Ecology* 17:707–709.

Dickersin, K. 1997. How important is publication bias? A synthesis of available data. *AIDS Education and Prevention* 9 (Supplement 1):15–21.

Dickersin, K. 2005 Publication bias: Recognizing the problem, understanding its origins and scope, and preventing harm. In *Publication Bias in Meta-Analysis: Prevention, Assessment and Adjustments*, edited by H. R. Rothstein, A. J. Sutton, and M. Borenstein, 11–34.. Wiley, Chichester, UK.

Diniz-Filho, J. A. F. 2001. Phylogenetic autocorrelation under distinct evolutionary processes. *Evolution* 55:1104–1109.

Ditlhogo, M.K.M., R. James, B. R. Laurence, and W. J. Sutherland. 1992. The effect of conservation management of reed beds. I. The invertebrates. *Journal of Applied Ecology* 29:265–276.

Dixon, P. M. 1993. The bootstrap and the jackknife: Describing the precision of ecological indices. In *Design and Analysis of Ecological Experiments*, edited by S. M. Scheiner and J. Gurevitch, 290–318. Chapman and Hall, New York.

Dodds, P. S., R. H. Rothman, and J. S. Weitz. 2001. Reexamination of the "3/4 law" of metabolism. *Journal of Theoretical Biology* 209:9–27.

Dominici, F., G. Parmigiani, R. L. Wolpert, and V. Hasselblad. 1999. Meta-analysis of migraine headache treatments : Combining information from heterogeneous designs. *Journal of the American Statistical Association* 94:16–28.

Downtown, W.J.S., W.J.R. Grant, and B. R. Loveys. 1987. Carbon dioxide enrichment increases yield of Valencia orange. *Australian Journal of Plant Physiology* 14:493–501.

Draper, D., J. S. Hodges, C. L. Mallows, and D. Pregibon. 1993. Exchangeability and data analysis. *Journal of the Royal Statistical Society Series A* 156:9–37.

Dubois, F. and F. Cezilly. 2002. Breeding success and mate retention in birds: A meta-analysis *Behavioral Ecology and Sociobiology* 52:357–364.

Dulvy, N. K. and N.V.C. Polunin. 2004. Using informal knowledge to infer human-induced rarity of a conspicuous reef fish. *Animal Conservation* 7:365–374.

DuMouchel, W. 1990. Bayesian meta-analysis. In *Statistical Methodology in the Pharmaceutical Sciences*, edited by D. A. Berry, 509–529. Dekker, New York.

DuMouchel, W. 1995. Meta-analysis for dose-response models. *Statistics in Medicine* 14:679–685.

DuMouchel, W. H. and J. E. Harris. 1983. Bayes and empirical Bayes methods for combining cancer experiments in man and other species. *Journal of the American Statistical Association* 78:293–308.

Duval, S. 2005. The trim and fill methods. In *Publication Bias in Meta-Analysis*, edited by H. R. Rothstein, A. J. Sutton, and M. Borenstein, 127–144. John Wiley, Chichester, UK.

Duval, S. and R. Tweedie. 2000a. A nonparametric 'trim and fill' method of accounting for publication bias in meta-analysis. *Journal of the American Statistical Association* 95:89–99.

Duval, S. and R. Tweedie. 2000b. Trim and fill: A simple funnel-plot-based method of testing and adjusting for publication bias in meta-analysis. *Biometrics* 56:455–463.

Dwan, K., D. G. Altman, J. A. Arnaiz, J. Bloom, A.-W. Chan, E. Cronin, E. Decullier, et al. 2008. Systematic review of the empirical evidence of study publication bias and outcome reporting bias. *PLoS ONE* 3(8):e3081. doi: 10.1371/journal.pone.0003081.

EBCTCG (Early Breast Cancer Trialists' Collaborative Group). 1992. Systemic treatment of early breast cancer by hormonal, cytotoxic, or immune therapy: 133 randomised trials involving 31,000 recurrences and 24,000 deaths among 75,000 women. *The Lancet* 339:1–15, 71–85.

Eddy, D. M. 1989. The confidence profile method: A Bayesian method for assessing health technologies. *Operations Research* 37:210–228.

Edgington, E. S. 1987. *Randomization Tests*, 2nd edition. Marcel Dekker, New York.

Edwards, P., M. Clarke, C. DiGuiseppi, S. Pratap, I. Roberts, and R. Wentz. 2002. Identification of randomized controlled trials in systematic reviews: Accuracy and reliability of screening records. *Statistics in Medicine* 21:1635–1640.

Efron, B. 1982. *The Jackknife, the Bootstrap, and Other Resampling Plans*, CBMS-NSF Monograph 38. Society of Industrial and Applied Mathematics, Philadelphia, PA.

Efron, B. and C. N. Morris. 1973. Stein's estimation rule and its competitors—an empirical Bayes approach. *Journal of the American Statistical Association* 68:34–38.

Efron, B. and C. N. Morris. 1977. Stein's paradox in statistics. *Scientific American* 236:119–127.

Efron, B. and R. Tibshirani. 1997. Improvements on cross-validation: The bootstrap method. *Journal of the American Statistical Association* 92:548–560.

Egger M., G. Davey Smith, and D. G. Altman. 2001. *Systematic Reviews in Health Care*. BMJ Books, London.

Egger, M., G. Davey Smith, and A. N. Phillips. 1997. Meta-analysis: principles and procedures. *British Medical Journal* 315:1533–1537.

Egger, M., G. Davey Smith, M. Schneider, and C. Minder. 1997. Bias in meta-analysis detected by a simple, graphical test. *British Medical Journal* 315:629–634.

Egger, E., T. Zellweger-Zähner, M. Schneider, C. Junker, C. Lengeler, and G. Antes. 1997. Language bias in randomised controlled trials published in English and German. *The Lancet* 350: 326–329.

Ellison, A. M. 2001. Exploratory data analysis and graphic display. In *Design and Analysis of Ecological Experiments*, edited by S. M. Scheiner and J. Gurevitch, 37–62, 2nd edition. Oxford University Press, Oxford.

Ellison, A. M. 2004. Bayesian inference in ecology. *Ecology Letters* 7:509–520.

Elton, C. S. 1958. *The Ecology of Invasions by Animals and Plants*. Methuen and Co, London.

Engels, E. A., C. H. Schmid, N. Terrin, I. Olkin, and J. Lau. 2000. Heterogeneity and statistical significance in meta-analysis: An empirical study of 125 meta-analyses. *Statistics in Medicine* 19:1707–1728.

Englund, G., O. Sarnell, and S. D. Cooper. 1999. The importance of data-selection criteria: Meta-analyses of stream predation experiments. *Ecology* 80:1132–1141.

Enquist, B. J., J. H. Brown, and G. B. West. 1998. Allometric scaling of plant energetics and population density. *Nature* 395:163–165.

EPA. 1992. Respiratory health effects of passive smoking: Lung cancer and other disorders. US Environmental Protection Agency, Washington, D.C. EPA/600/6-90/006F. http://www.epa.gov/ncea/ets/pdfs/acknowl.pdf.

Evidence-based Working Group. 1992. Evidence-based medicine: A new approach to teaching the practice of medicine. *Journal of the American Medical Association* 268:2420–2425.

Ewen, J. G., P. Cassey, and A. P. Møller. 2004. Facultative primary sex ratio variation: A lack of evidence in birds? *Proceedings of the Royal Society of London Series B* 271:1277–1282.

Eysenck, H. 1978. An exercise in mega-silliness. *American Psychologist* 33:519.

Eysenck, H. J. 1994. Systematic reviews: Meta-analysis and its problems. *British Medical Journal* 309:789–792.

Falconer, D. S. and T.F.C. Mackay. 1996. *Introduction to Quantitative Genetics*, 4th edition. Addison Wesley Longman Limited, London.

Fanelli, D. 2010. "Positive" results increase down the hierarchy of the sciences. *PloS ONE* 5(4):e10068. doi: 10.1371/journal.pone.0010068.

Farrell-Gray, C. C. and N. J. Gotelli. 2005. Allometric exponents support a 3/4-power scaling law. *Ecology* 86:2083–2087.

Fayed, S. A., M. D. Jennions, and P. R. Y. Backwell. 2008. What factors contribute to an ownership advantage? *Biology Letters* 4:143–145.

Fazey, I., J. A. Fazey, J. G. Salisbury, D. B. Lindenmayer, and S. Dovers. 2006. The nature and role of experiential knowledge for environmental conservation. *Environmental Conservation* 33:1–10.

Fazey, I., J. G. Salisbury, D. B. Lindenmayer, J. Maindonals, and J. Douglas. 2004. Can methods applied in medicine be used to summarize and disseminate conservation research? *Environmental Conservation* 31:190–198.

Feingold, A. 1988. Cognitive gender differences are disappearing. *American Psychologist* 43:95–103.

Felsenstein, J. 1985. Phylogenies and the comparative method. *American Naturalist* 125:1–15.

Felsenstein, J. 2008. Comparative methods with sampling error and within-species variation: Contrasts revisited and revised. *American Naturalist* 171:713–725.

Fergusson, D., K. Glass, B. Hutton, and S. Shapiro. 2005 Randomized controlled trials of aprotinin in cardiac surgery: Could clinical equipoise have stopped the bleeding? *Clinical Trials* 2:218–232.

Fern, E. F. and K. B. Monroe. 1996. Effect-size estimates: Issues and problems in interpretation. *Journal of Consumer Research* 23:89–105.

Fernandez-Duque, E. 1997. Comparing and combining data across studies: Alternatives to significance testing. *Oikos* 79:616–618.

Fernandez-Duque, E. and C. Valeggia. 1994. Meta-analysis: A valuable tool in conservation research. *Conservation Biology* 8:555–561.

Festa-Bianchet, M. 1996. Offspring sex ratio studies of mammals: Does publication depend upon the quality of the research or the direction of the results? *Ecoscience* 3:42–44.

Field, S. A., A. J. Tyre, N. Jonzen, J. R. Rhodes, and H. P. Possingham. 2004. Minimizing the cost of environmental management decisions by optimizing statistical thresholds. *Ecology Letters* 7:669–675.

Fischbach, L. A., K. J. Goodman, M. Feldman, and C. Aragaki. 2002. Sources of variation of Helicobacter pylori treatment success in adults worldwide: A meta-analysis. *International Journal of Epidemiology* 31:128–139.

Fisher, R. A. 1925. *Statistical Methods for Research Workers*. Oliver and Boyd, Edinburgh.

Fleiss, J. L. 1994. Measures of effect size for categorical data. In *The Handbook of Research Synthesis*, edited by H. Cooper and L. V. Hedges, 245–260. Russell Sage Foundation, New York.

Follmann, D., P. Elliott, I. Suh, and J. Cutler. 1992. Variance imputation for overviews of clinical trials with continuous response. *Journal of Clinical Epidemiology* 45:769–773.

Folstad, I. and A. J. Karter. 1992. Parasites, bright males, and the immunocompetence handicap. *American Naturalist* 139:603–622.

Ford, E. D. 2000. *Scientific Method for Ecological Research*. Cambridge University Press, Cambridge.

Ford, E. D. and H. Ishii. 2001. The method of synthesis in ecology. *Oikos* 93:153–160.

Forstmeier, W. and H. Schielzeth. 2011. Cryptic multiple hypothesis testing in linear models: Overestimated effect sizes and the winner's curse. *Behavioral Ecology and Sociobiology* 65: 47–55.

Foster, J. R. 2001. Statistical power in forest monitoring. *Forest Ecology and Management* 151:211–222.

Frank, S. A. 1990. Sex allocation theory for birds and mammals. *Annual Review of Ecology and Systematics* 21:13–55.

Freckleton, R. P., P. H. Harvey, and M. Pagel. 2002. Phylogenetic analysis and comparative data: A test and review of evidence. *American Naturalist* 160:712–726.

Furukawa, T. A., C. Barbui, A. Cipriani, P. Brambilla, and N. Watanabe. 2006. Imputing missing standard deviation can provide accurate results. *Journal of Clinical Epidemiology* 59:7–10.

Furukawa, T. A., D. L. Streiner, and S. Hori. 2000. Discrepancies among megatrials. *Journal of Clinical Epidemiology* 53:1193–1199.

Galbraith, R. F. 1984. Some applications of radial plots. *Journal of the American Statistical Association* 89:1232–1242.

Galbraith, R. F. 1988. A note on graphical presentation of estimated odds ratios from several clinical trials. *Statistics in Medicine* 7:889–894.

Garamszegi, L. Z., S. Calhim, N. Dochtermann, G. Hegyi, P. L. Hurd, C. Jorgensen, N. Kutsukake, et al. 2009. Changing philosophies and tools for statistical inferences in behavioral ecology. *Behavioral Ecology* 20:1365–1375.

Garant, D., L.E.B. Kruuk, T. A. Wilkin, R. H. McCleery, and B. C. Sheldon. 2005. Evolution driven by differential dispersal within a wild bird population. *Nature* 433:60–65.

Gardner, T. A., I. M. Côté, J. A. Gill, A. Grant, and A. R. Watkinson. 2003. Long-term region-wide declines in Caribbean corals. *Science* 301:958–960.

Garland, T., Jr., A. W. Dickerman, C. M. Janis, and J. A. Jones. 1993. Phylogenetic analysis of covariance by computer simulation. *Systematic Biology* 42:265–292.

Garland, T. and A. R. Ives. 2000. Using the past to predict the present: Confidence intervals for regression equations in phylogenetic comparative methods. *American Naturalist* 155:346–364.

Gates, S. 2002. Review of methodology of quantitative reviews using meta-analysis in ecology. *Journal of Animal Ecology* 71:547–557.

Gehr, B. T., C. Weiss, and F. Porzsolt. 2006. The fading of reported effectiveness. A meta-analysis of randomized controlled trials. *BMC Medical Research Methodology* 6:25.

Gelfand, A. E. and A.F.M. Smith. 1990. Sampling-based approaches to calculating marginal densities. *Journal of the American Statistical Association* 85:398–409.

Gelman, A. 2006. Prior distributions for variance parameters in hierarchical models. *Bayesian Analysis* 1:515–533

Gelman, A., J. B. Carlin, H. S. Stern, and D. B. Rubin. 2004. *Bayesian Data Analysis*, 2nd edition. Chapman and Hall, New York.

Gelman, A. and J. Hill. 2007. *Data Analysis Using Regression and Multilevel/Hierarchical Models*. Cambridge University Press, Cambridge.

Gelman, A. and D. B. Rubin. 1992. Inference from iterative simulation using multiple sequences (with discussion and rejoinder). *Statistical Science* 7:457–511.

Gerodette, T. 1987. A power analysis for detecting trends. *Ecology* 68:1364–1372.

Getty, T. 2006. Sexually selected traits are not similar to sporting handicaps. *Trends in Ecology & Evolution* 21:83–88.

Gilbert, D. 2006. *Stumbling Upon Happiness*. Knopf, New York.

Gilks, W. R., S. Richardson, and D. J. Spiegelhalter, editors. 1996. *Markov Chain Monte Carlo in Practice*. Chapman and Hall, New York.

Gilpin, A. R. 1993 Table for conversion of Kendall's tau to Spearman's rho within the context of measures of magnitude of effect for meta-analysis. *Educational and Psychological Measurement* 53:87–92.

Glass, G. V. 1976. Primary, secondary, and meta-analysis of research. *Educational Researcher* 5:3–8.

Glass, G. V., B. McGaw, and M. L. Smith. 1981. *Meta-analysis in Social Research*. SAGE Publications, Beverly Hills, CA.

Glass, G. V. and M. L. Smith. 1979. Meta-analysis of the relationship between class size and achievement. *Educational Evaluation and Policy Analysis* 1:2–16.

Glasziou, P. P. and S. L. Sanders. 2002. Investigating causes of heterogeneity in systematic reviews. *Statistics in Medicine* 21:1503–1511.

Glaves, D. J. and N. E. Haycock, editors. 2005. Defra Review of the Heather and Grass Burning Regulations and Code: Science Panel Assessment of the Effects of Burning on Biodiversity, Soils and Hydrology. DEFRA Conservation, Uplands and Rural Europe Division, Uplands Management Branch, London.

Glazier, D. S. 2005. Beyond the "3/4 power law" variation in the intra- and inter- specific scaling of metabolic rate in animals. *Biological Reviews* 80:1–52.

Gleser, L. J. and I. Olkin. 1994. Stochastically dependent effect sizes. In *The Handbook of Research Synthesis*, edited by H. Cooper and L. V. Hedges, 339–355. Russell Sage Foundation, New York.

Goldman, L. and A. R. Feinstein. 1979. Anticoagulants for myocardial infarction: The problem of pooling, drowning, and floating. *Annals of Internal Medicine* 90:92–94.

Goldstein, H. 1995. *Multilevel Statistical Models*. Edward Arnold, London.

Goldstein, H., J. Rasbash, I. Plewis, D. Draper, W. Browne, M. Yang, G. Woodhouse, and M. J. R. Healy. 1998. *A User's Guide to MLwiN*. Institute of Education, London.

Gontard-Danek, M. C. and A. P. Møller. 1999. The strength of sexual selection: A meta-analysis of bird studies. *Behavioral Ecology* 10:476–486.

Goodman, S. N. and J. A. Berlin. 1994. The use of predicted confidence intervals when planning experiments and the misuse of power when interpreting results. *Annals of Internal Medicine* 121:200–206.

Gotelli, N. J. and D. J. McCabe. 2002. Species co-occurrence: A meta-analysis of JM Diamond's assembly rules model. *Ecology* 83:2091–2096.

Gotzsche, P. C. 1987. Reference bias in reports of drug trials. *British Medical Journal* 295:654–656.

Grabe, S., L. M. Ward, and J. S. Hyde. 2008. The role of the media in body image concerns among women: A meta-analysis of experimental and correlational studies. *Psychological Bulletin* 134:460–476.

Grafen, A. 1989. The phylogenetic regression. *Philosophical Transactions of the Royal Society of London Series B* 326:119–157.

Greenland, S. 1987. Quantitative methods in the review of epidemiologic literature. *Epidemiologic Review* 9:1–30.

Greenland, S. 2005. Multiple-bias modeling for analysis of observational data. *Journal of the Royal Statistical Society Series A* 168:267–308.

Greenwald, A. G., R. Gonzalez, R. J. Harris, and D. Guthrie. 1996. Effect sizes and p-values: What should be reported and what should be replicated? *Psychophysiology* 33:175–188.

Greischar, M. A. and B. Koskella. 2007. A synthesis of experimental work on parasite local adaptation. *Ecology Letters* 10:418–434.

Griffin, A. S., B. C. Sheldon, and S. A. West. 2005. Cooperative breeders adjust offspring sex ratios to produce helpful helpers. *American Naturalist* 166:628–632.

Griffin, A. S. and S. A. West. 2003. Kin discrimination and the benefit of helping in cooperatively breeding vertebrates. *Science* 302:634–636.

Griffith, S. C., T. H. Parker, and V. A. Olson. 2006. Melanin- versus carotenoid-based sexual signals: Is the difference really so black and red? *Animal Behaviour* 71:749–763.

Groβ, J. 2003. *Linear Regression*. Springer, Berlin.

Grueber, C. E., S. Nakagawa, R. J. Laws, I. G. Jamieson. 2011. Multimodel inferences in ecology and evolution: Challenges and solutions. *Journal of Evolutionary Biology* 24: 699–711.

Guo, L. B. and R. M. Gifford. 2002. Soil carbon stocks and land use change: A meta-analysis. *Global Change Biology* 8:345–360.

Guo, Q. 2003. Plant abundance: The measurement and relationship with seed size. *Oikos* 101:639–642.

Gurevitch, J. 2006. Commentary on Simberloff (2006): Meltdowns, snowballs and positive feedback. *Ecology Letters* 8:919–921.

Gurevitch, J., P. S. Curtis, and M. H. Jones. 2001. Meta-analysis in ecology. *Advances in Ecological Research* 32:199–247.

Gurevitch, J. and L. V. Hedges. 1993. Meta-analysis: Combining the results of independent experiments. In *Design and Analysis of Ecological Experiments*, edited by S. M. Scheiner and J. Gurevitch, 373–398. Chapman and Hall, New York.

Gurevitch, J. and L. V. Hedges. 1999. Statistical issues in ecological meta-analysis. *Ecology* 80: 1142–1149.

Gurevitch, J. and K. Mengersen. 2010. A statistical view of synthesizing patterns of species richness along productivity gradients: Devils, forests, and trees. *Ecology* 91:2553–2560.

Gurevitch, J., J. A. Morrison, and L. V. Hedges. 2000. The interaction between competition and predation: A meta-analysis of field experiments. *American Naturalist* 155:435–453.

Gurevitch, J., L. L. Morrow, A. Wallace, and J. S. Walsh. 1992. A meta-analysis of competition in field experiments. *American Naturalist* 140:539–572.

Guyatt, G. H., A. D. Oxman, G. E. Vist, R. Kunz, Y. F. Ytter, P. Alonso-Coello, and H. Schunemann. 2008. Rating quality of evidence and strength of recommendations: GRADE: an emerging consensus on rating quality of evidence and strength of recommendations. *British Medical Journal* 336:924–926.

Haddock, C. K., D. Rindskopf, and W. R. Shadish. 1998. Using odds ratios as effect sizes for meta-analysis of dichotomous data: A primer on methods and issues. *Psychological Methods* 3:339–353.

Hadfield, J. 2009. MCMCglmm package for R. Available at http://cran.r-project.org/web/packages/MCMCglmm.

Hadfield, J. D. and S. Nakagawa. 2010. General quantitative genetic methods for comparative biology: Phylogenies, taxonomies, meta-analysis and multi-trait models for continuous and categorical characters. *Journal of Evolutionary Biology* 23:494–508.

Hahn, S., P. R. Williamson, and J. L. Hutton. 2002. Investigation of within-study selective reporting in clinical research: Follow-up of applications submitted to a local research ethics committee. *Journal of Evaluation in Clinical Practice* 8:353–359.

Hahn, S., P. R. Williamson, J. L. Hutton, P. Garner, and E. V. Flynn. 2000. Assessing the potential for bias in meta-analysis due to selective reporting of subgroups analyses within studies. *Statistics in Medicine* 19:3325–3336.

Halpern, B. S. 2003. The impact of marine reserves: Do reserves work and does reserve size matter? *Ecological Applications* 13:S117–S137.

Halpern, S. D. and J. A. Berlin. 2005. Beyond conventional publication bias: Other determinants of data suppression. In *Publication Bias in Meta-Analysis*, edited by H. R. Rothstein, A. J. Sutton, and M. Borenstein, 303–317. John Wiley, Chichester, UK.

Hamilton, G., F. Fielding, A. Chiffings, B. Hart, R. Johnstone, and K. Mengersen. 2007. Investigating the use of a Bayesian network to model the risk of Lyngbya majuscule bloom initiation in Deception Bay, Queensland. *Human and Ecological Risk Assessment* 13:1271–1287.

Hansen, T. F. 1997. Stabilizing selection and the comparative analysis of adaptation. *Evolution* 51:1341–1351.

Harbord, R. M., M. Egger, and J.A.C. Sterne. 2006. A modified test for small-study effects in meta-analyses of controlled trials with binary endpoints. *Statistics in Medicine* 25: 3443–3457.

Hardy, I.C.V., editor. 2002. *Sex Ratio: Concepts and Research Methods*. Cambridge University Press, Cambridge.

Harley, S. J. and R. A. Myers. 2001. Hierarchical Bayesian models of length-specific catchability of research trawl surveys. *Canadian Journal of Fisheries and Aquatic Sciences* 58:1569–1584.

Harley, S. J., R. A. Myers, N. Barrowman, K. Bowen, and R. Amiro. 2001. Estimation of research trawl survey catchability for biomass reconstruction of the eastern Scotian Shelf. Canadian Science Advisory Secretariat Research Documents 84.

Harley, S. J., R. A. Myers, and C. A. Field. 2004. Hierarchical models improve abundance estimates: Spawning biomass of hoki in Cook Strait, New Zealand. *Ecological Applications* 14: 1479–1494.

Harrison, F. 2011. Getting started with meta-analysis. *Methods in Ecology and Evolution* 2:1–10.

Harvey, P. H. and M. D. Pagel. 1991. *The Comparative Method in Evolutionary Biology*. Oxford University Press, Oxford.

Hauck, W. W. 1979. The large-sample variance of the Mantel-Haenszel estimator of a common odds ratio. *Biometrics* 35:817–819.

Hauck, W. W. 1989. Odds ratio inference from stratified samples. *Communications in Statistics* 18A:767–800.

Hawkins, D. M. and D. H. Olwell. 1997. *Cumulative Sum Charts and Charting for Quality Improvement*. Springer Verlag, New York.

Hayes, J. P. and R. J. Steidl. 1997. Statistical power analysis and amphibian population trends. *Conservation Biology* 11:273–275.

Heck, K. L., G. Hays, and R. J. Orth. 2003. Critical evaluation of the nursery role hypothesis for seagrass meadows. *Marine Ecology Progress Series* 253:123–136.

Hedges, L. V. 1981. Distribution theory for Glass's estimator of effect size and related estimators. *Journal of Educational Statistics* 6:107–128.

Hedges, L. V. 1983. A random effects model for effect sizes. *Psychological Bulletin* 93:388–395.

Hedges, L. V. 1984. Estimation of effect size under non-random sampling: The effects of censoring studies yielding statistically insignificant mean differences. *Journal of Educational Statistics* 9:61–85.

Hedges, L. V. 1989. An unbiased correction for sampling error in validity generalization studies. *Journal of Applied Psychology* 74:469–477.

Hedges, L. V. 1992 Modelling publication selection effects in meta-analysis. *Statistical Science* 7:246–255.

Hedges, L. V. 1994. Fixed effects models. In *The Handbook of Research Synthesis*, edited by H. Cooper and L. V. Hedges, 285–299. Russell Sage Foundation, New York.

Hedges, L. V., J. Gurevitch, and P. S. Curtis. 1999. The meta-analysis of response ratios in experimental ecology. *Ecology* 80:1150–1156.

Hedges, L. V. and I. Olkin. 1980. Vote-counting methods in research synthesis. *Psychological Bulletin* 88:359–369.

Hedges, L. V. and I. Olkin. 1985. *Statistical Methods for Meta-analysis*. Academic Press, Orlando, FL.

Hedges, L. V. and T. D. Pigott. 2001. The power of statistical tests in meta-analysis. *Psychological Methods* 6:203–217.

Hedges, L. V. and T. D. Pigott. 2004. The power of statistical tests for moderators in meta-analysis. *Psychological Methods* 9:426–445.

Hedges, L. V. and J. L. Vevea. 1996. Estimating effect sizes under publication bias: Small sample properties and robustness of a random effects selection model. *Journal of Educational and Behavioral Statistics* 21:299–332.

Hedges, L. V. and J. L. Vevea. 2005. Selection model approaches. In *Publication Bias in Meta-Analysis*, edited by H. R. Rothstein, A. J. Sutton, and M. Borenstein, 145–174.. John Wiley and Sons, Chichester, UK.

Helser, T. E. and H. Lai. 2004. A Bayesian hierarchical meta-analysis of fish growth: With an example for the North American largemouth bass, *Micropterus salmoides*. *Ecological Modelling* 178:399–416.

Hendry, A. P., T. J. Farrugia, and M. T. Kinnison. 2008. Human influences on the rate of phenotypic change in wild animal populations. *Molecular Ecology* 17:20–29.

Herbison, P., J. Hay-Smith, and W. J. Gillespie. 2006. Adjustment of meta-analyses on the basis of quality scores should be abandoned. *Journal of Clinical Epidemiology* 59:1249–1256.

Herre, E. A. 1985. Sex-ratio adjustment in fig wasps. *Science* 228:896–898.

Herre, E. A. 1987. Optimality, plasticity and selective regime in fig wasp sex-ratios. *Nature* 329:627–629.

Hersch, E. I. and P. C. Phillips. 2004. Power and potential bias in field studies of natural selection. *Evolution* 58:479–485.

Higgins, J.P.T. and S. Green, editors. 2011. *Cochrane Handbook for Systematic Reviews of Interventions. Version 5.1.0*. The Cochrane Collaboration. Available at http://handbook.cochrane.org.

Higgins, J.P.T. and S. G. Thompson. 2002. Quantifying heterogeneity in a meta-analysis. *Statistics in Medicine* 21:1539–1558.

Higgins, J.P.T., S. G. Thompson, J. J. Deeks, and D. G. Altman. 2003. Measuring inconsistency in meta-analyses. British Medical Journal 327:557–560.

Higgins, J.P.T., A. Whitehead, and M. Simmonds. 2011. Sequential methods for random-effects meta-analysis. *Statistics in Medicine* 30:903–921.

Hillebrand, H. 2008. Meta-analysis in ecology. In *Encyclopedia of Life Sciences (ELS)*. John Wiley, Chichester, UK. doi: 10.1002/9780470015902.a0003272.

Hillebrand, H., D. S. Gruner, E. T. Borer, M. E. S. Bracken, E. E. Cleland, J. J. Elser, W. S. Harpole, et al. 2007. Consumer versus resource control of producer diversity depends on ecosystem type and producer community structure. *Proceedings of the National Academy of Sciences of the United States of America* 104:10904–10909.

Hoeksema, J. D. and S. E. Forde. 2008. A meta-analysis of factors affecting local adaptation between interacting species. *American Naturalist* 171:275–290.

Hoenig, J. M. and D. M. Heisey. 2001. The abuse of power: The pervasive fallacy of power calculations for data analysis. *American Statistician* 55:19–24.

Holmes, C. T. and K. M. Mathews. 1984. The effects of non-promotion on elementary and junior high school pupils: A meta-analysis. *Review of Educational Research* 54:225–236.

Hopewell, S., S. McDonald, M. J. Clarke, and M. Egger. 2007. Grey literature in meta-analyses of randomized trials of health care interventions. *Cochrane Database of Systematic Reviews* 2: Art. No.: MR000010. doi: 0.1002/14651858.MR000010.pub3.

Houle, D., C. Pélabon, G. P. Wagner, and T. F. Hansen. 2011. Measurement and meaning in biology. *The Quarterly Review of Biology* 86:3–34.

Hozo, S. P., B. Djulbegovic, and I. Hozo. 2005. Estimating the mean and variance from the median, range, and the size of a sample. *BMC Medical Research Methodology* 5:13.

Hu, F. and J. V. Zidek. 1995. Incorporating relevant sample information using the likelihood. Technical Report No. 161. Department of Statistics, University of British Columbia. Vancouver, BC, Canada.

Huberty, A. F. and R. F. Denno. 2004. Plant water stress and its consequences for herbivorous insects: A new synthesis. *Ecology* 85:1383–1398.

Huedo-Medina, T. B., F. Sánchez-Meca, F. Marín-Martínez, and J. Botella. 2006. Assessing heterogeneity in meta-analysis: Q statistic or I^2 index? *Psychological Methods* 11:193–206.

Huffcutt, A. I. and W. Arthur. 1995. Development of a new outlier statistic for meta-analytic data. *Journal of Applied Psychology* 80:327–334.

Hughes, T. P., A. H. Baird, E. A. Dinsdale, V. J. Harriott, N. A. Moltschaniwskyj, M. S. Pratchett, J. E. Tanner, and B. L. Willis. 2002. Detecting regional variation using meta-analysis and large-scale sampling: Latitudinal patterns in recruitment. *Ecology* 83:436–451.

Hunt, J., R. C. Brooks, M. D. Jennions, M. J. Smith, C. L. Bentsen, and L. F. Bussière. 2004. High-quality male crickets invest heavily in sexual display but die young. *Nature* 432:1024–1027.

Hunter, J. E. and F. L. Schmidt. 1990. *Methods of Meta-analysis: Correcting Error and Bias in Research Findings*. SAGE Publications, Beverly Hills, CA.

Hunter, J. E. and F. L. Schmidt. 2004. *Methods of Meta-analysis: Correcting Error and Bias in Research Findings*, 2nd edition. SAGE Publications, Thousand Oaks, CA.

Hurlbert, S. H. 1994. Old shibboleths and new syntheses [Book review of *Design and Analysis of Ecological Experiments*, edited by S. M. Scheiner and J. Gurevitch]. *Trends in Ecology & Evolution* 9:495–496.

Hurlbert, S. H. 2004. On misinterpretations of pseudoreplication and related matters: A reply to Oksanen. *Oikos* 104:591–597.

Hutton, J. L. and P. R. Williamson. 2000. Bias in meta-analysis due to outcome variable selection within studies. *Applied Statistics* 49:359–370.

Hyatt, L. A., M. S. Rosenberg, T. G. Howard, G. Bole, W. Fang, J. Anastasia, K. Brown, et al. 2003. The distance dependence prediction of the Janzen-Connell hypothesis: A meta-analysis. *Oikos* 103:590–602.

Hyde, J. S., E. Fennema, and S. J. Lamon. 1990. Gender differences in mathematics performance: A meta-analysis. *Psychological Bulletin* 107:139–155.

Ioannidis, J.P.A. 1998. Effect of the statistical significance of results on the time to completion and publication of randomized efficacy trials. *Journal of the American Medical Association* 279:281–286.

Ioannidis, J.P.A. 2003. Genetic associations: False or true? *Trends in Molecular Medicine* 9:135–138.

Ioannidis, J.P.A. 2005a. Contradicted and initially stronger effects in highly cited clinical research. *Journal of the American Medical Association* 294:218–228.

Ioannidis, J.P.A. 2005b. Distinguishing biases from genuine heterogeneity: distinguishing artifactual from substantive effects. In *Publication Bias in Meta-Analysis*, edited by H. R. Rothstein, A. J. Sutton, and M. Borenstein, 287–302. John Wiley, Chichester, UK.

Ioannidis, J.P.A. 2005c. Why most published research findings are false. *PLOS Medicine* 2: e124. doi:10.1371/journal.pmed.0020124

Ioannidis, J.P.A. 2008. Why most discovered true associations are inflated. *Epidemiology* 19:640–647.

Ioannidis, J.P.A., J. C. Cappelleri, and J. Lau. 1998. Issues in comparisons between meta-analyses and large trials. *Journal of the American Medical Association* 279:1089–1093.

Ioannidis, J.P.A., J. C. Cappelleri, H. S. Sacks, and J. Lau. 1997. The relationship between study design, results, and reporting of randomized clinical trials of HIV infection. *Controlled Clinical Trials* 18:431–444.

Ioannidis, J.P.A. and J. Lau. 1997. The impact of high risk patients on the results of clinical trials. *Journal of Clinical Epidemiology* 50:1089–1098.

Ioannidis, J.P.A. and J. Lau. 1999. State of the evidence: Current status and prospects of meta-analysis in infectious diseases. *Clinical Infectious Diseases* 29:1178–1185.

Ioannidis, J.P.A. and J. Lau. 2001. Evolution of treatment effects over time: empirical insight from recursive meta-analyses. *Proceedings of the National Academy of Sciences of the United States of America* 98:831–836.

Ioannidis, J.P.A., E. E. Ntzani, T. A. Trikalinos, and D. G. Contopoulus-Ioannidis. 2001. Replication validity of genetic association studies. *Nature Genetics* 29:306–309.

Ioannidis, J.P.A., N. A. Patsopoulos, and H. R. Rothstein. 2008. Reasons or excuses for avoiding meta-analysis in forest plots. *British Medical Journal* 336:1413–1415.

Ioannidis, J.P.A. and T. A. Trikalinos. 2005. Early extreme contradictory estimates may appear in published research: the Proteus phenomenon in molecular genetics research and randomized trials. *Journal of Clinical Epidemiology* 58:543–549.

Ioannidis, J.P.A. and T. A. Trikalinos. 2007. The appropriateness of asymmetry tests for publication bias in meta-analysis: A large survey. *Canadian Medical Association Journal* 176:1091–1096.

Ives, A. R. and S. R. Carpenter. 2007. Stability and diversity of ecosystems. *Science* 317:58–62.

Iyengar, S. and J. B. Greenhouse. 1988. Selection models and the file drawer problem. *Statistical Science* 3:109–135.

Jablonski, L. M., X. Z. Wang, and P. S. Curtis. 2002. Plant reproduction under elevated CO_2 conditions: A meta-analysis of reports on 79 crop and wild species. *New Phytologist* 156:9–26.

Jackson, D. 2006. The power of the standard test for the presence of heterogeneity in meta-analysis. *Statistics in Medicine* 25:2688–2699.

Jadad, A. R., R. A. Moore, D. Carroll, C. Jenkinson, D. J. M. Reynolds, D. J. Gavaghan, and H. J. McQuay. 1996. Assessing the quality of reports of randomized clinical trials: Is blinding necessary? *Controlled Clinical Trials* 17:1–12.

James, A., S. Low Choy, and K. Mengersen. 2010. Elicitator: An expert elicitation tool for ecology. *Environmental Modelling and Software* 25:129–145.

James, W. and C. Stein. 1961. Estimation with quadratic loss. *Proceedings of the Fourth Berkeley Symposium on Mathematical Statistics and Probability* 1:311–319.

Janzen, D. H. 1970. Herbivores and the number of tree species in tropical forests. *American Naturalist* 104:501–528.

Järvinen, A. 1991. A meta-analytic study of the effects of female age on laying-date and clutch-size in the great tit *Parus major* and the pied flycatcher *Ficedula hypoleuca*. *Ibis* 133:62–67.

Jeng, G. T., J. R. Scott, and L. F. Burmeister. 1995. A comparison of meta-analytic results using literature vs. individual patient data. *Journal of the American Medical Association* 274:830–836.

Jennions, M.D. and H. Kokko. 2010. Sexual selection. In *Evolutionary Behavioral Ecology*, edited by D. F. Westneat and C. W. Fox, 343–364. Oxford, Oxford University Press.

Jennions, M. D. and A. P. Møller. 2002a. Publication bias in ecology and evolution: An empirical assessment using the "trim and fill" method. *Biological Reviews* 77:211–222.

Jennions, M. D. and A. P. Møller. 2002b. Relationships fade with time: A meta-analysis of temporal trends in publication in ecology and evolution. *Proceedings of the Royal Society of London Series B* 269:43–48.

Jennions, M. D. and A. P. Møller. 2003. A survey of the statistical power of research in behavioral ecology and animal behavior. *Behavioral Ecology* 14:438–455.

Jennions, M. D., A. P. Møller, and J. Hunt. 2004. Meta-analysis can "fail": reply to Kotiaho and Tomkins. *Oikos* 104:191–193.

Jennions, M. D., A. P. Møller, and M. Petrie. 2001. Sexually selected traits and adult survival: A meta-analysis. *Quarterly Review of Biology* 76:3–36.

Johnson, D. H. 1999. The insignificance of significance testing. *Journal of Wildlife Management* 63:763–772.

Johnson, D. W. and P. S. Curtis. 2001. Effects of forest management on soil C and N storage: Meta analysis. *Forest Ecology and Management* 140:227–238.

Johnson, J. B. and K. S. Omland. 2004. Model selection in ecology and evolution. *Trends in Ecology & Evolution* 19:101–108.

Jones, A. P., R. D. Riley, P. R. Williamson, and A. Whitehead. 2009. Meta-analysis of individual patient data versus aggregate data from longitudinal clinical trials. *Clinical Trials* 8:18–27.

Jones, A. T., W. N. Venables, J. B. Dry, and J. T. Wiskich. 1994. Random effects and variances: A synthesis of nonlinear regression analyses of mitochondrial electron transport. *Biometrika* 81:219–235.

Jüni, P., F. Holenstein, J.A.C. Sterne, C. Bartlett, and M. Egger. 2002. Direction and impact of language bias in meta-analyses of controlled trials: Empirical study. *International Journal of Epidemiology* 31:115–123.

Jüni, P., A. Witschi, R. Bloch, and M. Egger. 1999. The hazards of scoring the quality of clinical trials for meta-analysis. *Journal of the American Medical Association* 282:1054–1060.

Kalaian, H. A. and S. W. Raudenbush. 1996. A multivariate mixed linear model for meta-analysis. *Psychological Methods* 1:227–235.

Kalcounis-Rüppell, M. C., J. M. Psykkakis, and R. M. Brigham. 2005. Tree roost selection by bats: An empirical synthesis using meta-analysis. *Wildlife Society Bulletin* 33:1123–1132.

Kampichler, C. and A. Bruckner. 2009. The role of microarthropods in terrestrial decomposition: A meta-analysis of 40 years of litterbag studies. *Biological Reviews* 84:375–389.

Kelley, G. A., K. S. Kelley, and Z. Vu Tran. 2004. Retrieval of missing data for meta-analysis: A practical example. *International Journal of Technology Assessment in Health Care* 20:296–299.

Kelly, C. D. 2006. Replicating empirical research in behavioral ecology: How and why it should be done but rarely ever is. *Quarterly Review of Biology* 81:221–236.

Kelly, C. D., and M. D. Jennions. 2011. Sexual selection and sperm quantity: Meta-analyses of strategic ejaculation. *Biological Reviews* 86:863–884.

Kendall, M. G. 1970. Rank *Correlation Methods*, 4th edition. Charles Griffin, London.

Kerlikowske, K., D. Grady, S. M. Rubin, C. Sandrock, and V. L. Ernster. 1995. Efficacy of screening mammography: A meta-analysis. *Journal of the American Medical Association* 273:149–154.

Khan, K. S., R. Kunz, J. Kleijnen, and G. Antes. 2003. *Systematic Reviews to Support Evidence-based Medicine: How to Apply Findings of Healthcare Research.* Royal Society of Medicine Press, London.

Kirk, R. E. 2007. Effect magnitude: A different focus. *Journal of Statistical Planning and Inference* 137:1634–1646.

Kjaergard, L. L. and C. Gloud. 2002. Citation bias of hepato-biliary randomized clinical trials. *Journal of Clinical Epidemiology* 55:407–410.

Knapczyk, F. N. and J. K. Conner. 2007. Estimates of the average strength of natural selection are not inflated by sampling error or publication bias. *American Naturalist* 170:501–508.

Kokko, H. 1997. Evolutionarily stable strategies of age-dependent sexual advertisement. *Behavioral Ecology and Sociobiology* 41:99–107.

Kokko, H., M. D. Jennions, and R. C. Brooks. 2006. Unifying and testing models of sexual selection. *Annual Reviews of Ecology, Evolution and Systematics* 37:43–66.

Kokko, H., A. Lopez-Sepulcre, and L. J. Morrell. 2006. From hawks and doves to self-consistent games of territorial behavior. *American Naturalist* 167:901–912.

Koricheva, J. 2002. Meta-analysis of sources of variation in fitness costs of plant antiherbivore defenses. *Ecology* 83:176–190.

Koricheva, J. 2003. Non-significant results in ecology: a burden or a blessing in disguise? *Oikos* 102:397–401.

Koricheva, J., S. Larsson, and E. Haukioja. 1998. Insect performance on experimentally stressed woody plants: A meta-analysis. *Annual Review of Entomology* 43:195–216.

Koricheva, J., H. Nykänen, and E. Gianoli. 2004. Meta-analysis of trade-offs among plant antiherbivore defenses: Are plants jacks-of-all-trades, masters of all? *American Naturalist* 163:E64–E75.

Kotiaho, J. S. and J. L. Tomkins. 2002. Meta-analysis, can it ever fail? *Oikos* 96:551–553.

Kozlowski, J., M. Konarzewski, and A. T. Gawelczyk. 2003. Intraspecific body size optimization produces intraspecific allometries. In *Macroecology: Concepts and Consequences*, edited by T. M. Blackburn and K. J. Gaston, 299–320. Blackwell Science, Oxford.

Krasnov, B., V. Maxim, N. Korallo-Vinarskaya, D. Mouillot, and R. Poulin. 2009. Inferring associations among parasitic gamasid mites from census data. *Oecologia* 160:175–185.

Kraus, S. J. 1995. Attitudes and the prediction of behavior: A meta-analysis of the empirical literature. *Personality and Social Psychology Bulletin* 21:58–75.

Krist, M. 2010. Egg size and offspring quality: A meta-analysis in birds. *Biological Reviews* 86:692–716.

Kruuk, L.E.B., T. H. Clutton-Brock, S. D. Albon, J. M. Pemberton, and F. E. Guiness. 1999. Population density affects sex ratio variation in red deer. *Nature* 399:459–461.

Kuhn, T. S. 1970. *The Structure of Scientific Revolutions*. University of Chicago Press, Chicago.

Kuhnert, P. M., T. G. Martin, and S. P. Griffiths. 2010. A guide to eliciting and using expert knowledge in Bayesian ecological models. *Ecology Letters* 13:900–914.

Kuhnert, P. M., T. G. Martin, K. Mengersen, and H. P. Possingham. 2005. Assessing the impacts of grazing levels on bird density in woodland habitat: A Bayesian approach using expert opinion. *Environmetrics* 16:717–747.

Kulinskaya, E. and J. Koricheva. 2010. Use of quality control charts for detection of outliers and temporal trends in cumulative meta-analysis. *Research Synthesis Methods* 1: 297–307.

Kulmatiski, A., K. H. Beard, J. R. Stevens, and S. M. Cobbold. 2008. Plant-soil feedbacks: A meta-analytical review. *Ecology Letters* 11:980–999.

Kuss, O. and A. Koch. 1996. Meta-analysis macros for SAS. *Computational Statistics and Data Analysis* 22:325–333.

L'Abbé, K. A., A. S. Detsky, and K. O'Rourke. 1987. Meta-analysis in clinical research. *Annals of Internal Medicine* 107:224–233.

Lajeunesse, M. J. 2009. Meta-analysis and the comparative phylogenetic method. *American Naturalist* 174:369–381.

Lajeunesse, M. J. 2011a. *phyloMeta*: A program for phylogenetic comparative analyses with meta-analysis. *Bioinformatics* 27:2603–2604.

Lajeunesse, M. J. 2011b. On the meta-analysis of response ratios for studies with correlated and multi-group designs. *Ecology* 92:2049–2055.

Lajeunesse, M. J. and M. R. Forbes. 2002. Host range and local parasite adaptation. *Proceedings of the Royal Society of London Series B* 269:703–710.

Lajeunesse, M. J. and M. R. Forbes. 2003. Variable reporting and quantitative reviews: A comparison of three meta-analytical techniques. *Ecology Letters* 6:448–454.

Lambert, P. C., A. J. Sutton, K. R. Abrams, and D. R. Jones. 2002. A comparison of summary patient-level covariates in meta-regression with individual patient data meta-analysis. *Journal of Clinical Epidemiology* 55:86–94.

Lan, K. K., W. F. Rosenberger, and J. M. Lachin. 1993. Use of spending functions for occasional or continuous monitoring of data in clinical trials. *Statistics in Medicine* 12:2219–2231.

Lan, K.K.G., M. Hu, and J. C. Cappelleri. 2003. Applying the law of iterated logarithm to cumulative meta-analysis of continuous endpoint. *Statistica Sinica* 13:1135–1145.

Landman, J. T. and R. W. Dawes. 1982. Psychotherapy outcome: Smith and Glass' conclusions stand up under scrutiny. *American Psychologist* 37:504–516.

Lau, J., E. M. Antman, J. Jimenez-Silva, B. Kupelnick, F. Mosteller, and T. C. Chalmers. 1992. Cumulative meta-analysis of therapeutic trials for myocardial infarction. *New England Journal of Medicine* 327:248–254.

Lau, J., J.P.A. Ioannidis, and C. H. Schmid. 1998. Summing up evidence: One answer is not always enough. *The Lancet* 351:123–127.

Lau, J., J.P.A. Ioannidis, N. Terrin, C. H. Schmid, and I. Olkin. 2006. The case of the misleading funnel plot. *British Medical Journal* 333:597–600.

Lau, J., C. H. Schmid, and T. C. Chalmers. 1995. Cumulative meta-analysis of clinical trials builds evidence for exemplary medical care. *Journal of Clinical Epidemiology* 48:45–57.

Leeflang, M. M., J. J. Deeks, C. Gatsonis, P. M. Bossuyt, and C.D.T.A.W. Group. 2008. Systematic reviews of diagnostic test accuracy. *Annals of Internal Medicine* 149:889–897.

Legendre, P. 2000. Comparison of permutation methods for the partial correlation and partial Mantel tests. *Journal of Statistical Computation and Simulation* 67:37–73.

Leimar, O. 1996. Life history analysis of the Trivers and Willard problem. *Behavioral Ecology* 7:316–325.

Leimu, R. and J. Koricheva. 2004. Cumulative meta-analysis: A new tool for detection of temporal trends and publication bias in ecology. *Proceedings of the Royal Society of London Series B* 271:1961–1966.

Leimu, R. and J. Koricheva. 2005a. Does scientific collaboration increase the impact of ecological articles? *Bioscience* 55:438–443.

Leimu, R. and J. Koricheva. 2005b. What determines the citation frequency of ecological papers? *Trends in Ecology & Evolution* 20:28–32.

Leishman, M. R. and B. R. Murray. 2001. The relationship between seed size and abundance in plant communities: Model predictions and observed patterns. *Oikos* 94:151–161.

LeLorier, J., G. Grégoire, A. Benhaddad, J. Lapierre, and F. Derderian. 1997. Discrepancies between meta-analyses and subsequent large randomized, controlled trials. *New England Journal of Medicine* 337:536–542.

Levey, D. J., R. S. Duncan, and C. F. Levins. 2007. Use of dung as a tool by burrowing owls. *Nature* 431:39.

Levine, J. 2001. Trial assessment procedure scale (TAPS). In *Guide to Clinical Trials*, edited by B. Spilker, 780–786. Raven Press, New York.

Levine, J. M., P. B. Adler, and S. G. Yelenik. 2004. A meta-analysis of biotic resistance to exotic plant invasions. *Ecology Letters* 7:975–989.

Levitt, S. D. and S. J. Dubner. 2005. *Freakonomics: A Rogue Economist Explores the Hidden Side of Everything*. Harper Collins, New York.

Lewis, S. and M. Clarke. 2001. Forest plots: Trying to see the wood and the trees. *British Medical Journal* 322:1479–1480.

Light, R. J. and D. B. Pillemer. 1984. *Summing up; The science of Reviewing Research*. Harvard University Press, Cambridge.

Light, R. J., J. D. Singer, and J. B. Willett. 1994. The visual presentation and interpretation of meta-analyses. In *The Handbook of Research Synthesis*, edited by H. Cooper and L. V. Hedges, 439–453. Russell Sage Foundation, New York.

Lipsey, M. W. 2003. Those confounded moderators in meta-analysis: Good, bad, and ugly. *Annals of the American Academy of Political and Social Science* 587:69–81.

Lipsey, M. W. and B. Wilson. 1993. The efficacy of psychological, educational, and behavioural treatment: Confirmation from meta-analysis. *American Psychologist* 48:1181–1209.

Lipsey, M. W. and D. B. Wilson. 2001. *Practical Meta-analysis*. SAGE Publications, Thousand Oaks, CA.

Littell, R. C., G. A. Milliken, W. W. Stroup, and R. R. Wolfinger. 1996. *SAS System for Mixed Models*. SAS Institute, Cary, NC.

Little, R.J.A. and D. B. Rubin. 2000. Causal effects in clinical and epidemiological studies via potential outcomes: Concepts and analytical approaches. *Annual Review of Public Health* 21:121–145.

Little, R.J.A. and D. B. Rubin. 2002. *Statistical Analysis with Missing Data*, 2nd edition. John Wiley and Sons, New York.

Liu, H. and P. Stiling. 2006. Testing the enemy release hypothesis: A review and meta-analysis *Biological Invasions* 8:1535–1545.

Lively, C. M., M. F. Dybdahl, J. Jokela, E. E. Osnas, and L. F. Delph. 2004. Host sex and local adaptation by parasites in a snail-trematode interaction. *American Naturalist* 164:S6–S18.

Lortie, C. J., L. W. Aarssen, A. E. Budden, J. Koricheva, R. Leimu, and T. Tregenza. 2007. Publication bias and merit in ecology. *Oikos* 116:1247–1253.

Lortie, C. J. and R. M. Callaway. 2006. Re-analysis of meta-analysis: Support for the stress-gradient hypothesis. *Journal of Ecology* 94:7–16.

Lortie, C. J. and A. Dyer. 1999. Over-interpretation: Avoiding the stigma of non-significant results. *Oikos* 87:183–185.

Losos, J. 1996. Phylogenetic perspectives on community ecology. *Ecology* 77:1344–1354.

Low Choy, S., J. Murray, A. James, and K. Mengersen. 2010. Indirect elicitation from ecological experts: From methods and software to habitat modelling and rock-wallabies. In *The Oxford Handbook of Applied Bayesian Analysis*, edited by A. O'Hagan and M. West, 511–544. Oxford University Press, Oxford.

Low Choy, S. J., R. A. O'Leary, and K. L. Mengersen. 2009. Elicitation by design in ecology: Using expert opinion to inform priors for Bayesian statistical models. *Ecology* 90:265–277.

Lu, G., A. E. Ades, A. J. Sutton, N. J. Cooper, A. H. Briggs, and D. M. Caldwell. 2007. Meta-analysis of mixed treatment comparisons at multiple follow-up times. *Statistics in Medicine* 26:3681–3699.

Lunn, D. J., A. Thomas, N. Best, and D. Spiegelhalter. 2000. WinBUGS—a Bayesian modelling framework: Concepts, structure, and extensibility. *Statistics and Computing* 10:325–337.

Lunn, D., D. Spiegelhalter, A. Thomas, and N. Best. 2009. The BUGS project: Evolution, critique and future directions (with discussion). *Statistics in Medicine* 28:3049–3082

Lynn, M. R. 1989. Meta-analysis: Appropriate tool for the integration of nursing research? *Nursing Research* 38:302–305.

Ma, J., W. Liu, A. Hunter, and W. Zhang. 2008. Performing meta-analysis with incomplete statistical information in clinical trials. *BMC Medical Research Methodology* 8:56.

Macaskill, P. 2004. Empirical Bayes estimates generated in a hierarchical summary ROC analysis agreed closely with those of a full Bayesian analysis. *Journal of Clinical Epidemiology* 57:925–932.

MacKenzie, B. R., R. A. Myers, and K. G. Bowen. 2003. Spawner-recruit relationships and fish stock carrying capacity in aquatic ecosystems. *Marine Ecology Progress Series* 248:209–220.

Maddison, W. 1989. Reconstructing character evolution on polytomous cladograms. *Cladistics* 5:365–377.

Maddison, W. 1990. A method for testing the correlated evolution of two binary characters: Are gains or losses concentrated on branches of a phylogenetic tree? *Evolution* 44:539–557.

Maestre, F. T., F. Valladares, and J. F. Reynolds. 2005. Is the change of plant-plant interactions with abiotic stress predictable? A meta-analysis of field results in arid environments. *Journal of Ecology* 93:748–757.

Maestre, F. T., F. Valladares, and J. F. Reynolds. 2006. Does one model fit all? A reply to Lortie and Callaway. *Journal of Ecology* 94:17–22.

Manly, B. F. J. 1997. *Randomization, Bootstrap and Monte Carlo Methods in Biology*. Chapman and Hall, London.

Mann, C. 1990. Meta-analysis in the breech. *Science* 249:476–480.

Mann, T. 2005. *The Oxford Guide to Library Research*. Oxford University Press, Oxford.

Manson, F. J., N. R. Loneragan, B. D. Harch, G. A. Skilleter, and L. Williams. 2005. A broad-scale analysis of links between coastal fisheries production and mangrove extent: A case-study for northeastern Australia. *Fisheries Research* 74:69–85.

Mantel, N. 1967. The detection of disease clustering and a generalized regression approach. *Cancer Research* 27:209–220.

Mantel, N. and W. Haenszel. 1959. Statistical aspects of the analysis of data from retrospective studies of disease. *Journal of the National Cancer Institute* 22:719–748.

Mantel, N. and R. S. Valand. 1970. A technique of nonparametric multivariate analysis. *Biometrics* 26:547–558.

Mapstone, B. 1995. Scalable decision rules for environmental impact studies: Effect size, type 1, and type 2 errors. *Ecological Applications* 5:401–410.

Marín-Martínez, F. and J. Sánchez-Meca. 1999. Averaging dependent effect sizes in meta-analysis: A cautionary note about procedures. *Spanish Journal of Psychology* 2:32–38.

Markow, T. A. and G. M. Clarke. 1997. Meta-analysis of the heritability of developmental stability: A giant step backward. *Journal of Evolutionary Biology* 10:31–37.

Marsh, D. M. 2001. Fluctuations in amphibian populations: A meta-analysis. *Biological Conservation* 101:327–335.

Martin, T. G., P. M. Kuhnert, K. Mengersen, and H. P. Possingham. 2005. The power of expert opinion in ecological models using Bayesian methods: Impact of grazing on birds. *Ecological Applications* 15:266–280.

Martínez-Abraín, A. 2008. Statistical significance and biological relevance: A call for a more cautious interpretation of results in ecology. *Acta Oecologica* 34:9–11.

Martins, E. P. 1994. Estimating the rate of phenotypic evolution from comparative data. *American Naturalist* 144:193–209.

Martins, E. P. 2000. Adaptation and comparative method. *Trends in Ecology & Evolution* 15:296–299.

Martins, E. P. and T. Garland. 1991. Phylogenetic analyses of the correlated evolution of continuous characters: A simulation study. *Evolution* 45:534–557.

MathWorks. 2007. *MATLAB*. MathWorks, Natick, MA.

Mayr, E. 1963. *Animal Species and Evolution*. Harvard University Press, Cambridge.

McAuley, L., B. Pham, P. Tugwell, and D. Moher. 2000. Does the inclusion of grey literature influence estimates of intervention effectiveness reported in meta-analyses? *The Lancet* 356:1228–1231.

McCarthy, J. M., C. L. Hein, J. D. Olden, and M. J. Vander Zanden. 2006. Coupling long-term studies with meta-analysis to investigate impacts of non-native crayfish on zoobenthic communities. *Freshwater Biology* 51:224–235.

McCarthy, M. A. 2007. *Bayesian Methods for Ecology*. Cambridge University Press, Cambridge.

McDaniel, M. A., H. R. Rothstein, and D. Whetzel. 2006. Publication bias: A case study of four test vendor manuals. *Personnel Psychology* 59:927–953.

McIntosh, M. W. 1996. The population risk as an explanatory variable in research synthesis of clinical trials. *Statistics in Medicine* 15:1713–1728.

McIver, J. D., R.E.J. Boerner, and S. C. Hart. 2008. The national fire and fire surrogate study: Ecological consequences of alternative fuel reduction methods in seasonally dry forests. *Forest Ecology and Management* 255:3075–3080.

McNab, B. K. 1997. On the utility of uniformity in the definition of basal rate of metabolism. *Physiological Zoology* 70:718–720.

Meinick, D. J. and G. A. Hoelzer. 1994. Patterns of speciation and limits to phylogenetic resolution. *Trends in Ecology & Evolution* 9:104–107.

Meunier, J., S. A. West, and M. Chaupisat. 2008. Split sex ratios in the social Hymenoptera: A meta-analysis. *Behavioral Ecology* 19:382–390.

Meunier, J., S. F. Pinto, T. Burri, A. Roulin. 2011. Eumelanin-based coloration and fitness parameters in birds: A meta-analysis. *Behavioral Ecology and Sociobiology* 65: 559–567.

Meydani, S. N., J. Lau, G. E. Dallal, and M. Meydani. 2005. High-dosage vitamin E supplementation and all-cause mortality. *Annals of Internal Medicine* 143:153.

Micheli, F., B. S. Halpern, L. W. Botsford, and R. R. Warner. 2004. Trajectories and correlates of community change in no-take marine reserves. *Ecological Applications* 14:1709–1723.

Mickenautsch, S. 2010. Systematic reviews, systematic error and the acquisition of clinical knowledge. *BMC Medical Research Methodology* 10:53.

Miller, E. R., R. Pastor-Barriuso, D. Dalal, R. A. Riemersma, L. J. Appel, and E. Guallar. 2005. Meta-analysis: High-dosage vitamin E supplementation may increase all-cause mortality. *Annals of Internal Medicine* 142:37–46.

Miller, N. and V. E. Pollock. 1994. Meta-analytic synthesis for theory development. In *The Hand-book of Research Synthesis*, edited by H. Cooper and L. V. Hedges, 457–485. Russell Sage Foundation, New York.

Milsom, T. P., S. D. Langton, W. K. Parkin, S. Peel, J. D. Bishop, J. D. Hart, and N. P. Moore. 2000. Habitat models of bird species' distribution: An aid to the management of coastal grazing marshes. *Journal of Applied Ecology* 37:206–727.

Mittelbach, G. G., C. F. Steiner, S. M. Scheiner, K. L. Gross, H. L. Reynolds, R. B. Waide, M. R. Willig, S. I. Dodson, and L. Gough. 2001. What is the observed relationship between species richness and productivity? *Ecology* 82:2381–2396.

Miyazawa, K. and M. J. Lechowicz. 2004. Comparative seedling ecology of eight North American Spruce (*Picea*) species in relation to their geographic ranges. *Annals of Botany* 94:635–644.

Moala, F. A. and A. O'Hagan. 2010. Elicitation of multivariate prior distributions: A nonparametric Bayesian approach. *Journal of Statistical Planning and Inference* 140:1635–1655.

Moher, D., D. J. Cook, S. Eastwood, I. Olkin, D. Rennie, and D. F. Stroup. 1999. Improving the quality of reports of meta-analyses of randomised controlled trials: The QUOROM statement. *The Lancet* 354:1896–1900.

Moher, D., A. R. Jadad, G. Nichol, M. Penman, P. Tugwell, and S. Walsh. 1995. Assessing the quality of randomized controlled trials: An annotated bibliography of scales and checklists. *Controlled Clinical Trials* 16:62–73.

Moher, D., A. R. Jadad, and P. Tugwell. 1996. Assessing the quality of randomized controlled trials: Current issues and future directions. *International Journal of Technology Assessment in Health Care* 12:195–208.

Moher, D., A. Liberati, J. Tetzlaff, D. G. Altman. 2009. Preferred reporting items for systematic reviews and meta-analyses: The PRISMA statement. *Annals of Internal Medicine* 151:264–269.

Moher, D., K. F. Schulz, and D. G. Altman. 2001. The CONSORT statement: Revised recommendations for improving the quality of reports of parallel-group randomized trials. *Annals of Internal Medicine* 134:657–662.

Molbo, D., C. A. Machado, J. G. Sevenster, L. Keller, and E. A. Herre. 2003. Cryptic species of fig-pollinating wasps: Implications for the evolution of the fig-wasp mutualism, sex allocation, and precision of adaptation. *Proceedings of the National Academy of Sciences of the United States of America* 100:5867–5872.

Moles, A. T., A. D. Falster, M. R. Leishman, and M. Westoby. 2004. Small-seeded species produce more seeds per square metre of canopy per year, but not per individual per lifetime. *Journal of Ecology* 92:384–396.

Møller, A. P. and R. V. Alatalo. 1999. Good-genes effects in sexual selection. *Proceedings of the Royal Society of London Series B* 266:85–91.

Møller, A. P. and M. D. Jennions. 2001. Testing and adjusting for publication bias. *Trends in Ecology & Evolution* 16:580–586.

Møller, A. P. and M. D. Jennions. 2002. How much variance can be explained by ecologists and evolutionary biologists? *Oecologia* 132:492–500.

Møller, A. P. and R. Thornhill. 1998. Bilateral symmetry and sexual-selection: A meta-analysis. *American Naturalist* 151:174–192.

Møller, A. P., R. Thornhill, and S. W. Gangestad. 2005. Direct and indirect tests for publication bias: Asymmetry and sexual selection. *Animal Behaviour* 70:497–506.

Moreno S. G., A. J. Sutton, A. E. Ades, T. D. Stanley, K. R. Abrams, J. L. Peters and N. J. Cooper. 2009. Assessment of regression-based methods to adjust for publication bias through a comprehensive simulation study. *BMC Medical Research Methodology* 9: 2.

Mosqueira, I., I. M. Côté, S. Jennings, and J. D. Reynolds. 2000. Conservation benefits of marine reserves for fish populations. *Animal Conservation* 4:321–332.

Mullen, B., P. Muellerleile, and B. Bryant. 2001. Cumulative meta-analysis: A consideration of indicators of sufficiency and stability. *Personality and Social Psychology Bulletin* 27:1450–1462.

Mumby, P. J., A. J. Edwards, J. E. Arias-Gonzalez, K. C. Lindeman, P. G. Blackwell, A. Gall, M. I. Gorczynska, et al. 2004. Mangroves enhance the biomass of coral reef fish communities in the Caribbean. *Nature* 427:533–536.

Muncer, S. J., M. Craigie, and J. Holmes. 2003. Meta-analysis and power: Some suggestions for the use of power in research synthesis. *Understanding Statistics* 2:1–12.

Muncer, S. J., S. Taylor, and M. Craigie. 2002. Power dressing and meta-analysis: Incorporating power analysis into meta-analysis. *Journal of Advanced Nursing* 38:274.

Murray, B. R., B. P. Kelaher, G. C. Hose, and W. F. Figueira. 2005. A meta-analysis of the inter-specific relationship between seed size and plant abundance within local communities. *Oikos* 110:191–194.

Murtaugh, P. A. 2002. Journal quality, effect size, and publication bias in meta-analysis. *Ecology* 83:1162–1166.

Myers, R. A., J. K. Baum, T. D. Shepherd, S. P. Powers, and C. H. Peterson. 2007. Cascading effects of the loss of apex predatory sharks from a coastal ocean. *Science* 315:1846–1850.

Myers, R. A. and B. Worm. 2003. Rapid worldwide depletion of predatory fish communities. *Nature* 423:280–283.

Myers, R. A. and B. Worm. 2005. Extinction, survival or recovery of large predatory fishes. *Philosophical Transactions of the Royal Society of London Series B* 360:13–20.

Nakagawa, S. 2004 A farewell to Boneferroni: The problem of how low statistical power and publication bias. *Behavioral Ecology* 15:1044–1045.

Nakagawa, S. and I. C. Cuthill. 2007. Effect size, confidence interval and statistical significance: A practical guide for biologists. *Biological Reviews* 82:591–605.

Nakagawa, S., N. Ockendon, D.O.S. Gillespie, B. J. Hatchwell, and T. Burke. 2007. Assessing the function of how house sparrows' bib size using a flexible meta-analysis method. *Behavioral Ecology* 18:831–840.

Nakagawa, S., N. Ockendon, D.O.S. Gillespie, B. J. Hatchwell, and T. Burke. 2011. Erratum. *Behavioral Ecology* 22:445–446.

Nam, I.-S., K. L. Mengersen, and P. Garthwaite. 2003. Multivariate meta-analysis. *Statistics in Medicine* 22:2309–2333.

Neff, B. D. and J. D. Olden. 2006. Is peer review a game of chance? *Bioscience* 56:333–340.

Newton, A. C., G. B. Stewart, A. Diaz, D. Golicher, and A. S. Pullin. 2007. Bayesian Belief Networks as a tool for evidence-based conservation management. *Journal for Nature Conservation* 15:144–160.

Newton, A. C., G. B. Stewart, G. Myers, A. Diaz, S. Lake, J. M. Bullock, and A. S. Pullin. 2009. Impacts of grazing on lowland heathland in north-west Europe. *Biological Conservation* 142:935–947.

Norby, R. J., M. F. Cotrufo, P. Ineson, E. G. O'Neill, and J. G. Canadell. 2001. Elevated CO_2, litter chemistry, and decomposition: a synthesis. *Oecologia* 127:153–165.

Normand, S. L. 1995. Meta-analysis software: A comparative review. *American Statistician* 49:298–309.

Normand, S. L. 1999. Tutorial in biostatistics meta-analysis: Formulating, evaluating, combining, and reporting. *Statistics in Medicine* 18:321–359.

Novick, M. R., P. H. Jackson, D. T. Thayer, and N. S. Cole. 1972. Estimating multiple regressions in m groups: A cross-validation study. *British Journal of Mathematical and Statistical Psychology* 25:33–50.

Nykänen, H. and J. Koricheva. 2004. Damage-induced changes in woody plants and their effects on insect herbivore performance: A meta-analysis. *Oikos* 104:247–268.

O'Hagan, A., C. E. Buck, A. Daneshkhah, R. Eiser, P. Garthwaite, D. Jenkinson, J. Oakley, and T. Rakow. 2006. *Uncertain Judgments: Eliciting Experts' Probabilities.* John Wiley and Sons, Chichester, UK.

O'Hagan, A. and J. Forster. 2004. *Kendall's Advanced Theory of Statistics. Volume 2B: Bayesian Inference,* 2nd edition. Oxford University Press, New York.

O'Hara, R. B. 2005. The anarchist's guide to ecological theory. Or, we don't need no stinkin' laws. *Oikos* 110:390–393.

O'Leary, R. A., S. Low Choy, J. Murray, M. Kynn, R. Denham, T. Martin, and K. Mengersen. 2009. Comparison of three expert elicitation methods for logistic regression on predicting the presence of the threatened brush-tailed rock-wallaby *Petrogale penicillata. Environmetrics* 20:379–398.

Ogilvie, D., D. Fayter, M. Petticrew, A. Sowden, S. Thomas, M. Whitehead, and G. Worth. 2008. The harvest plot: A method for synthesising evidence about the differential effects of interventions. *BMC Medical Research Methodology* 8:8.

Oksanen, L. 2003. Logic of experiments in ecology: Is pseudoreplication a pseudoissue? *Oikos* 94:27–38.

Oliver, M. B. and J. S. Hyde. 1993. Gender differences in sexuality: A meta-analysis. *Psychological Bulletin* 114:29–51.

Orwin, R. G. 1983. A failsafe N for effect size in meta-analysis. *Journal of Educational Statistics* 8:157–159.

Orwin, R. G. and D. S. Cordray. 1985. Effects of deficient reporting on meta-analysis: A conceptual framework and reanalysis. *Psychological Bulletin* 97:134–147.

Osenberg, C. W., O. Sarnelle, and S. D. Cooper. 1997. Effect size in ecological experiments: The application of biological models in meta-analysis. *American Naturalist* 150:798–812.

Osenberg, C. W., O. Sarnelle, S. D. Cooper, and R. D. Holt. 1999. Resolving ecological questions through meta-analysis: Goals, metrics, and models. *Ecology* 80:1105–1117.

Osenberg, C. W. and C. M. St. Mary. 1998. Meta-analysis: Synthesis or statistical subjugation? *Integrative Biology* 1:37–41.

Oyserman, D., H. M. Coon, and M. Kemmelmeier. 2002. Rethinking individualism and collectivism: Evaluation of theoretical assumptions and meta-analyses. *Psychological Bulletin* 128:3–72.

Pagel, M. 1994. Detecting correlated evolution on phylogenies: A general method for the comparative analysis of discrete characters. *Proceedings of the Royal Society of London Series B* 255:437–445.

Pagel, M. 1997. Inferring evolutionary processes from phylogenies. *Zoologica Scripta* 26:331–348.

Paine, R. T. 2002. Advances in ecological understanding: By Kuhnian revolution or conceptual evolution. *Ecology* 83:1553–1559.

Palmer, A. R. 1999. Detecting publication bias in meta-analyses: A case study of fluctuating asymmetry and sexual selection. *American Naturalist* 154:220–233.

Palmer, A. R. 2000. Quasireplication and the contract of error: Lessons from sex ratios, heritabilities and fluctuating asymmetry. *Annual Reviews of Ecology and Systematics* 31:441–480.

Parker, J. D., D. E. Burkepile, and M. E. Hay. 2006. Opposing effects of native and exotic herbivores on plant invasions. *Science* 311:1459–1461.

Parmesan, C. and G. Yohe. 2003. A globally coherent fingerprint of climate change impacts across natural systems. *Nature* 421:37–42.

Patsopoulos, N. A., A. A. Analatos, and J.P.A. Ioannidis. 2005. Relative citation impact of various study designs in the health sciences. *Journal of the American Medical Association* 293:2362–2366.

Patterson, H. D. and R. Thompson. 1971. Recovery of inter-block information when block sizes are unequal. *Biometrika* 58:545–554.

Paul, P. A., P. E. Lipps, and L. V. Madden. 2005. Relationship between visual estimates of Fusarium head blight intensity and deoxynivalenol accumulation in harvested wheat grain: A meta-analysis. *Phytopathology* 95:1225–1236.

Paul, P. A., P. E. Lipps, and L. V. Madden. 2006. Meta-analysis of regression coefficients for the relationship between Fusarium head blight and deoxynivalenol content of wheat. *Phytopathology* 96:951–961.

Pearson, E. S. 1932. The percentage limits for the distribution of range in samples from a normal population. *Biometrika* 24:404–417.

Pearson, K. 1904. Report on certain enteric fever inoculation statistics. *British Medical Journal* 2:1243–1246.

Pen, I. and F. J. Weissing. 2000. Sex ratio optimization with helpers at the nest. *Proceedings of the Royal Society of London Series B* 267:539–544.

Peterman, R. M. 1990a. Statistical power analysis can improve fisheries research and management. *Canadian Journal of Fisheries and Aquatic Science* 47:2–15.

Peterman, R. M. 1990b. The importance of reporting statistical power: The forest decline and acidic deposition example. *Ecology* 71:2024–2027.

Peterman, R. M. 1995. Statistical power of methods of meta-analysis. *Trends in Ecology & Evolution* 10:460.

Peters, J. and K. Mengersen. 2008. Selective reporting of adjusted estimates in observational epidemiology studies: Reasons and implications for meta-analyses. *Evaluation and the Health Professions* 31:370–389.

Peters, J., A. Sutton, D. R. Jones, K. R. Abrams, and L. Rushton. 2008. Contour-enhanced meta-analysis funnel plots help distinguish publication bias from other causes of asymmetry. *Journal of Clinical Epidemiology* 61:991–996

Peters, J. L., A. J. Sutton, D. R. Jones, K. R. Abrams, and L. Rushton. 2006. Comparison of two methods to detect publication bias in meta-analysis. *Journal of the American Medical Association* 8:676–680.

Peters, J. L., A. J. Sutton, D. R. Jones, K. R. Abrams, and L. Rushton. 2007. Performance of the trim and fill method in the presence of publication bias and between-study heterogeneity. *Statistics in Medicine* 26:4544–4562.

Peterson, R. A. 2000. A meta-analysis of variance accounted for and factor loadings in exploratory factor analysis. *Marketing Letters* 11:261–275.

Peto, R. 1985. Discussion of "On the allocation of treatments in sequential medical trials" by J. Bather. *International Statistical Review* 53:1–13.

Petticrew, M. and H. Roberts. 2006. *Systematic Reviews in the Social Sciences: A Practical Guide.* Blackwell, Oxford, UK.

Pham, B., R. Platt, L. McAuley, T. P. Klassen, and D. Moher. 2001. Is there a 'best' way to detect and minimize publication bias? An empirical evaluation. *Evaluation and the Health Professions* 24:109–125.

Philbrook, H. T., N. Barrowman, and A. X. Garg. 2007. Imputing variance estimates do not alter the conclusions of a meta-analysis with continuous outcomes: A case study of changes in renal function after kidney donation. *Journal of Clinical Epidemiology* 60:228–240.

Piatelli-Palmarini, M. 1994. *Inevitable Illusions: How Mistakes of Reason Rule Our Minds.* John Wiley and Sons, New York.

Pigott, T. D. 1994. Methods for handling missing data in research synthesis. In *The Handbook of Research Synthesis*, edited by H. Cooper and L. V. Hedges, 163–176 . Russell Sage Foundation, New York.

Pigott, T. D. 2001. Missing predictors in models of effect size. *Evaluation & the Health Professions* 24:277–307.

Pogue, J. M. and S. Yusuf. 1997. Cumulating evidence from randomized trials: Utilizing sequential monitoring boundaries for cumulative meta-analysis. *Controlled Clinical Trials* 18:580–593.

Poole, C. and S. Greenland. 1999. Random-effects meta-analyses are not always conservative. *American Journal of Epidemiology* 150:469–475.

Poulin, R. 1995. Clutch size and egg size in free-living parasitic copepods: A comparative analysis. *Evolution* 49:325–336.

Poulin, R. 2000. Manipulation of host behaviour by parasites: a weakening paradigm? *Proceedings of the Royal Society of London Series B* 267:787–792.

Provost, J., G. Berthomieu, and P. Morel. 2000. Low-frequency p-and g-mode solar oscillations. *Astronomy and Astrophysics* 353:775–785.

Puetz, T. W., P. J. O'Connor, and R. K. Dishman. 2006. Effect of chronic exercise on feelings of energy and fatigue: a quantitative synthesis. *Psychological Bulletin* 132:866–876.

Pullin, A. S. and T. M. Knight. 2003. Support for decision making in conservation practice: An evidence-based approach. *Journal for Nature Conservation* 11:83–90.

Pullin, A. S., T. M. Knight, D. A. Stone, and K. Charman. 2004. Do conservation managers use scientific evidence to support their decision-making? *Biological Conservation* 119:245–252.

Pullin, A. S. and G. B. Stewart. 2006. Guidelines for systematic review in conservation and environmental management. *Conservation Biology* 20:1647–1656.

Purvis, A. and T. Garland. 1993. Polytomies in comparative analyses of continuous characters. *Systematic Biology* 42:569–575.

Quinn, G. P. and M. J. Keough. 2002. *Experimental Design and Data Analysis for Biologists*. Cambridge University Press, Cambridge.

Qvarnström, A., T. Part, and B. C. Sheldon. 2000. Adaptive plasticity in mate preference linked to differences in reproductive effort. Nature 405:344–347.

Rasmussen, S. A., S. Y. Chu, S. Y. Kim, C. H. Schmid, and J. Lau. 2008. Maternal obesity and risk of neural tube defects: A metaanalysis. *American Journal of Obstetrics & Gynecology* 198:611–619.

Raudenbush, S., B. Becker, and H. Kalaian. 1988. Modeling multivariate effect sizes. *Psychological Bulletin* 103:111–120.

Raudenbush, S. W. 1994. Random effects models. In *The Handbook of Research Synthesis*, edited by H. Cooper and L. V. Hedges, 301–321 . Russell Sage Foundation, New York.

Ravnskov, U. 1992. Cholesterol lowering trials in coronary heart disease: Frequency of citation and outcome. *British Medical Journal* 305:15–19.

Reed, D. H. and R. Frankham. 2001. How closely correlated are molecular and quantitative measures of genetic variation? A meta-analysis. *Evolution* 55:1095–1103.

Reed, J. G. and P. M. Baxter. 2003. *Library Use: Handbook for Psychology*, 3rd edition. American Psychological Association, Washington, DC.

Ren, C., G. M. Williams, L. Morawska, K. Mengersen, and S. Tong. 2007. Ozone modifies associations between temperature and cardiovascular mortality—analysis of the NMMAPS data. *Occupational and Environmental Medicine* 65:255–260.

Reynolds, J. D. and J. C. Jones. 1999. Female preference for preferred males is reversed under low oxygen conditions in the common goby (*Pomatoschistus microps*). *Behavioral Ecology* 10:149–154.

Reynolds, M. R., H. E. Burhart, and R. F. Daniels. 1981. Procedures for statistical validation of stochastic simulation models. *Forest Science* 27:349–364.

Reznick, D. N., L. Nunney, and A. Tessier. 2000. Big houses, big cars, superfleas and the costs of reproduction. *Trends in Ecology & Evolution* 15:421–425.

Ricciardi, A. and J. M. Ward. 2006. Comment on opposing effects of native and exotic herbivores on plant invasions. *Science* 313:298a.

Richards, T. A. and D. Bass. 2005. Molecular screening of free-living microbial eukaryotes: Diversity and distribution using a meta-analysis. *Current Opinions in Microbiology* 8:240–252.

Ricklefs, R. E. and C. D. Cadena. 2008. Heritability of longevity in captive populations of non-domesticated mammals and birds. *Journals of Gerontology A* 63:435–446.

Ricklefs, R. E. and J. M. Starck. 1996. The application of phylogenetically independent contrasts: A mixed progress report. *Oikos* 77:167–172.

Ridley, J., N. Kolm, R. P. Freckleton, and M. J. G. Gage. 2007. An unexpected influence of widely used significance thresholds on the distribution of reported P-values. *Journal of Evolutionary Biology* 20:1082–1089.

Rief, W. and S. G. Hofmann. 2009. The missing data problem in meta-analysis. *Archives of General Psychiatry* 65:238.

Riek, A. 2008. Relationship between field metabolic rate and body weight in mammals: Effect of the study. *Journal of Zoology* 276:187–194.

Riley, R. D. 2009. Multivariate meta-analysis: the effect of ignoring within-study correlation. *Journal of the Royal Statistical Society Series A* 172:789–811.

Riley, R. D., K. R. Abrams, A. J. Sutton, P. C. Lambert, and J. R. Thompson. 2007. Bivariate random-effects meta-analysis and the estimation of between-study correlation. *BMC Medical Research Methodology* 7:3.

Riley, R. D., P. C. Lambert, and G. Abo-Zakl. 2010. Meta-analysis of individual participant data: Rationale, conduct and reporting. *British Medical Journal* 34:c221.

Riley, R. D., M. C. Simmonds, and M. P. Look. 2007. Evidence synthesis combining individual patient data and aggregate data: A systematic review identified current practice and possible methods. *Journal of Clinical Epidemiology* 60:431–439.

Riley, R. D. and E. W. Steyerberg. 2010. Meta-analysis of a binary outcome using individual participant data and aggregate data. *Research Synthesis Methods* 1:2–19.

Riley, R. D., A. J. Sutton, K. R. Abrams, and P. C. Lambert. 2004. Sensitivity analyses allowed more appropriate and reliable meta-analysis conclusions for multiple outcomes when missing data was present. *Journal of Clinical Epidemiology* 57:911–924.

Riley, R. D., J. R. Thompson, and K. R. Abrams. 2008. An alternative model for bivariate random-effects meta-analysis when the within-study correlations are unknown. *Biostatistics* 9:172–186.

Robert, C. P. and G. Casella. 2010. *Monte Carlo Statistical Methods*. 2nd edition. Springer Verlag, New York.

Roberts, M. L., K. L. Buchanan, and M. R. Evans. 2004. Testing the immunocompetence handicap hypothesis: A review of the evidence. *Animal Behavior* 68:227–239.

Roberts, P. D., G. B. Stewart, and A. S. Pullin. 2006. Are review articles a reliable source of evidence to support conservation and environmental management? A comparison with medicine. *Biological Conservation* 132:409–423.

Robins, J. M., S. Greenland, and N. E. Breslow. 1986. A general estimator for the variance of the Mantel-Haenszel odds ratio. *American Journal of Epidemiology* 124:719–723.

Roff, D. A. 2002. *Life History Evolution*. Sinaeur and Associates, Massachusetts.

Rohlf, F. J. 2001. Comparative methods for the analysis of continuous variables: Geometric interpretations. *Evolution* 55:2143–2160.

Rong, J., E. A. Feltus, V. N. Waghmare, G. J. Pierce, P. W. Chee, X. Draye, Y. Saranga, et al. 2007. Meta-analysis of polyploid cotton QTL shows unequal contributions of subgenomes to a complex network of genes and gene clusters implicated in lint fiber development. *Genetics* 176:2577–2588.

Root, T. L., J. T. Price, K. R. Hall, S. H. Schneider, C. Rosenzweig, and J. A. Pounds. 2003. Fingerprints of global warming on wild animals and plants. *Nature* 421:57–60.

Rosenberg, M.S. 2000. The Comparative Claw Morphology, Phylogeny, and Behavior of Fiddler Crabs (Genus *Uca*). PhD diss., Stony Brook University, Stony Brook, New York.

Rosenberg, M.S. 2001. The systematics and taxonomy of fiddler crabs: A phylogeny of the genus *Uca*. *Journal of Crustacean Biology* 21:839–869.

Rosenberg, M. S. 2005. The file-drawer problem revisited: A general weighted method for calculating fail-safe numbers in meta-analyses. *Evolution* 59:464–468.

Rosenberg, M. S., D. C. Adams, and J. Gurevitch. 2000. *MetaWin: Statistical Software for Meta-analysis*. Sinauer Associates, Sunderland, MA.

Rosenfeld, R. M. and J. C. Post. 1992. Meta-analysis of antibiotics for the treatment of otitis media with effusion. *Otolaryngology—Head and Neck Surgery* 106:378–386.

Rosenthal, R. 1979. The "file drawer problem" and tolerance for null results. *Psychological Bulletin* 86:638–641.

Rosenthal, R. 1991. *Meta-analytic Procedures for Social Research*. SAGE Publications, Newbury Park, CA.

Rosenthal, R. 1994. Parametric measures of effect size. In *The Handbook of Research Synthesis*, edited by H. Cooper and L. V. Hedges, 231–244. Russell Sage Foundation, New York.

Rosenthal, R. and M. DiMatteo. 2001. Meta-analysis: Recent developments in quantitative methods in literature reviews. *Annual Review of Psychology* 52: 59–82.

Rosenthal, R. and R. L. Rosnow. 1985. *Contrast Analysis: Focused Comparisons in the Analysis of Variance*. Cambridge University Press, New York.

Rosenthal, R. and R. L. Rosnow. 1991. *Essentials of Behavioral Research: Methods and Data Analysis*, 2nd edition. McGraw-Hill, New York.

Rosenthal, R., R. L. Rosnow, and D. B. Rubin. 2000. *Contrast and Effect Sizes in Behavioral Research: A Correlational Approach*. Cambridge University Press, Cambridge.

Rosenthal, R. and D. Rubin. 1978. Interpersonal expectancy effects: The first 345 studies. *Behavioral and Brain Sciences* 3:377–415.

Rosenthal, R. and D, Rubin. 1982. Comparing effect sizes of independent studies. *Psychological Bulletin* 92:500–504.

Roth, P. L. 2008. Software review: Hunter-Schmidt meta-Analysis programs 1.1. *Organizational Research Methods* 11:192–196.

Rothstein, H. R. and S. Hopewell. 2009. Grey literature. In *The Handbook of Research Synthesis and Meta-analysis*, 2nd edition, edited by H. Cooper, L. V. Hedges, and J. C. Valentine, 103–125. Russell Sage Foundation, New York.

Rothstein, H. R., A. Sutton, and M. Borenstein, editors. 2005. *Publication Bias in Meta-analysis*. John Wiley, Chichester, UK.

Rubin, D. B. 1984. Bayesianly justifiable and relevant frequency calculations for the applied statistician. *Annals of Statistics* 12:1151–1172.

Rubin, D. B. and N. Schenker. 1991. Multiple imputation in health-care databases: An overview and some applications. *Statistics in Medicine* 10:585–598.

Rustad, L. E., J. L. Campbell, G. M. Marion, R. J. Norby, M. J. Mitchell, A. E. Hartley, J.H.C. Cornelissen, J. Gurevitch, and GCTE-NEWS. 2001. A meta-analysis of the response of soil respiration, net nitrogen mineralization, and aboveground plant growth to experimental ecosystem warming. *Oecologia* 126:543–562.

Rutter, C. M. and C. A. Gatsonis. 2001. A hierarchical regression approach to meta-analysis of diagnostic test accuracy evaluations. *Statistics in Medicine* 20:2865–2884.

Sackett, D. L., P. Glasziou, and I. Chalmers. 1997. Meta-analysis may reduce imprecision, but it can't reduce bias. Unpublished commentary commissioned by the New England Journal of Medicine.

Sackett, D. L., W. M. Rosenberg, J. A. Gray, R. B. Haynes, and W. S. Richardson. 1996. Evidence based medicine: What it is and what it isn't. *British Medical Journal* 312:71–72.

Saikkonen, K., P. Lehtonen, M. Helander, J. Koricheva, and S. H. Faeth. 2006. Model systems in ecology: Dissecting the endophyte-grass literature. *Trends in Plant Science* 11:428–433.

Salanti, G. and J.P.T. Higgins. 2008. Meta-analysis of genetic association studies under different inheritance models using data reported as merged genotypes. *Statistics in Medicine* 27:764–777.

Salanti, G., J.P.T. Higgins, A. E. Ades, and J.P.A. Ioannidis. 2008. Evaluation of networks of randomized trials. *Statistical Methods in Medical Research* 17:279–301.

Salanti, G., F. K. Kawoura, and J. P. A. Ioannidis. 2008. Exploring the geometry of treatment networks. *Annals of Internal Medicine* 148:544–553.

Sanderson, S., L. D. Tatt, and J.P.T. Higgins. 2007. Tools for assessing quality and susceptibility to bias in observational studies in epidemiology: A systematic review and annotated bibliography. *International Journal of Epidemiology* 36:666–676.

Santos, E.S.A., D. Scheck, and S. Nakagawa. 2011. Dominance and plumage traits: Meta-analysis and metaregression analysis. *Animal Behavior* 82:3–19.

Savage, L. J. 1954. *The Foundations of Statistics.* Wiley and Sons, New York.

Savage, V. M., J. F. Gillooly, W. H. Woodruff, G. B. West, A. P. Allen, B. J. Enquist, and A. C. Brown. 2004. The predominance of quarter-power scaling in biology. *Functional Ecology* 18:257–282.

Sax, D. F., J. J. Stachowicz, J. H. Brown, J. F. Bruno, M. N. Dawson, S. D. Gaines, R. K. Grosberg, et al. 2007. Ecological and evolutionary insights from species invasions. *Trends in Ecology & Evolution* 22:465–471.

Schafer, J. L. 1997. *Analysis of Incomplete Multivariate Data.* Chapman and Hall, London.

Scheiner, S. M., S. B. Cox, M. Willig, G. G. Mittelbach, C. Osenberg, and M. Kaspari. 2000. Species richness, species-area curves and Simpson's paradox. *Evolutionary Ecology Research* 2:791–802.

Scherer, R. W., P. Langenberg, and E. von Elm. 2007. Full publication of results initially presented in abstracts. *Cochrane Database of Methodology Reviews* Issue 2, Art. No.: MR000005. doi: 10.1002/14651858.MR000005.pub3.

Schino, G. 2007. Grooming and agonistic support: A meta-analysis of primate reciprocal altruism. *Behavioral Ecology* 18:115–120.

Schmid, C. H. and E. N. Brown. 2000. Bayesian Hierarchical Models. In *Methods in Enzymology,* Volume 321: Numerical Computer Methods, Part C, edited by M. L. Johnson and L. Brand, 305–330. Academic Press, New York.

Schmid, C. H., M. Landa, T. H. Jafar, I. Giatras, T. Karim, M. Reddy, P. C. Stark, and A. S. Levey. 2003. Constructing a database of individual clinical trials for longitudinal analysis. *Controlled Clinical Trials* 24:324–340.

Schmid, C. H., J. Lau, M. W. McIntosh, and J. C. Cappelleri. 1998. An empirical study of the effect of the control rate as a predictor of treatment efficacy in meta-analysis of clinical trials. *Statistics in Medicine* 17:1923–1942.

Schmid, C. H., P. C. Stark, J. A. Berlin, P. Landais, and J. Lau. 2004. Meta-regression detected associations between heterogeneous treatment effects and study-level, but not patient-level, factors. *Journal of Clinical Epidemiology* 57:683–697.

Schmidt, F. L. 1992. What do data really mean? Research findings, meta-analysis, and cumulative knowledge in psychology. *American Psychologist* 47:1173–1181.

Schmidt, F. L. and J. E. Hunter. 1977. Development of a general solution to the problem of validity generalization. *Journal of Applied Psychology* 62:529–540.

Schmitz, O. J., P. A. Hambäck, and A. P. Beckerman. 2000. Trophic cascades in terrestrial systems: A review of the effects of carnivore removals on plants. *American Naturalist* 155:141–153.

Schulz, K. F., I. Chalmers, R. J. Hayes, and D. G. Altman. 1995. Empirical evidence of bias. Dimensions of methodological quality associated with estimates of treatment effects in controlled trials. *Journal of the American Medical Association* 273:408–412.

Schwarzer, G., G. Antes, and M. Schumacher. 2002. Inflation of type I error rate in two statistical tests for the detection of publication bias in meta-analyses with binary outcomes. *Statistics in Medicine* 21:2465–2477.

Searles, P. S., S. D. Flint, and M. M. Caldwell. 2001. A meta-analysis of plant field studies simulating stratospheric ozone depletion. *Oecologia* 127:1–10.

Shadish, W. R., T. Cook, and D. Campbell. 2002. *Experimental and Quasi-Experimental Designs for Generalized Causal Inference.* Houghton-Mifflin, New York.

Sheldon, B. C. and S. A. West. 2004. Maternal dominance, maternal condition, and offspring sex ratio in ungulate mammals. *American Naturalist* 163:40–54.

Sheridan, P. and C. Hays. 2003. Are mangroves nursery habitat for transient fishes and decapods? *Wetlands* 23:449–458.

Sherman, L., D. Farrington, B. Welsh, and D. Mackenzie. 2002. *Evidence Based Crime Prevention.* Routledge, New York.

Sheu, C.-F. and S. Suzuki. 2001. Meta-analysis using linear mixed models. *Behavior Research Methods, Instruments & Computers* 33:102–107.

Sih, A., P. Crowley, M. McPeek, J. Petranka, and K. Strohmeier. 1985. Predation, competition, and prey communities—a review of field experiments. *Annual Review of Ecology and Systematics* 16:269–311.

Silk, J. B. and G. R. Brown. 2008. Local resource competition and local resource enhancement shape primate birth sex ratios. *Proceedings of the Royal Society of London Series B* 275:1761–1765.

Silk, J. B., E. Willoughby, and G. R. Brown. 2005. Maternal rank and local resource competition do not predict birth sex ratios in wild baboons. *Proceedings of the Royal Society of London Series B* 272:859–864.

Simberloff, D. 2006a. Invasional meltdown 6 years later: Important phenomenon, unfortunate metaphor or both? *Ecology Letters* 8:912–919.

Simberloff, D. 2006b. Rejoinder to Simberloff (2006): Don't calculate effect sizes; study ecological effects. *Ecology Letters* 8:921–922.

Simmonds, M. C. and J.P.T. Higgins. 2007. Covariate heterogeneity in meta-analysis: Criteria for deciding between meta-regression and individual patient data. *Statistics in Medicine* 26:2982–2999.

Simmonds, M. C., J.P.T. Higgins, L. A. Steward, J. F. Tierney, M. J. Clarke, and S. G. Thompson. 2005. Meta-analysis of individual patient data from randomized trials: A review of methods used in practice. *Clinical Trials* 2:209–217.

Simmons, L. W. 2005. The evolution of polyandry: Sperm competition, sperm selection, and offspring viability. *Annual Review of Ecology, Evolution and Systematics* 36:125–146.

Simmons, L. W., J. L. Tomkins, J. S. Kotiaho, and J. Hunt. 1999. Fluctuating paradigm. *Proceedings of the Royal Society of London Series B* 266:593–595.

SIOP. 2003. *Principles for the Validation and Use of Personnel Selection Procedures*, 4th edition. Society for Industrial and Organizational Psychology. Bowling Green, OH. Available at www.siop.org.

Slavin, R. E. 1986. Best-evidence synthesis: An alternative to meta-analytic and traditional reviews. *Educational Researcher* 15:5–11.

Smart, C. R., R. E. Hendrick, J. H. Rutledge, and R. A. Smith. 1995. Benefit of mammography screening in women ages 40 to 49 years: Current evidence from randomized controlled trials. *Cancer* 75:1619–1626.

Smart, R. G. 1964. The importance of negative results in psychological research. *Canadian Psychologist* 5:225–232.

Smith, B. R. and D. T. Blumstein. 2008 Fitness consequences of personality: A meta-analysis. *Behavioral Ecology* 19:448–455.

Smith, M. L. and G. V. Glass. 1977. Meta-analysis of psychotherapy outcome studies. *American Psychologist* 32:752–760.

Smith, R. K., A. S. Pullin, G. B. Stewart, and W. J. Sutherland. 2010. Effectiveness of predator removal for enhancing bird populations. *Conservation Biology* 24:820–829.

Smouse, P. E., J. C. Long, and R. R. Sokal. 1986. Multiple regression and correlation extension of the Mantel test of matrix correspondence. *Systematic Zoology* 35:627–632.

Sohn, S. Y. 2000. Multivariate meta-analysis with potentially correlated marketing study results. *Naval Research Logistics* 47:500–510.

Soininen, J., R. McDonald, and H. Hillebrand. 2007. The distance decay of similarity in ecological communities. *Ecography* 30:3–12.

Sokal, R. R. and F. J. Rohlf. 1995. *Biometry: The Principles and Practice of Statistics in Biological Research*, 3rd edition. W. H. Freeman, New York.

Song, F. 1999. Exploring heterogeneity in meta-analysis: Is the L'Abbé plot useful? *Journal of Clinical Epidemiology* 52:725–730.

Song, F., A. Easterwood, S. Gilbody, L. Duley, and A. Sutton. 2000. Publication and other selection biases in systematic reviews. *Health Technology Assessment* 4:1–115.

Song, F. and S. Gilbody. 1998. Bias in meta-analysis detected by a simple, graphical test: Increase in studies of publication bias coincided with increasing use of meta-analysis. *British Medical Journal* 316:471.

Song, F., K. S. Khan, J. Dinner, and A. J. Sutton. 2002. Asymmetric funnel plots and publication bias in meta-analyses of diagnostic accuracy. *International Journal of Epidemiology* 31:88–95.

Spiegelhalter, D. J., L. S. Freedman, and M.K.B. Parmar. 1994. Bayesian approaches to randomized trials (with discussion). *Journal of the Royal Statistical Society Series A* 157:357–416.

Spiegelhalter, D. J., A. Thomas, N. Best, and D. Lunn. 2003. *WinBUGS User Manual*, Version 1.4. MRC Biostatistics Unit, Cambridge, UK.

Spiegelhalter, D. J., K. R. Abrams, and J. P. Myles. 2004. *Bayesian Approaches to Clinical Trials and Health-Care Evaluation*. Wiley, New York.

Stampfer, M. J. and G. A. Colditz. 1991. Estrogen replacement therapy and coronary heart disease: A quantitative assessment of the epidemiologic evidence. *Preventive Medicine* 20:47–63.

Steidl, R. J. and L. Thomas. 2001. Power analysis and experimental design. In *Design and Analysis of Ecological Experiments*, edited by S. M. Scheiner and J. Gurevitch, 14–36, 2nd edition. Oxford University Press, Oxford.

Sterling, T. D. 1959. Publication decisions and their possible effects on inferences drawn from test of significance—or vice-versa. *Journal of the American Statistical Association* 54:30–34.

Sterne, J.A.C. and M. Egger. 2005. Regression methods to detect publication and other bias in meta-analysis. In *Publication Bias in Meta-Analysis*, edited by H. R. Rothstein, A. Sutton, and M. Borenstein, 99–110. John Wiley, Chichester, UK.

Stern, J. M. and R. J. Simes. 1997. Publication bias: Evidence of delayed publication in a cohort study of clinical research projects. *British Medical Journal* 315:640–645.

Sterne, J.A.C., B. J. Becker, and M. Egger. 2005. The funnel plot. In *Publication Bias in Meta-Analysis*, edited by H. R. Rothstein, A. Sutton, and M. Borenstein, 76–98. John Wiley, Chichester, UK.

Sterne, J.A.C., M. Egger, and G. Davey Smith. 2001. Investigating and dealing with publication and other biases in meta-analysis. *British Medical Journal* 323:101–105.

Sterne, J.A.C., D. Gavaghan, and M. Egger. 2000. Publication and related biases in meta-analysis: Power of statistical tests and prevalence in the literature. *Journal of Clinical Epidemiology* 53:1119–1129.

Sterne, J.A.C., R. Harris, R. Harbord, and T. J. Steichen. 2007. User-written packages for meta-analysis in Stata. Available at http://www.stata.com/support/faqs/stat/meta.html.

Sterne J.A.C., R. Harris, R. Harbord, and T. J. Steichen. 2009. *Meta-Analysis in Stata: An Updated Collection from the Stata Journal*. Stata Press, College Station, TX.

Sterne, J.A.C., P. Jüni, K. F. Schulz, D. G. Altman, and M. Egger. 2002. Statistical methods for assessing the influence of study characteristics on treatment effects in 'meta-epidemiological' research. *Statistics in Medicine* 21:1513–1524.

Stewart, G. B. 2010. Meta-analysis in applied ecology. *Biology Letters* 6:78–81.

Stewart, G. B., H. R. Bayliss, D. A. Showler, W. J. Sutherland, and A. S. Pullin. 2009. Effectiveness of engineered in-stream structure mitigation measures to increase salmonid abundance: A systematic review. *Ecological Applications* 19:931–941.

Stewart, G. B., C. F. Coles, and A. S. Pullin. 2005. Applying evidence-based practice in conservation management: Lessons from the first systematic review and dissemination projects. *Biological Conservation* 126:270–278.

Stewart, G. B., E. S. Cox, M. G. Le Duc, R. J. Pakeman, A. S. Pullin, and R. H. Marrs. 2008. Control of bracken across the UK: Meta-analysis of a multi-site study. *Annals of Botany* 101:957–970.

Stewart, G. B. and A. S. Pullin. 2008. The relative importance of grazing stock type and grazing intensity for conservation of mesotrophic pasture. *Journal of Nature Conservation* 16:175–185.

Stewart, G. B., A. S. Pullin, and C. F. Coles. 2007. Poor evidence-base for assessment of windfarm impacts on birds. *Environmental Conservation* 34:1-11.

Stewart, G. B., A. S. Pullin, and C. Tyler. 2007. The effectiveness of asulam for bracken (*Pteridium aquilinium*) control in the U.K.: A meta-analysis. *Environmental Management* 40:447–460.

Stewart, L. A. and M. K. Parmar. 1993. Meta-analysis of the literature or of individual patient data: is there a difference? *The Lancet* 341:418–422.

Stewart, L. A. and J. F. Tierney. 2002. To IPD or not to IPD?: Advantages and disadvantages of systematic reviews using individual patient data. *Evaluation and the Health Professions* 25:76–97

Stigler, S. M. 1990. *The History of Statistics: The Measurement of Uncertainty before 1900.* Harvard University Press, Cambridge, MA.

Stijnen, T. 2000. Tutorial in biostatistics. Meta-analysis: formulating, evaluating, combining and reporting by S.-L. T. Normand. *Statistics in Medicine* 19:159–161.

Stoehr, A. M. 1999. Are significance thresholds appropriate for the study of animal behaviour? *Animal Behavior* 57:F22–F25.

Strauss, S. Y., J. A. Lau, T. W. Schoener, and P. Tiffin. 2008. Evolution in ecological field experiments: Implications for effect size. *Ecology Letters* 11:199–207.

Stroup, D. F., J. A. Berlin, S. C. Morton, I. Olkin, G. D. Williamson, D. Rennie, D. Moher, B. J. Becker, T. A. Sipe, and S. B. Thacker. 2000. Meta-analysis of observational studies in epidemiology: A proposal for reporting. *Journal of the American Medical Association* 283:2008–2012.

Surowiecki, J. 2004. *The Wisdom of Crowds*. Little, Brown, New York.

Sutherland, W. J., A. S. Pullin, P. M. Dolman, and T. M. Knight. 2004. The need for evidence-based conservation. *Trends in Ecology & Evolution* 19:305–308.

Sutton, A. J. and K. R. Abrams. 2001. Bayesian methods in meta-analysis and evidence synthesis. *Statistical Methods in Medical Research* 10:277–303.

Sutton, A. J., K. R. Abrams, D. R. Jones, T. A. Sheldon, and F. Song. 2000. *Methods for Meta-Analysis in Medical Research*. Wiley and Sons, Chichester, UK.

Sutton, A. J., N. J. Cooper, D. R. Jones, P. C. Lambert, J. R. Thompson, and K. R. Abrams. 2007. Evidence-based sample size calculations based upon updated meta-analysis. *Statistics in Medicine* 26:2479–2500.

Sutton, A. J., S. J. Duval, R. Tweedie, K. R. Abrams, and D. R. Jones. 2000. Empirical assessment of effect of publication bias on meta-analyses. *British Medical Journal* 320:1574–1577.

Sutton, A. J. and T. D. Pigott. 2005. Bias in meta-analysis induced by incompletely reported studies. In *Publication Bias in Meta-Analysis*, edited by H. R. Rothstein, A. Sutton, and M. Borenstein, 223–239. John Wiley, Chichester, UK.

Svenning, J. C. and S. J. Wright. 2005. Seed limitation in a Panamanian forest. *Journal of Ecology* 93:853–862.

Taborsky, M. 2009. Biased citation practice and taxonomic parochialism. *Ethology* 115:105–111.

Taleb, N. N. 2007. *The Black Swan*. Random House, New York.

Tang, J. L. and J.L.Y. Liu. 2000. Misleading funnel plot for detection of bias in meta-analysis. *Journal of Clinical Epidemiology* 53:477–484.

Taveggia, T. C. 1974. Resolving research controversy through empirical cumulation: Toward reliable sociological knowledge. *Sociological Methods & Research* 2:395–407.

Taylor, B. L. and T. Gerodette. 1993. The uses of statistical power in conservation biology: The vaquita and northern spotted owl. *Conservation Biology* 7:489–500.

Terrell, C. D. 1982 Table for converting the point biserial to the biserial. *Educational and Psychological Measurement* 42:983–986.

Terrin, N., C. H. Schmid, and J. Lau. 2005. In an empirical evaluation of the funnel plot, researchers could not visually identify publication bias. *Journal of Clinical Epidemiology* 58:894–901.

Terrin, N., C. H. Schmid, J. Lau, and I. Olkin. 2003. Adjusting for publication bias in the presence of heterogeneity. *Statistics in Medicine* 22:2113–2126.

Thomas, L. and F. Juanes. 1996. The importance of statistical power analysis: An example from animal behaviour. *Animal Behavior* 52:856–859.

Thompson, S. G. and J.P.T. Higgins. 2002. How should meta-regression analyses be undertaken and interpreted. *Statistics in Medicine* 21:1559–1573.

Thompson, S. G., R. M. Turner, and D. E. Warn. 2001. Multilevel models for meta-analysis, and their application to absolute risk differences. *Statistical Methods in Medical Research* 10:375–392.

Thornhill, R., A. P. Møller, and S. Gangestad. 1999. The biological significance of fluctuating asymmetry and sexual selection: A reply to Palmer. *American Naturalist* 154:234–241.

Timi, J. T. and R. Poulin. 2007. Different methods, different results: Temporal trends in the study of nested subset patterns in parasite communities. *Parasitology* 135:131–138.

Timm, N. H. 1999. Testing multivariate effect sizes in multiple-endpoint studies. *Multivariate Behavioral Research* 34:132–145.

Tomkins, J. L. and J. S. Kotiaho. 2004. Publication bias in meta-analysis: Seeing the wood for the trees. *Oikos* 104:194–196.

Tonhasca, A. and D. N. Byrne. 1994. The effects of crop diversification on herbivorous insects: A meta-analysis approach. *Ecological Entomology* 19:239–244.

Toro, M. A. and A. Caballero. 2005. Characterization and conservation of genetic diversity in subdivided populations. *Philosophical Transactions of the Royal Society of London Series B* 360:1367–1378.

Torres-Vila, L. M. and M. D. Jennions. 2005. Male mating history and female fecundity in the Lepidoptera: Do male virgins make better partners? *Behavioral Ecology and Sociobiology* 57:318–326.

Torres-Vila, L. M., M. C. Rodrígues-Molina, and M. D. Jennions. 2004. Polyandry and fecundity in the Lepidoptera: Can methodological and conceptual approaches bias outcomes? *Behavioral Ecology and Sociobiology* 55:315–324.

Toth, G. B. and H. Pavia. 2007. Induced herbivore resistance in seaweeds: A meta-analysis. *Journal of Ecology* 95:425–434.

Traill, L. W., C. J. A. Bradshaw, and B. W. Brook. 2007. Minimum viable population size: A meta-analysis of 30 years of published estimates. *Biological Conservation* 139:159–166.

Treadwell, J. R., S. J. Tregar, J. T. Reston, and C. M. Turkelson. 2007. A system for rating the stability and strength of medical evidence. *BMC Medical Research Methodology* 6:52.

Tregenza, T. and N. Wedell. 1997. Natural selection bias? *Nature* 386:234.

Tregenza, T. and N. Wedell. 1998. Benefits of multiple mates in the cricket *Gryllus bimaculatus*. *Evolution* 52:1726–1730.

Trikalinos, T. A., R. Churchill, M. Ferri, S. Leucht, A. Tuunainen, K. Wahlbeck, and J.P.A. Ioannidis. 2004. Effect sizes in cumulative meta-analyses of mental health randomized trials evolved over time. *Journal of Clinical Epidemiology* 57:1124–1130.

Trikalinos, T. A. and J.P.A. Ioannidis. 2005. Assessing the evolution of effect sizes over time. In *Publication Bias in Meta-Analysis*, edited by H. R. Rothstein, A. Sutton, and M. Borenstein, 241–259. John Wiley, Chichester, UK.

Trivers, R. L. and D. E. Willard. 1973. Natural selection on parental ability to vary the sex ratio of offspring. *Science* 179:90–92.

Tufte, E. R. 1990. *Envisioning Information*. Graphic Press, Cheshire, CT.

Tufte, E. R. 2001. *The Visual Display of Quantitative Information*, 2nd edition. Graphic Press, Cheshire, CT.

Tukey, J. W. 1972. Some graphic and semigraphic displays. In *Statistical Papers in Honor of George W. Snedecor*, edited by T. A. Bancroft, 293–316. The Iowa State University Press, IA.

Turner, E. H., A. M. Matthews, E. Linardatos, R. A. Tell, and R. Rosenthal. 2008. Selective publication of antidepressant trials and its influence on apparent efficacy. *New England Journal of Medicine* 358:252–260.

Turner, H., R. Boruch, J. Lavenberg, J. Schoeneberger, and D. de Moya. 2004. Electronic registers of trials. Paper presented at Fourth Annual Campbell Collaboration Colloquium: A First Look at the Evidence, February 2004, Washington DC.

Turner, R. M., R. Z. Omar, and S. G. Thompson. 2006. Modelling multivariate outcomes in hierarchical data, with application to cluster randomized trials. *Biometrical Journal* 48:333–345.

Turner, R. M., R. Z. Omar, M. Yang, H. Goldstein, and S. G. Thompson. 2000. A multilevel model framework for meta-analysis of clinical trials with binary outcomes. *Statistics in Medicine* 19:3417–3432.

Tyler, C., A. S. Pullin, and G. B. Stewart. 2006. Effectiveness of management interventions to control invasion by *Rhododendron ponticum*. *Environmental Management* 37:513–522.

Tylianakis, J. M., R. K. Didham, J. Bascompte, and D. A. Wardle. 2008. Global change and species interactions in terrestrial ecosystems. *Ecology Letters* 11:1351–1363.

Vacha-Haase, T. 1998. Reliability generalization: Exploring variance in measurement error affecting score reliability across studies. *Educational and Psychological Measurement* 58:6–20.

Vacha-Haase, T., R. K. Henson, and J. Caruso. 2002. Reliability generalization: Moving toward improved understanding and use of score reliability. *Educational and Psychological Measurement* 62:562–569.

Valentine, J. C., T. D. Pigott, and H. R. Rothstein. 2010. How many studies do you need?: A primer on statistical power in meta-analysis. *Journal of Educational and Behavioral Statistics* 35:215–247.

van Houwelingen, H., K. Zwinderman, and T. Stijnen. 1993. A bivariate approach to meta-analysis. *Statistics in Medicine* 12:2272–2284.

van Houwelingen, H. C. 1995. Meta-Analysis: Methods, limitations and applications. *Biocybernetics and Biomedical Engineering* 15:53–61.

van Houwelingen, H. C., L. R. Arends, and T. Stijnen. 2002. Advanced methods in meta-analysis: Multivariate approach and meta-regression. *Statistics in Medicine* 21:589–624.

van Ijzendoom, M. H., F. Juffer, and C. W. Klein-Poelhuis. 2005. Adoption and cognitive development: A meta-analytic comparison of adopted and nonadopted children's IQ and school performance. *Psychological Bulletin* 131:301–316.

van Zandt, P. A. and S. Mopper. 1998. A meta-analysis of adaptive deme formation in phytophagous insect populations. *American Naturalist* 152:595–604.

VanderWerf, E. 1992. Lack's clutch size hypothesis: an examination of the evidence using meta-analysis. *Ecology* 73:1699–1705.

Vehviläinen, H., J. Koricheva, and K. Ruohomäki. 2007. Tree species diversity influences herbivore abundance and damage: Meta-analysis of long-term forest experiments. *Oecologia* 152:287–298.

Verdú, M. and A. Traveset. 2004. Bridging meta-analysis and the comparative method: A test of seed size effect on germination after frugivores' gut passage. *Oecologia* 138:414–418.

Verdú, M. and A. Traveset. 2005. Early emergence enhances plant fitness: A phylogenetically controlled meta-analysis. *Ecology* 86:1385–1394.

Vevea, J. L. and L. V. Hedges. 1995. A general linear model for estimating effect size in the presence of publication bias. *Psychometrika* 60:419–435.

Vevea, J. L. and C. W. Woods. 2005. Publication bias in research synthesis: Sensitivity analysis using a priori weight functions. *Psychological Methods* 10:428–443.

Viechtbauer, W. 2006. *mima: An S-Plus/R Function to Fit Meta-analytic Mixed-, Random-, and Fixed-Effects Models. Computer Software and Manual*. Available from http://www.wvbauer.com/.

Viechtbauer, W. 2010. Conducting meta-analyses in R with the metafor package. *Journal of Statistical Software* 36:3. Available at http://www.jstatsoft.org/v36/i03/paper.

Vila, M., M. Williamson, and M. Lonsdale. 2004. Competition experiments on alien weeds with crops: Lessons for measuring plant invasion impact? *Biological Invasions* 6:59–69.

Villar, J., G. Carroli, and J. M. Belizan. 1995. Predictive ability of meta-analysis of randomised controlled trials. *The Lancet* 345:772–776.

von Ende, N. 2001. Repeated-measures analysis: growth and other time dependent measures. In *Design and Analysis of Ecological Experiments*, 2nd edition, edited by S. M. Scheiner and J. Gurevitch, 134–157. Oxford University Press, New York.

Wacholder, S., S. Chancock, M. Garcia-Closas, L. El Ghormli, and N. Rothman. 2004. Assessing the probability that a positive report is false: An approach for molecular epidemiology studies. *National Cancer Institute* 96:434–442.

Wachter, K. 1988. Disturbed by Meta-Analysis? *Science* 241:1407–1408.

Walker, M. D., C. H. Wahren, R. D. Hollister, G.H.R. Henry, L. E. Ahlquist, J. M. Alatalo, M. S. Bret-Harte, et al. 2006. Plant community responses to experimental warming across the tundra biome. *Proceedings of the National Academy of Sciences of the United States of America* 103:1342–1346.

Wallace B. C., I. J. Dahabreh, T. A. Trikalinos, J. Lau, P. Trow, and C. H. Schmid. 2012. Closing the gap between methodologists and end-users: R as a computational back-end. *Journal of Statistical Software* 49:5. Available at http://www.jstatsoft.org/v49/i05/paper.

Wallace, B. C., C. H. Schmid, J. Lau, and T. A. Trikalinos. 2009. MetaAnalyst: Software for meta-analysis of binary, continuous and diagnostic data. *BMC Medical Research Methodology*, 9:80.

Walsh, H. E., M. G. Kidd, T. Moum, and V. L. Friesen. 1999. Polytomies and the power of phylogenetic inference. *Evolution* 53:932–937.

Walsh, J. E. 1947. Concerning the effect of intraclass correlation on certain significance tests. *Annals of Mathematical Statistics* 18:88–96.

Walter, S. D. and X. Yao. 2007. Effect sizes can be calculated for studies reporting ranges for outcome variables in systematic reviews. *Journal of Clinical Epidemiology* 60:849–852.

Wan, S. Q., D. F. Hui, and Y. Q. Luo. 2001. Fire effects on nitrogen pools and dynamics in terrestrial ecosystems: A meta-analysis. *Ecological Applications* 11:1349–1365.

Wang, M. C. and B. J. Bushman. 1999. *Integrating Results through Meta-analytic Review using SAS Software*. SAS Institute, Cary, NC.

Wang, X. 2006. Approximating Bayesian inference by weighted likelihood method. *Canadian Journal of Statistics* 34:279–298.

Wang, X. and J. V. Zidek. 2005. Selecting likelihood weights by cross-validation. *Annals of Statistics* 33:463–500.

Waring, G. L. and N. S. Cobb. 1992. The impact of plant stress on herbivore population dynamics. In *Insect-Plant Interactions*, Volume 4, edited by E. A. Bernays, 167–226. CRC-Press, Boca Raton, FL.

Watt, A. D. 1994. The relevance of the stress hypothesis to insects feeding on tree foliage. In *Individuals, Populations and Patterns in Ecology*, edited by S. R. Leather, A. D. Watt, N. J. Mills, and K. F. A. Walters, 73–85. Intercept, Andover, UK.

Webb, C. O. and M. J. Donoghue. 2004. Phylomatic: Tree assembly for applied phylogenetics. *Molecular Ecology News* 5:181–183.

Weintraub, I. 2000. The role of grey literature in the sciences. Available at: http://library.brooklyn .cuny.edu/access/greyliter.htm. Accessed on July 31, 2006.

Weisz, J. 2004. *Psychotherapy for Children and Adolescents: Evidence-Based Treatments and Case Examples*. Cambridge University Press, Cambridge, UK.

West, G. B., J. H. Brown, and B. J. Enquist. 1997. A general model for the origin of allometric scaling laws in biology. *Science* 276:122–126.

West, G. B., W. H. Woodruff, and J. H. Brown. 2002. Allometric scaling of metabolism from molecules and mitochondria to cells and mammals. *Proceedings of the National Academy of Sciences of the United States of America* 99:2473–2478.

West, S. A. 2009. *Sex Allocation*. Princeton University Press, Princeton, NJ.

West, S. A. and B. C. Sheldon. 2002. Constraints in the evolution of sex ratio adjustment. *Science* 295:1685–1688.

West, S. A., D. M. Shuker, and B. C. Sheldon. 2005. Sex-ratio adjustment when relatives interact: A test of constraints on adaptation. *Evolution* 59:1211–1228.

Westoby, M., M. R. Leishman, and J. M. Lord. 1995a. Issues of interpretation following phylogenetic correction. *Journal of Ecology* 83:892–893.

Westoby, M., M. R. Leishman, and J. M. Lord. 1995b. Further remarks on phylogenetic correction. *Journal of Ecology* 83:727–729.

Westoby, M., M. R. Leishman, and J. M. Lord. 1995c. On misinterpreting the "phylogenetic correction." *Journal of Ecology* 83:531–534.

Wetterslev, J., K. Thorlund, J. Brok, and C. Gluud. 2008. Trial sequential analysis may establish when firm evidence is reached in cumulative meta-analysis. *Journal of Clinical Epidemiology* 61:64–75.

White, C. R., T. M. Blackburn, and R. S. Seymour. 2009. Phylogenetically informed analysis of the allometry of mammalian basal metabolic rate supports neither geometric nor quarter-power scaling. *Evolution* 63:2658–2667.

White, C. R., P. Cassey, and T. M. Blackburn. 2007. Allometric exponents do not support a universal metabolic allometry. *Ecology* 88:315–323.

White, C. R. and R. S. Seymour. 2003. Mammalian basal metabolic rate is proportional to body mass 2/3. *Proceedings of the National Academy of Sciences of the United States of America* 100:4046–4049.

White, H. D. 1994. Scientific communication and literature retrieval. In *The Handbook of Research Synthesis*, edited by H. Cooper and L. V. Hedges, 41–55. Russell Sage Foundation, New York.

Whitehead, A. 2002. *Meta-analysis of Controlled Clinical Trials*. John Wiley and Sons, West Sussex, UK.

Whitehead, A., R. Z. Omar, J.P.T. Higgins, E. Savaluny, R. M. Turner, and S. G. Thompson. 2001. Meta-analysis of ordinal outcomes using individual patient data. *Statistics in Medicine* 20:2243–2260.

Whittaker, R. J. 2010. Meta-analyses and mega-mistakes: Calling time on meta-analysis of the species richness–productivity relationship. *Ecology* 91:2522–2533.

Whittingham, M. J., P. A. Stephens, R. B. Bradbury, and R. P. Freckleton. 2006. Why do we still use stepwise modelling in ecology and behaviour. *Journal of Animal Ecology* 75:1182–1189.

Whymper, E. 1986. *Scrambles Amongst the Alps*. Webb and Bower, London.

Wiebe, N., B. Vandermeer, R. W. Platt, T. P. Klassen, D. Moher, and N. J. Barrowman. 2006. A systematic review identifies a lack of standardization in methods for handling missing variance data. *Journal of Clinical Epidemiology* 59:342–353.

Wikström, N., V. Savolainen, and M. W. Chase. 2001. Evolution of the angiosperms: Calibrating the family tree. *Proceedings of the Royal Society of London Series B* 268:2211–2220.

Williams, R. L., T. C. Chalmers, K. C. Stange, F. T. Chalmers, and S. J. Bowlin. 1993. Use of antibiotics in preventing recurrent acute otitis media and in treating otitis media with effusion: A meta-analytic attempt to resolve the brouhaha. *Journal of the American Medical Association* 270:1344–1351.

Wilson, E. O. 1975. *Sociobiology: The New Synthesis*. The Belknap Press of the Harvard University Press, Cambridge, MA.

Wolf, F. M. 1986. *Meta-analysis: Quantitative Methods for Research Synthesis*. SAGE Publications, Newbury Park, CA.

Wolpert, R. and K. Mengersen. 2004. Adjusted likelihoods for synthesizing empirical evidence from studies that differ in quality and design: Effects of environmental tobacco smoke. *Statistical Science* 19:450–471.

Wong, B.B.M. and H. Kokko. 2005. Is science as global as we think? *Trends in Ecology & Evolution* 20:475–476.

Wood, J. 2008. Methodology for dealing with duplicate study effects in a meta-analysis. *Organizational Research Methods* 11:79–95.

Wood, L., M. Egger, L. L. Gluud, K. F. Schulz, P. Jüni, D. G. Altman, C. Gluud, R. M. Martin, A. J. Wood, and J. A. C. Sterne. 2008. Empirical evidence of bias in treatment effect estimates in controlled trials with different interventions and outcomes: Meta-epidemiological study. *British Medical Journal* 336:601–605.

Woodhouse, G., editor. 1996. *Multilevel Modelling Applications: A Guide for Users of MLn*. Institute of Education, University of London, London.

Wooster, D. 1994. Predator impacts on stream benthic prey. *Oecologia* 99:7–15.

Worm, B., H. K. Lotze, H. Hillebrand, and U. Sommer. 2002. Consumer versus resource control of species diversity and ecosystem functioning. *Nature* 417:848–851.

Worm, B. and R. A. Myers. 2004. Managing fisheries in a changing climate—no need to wait for more information: Industrialized fishing is already wiping out stocks. *Nature* 429:15.

Yang, M., H. Goldstein, and J. Rasbash. 1996. *MLn Macros for Advanced Multilevel Modelling*, Version 1.1. Institute of Education, University of London, London.

Young, C. and R. Horton. 2005. Putting clinical trials into context. *The Lancet* 366:107–108.

Yuan, Y. and R.J.A. Little. 2009. Meta-analysis of studies with missing data. *Biometrics* 65:487–496.

Yusuf, S., R. Peto, J. Lewis, R. Collins, and P. Sleight. 1985. Beta blockade during and after myocardial infarction: An overview of the randomized trials. *Progress in Cardiovascular Diseases* 27:335–371.

Zamora J., V. Abraira, A. Muriel, K. Khan, and A. Coomarasamy. 2006. Meta-DiSc: A software for meta-analysis of test accuracy data. *BMC Medical Research Methodology* 6:31.

Zarin, D. A., T. Tse, and N. C. Ide. 2005. Trial registration at ClinicalTrials.gov between May and October 2005. *New England Journal of Medicine* 353:2779–2787.

Zvereva, E. L., E. Toivonen, and M. V. Kozlov. 2008. Changes in species richness of vascular plants under the impact of air pollution: A global perspective. *Global Ecology and Biogeography* 17:305–319.

LIST OF CONTRIBUTORS

Isabelle M. Côté
Department of Biological Sciences
Simon Fraser University
Burnaby, BC
V5A 1S6
Canada

Peter S. Curtis
Department of Evolution, Ecology
  & Organismal Biology
The Ohio State University
318 W. 12th Avenue
Columbus, OH 43210
US

Jessica Gurevitch
Department of Ecology & Evolution
Stony Brook University
Stony Brook, NY 11794-5245
US

Michael D. Jennions
Evolution, Ecology & Genetics
Research School of Biology
The Australian National University
Canberra, ACT 0200
Australia

Julia Koricheva
School of Biological Sciences
Royal Holloway, University of London
Egham, Surrey, TW20 0EX
UK

Marc J. Lajeunesse
Department of Integrative Biology
University of South Florida
SCA 110, 4202 East Fowler Avenue
Tampa, Fl 33620-5200
US

Joseph Lau
Center for Evidence Based Medicine and
  Department of Biostatistics
University Box GS121-8
121 South Main Street
Brown University
Providence, RI 02912
US

(formerly of Tufts Medical Center, Institute
  for Clinical Research and Health Policy
  Studies, Boston, MA, US)

Christopher J. Lortie
Department of Biology
York University
4700 Keele Street, Toronto, Ontario
M6S 2E2
Canada

Kerrie Mengersen
School of Mathematical Sciences
Queensland University of Technology
GPO Box 2434
Brisbane, Qld 4001
Australia

Michael S. Rosenberg
School of Life Sciences
Arizona State University
PO Box 874501
Tempe, AZ 85287-4501
US

Hannah R. Rothstein
Zicklin School of Business
Baruch College
The City University of New York
1 Bernard Baruch Way
New York, NY 10010
US

Christopher H. Schmid
Center for Evidence Based Medicine and
  Department of Biostatistics
University Box GS121-8
121 South Main Street
Brown University
Providence, RI 02912
US
(formerly of Tufts Medical Center, Institute
  for Clinical Research and Health Policy
  Studies, Boston, MA, US)

Gavin B. Stewart
Centre for Reviews and Dissemination
University of York
York, YO10 5DD
UK

SUBJECT INDEX

aggregation bias, 32. *See also* ecological bias

AGORA, 427

AGRICOLA, 39, 442

Akaike's information criterion (AIC) for model comparison, 180, 292–293, 296–297, 308, 402

allometric scaling, 368, 382, 386, 390–391

analysis of covariance (ANCOVA), 22, 98, 100, 122, 200

analysis of variance (ANOVA), 11, 22, 77, 98, 114, 115, 117, 197, 293; and factorial model, 98, 124; and fixed–effects model, 94; vs. meta-analysis, 94, 315–316; nested, 118; one-way, 113, 187, 200, 201, 369; and partitioning of heterogeneity, 114, 358; and restricted maximum likelihood approach (REML), 128; two-way, 200

apples and oranges dilemma, 12, 29, 284–285, 318, 319, 427. *See also* meta-analysis: criticism of

Bayes' rule, 103, 145–146, 152

Bayesian analysis, 102–104, 145–155; advantages of, 120, 145; approach to inference in, 100–104, 118, 122, 145–147, 152, 168; ecological examples of, 104, 307–309; vs. frequentist approach, 90–92, 152; incorporation of expert opinions in, 12, 32; for multivariate models, 282; software for, 103, 155, 176, 183–186, 189–190; testing for heterogeneity in, 105; updating in, 106, 149–152; WinBUGS codes for, 169–173, 278–281; worked examples of, 155–168, 263–264

Bayesian belief networks, 29, 80

Begg and Mazumdar test for publication bias, 177, 180, 220

behavior, 46, 240, 284, 365, 376, 382, 395, 408; dispersal, 48; effects of parasites on host, 238, 241; feeding, 63

behavioral ecology, 374

Bergmann's rule, 383, 385

between-study variance, 93, 97, 106, 363; in Bayesian analysis, 103, 145, 146, 149; Der-Simonian and Laird approximation of, 125, 126, 133, 136; Hedges approximation of, 125, 133, 136; and heterogeneity, 112; in maximum likelihood estimation, 127, 132, 135; in methods of moments estimation, 101; and standard error, 353; unknown, 131–132, 148; vs. within-study variance, 94, 104, 316. *See also* tau squared (τ^2)

bias, 16, 35, 147, 262, 326; aggregation, 32; cognitive, 365; in data extraction, 56–57; dissemination, 197, 210–211, 215; ecological, 32, 305–306, 423; experimental, 59; due to imputation, 203, 205–206; due to missing data, 160, 195, 198, 203, 274, 283; due to multiple testing, 331; due to nonindependence of effect size estimates, 257, 271; in meta-analyses of multisite studies, 315; phylogenetic, 284, 289, 292, 293, 294; publication, 23, 84, 197, 198, 206, 326, 343 (*see also* publication bias); reporting, 225, 274; research, 228–229, 241, 285, 315, 328, 402; sampling, 365, in search strategy, 18, 27, 34, 38, 42, 44, 48, 327, 420; selection, 40, 49, 54, 300, 326, 328; sex, 389, 394; and study quality, 59; submission, 214–215; time-lag, 240–241; in vote counting and narrative reviews, 6, 349, 364, 365

bibliographic databases. *See* reference databases

binary data, 31, 61, 76, 102, 128, 137, 148, 164, 304, 418; Bayesian analysis of, 311; mixed models for, 281; software for meta-analysis of, 175, 176, 180–186; transformations for, 261

binomial distribution, 74, 76, 78, 79, 82, 91, 125, 137, 144, 164; logit transformation of, 81; meta-analysis of data with, 135, 138

biodiversity, 30, 75; loss of, xi, 13, 218, 242, 314, 415; maintenance of, 382; management of, 420

Biological Abstracts, 39

biological rule, 17, 383

biological significance, 33, 336, 337; vs. statistical significance, 63

BIOSIS Previews, 39

birds, 18, 46, 72, 77, 83, 101, 138, 216, 228, 238, 239, 241, 257, 285, 364, 381; metabolic rate of, 390–391; predator effects on, 421; sex ratios in, 394; sexual dimorphism in, 399; wind farm impact on, 30, 47, 48, 325, 327

Bland-Altman plot, 341, 347